30th International Baltic Road Conference (IBRC 2021)

IOP Conference Series: Materials Science and Engineering
Volume 1202

Riga, Latvia
23 - 24 August 2021

ISBN: 978-1-7138-4661-1
ISSN: 1757-8981

Printed from e-media with permission by:

Curran Associates, Inc.
57 Morehouse Lane
Red Hook, NY 12571

Some format issues inherent in the e-media version may also appear in this print version.

Institute of Physics
Dirac House, Temple Back
Bristol BS1 6BE UK

Phone: 44 1 17 929 7481
Fax: 44 1 17 920 0979

techtracking@iop.org

Additional copies of this publication are available from:

Curran Associates, Inc.
57 Morehouse Lane
Red Hook, NY 12571 USA
Phone: 845-758-0400
Fax: 845-758-2633
Email: curran@proceedings.com
Web: www.proceedings.com

TABLE OF CONTENTS

Preface for the Proceedings of the 30th International Baltic Road Conference, August 23 – 24, 2021

These proceedings contain full reports submitted in an open call to the 30th International Baltic Road Conference held virtually on August 23 and 24, 2021.

The conditions for including the reports in these proceedings were the following:

1. Full reports were submitted before the formal deadline of March 15, 2021,
2. Authors of reports have expressed their consent to include their reports in these proceedings.

Call for papers was launched in the second half of 2020, and until the formal deadline 50 full reports were received. The reports were reviewed by the scientific committee, consisting of:

1. Dr. Andrus Aavik, Tallinn University of Technology, Estonia,
2. Dr. Artu Ellmann, Tallinn University of Technology, Estonia,
3. Dr. Dago Antov, Tallinn University of Technology, Estonia,
4. Dr. Ainārs Paeglītis, Riga Technical University, Latvia,
5. Dr. Atis Zariņš, Riga Technical University, Latvia,
6. Dr. Juris Smirnovs, Riga Technical University, Latvia,
7. Dr. Viktors Haritonovs. Riga Technical University, Latvia,
8. M Eng Jānis Barbars, "Latvian State Roads" Latvia,
9. Dr. Alfredas Laurinavičius, Vilnius Gedimino Technical University, Lithuania,
10. Dr., Audrius Vaitkus, Vilnius Gedimino Technical University, Lithuania,
11. Dr. Daiva Žilionienė, Vilnius Gedimino Technical University, Lithuania,
12. Dr. Donatas Čygas, Vilnius Gedimino Technical University, Lithuania,
13. Dr. Viktoras Vorobjovas, Vilnius Gedimino Technical University, Lithuania.

The reports are categorised according to seven themes of the conference:

- Future Mobility, Strategic Planning & Road Financing
- Road Construction & Innovative Materials
- Bridges
- Traffic Safety
- Road Routine Maintenance
- Smart Road Solutions & ITS
- Environment, Climate Change & Energy Efficiency

The conference itself was organised as a fully virtual event with the help of a special on-line conference platform. The organiser - the Baltic Road Association - decided to organise a virtual conference because of the global Covid-19 pandemic in order to reach its target audience – road authorities and experts of the Baltic States, since there were severe restrictions for travelling and public gatherings in place in the Baltic States in August, 2021. International Baltic Road Conferences are important formal milestones when the chairing of the Baltic Road Association is transferred from one Baltic State to another, and this transfer is performed every fourth year usually in the end of August. Additional argument for not postponing the Conference for a year was that there were already a number of similar competing international events planned for 2022.

The conference was broadcast from Riga, Latvia on August 23 and 24, 2021 and attracted 289 delegates: 127 from Latvia, 51 from Lithuania, 48 from Estonia and 63 from other countries.

The number of conference reports reached 77: 9 keynote presentations, 6 dedicated presentations in dedicated sessions and 62 presentations in technical sessions. The number of presentations exceeds the number of reports included in these proceedings since there were few full reports not submitted before the formal deadline and since the organiser at a later moment decided to include few extra presentations because of the relevance of their topics.

The conference programme consisted of 2 keynote sessions with invited speakers in the beginning and the end of the conference, 2 dedicated sessions with invited speakers and 14 technical sessions for the speakers who participated in the open call for scientific reports. 4 parallel sessions were broadcast simultaneously, and each speaker had 15 minutes allocated for presentation. Time for questions & answers was allocated at the end of each session and they were monitored by the moderator of each session with the help of a chat function available in the conference platform. Rating function was included, and the participants could vote for the most interesting presentation. The participants could also tag the themes of their interests and update their personal profiles in order to promote interaction.

Broadcasting of the conference was performed by the conference centre of Radisson Blu Latvia Hotel, and virtual conference platform was provided by mitto.me .

In general, virtual format of the event proved to be successful and earned compliments from the participants.

IOP Conf. Series: Materials Science and Engineering 1202 (2021) 011002 doi:10.1088/1757-899X/1202/1/011002

Peer review declaration

All conference organisers/editors are required to declare details about their peer review. Therefore, please provide the following information:

- **Type of peer review: Single-blind / Double-blind / Triple-blind / Open / Other (please describe)**

  Each paper was blind peer-reviewed by two reviewers. Members of the Conference scientific committee served as reviewers.

- **Conference submission management system:**

  EasyChair platform (easychair.org) was used for the management of the Conference papers.

- **Number of submissions received:**

  Initially the Conference organiser received 86 abstracts of scientific papers.

- **Number of submissions sent for review:**

  82 submitted reports were sent for review.

- **Number of submissions accepted:**

  50 papers were accepted.

- **Acceptance Rate (Number of Submissions Accepted / Number of Submissions Received X 100):**

  (48 / 86) x 100 = 56 %

- **Average number of reviews per paper:**

  Two.

- **Total number of reviewers involved:**

  Thirteen reviewers from 3 countries:
  1. Alfredas Laurinavičius, Dr.sc.ing., Vilnius Gedemina Technical University, Lithuania
  2. Andrus Aavik, Dr.sc.ing., Tallinn University of Technology, Estonia
  3. Artu Ellmann, Dr.sc.ing., Tallinn University of Technology, Estonia
  4. Atis Zarins, Dr.sc.ing., Riga Technical University, Latvia
  5. Audrius Vaitkus, Dr.sc.ing., Vilnius Gedemino Technical University, Lithuania
  6. Dago Antov, Dr.sc.ing., Tallinn University of Technology, Estonia
  7. Daiva Žilionienė, Dr.sc.ing., Vilnius Gedemino Technical University, Lithuania

8. Donatas Čygas, Dr.sc.ing., Vilnius Gedemino Technical University, Lithuania
9. Juris Smirnovs, Dr.sc.ing., Riga Technical University, Latvia
10. Viktoras Vorobjovas, Dr.sc.ing., Vilnius Gedemino Technical University, Lithuania
11. Viktors Haritonovs, Dr.sc.ing., Riga Technical University, Latvia
12. Ainars Paeglitis, Dr.sc.ing. Riga Technical University, Latvia
13. Janis Barbars, M Eng., "Latvian State Roads", Latvia

- **Any additional info on review process:**

Summaries / abstracts of the submitted papers were also reviewed in preliminary evaluation.

During evaluation the following factors were considered:
1. Whether thehe author is familiar with the existing state of research,
2. Whether the topic is relevant to the scope of 30IBRC and corresponds to one of the seven main themes of the Conference,
3. Whether the report his is a new and original contribution,
4. Whether the title is appropriate,
5. Whether thehe abstract and keywords are adequate,
6. Whether the presentation of material is logical and technically correct,
7. Whether the interpretations and conclusions are sound and justified by the results,
8. Whether the writing style / English is clear and understandable,
9. Whether the paper is of the right length,
10. And finally, whether the references are adequate.

After the evaluaton the following recommendations were expressed towards each report :
- Acceptable without any changes
- Acceptable with modifications
- Unacceptable

- **Contact person for queries:**

Martins Dambergs martins.dambergs@lvceli.lv

Please submit this form along with the rest of your files on the submission date written in your publishing agreement.

The information you provide will be published as part of your proceedings.

EUROPEAN ITS PLATFORM: RETHINKING ITS

Orestis GIAMARELOS[1], Tobias REIFF[1]

[1]*Section Traffic Management and Road Maintenance, Federal Highway Research Institute, Bergisch Gladbach, Germany*

Corresponding author: Giamarelos, Orestis, giamarelos@bast.de

Abstract. The expectations placed on the existing European transport infrastructure are increasing due to the growing mobility needs of the population on the one hand and the rising volume of freight traffic on the other. The aim of the European ITS Platform is to increase the efficiency of the TEN-T core network through a better use of the existing infrastructure by implementing Intelligent Transport Systems (ITS). It fosters interoperability and the development of uniform technical standards by monitoring and disseminating the results delivered by the five CEF ITS Road Corridor Projects.
One major outcome is the drafting of the "Reference Handbook for harmonized ITS Core Service Deployment in Europe". This digital freely available handbook comprises a series of guidance and advice for use by road authorities and operators to support them in the development of their strategic approach, design development, deployment, installation and operation of ITS and remain compliant with European legislation. Its key features include: Deployment Guidance for 14 different ITS Services, including DATEX reference profiles, Guidance on the provision and use of traffic information provided under the European Directive (2010/40/EU), C-ITS developments, Knowledge gained from the collection of 100 Best Practices etc.

Keywords: ITS, deployment guidance, best practice, reference handbook, digitalisation, traffic management, traveller information services, freight and logistics, C-ITS, DATEX.

Introduction

The European ITS Platform (short: *EU EIP*), a project co-financed by the Connecting Europe Facility of the European Union, is the follow up of projects already supported by the European Commission TEN-T programme named "European ITS Platform (EIP)" (2013-2015) and "European ITS Platform+ (EIP+)" (2014 - 2015). It is the place where National Ministries, Road Authorities, Road Operators and partners from the private and public sectors cooperate in order to foster, accelerate and optimize current and future ITS deployments in Europe in a harmonized way. Ensuring continuity of high-quality services for European end-users requires the creation of a proper environment for the harmonization of existing and future ITS Services. The European ITS Platform brings together the majority of the European key players, cooperating to establish an open "forum", aiming at providing valid contribution for the future strategy and policy recommendation for better development of ITS service along European road Corridors.

To increase the efficiency of the TEN-T Core Network Corridors, it is mandatory to encourage the development of an integrated trans-European network and a better use of the existing infrastructures by employing intelligent transport systems as well as uniform technical standards. Interoperability must be discussed, designed, tested and finally deployed on the basis of the evolution of technology, standards, specifications and open interfaces. By monitoring, processing, evaluating and disseminating results delivered by the five ITS Road Corridor projects: Arc Atlantique, Crocodile, NEXT-ITS, MedTIS and URSA MAJOR, the Works projects that are co-funded by the European Commission within the CEF MAP ITS Call 2014, the European ITS Platform can be considered as the technical European ITS "Knowledge Management Centre", contributing significantly to the most effective use of ITS standards and specifications. EU EIP includes also the evaluation of the results provided by the ITS Road Corridor projects in order to consolidate a harmonized and substantiated impact evaluation of the socio-economic benefits of ITS services.

Partners from 15 member states participate in EU EIP. Figure 1 presents their main tasks covered by various activities, working groups and task forces, as well as their major achievements in this framework. A significant achievement of the project, led by the "Monitoring and Dissemination" activity, is the publication of the "Reference Handbook for harmonised core ITS service deployment in Europe", which is presented in this paper.

Figure 1. The European ITS Platform in one figure.

Reference Handbook for harmonised core ITS service deployment in Europe

The “Reference Handbook for harmonized core ITS service deployment in Europe” (short: Reference Handbook) constitutes an essential basis for a harmonised and cross-border implementation of ITS services. It comprises a series of guidance and advice for use by road authorities and operators to support them in the development of their strategic approach, design development, deployment, installation and operation of ITS, while remaining compliant with European legislation. The purpose of this guidance is to assist Member States in taking a broadly similar approach so wider European added value can be achieved, whilst at the same time delivering the needs of individual Member States.

The content of the Reference Handbook has been written by ITS practitioners and experts in the field of Traffic and Traveller Information Services, Traffic Management Services and Freight & Logistics Services from across Europe, working for Road and Transport Authorities and Road Operators, in cooperation with the European Commission. The main part of its content had originally been drafted and adopted by Member States in 2012 as separate Deployment Guidelines, with an update of the included Best Practice examples in 2015, under the EasyWay and EIP programmes. The current handbook is now published as a major revision to reflect changing requirements in a time of radical technological change.

There were several reasons that made the update of the Deployment Guidelines a necessity. A main reason was the entry into force in the coming years of Delegated Regulations affecting many of the included ITS services and this made a revision of the ITS Deployment Guidelines essential, in order to maintain compliance with the regulations. Furthermore, updating all service descriptions to the state of the art, including latest developments in digitalisation and C-ITS, as well as incorporating knowledge gained from the collection of Best Practices and the results of the ITS Corridors and the EU EIP activities (e.g. Quality Package, NAPs, KPIs, Interfaces for Data Exchange etc.) is fundamental, in order to provide guidance to the ITS Community that is up-to-date, comprehensive and complete in all aspects. Making the information easier to reach and easier to understand and assimilate was another goal of the planned update, eliminating redundancies and merging all previously separate documents into one complete Reference Handbook. The major new features that the Reference Handbook introduces in order to overcome these challenges are presented in Figure 2.

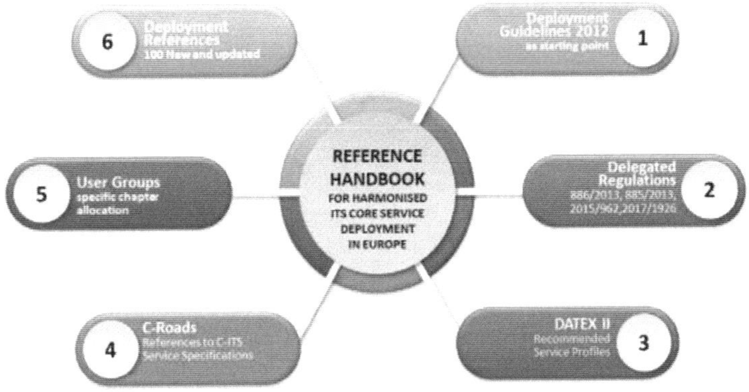

Figure 2. The major new features introduced in the Reference Handbook.

1. The Conceptual Setup – Deployment Guidelines 2012 as starting point

The original set of 19 separate documents (Easyway & European ITS Platform, 2012-2015) is now consolidated into one handbook, eliminating repetitions and redundancies and having all knowledge on ITS core service deployment in one document.

The conceptual setup is visualised in Figure 3. It begins with a common part which is valid for all ITS Services and is followed by three blocks: Traffic and Traveller Information Services, Traffic Management Services and Freight and logististic Services. Within these blocks there is a separate chapter for every ITS Service, e.g. Variable Speed Limits, Ramp Metering, Intelligent and Secure Truck Parking etc. The full list of ITS services included in the Reference Handbook is presented in Table 1.

Following these three blocks, there is a common annex with additional information for the interested users, covering the operating environments, presenting all deployment references collected grouped by ITS core service, as well as practical checklists to assist with service deployment.

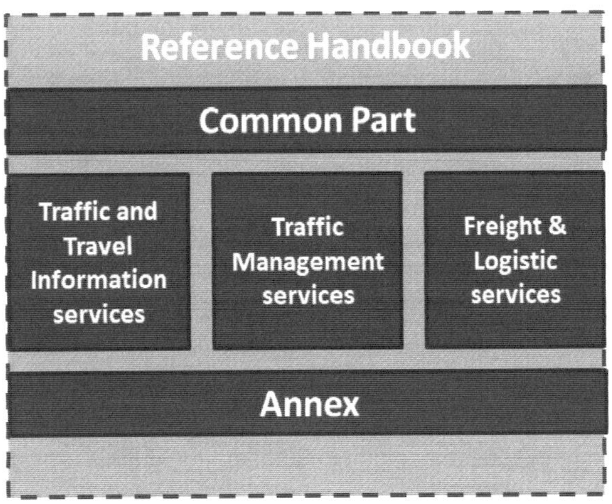

Figure 3. The conceptual setup.

Table 1. The ITS core services covered in the Reference Handbook.

Traffic and Traveller Information Services	Traffic Management Services	Freight & Logistic Services
- Forecast and Real Time Event Information - Traffic Condition and Travel Time Information Service - Speed Limit Information - Weather Information Service - Co-modal Traveller Information Services	- Dynamic lane management - Variable speed limits - Ramp metering - Hard shoulder running - Incident warning and management - HGV overtaking ban - Traffic Management Plans for Corridors and Networks	- Intelligent and Secure Truck Parking - Access to Abnormal Goods Transport Regulations

The core and the primary objective of the Reference Handbook is the harmonisation of the ITS service deployment throughout Europe. For each of the European ITS Core Services, requirements and advice are formulated from a pan-European perspective in such a way that:

- functional, organisational and technical interoperability between the ITS services is achieved,
- the end user can perceive and use the services offered in the same or at least a similar way (common look and feel),
- uniform implementation and evaluation benchmarks for the deployment of ITS Core services are available to the acting road operators when they intend to implement a new ITS service or improve an existing one.

The requirements and advice are formed with the keywords "MUST", "SHOULD" and "MAY". These are to be interpreted as described in RFC 2119 (Bradner, 1997). In order to claim compliance, a certain deployment must follow these rules:

- MUST / MUST NOT: An absolute requirement/prohibition. In case of non-fulfillment, only insurmountable reasons can be stated (e.g. legal regulations).
- SHOULD / SHOULD NOT: A strong suggestion. Non-fulfillment must be supported by very clearly described and for third parties comprehensible and traceable reasons.
- MAY: These elements are optional.

2. Delegated Regulations

The second major development is ensuring compliance to the Delegated Regulations under the ITS Directive. The relevant Delegated Regulations are shown in Table 2. The experts of the European ITS Platform checked the data requirements resulting from these Delegated Regulations and formulated relevant requirements for all affected ITS Services accordingly.

The practical gain is that users of the Reference Handbook can see easily for a deployed ITS service which data elements must be provided to the NAPs, in order to be compliant to the Delegated Regulations.

Table 2. The Delegated Regulations under the ITS Directive.

(EU) 885/2013	The provision of information services for safe and secure parking places for trucks and commercial vehicles.
(EU) 886/2013	The provision, where possible, of road safety-related minimum universal traffic information free of charge to users.
(EU) 2015/962	The provision of EU-wide real-time traffic information services.
(EU) 2017/1926	The provision of EU-wide multimodal travel information services.

3. DATEX II – Recommended Service Profiles

This innovation is the product of the cooperation of the Monitoring & Dissemination activity of EU EIP and the DATEX II Program Support Action, also co-funded by the European Union, with the support of the EU EIP activity on "Liaison and harmonization on interfaces for data exchange".

The experts involved in this task created a Recommended Service Profile for every ITS service included in the Reference Handbook, assisting this way the user with service deployment. All created profiles are available free of charge to all interested users and can be accessed through the Schema Generation Tool on the DATEX II webpage (DATEX II, 2020). The linkage to the Recommended Service Profiles is realized with the help of the newly introduced interface requirements and information provision standards which are provided for every service.

Figure 4 showcases the developed Recommended Service Profiles in the Schema Generation Tool of DATEX II PSA.

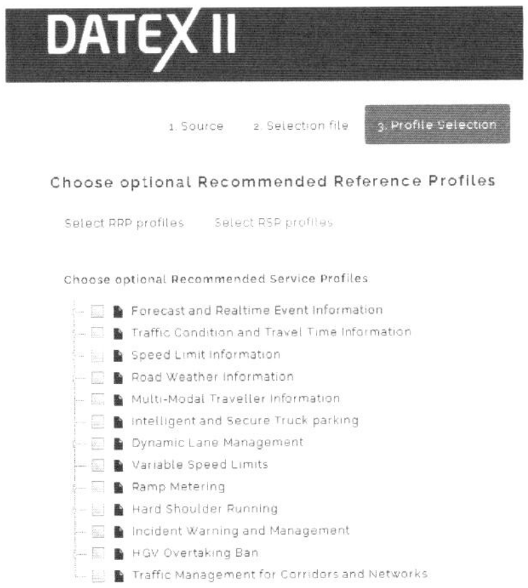

Figure 4. The DATEX II schema generation tool.

4. C-ROADS – References to C-ITS Service Specifications

In cooperation with the C-Roads project, the experts checked if C-ITS Services can support the ITS core services. A mapping of the ITS and C-ITS services has taken place in the development process of the Reference Handbook, in order to identify the services that support and relate to each other. In the new "IF2 requirements" that have been introduced in the handbook, the data elements needed for the C-ITS use case are listed. In addition, the C-Roads specification for this use case is referenced in the "information provision standard" sub-chapter for each ITS service. The references link to the latest C-Roads Release document (C-Roads, 2021).

5. User Groups

An analysis of the prospective users of the Reference Handbook and their needs took place in the drafting process of the handbook. The reason is that in such an extensive document it is especially important to ensure that all interested users can find the information they need. This analysis led to the definition of three main user groups of the handbook (Strategic Bodies, Implementation Managers and Expert Engineers). The next step was to allocate the chapters of the handbook that are of relevance to each group. This way each user group can have a tailor-made version of the document based on the needs of its members.

IOP Conf. Series: Materials Science and Engineering 1202 (2021) 012001 doi:10.1088/1757-899X/1202/1/012001

6. Deployment References

The collection of Deployment References is another task of the Monitoring and Dissemination activity of EU EIP that significantly contributed to the development of the Reference Handbook. The collection of ITS deployments from all over Europe started in the summer of 2017. The Deployment References come for the most part from the 5 CEF ITS Corridors but also include national projects from various member states. The 5 CEF ITS Road Corridor projects, which cover most member states of the European Union, are illustrated in the map in Figure 5. The Cross Corridor Cooperation task within the Monitoring and Dissemination activity, which was set up to enhance knowledge exchange and cooperation among the CEF ITS Corridors, also significantly facilitated the process and paved the way for this stream of Deployment References.

A standardized template was developed in order to collect these deployments. A sample page of this template is shown in Figure 6. It includes information on the ITS deployments, starting from the location of the ITS Service, the objectives, the budget, connection to different systems, benefits and lessons learnt from each deployment.

At the end of this task, approximately 100 deployment references have been collected, providing significant knowledge for the ITS service sections of the Handbook but also providing examples of ITS deployments to the interested users. All collected deployment references are provided in full form in an annex of the handbook.

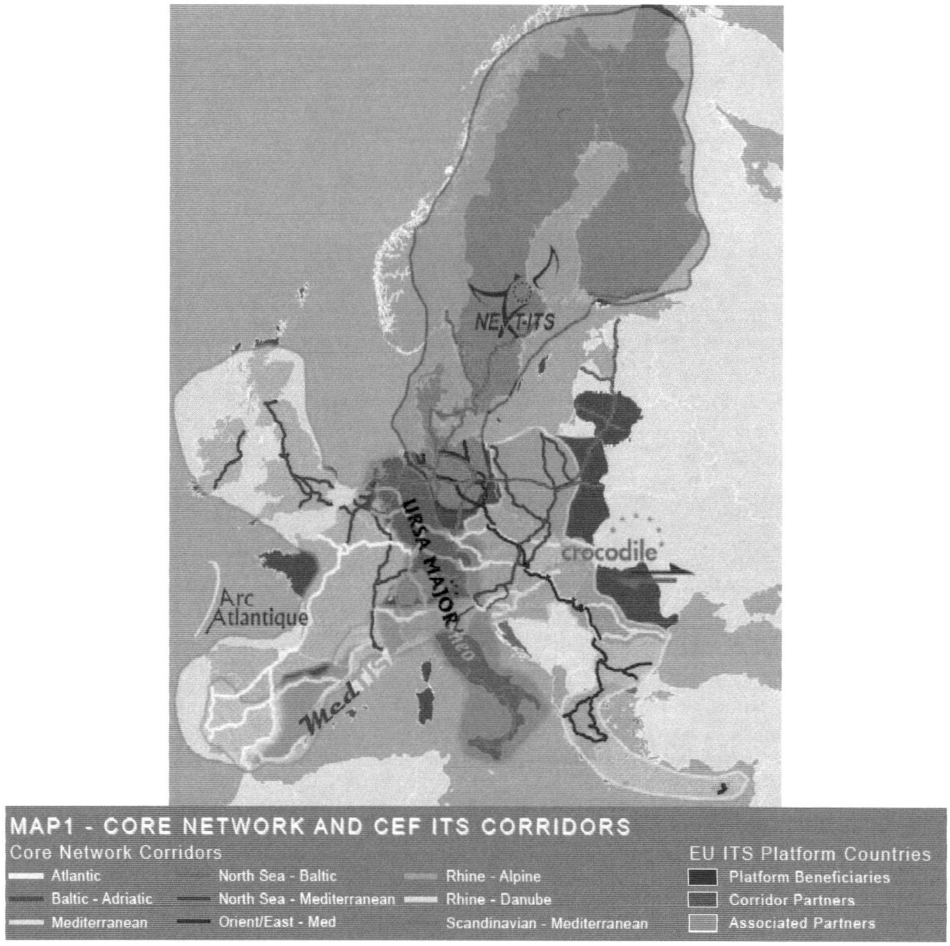

Figure 5. Map displaying the 5 CEF ITS Corridors: Arc Atlantique, Crocodile, MedTIS, NEXT-ITS and URSA MAJOR.

	defined to shorten the red phase or even to deactivate the traffic lights. If applicable, the number of vehicles allowed to pass can be increased from one to two. But in practice, algorithms ensure deactivation of RM traffic lights only in cases of heavy congestion on main carriageway and queuing vehicles on the entering ramp up to the detector requesting yellow. (see illustration #3)
Lessons learnt / factor of success / topics considered as good practice (technical, legal, organisational, financial)	Main effect of ramp metering in highly frequented parts of the motorway network is fragmenting vehicle platoons on the ramp. This ensures traffic flow on main carriageway. The original ALINEA algorithm is often too strong to reach this aim and needs modification in order to shorten the red phase.
Impacts assessment / results (if available)	Evaluation studies confirm the effectiveness of implemented systems. Occurrence of gridlocks and duration of congestions have been notably reduced. The speed level on main carriageway during rush hours has been increased when systems are in operation.

ILLUSTRATIONS

Installation with dynamic message about number of vehicles allowed to access (motorway A4, junction Moitzfeld)

Figure 6. A sample page of a collected Deployment Reference template.

Conclusion – Outlook

The European ITS Platform provides the necessary basis for National Ministries, Road Authorities, Road Operators and partners from the private and public sectors to cooperate in order to foster, accelerate and optimize current and future ITS deployments in Europe in a harmonized way and ensuring that in this process and in this new era of digitalisation no one is left behind.

The Reference Handbook for harmonised core ITS service deployment in Europe is a major outcome of this cooperation. It has been created by ITS experts and practitioners and refined in a commentary process by member states' experts. It constitutes an essential basis for a harmonised and cross-border implementation of ITS services. Thus, it is a powerful tool in the effort to master the ever-increasing challenges the European transport infrastructure faces. It makes a significant contribution to maintain a competitive European economy and achieve the three overarching objectives of efficiency, safety and the environment.

Finally, connected mobility and autonomous driving will bring new opportunities for road operators and road authorities but also significant new challenges, which will require a joint and coordinated effort. A European activity which offers a platform for a practical exchange of experiences and lessons learnt, as well as the development of harmonisation activities for ITS will continue to play an important role in the common effort to make the transport system safer, more efficient and environmentally friendly. The achievements of the European ITS Platform and the experience gained from the cooperation among the EU Member States on this matter in the framework of EU EIP and the preceding related projects constitutes a very promising sign for the future.

Funding

The European ITS Platform (EU EIP) is a project co-funded by the Connected Europe Facility of the European Union.

References

Bradner, S. (1997). *Key words for use in RFCs to Indicate Requirement Levels.* Internet Engineering Task Force – Network Working Group. https://www.ietf.org/rfc/rfc2119.txt
C-Roads (2021), *Common C-ITS Service and Use Case Definitions.* https://www.c-roads.eu
DATEX Group (2020), *Schema Generation Tool.* https://webtool.datex2.eu
Easyway & European ITS Platform (2012-2015). *ITS Deployment Guidelines 2012-2015.* https://dg.its-platform.eu/DGs2012

CODEC: CONNECTED DATA FOR ROAD INFRASTRUCTURE ASSET MANAGEMENT

Sukalpa BISWAS[1], John PROUST[1], Tadas ANDRIEJAUSKAS[1], Alex WRIGHT[1], Carl VAN GEEM[2], Darko KOKOT[3], António ANTUNES[4], Vânia MARECOS[4], José BARATEIRO[4], Shubham BHUSARI[5], Uros JOVANOVIC[6]

[1]*TRL Ltd, Wokingham, UK*
[2]*BRRC, Brussels, Belgium*
[3]*ZAG, Ljubljana, Slovenia*
[4]*LNEC, Lisbon, Portugal*
[5]*RHDHV, Amersfoort, Netherlands*
[6]*BEXEL, Ljubljana, Slovenia*
Corresponding author: Sukalpa Biswas, sbiswas@trl.co.uk

Received 15 March 2021; accepted date

Abstract. Road infrastructure asset management is rapidly transforming into a digital environment where data accessibility, effective integration and collaboration and accessibility from different sources and assets are key. However, current asset management processes are not yet fully integrated or linked, and there are incompatibilities between various systems and platforms that limit the ability to integrate asset management with BIM. The CoDEC project has sought to understand the current status of information management for assets, including inventory, condition and new data sources such as sensors and scanning systems, to identify the challenges and needs for linking and integrating different data sets to support effective asset management. As a result, CoDEC has developed a data dictionary framework to help link/integrate static and dynamic data for the "key" infrastructure assets (road pavements, bridges, tunnels). This will enable BIM and Asset Management Systems (AMS) to exchange data and help optimise and integrate data management across systems and throughout the different asset lifecycle phases, from build to operation. This work will be followed up with three pilot projects to demonstrate the feasibility of integrating asset data from various sources through linked data / semantic web technology to build the connection between AMS and BIM platforms.

Keywords: Asset management, BIM, Data dictionary, Linked data, Ontology, Asset data

Introduction

Building Information Modelling (BIM) is an information management process that has the potential to support asset management from concept to the end of life. The process is designed so that asset information can be generated, captured, maintained and used efficiently and effectively to optimise asset management. However, to date BIM processes have focused on the information gathered during the construction phase of the asset and typically do not address the data requirements required during the operational phase. There is a gap in defining the information requirements for the operational phase, and in how to define/accommodate these across BIM systems. The CEDR Transnational Research Programme funded CoDEC project (Connected Data for Effective Collaboration) had aimed to address this gap by establishing a better understanding of how BIM principles could be applied within the European highways industry to manage asset data during the operational phase. In particular, CoDEC has developed a specification to support the establishment of connections between asset management systems and BIM platforms - to make best use of legacy data and sensor/scanner data provided by new technologies. CoDEC has developed a "Data Dictionary" for infrastructure assets that could form the basis for data structures to support integration between different data management systems, and thereby improve the flow of asset data. Ultimately, the project aims to assist Road Authorities make efficient and effective use of data to support asset management.

1. Current state of using data dictionaries for asset management

A review of existing asset management and BIM practice within European National Road Administrations (NRA) was carried out by initially circulating a survey to a wide group of stakeholders. This was followed up with interviews with experts, which included detailed discussion on legacy and new data types (sensor data), data dictionaries, integration between management systems and the use of BIM platforms for highway assets (roads, bridges and tunnels). The review and consultation aimed to provide insight into the current situation within NRAs regarding data within asset management and their roadmaps for integrating asset management data with BIM, to establish the foundations for the development of the data dictionary.

1.1. Legacy data

Within the context of CoDEC we consider legacy data to be data that provides key information on the assets that fall under the responsibility of the NRA, and is the primary dataset applied for asset management. The data can be applied for the purposes of inventory (what is there), condition (how is it performing) and operation (how is it being used). The condition and operational aspect are commonly more challenging with respect to data collection and management.

The review found that the condition of assets is commonly monitored, but management of the condition data itself is disparate. The data are often stored in separate databases managed by individual NRA departments responsible for particular assets. These specialist data are often only accessible to members of the department and are shared with colleagues in other departments on request. Condition and maintenance management information is shared in various formats such as reports and maps. For pavements many NRAs use dedicated commercial Pavement Management Systems (PMS). For bridges, Bridge Management Systems (BMS) may be developed in-house or commercially sourced. Technology equipment installed on the network is again commonly managed by separate departments. Major tunnels are typically managed individually.

Operational asset management data typically includes data from real-time monitoring (cameras), traffic counters, traffic load sensors (via Weigh in Motion, WIM), bridge monitor sensors, monitoring of ventilation systems in tunnels etc. These sensors and devices are all assets that form part of the larger road, bridge or tunnel asset, and hence of the asset network as a whole. This data is typically used for local or network operational management by particular NRA departments, although some of the data is made available to the public (e.g., traffic counts).

1.2. Sensor and scanning data

Emerging technologies are opening up opportunities for NRAs to improve asset data collection, analysis and management. The review identified many technologies that NRAs are starting to use to support management of the network. In particular, a variety of different sensor/scanning technologies and data processing techniques are now available to NRAs, which CoDEC has structured into seven technology families:

- Embeddable and Fixed Sensors (e.g. Weather, WIM, loops, fire detectors, visibility meters, etc.).
- Airborne, terrestrial, and mobile type LiDAR technologies for monitoring highways and structures.
- Satellite Data Monitoring (satellite and aerial imagery, InSAR) and Unmanned Aerial Vehicles (UAVs) for inventory and monitoring of highways infrastructure.
- Internet of Things (IoT) and Connected Sensor Networks.
- Probe Vehicles and systems within Autonomous Vehicles that provide vehicle sensor data that can be used for highways asset management (crowdsourcing).
- Smartphones (crowdsourcing).
- Advanced Data Processing techniques (artificial intelligence, machine learning algorithms, etc.) to extract information on asset condition from images, video footage.

It was found that the integration of data from these new sensors and scanning technologies is not yet as mature as the technologies themselves. Although most of the technologies reviewed have a Technology Readiness Level (TRL) above 6, use cases are typically in the pilot/demonstration stage and are not implemented network wide. However, the extent and maturity of the sensor and scanning technology families differ depending on the asset type. For example, embeddable and fixed sensors, including IoT sensors and sensor networks, are relatively common in structures (bridges and tunnels), while vehicle-based sensing technologies and laser scanning are more common for carriageways.

1.3. Asset management and BIM

Many NRAs use BIM during the design and construction phases of large projects. After construction, the BIM model is sometimes delivered as a detailed "as-built" model, but little further use is made during the remainder of the operational life cycle of the asset (this is more commonly the case for pavement assets than structures such as bridges or tunnels). However, long-term asset management is still typically carried out in a traditional way using dedicated asset management systems:

- Pavement Management Systems (PMS) contain data about the road network, structural information about the layers of the road, traffic density and the current condition (e.g. using condition parameters that can be converted into indicators) of individual lengths of pavement
- Bridge Management Systems (BMS) contain bridge inspection data on each component (usually standardized, e.g. following PIARC recommendations) and data provided by sensors and advanced inspection techniques.
- Tunnel management systems contain data on the tunnel structure itself and the electronic equipment used to support tunnel operation.

Although a few initiatives across Europe have attempted to define standard formats for the data (e.g. AM4INFRA (Marcovaldi, Biccellari, 2018)) contained in these systems in general, the system of data management is very much tailored to the individual AMS within the NRA. Little work has been done to create a standardized "Data Dictionary" for each type of asset. Our review showed that attempts have been made by Highways England through its Asset Data Management Manual (ADMM) (Highways England, 2020), and the Lithuanian NRA has a Lithuanian State Road Information System (LAKIS). However, few other publicly-available data dictionaries were identified in the states contained in the review (Norway, Sweden and Germany), or is other sources providing information (France, Australia/New Zealand). It was found that whilst other countries do not have data dictionaries as such, some NRAs have developed Object Type Libraries (OTL), including Netherlands, Flanders and Finland. Most OTLs/data dictionaries include roads, bridges and tunnels. As noted above, the gap in the availability, extent and content of data dictionaries for highway assets presents a barrier to the exploitation of this data, in particular within the BIM environment.

2. Developing a Data dictionary

The purpose of a data dictionary is to provide a registry of detailed information about the contents of a dataset or database, such as the names of objects, their properties, data types or formats, and text descriptions. A data dictionary thus provides a common schema for anyone needing to use the data it contains and is therefore critical in aligning multiple stakeholders across different organisations, disciplines, and use cases.

One of the key objectives of the CoDEC project is to develop a data dictionary for major highway assets which is both general enough to be widely applicable across various NRAs, while also being specific enough to be operationally useful in practical terms.

The CoDEC Data Dictionary covers three key highway civil assets (pavements, bridges and tunnels), as well as preliminary entries for supporting assets/systems such as lighting, fire-fighting, and drainage. CoDEC has considered both legacy data for these assets, and data from new sources such as sensors and scanners that offer the potential to transform NRAs' future ability to manage highway assets.

The CoDEC Data Dictionary has been developed by building on previous work carried out in AM4INFRA to develop a data dictionary on tunnels and bridges, the Highways England UK-ADMM data dictionary (Highways England, 2020), the Data Standard for Road Management and Investment in Australia and New Zealand (DSRMI, for tunnels) (Austroads, 2019) and ifcRoad (buildingSMART, 2020).

2.1. Data Dictionary Structure

Assets vs components

Infrastructure assets are complex 'objects', with many interdependent elements making up what one might call "an asset". There is not always a clear delineation between what can be called an asset, and what is only a part of an asset – in fact, the delineation depends largely on the viewpoint of the person(s) making the judgement (e.g.: is a single kerbstone an asset? Or is it only a component of a whole kerb – and should this whole kerb then be considered an asset? Or, is the kerb a component of the roadway, and the roadway is the asset?). Further, when trying to define an asset based on its constituent parts, it is important understand the level of detail required – an asset like a tunnel could be described down to its individual nuts and bolts, or more simply in broad terms covering only major components.

The issue of defining an asset by its parts is made even more complex when considering that certain 'parts' of an asset are not really physical, but rather are abstractions used in order to describe and manage an asset – a key example of this is a road 'section' (sometimes called a road 'link'), which is simply a way of longitudinally delineating a length of road using defined start and end points. Although the road section of course corresponds to a real, physical road, the actual physical asset is not in reality 'sectioned' in this way – this abstraction is simply used in order to better manage and understand the road, since the road network must often be considered in separate 'sections' for practical reasons.

For CoDEC the guiding principle applied to address the challenge of defining and delineating assets and their components was to consider the ultimate application: *the need to develop a data dictionary that will support the management of that asset* (taking into account the need to accommodate the needs of sensors and sensor data). The information from the review and the experience and knowledge of the team in infrastructure asset management (i.e.: how assets are actually managed, practically), was brought together to draw the necessary judgements on: (1) what constitutes "an asset" vs the components of that asset, and (2) the level of detail needed to adequately describe that asset for the purposes of management.

It was decided to use a two-tier system for describing assets and their constituent parts, and that this system should be the same for all asset types, in order to maintain consistency (especially in terms of level of detail). Two types of objects were defined: Entities, and Elements: 'Entities' correspond to what are considered as assets; 'Elements' correspond to what are considered as components if those assets.

The terminology is based on ISO 12006-2:2001 (International Organization for Standardization, 2001), which defines the terms 'construction entity' and 'element'. The decision to maintain a consistent definition across all assets presented challenges, since different asset types may have very different levels of complexity in terms of their constituent parts. For example, a road pavement can be broken down into constituent parts in a relatively simple way while retaining a reasonable level of detail, whereas a bridge may require a more complex breakdown to adequately describe its components. Essentially, there are pros and cons either to maintaining a consistent approach or to allowing the approach to vary by asset type. CoDEC decided to prioritise consistency in this case.

The 'classification' layer

It is not strictly a requirement of a data dictionary that the objects within it are classified, or organised, in any particular way. Indeed, when considering only the machine-readable applications of a data dictionary classification is not necessary – it primarily functions as an aid to human-readability. Despite the ultimate goal of using the data dictionary in a machine-readable environment, it was decided that the human-readability was important (especially since it is intended to be widely-shared), and so a classification layer was included.

A two-tier classification system for infrastructure asset entities (class and sub-class) based on *UniClass 2015* (NBS, 2015) was used. So, for example, a bridge would be classified as: Entity Class = "Structures", and Entity Sub-Class = "Bridges". For sensors, the only classification which was meaningful within the context of our data dictionary was whether the sensor was a fixed-location or mobile sensor – therefore only one classification tier for sensors was used, with only these two options. Object properties were classified using a single classification tier: for asset entity/element properties we used a classification based on *Omniclass* (Construction Specifications Institute, 2017); and for sensor properties we used a classification based on *SensorML* (Open Geospatial Consortium, 2020).

A special case - sensors

A key aim of CoDEC is to consider sensors and the data they provide, which are increasingly used to support infrastructure asset management. It was decided that the data dictionary should not be simply a 'directory' of different sensor types and their properties, since this would quickly become an impossible task due to the (ever-increasing) number of different sensors available, and the detailed properties of each one. Hence, CoDEC considered it more useful to develop a set of "general" property sets and definitions which would apply to any type of sensor. The data dictionary is hence future-proofed and will remain useful in a wide variety of use cases.

Therefore sensors were not considered as 'assets' in themselves, but rather as separate objects. The data dictionary focused on identifying the property sets which would apply across in general to types of sensor. However, when considering sensors CoDEC considered it necessary to develop different property sets for sensors that have fixed-locations and those that are mobile. A key reason for this is to address differences in the approach taken to referencing the location of fixed and mobile sensors. In addition there can be differences in how sensors are defined - for example, one can consider an array (or network) of multiple fixed-location sensors but this does not apply to mobile sensors. Therefore, sensors CoDEC has placed sensors in their own dedicated section of the data dictionary, separate from asset entities and elements.

2.2. Data Dictionary Content

Once the structure was decided, the 'content' of the data dictionary was developed – i.e. identifying and defining the assets (entities), components (elements), and their properties; the sensor types and their properties; and the properties of the sensor data.

From our review of existing data dictionaries developed for the purpose of highways asset management it became clear that there were very few relevant examples available in the public domain. Ultimately, CoDEC identified two potential sources: the data dictionary developed as part of the AM4INFRA project (Marcovaldi, Biccellari, 2018), and Highways England's (English NRA) Asset Data Management Manual (ADMM) (Highways England, 2020)). However, a detailed review of both of these sources concluded that neither fully met needs of the CoDEC project as the AM4INFRA dictionary tends to a high/theoretical level, whilst the ADMM is set at a lower level with specific focus on Highways England's operating context. However, CoDEC drew on these to design a Data Dictionary that fall within the space between them - i.e. detailed enough for practical application, yet generic enough to be applicable across the many operating contexts found in European NRAs.

Further review, combined with workshops undertaken within the CoDEC consortium (drawing on the practical expertise of members to translate knowledge of asset management and asset data directly into the data dictionary), and assessment of real-world datasets was used to inform the content of the data dictionary. This included considering examples of sensor-produced datasets to identify the general properties shared across sensors, which provided key input to the 'Sensor' and 'Sensor Data' sections of the dictionary. Further workshops were held in which the data dictionary content was presented and discussed with representatives from CEDR NRAs to validate the approach and the content. An extract from the data dictionary is shown in Figure 1 (the figure is truncated to fit, and as such does not show all fields). This shows an example of the content and structure for the 'Lane' element (part of the 'Road section' entity).

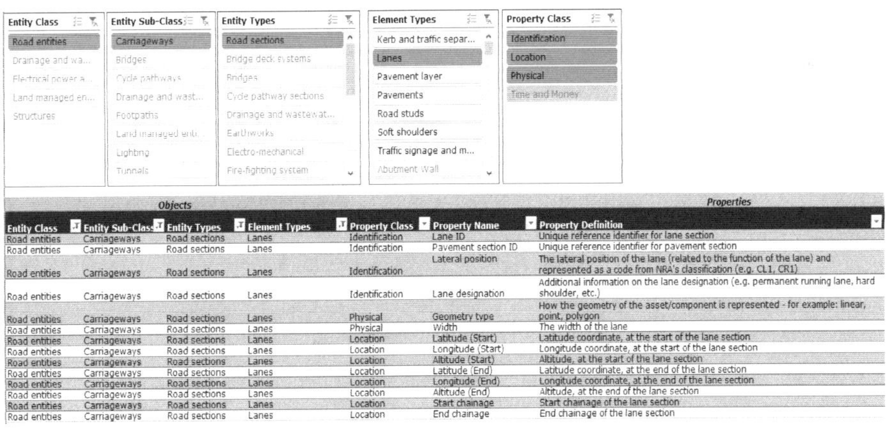

Objects				Properties		
Entity Class	Entity Sub-Class	Entity Types	Element Types	Property Class	Property Name	Property Definition
Road entities	Carriageways	Road sections	Lanes	Identification	Lane ID	Unique reference identifier for lane section
Road entities	Carriageways	Road sections	Lanes	Identification	Pavement section ID	Unique reference identifier for pavement section
Road entities	Carriageways	Road sections	Lanes	Identification	Lateral position	The lateral position of the lane (related to the function of the lane) and represented as a code from NRA's classification (e.g. CL1, CR1)
Road entities	Carriageways	Road sections	Lanes	Identification	Lane designation	Additional information on the lane designation (e.g. permanent running lane, hard shoulder, etc.)
Road entities	Carriageways	Road sections	Lanes	Physical	Geometry type	How the geometry of the asset/component is represented - for example: linear, point, polygon
Road entities	Carriageways	Road sections	Lanes	Physical	Width	The width of the lane
Road entities	Carriageways	Road sections	Lanes	Location	Latitude (Start)	Latitude coordinate, at the start of the lane section
Road entities	Carriageways	Road sections	Lanes	Location	Longitude (Start)	Longitude coordinate, at the start of the lane section
Road entities	Carriageways	Road sections	Lanes	Location	Altitude (Start)	Altitude, at the start of the lane section
Road entities	Carriageways	Road sections	Lanes	Location	Latitude (End)	Latitude coordinate, at the end of the lane section
Road entities	Carriageways	Road sections	Lanes	Location	Longitude (End)	Longitude coordinate, at the end of the lane section
Road entities	Carriageways	Road sections	Lanes	Location	Altitude (End)	Altitude, at the end of the lane section
Road entities	Carriageways	Road sections	Lanes	Location	Start chainage	Start chainage of the lane section
Road entities	Carriageways	Road sections	Lanes	Location	End chainage	End chainage of the lane section

Figure 1: Extract from the CoDEC data dictionary showing the entries under the 'Lane' element

3. Linked Data Methodologies and Tools

Linked Data and Semantic Web methodologies use ontologies to structure and share data. An ontology can be defined as a "formal, explicit specification of a shared conceptualization" (Studer et al., 1998), meaning that concepts, their constraints, and their relationships are encoded in a way that is systematically structured, explicit and machine-readable. This allows ontologies to be used to integrate and retrieve information, obtain semantically enhanced content, and to support knowledge management.

The Resource Description Framework (RDF), RDF Schema (RDFS) and the Ontology Web Language (OWL) were developed by the World Wide Web Consortium (W^3C). The RDF provides the basis for the "creation, exchange and use of annotations on the web" (Pan, Horrocks, 2009), using statements in the form of triples (subject, property, object). RDFS introduces class and hierarchy concepts, while OWL provides additional vocabulary and expressiveness (e.g. disjointedness, cardinality, object and data properties).

The CoDEC Data Dictionary draws on these concepts to provide a shared vocabulary and common language to enable the integration and sharing of data between different management systems. Linked data technologies, such as ontologies, are used to encode asset and sensor data in a formal, comprehensible, and explicit way.

3.1. Methodology for developing Linked Data

The Data Dictionary describes asset data using concepts, relationships and properties about highway assets, sensors, and sensor data. As noted above, the CoDEC ontology has been developed to build on the EUROTL framework (INTERLINK project, 2018). As a first step each Data dictionary concept or relationship was mapped to an existing class or property in EUROTL, as shown in Table 1. Properties can be defined either as an object property or data property, meaning a semantic relation between classes (for objects) or between the class and data (e.g. strings or numbers). Where a mapping is not present in the EUROTL CoDEC has created a new class or property (the CoDEC ontology has been developed using Stanford's Protégé (Musen, 2015)). However, this is always a sub-class of an EUROTL concept. This means that new CoDEC classes are an extension of the previous and ensures interoperability between the two ontologies.

As an example, the Bridge concept already exists in the EUROTL Framework (AM4INFRA:Bridge (Marcovaldi, Biccellari, 2018)). However the concept of a Structural Element (or equivalent) is not found in EUROTL. Hence, a new Structural Element class was created in the CoDEC ontology, as a sub-class of the already existing EUROTL concept EurOTL:PhysicalObject.

Table 1. Example of data dictionary to ontology mapping

Data Dictionary			**Ontology**		
Property	Description	Format	Domain	Object/Data Property	Range
Bridge ID	The unique reference identifier for bridge	String	bridgeID	is-a	Bridge
Bridge name	The name of the bridge	String	bridgeID	rdfs:label	xsd:string
Environment	Classification of surrounding environment (e.g. Rural/Urban)	String	bridgeID	inEnvironment	xsd:string
Region/District/Area	Relevant geographical situation	String	bridgeID	prov:atLocation	eurotl:LocationByIdentifier
Owner	Owner of the asset	String	bridgeID	hasOwner	prov:Agent (person or Org.)

3.2. API

An API has been developed by the CoDEC project to support data linking between systems. This is based on the principle of OpenAPI, as shown in Figure 2. On the bottom layer, the data integration environment is composed of a set of aligned ontologies, namely the EUROTL framework, the CoDEC ontology (based on the data dictionary) and a sensor ontology (the Sensor Network Ontology (Open Geospatial Consortium, 2020)). In order to manage the complexity of the linked data environment and create a separate layer that can be used without interfering with other layers, CoDEC exposes a set of REST services. These services are responsible for communicating with the linked data environment (typically through a set of SPARQL queries), and can be used by any application, as long as it has permission to access both services and data.

This layered approach has several advantages, the most critical one being the separation provided by multiple layers, which allows modification of the linked data structures without affecting the normal behaviour of external applications, as they "just need" to know how to call the services (their inputs and outputs).

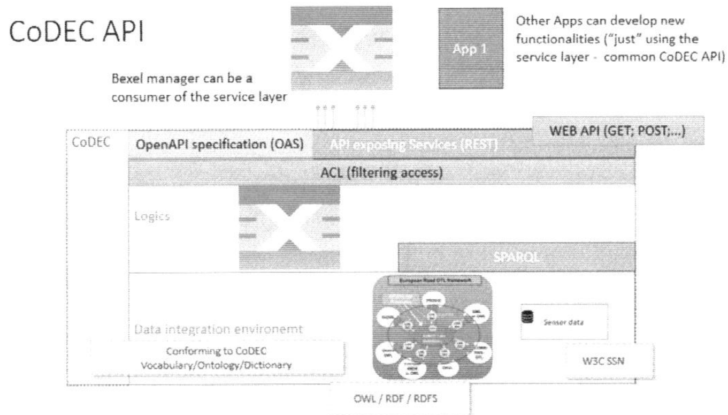

Figure 2. CoDEC API overview

3.3. Linked Data structure

Once the CoDEC ontology is finalized, with all the required classes and relationships, the next step is to integrate real data (namely, the data required for the pilot projects – see the next section for more details). The ontology is populated with data instances and made available in a linked data environment in order to be retrieved by external applications (via the API). The linked data environment that CoDEC uses is GraphDB, a database that follows RDF and SPARQL (query language for ontologies) specifications. Figure 3 presents a sample linked data structure for the fictional example of data for a bridge and a pavement.

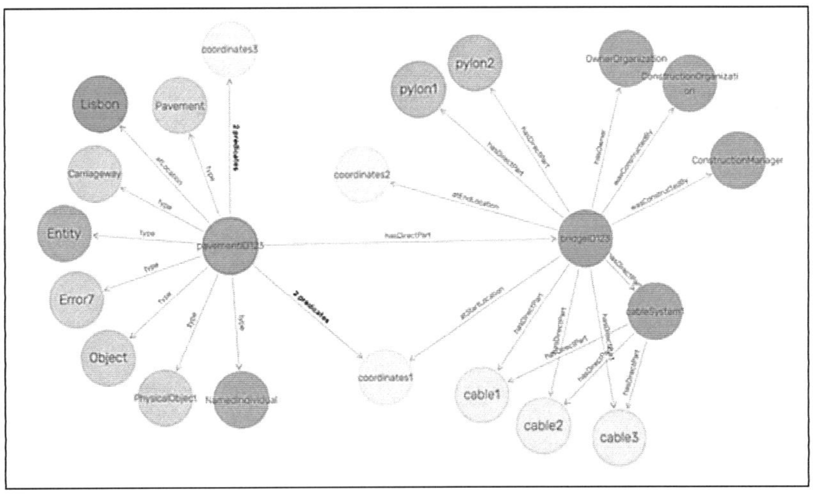

Figure 3. Sample of linked data using CoDEC ontology for a bridge and a pavement.

4. Demonstrating the concept

CoDEC will demonstrate the application of the data dictionary and ontology developed during the project, using a linked data approach, through three pilot projects. The pilot projects aim to:

– Demonstrate the connectivity between the Data Dictionary and EUROTL.
– Demonstrate visualisation of the integrated data (including sensor/scanned data sets) in an asset

IOP Conf. Series: Materials Science and Engineering 1202 (2021) 012002 doi:10.1088/1757-899X/1202/1/012002

management system.
- Demonstrate the ability to link data between BIM models, GIS and Asset Management Systems.

The projects will include the development of tools (including an Application Programming Interface, API via the Bexel Manager and the use of 3D visualization software) to demonstrate linking data between BIM platforms and asset management systems.

The pilot projects are being conducted with the support of three implementation partners: Agentschap Wegen & Verkeer (AWV, Belgian (Flemish) NRA) (Project 1); Rijkswaterstaat (RWS, Dutch NRA) (Project 2) and FTIA (Finnish NRA) (Project 3):

- Pilot project 1 – **Tunnel** - to show how the BIM model of a Tunnel can be enriched. It will link the BIM OTL to tunnel sensor data (such as thermal, visual sensors etc.) and provide an advanced 3D visualisation of this linked model.
- Pilot project 2 – **Bridge** - to integrate data from multiple bridge specific sensors, and demonstrate software tools and guidelines for linking, preparing and visualizing bridge data.
- Pilot project 3 – **Highway** – to link data between BIM models and a road network within a GIS (Geographic Information System) environment, requiring accurate spatial mapping between the two.

At the time of writing this paper all three pilot projects are still in progress. The section below provides a description of pilot project 3 in order to give an example of the way in which the CoDEC ontology is being applied in a practical use case.

4.1. Pilot project 3 - Linking BIM & GIS for Highway Asset Management

Whilst BIM models are often created for the design/construct phase, highways are primarily managed during the operational phase using GIS-based Asset Management Systems (AMS). Likewise, LiDAR scans – increasingly being used to gather physical data about highways assets – are often translated into BIM models rather than GIS-based information. In all of these cases, BIM models often hold information useful for asset management which is not being made available in the GIS-based AMS where these assets are managed. In other words, BIM models represent a new source of data which can be used to enrich (and/or complement) the data held within legacy systems – this is a particular focus of the CoDEC project. Pilot project 3 aims to address this issue by providing methods to link data between the BIM and GIS-based environments.

This pilot project focuses on the carriageway pavement as the key asset. Although some of the methods developed are specific to the challenges of linking pavement asset data, the general approach is more widely applicable to other non-linear highways assets. The project has access to multiple BIM models of sections of the road network within TRL's Smart Mobility Living Lab, located in the London Borough of Greenwich, UK, as well as a GIS-format network model of the whole of the borough's road network (provided by the local roads authority). Within the BIM models, the pavement is represented as a series of 2D 'slabs' of ~10m length. The challenge for this pilot project is to derive useful information from the geometry of these slabs, and then link that information with the corresponding location (i.e.: 10m subsection) within the GIS network model – an example is given in Figure below. The pilot project team has had to take innovative approaches to tackling both of these challenges and have developed solutions which will be disseminated in future CoDEC communications.

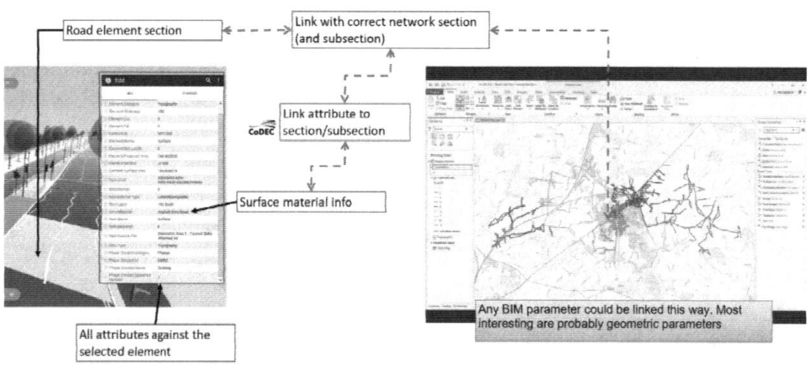

Figure 4: Example illustrating the linking of data for a particular subsection of pavement between the BIM model and GIS

The CoDEC ontology (described above) is being used to provide a structuring framework for organizing and linking the relevant data for this pilot, and so far is proving successful in that task. The linking methodology is summarized in Figure below. The pilot project has provided a practical opportunity to test the content and structure of the ontology (which is based on the data dictionary) – this has directly resulted in improvements and additions to the ontology.

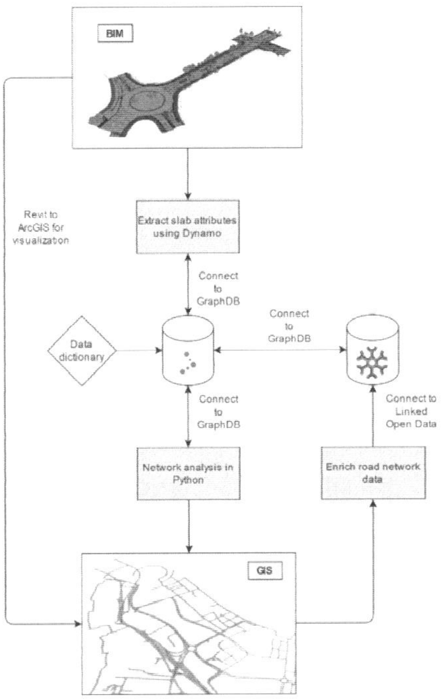

From BIM-csv export to Linked Database

1. Read the csv file in Python.
2. Convert the longitude and latitude values to British National Grid coordinates
3. Assign the coordinates to the corresponding slab object.
4. Create a convex hull around the slab object.
5. Read the vertices of the convex hull and store them in a GeoSPARQL Polygon object in Well Known Text (WKT) format.
6. Build a SPARQL query to insert the slab data to the Linked Database.
7. Execute the query using the GraphDB API.

From Linked Database to GIS

1. Build a SPARQL query to read the slab data from the Linked Database.
2. Execute the query using the GraphDB API.
3. Convert the slab geometries to ArcGIS Polygon features.
4. Intersect the slab features with the road network features.
5. Measure the start and end chainage of the intersect result on the network.

From GIS to Linked Database

1. Store the network information acquired through the GIS analysis to the slab object in Python.
2. Build a SPARQL query to insert the slab data to the Linked Database.
3. Execute the query using the GraphDB API.

Figure 5: Steps involved in linking data between the BIM models and the GIS using the CoDEC-developed tools and schema

Conclusions

Consultation with European NRAs showed many NRAs use BIM during the design and construction phases of large projects and sometimes stored as a detailed "as-built" model, but little further use is made during the remainder of the operational life cycle of the asset. Long-term asset management is still typically carried out using dedicated asset management systems. Lack of the interoperability of the data types, their standards, and isolated data management/storage principles act as barriers for effective data exchange.

It is clear that new data sources (from new sensor and scanning technologies) hold huge potential for improving the way that we manage highways assets – indeed, it is absolutely necessary that highways asset management adapts to capitalize on the increasing pace of the digital revolution. However, it is important to take a methodical and structured approach to dealing with the underlying data – the increasing volume, variety and velocity of the data demands it.

The CoDEC project (although not yet complete) is developing and demonstrating practical ways to structure asset-related data in such a way as to make them operationally useful. The CoDEC approach relies in large part on having a methodically developed framework for the data (the data dictionary structure), and translating this into a machine-readable framework (the ontology), in order to provide a means of making data more interoperable – a step on the journey to the ultimate goal of making data seamlessly available when and where it is needed.

It is important to note that it has taken significant effort to get this far, and there is still further to go within the scope of the project. However, CoDEC already shown that in ongoing pilot projects, with an initial investment of effort, it is possible to create a data structure which, when applied using the right methods, unlocks great potential for enriching legacy data, systems, and ways of working. When used within a digital-led organizational context, there is huge potential for such an approach to unlock value from (and add value to) asset data, ultimately leading to multiple benefits in terms of increased asset knowledge (and therefore better asset management), more efficient use of data (and therefore lower costs), and more effective use of data (and therefore greater benefits).

Acknowledgements

The CoDEC project would like to thank all the experts who were involved in stakeholder consultation activities by sharing their knowledge, experience and providing suggestions for the project.

Funding

The CoDEC project was carried out as part of the CEDR Transnational Road Research Programme Call 2018: BIM. The funding for the research was provided by the national road administrations of Austria, Belgium – Flanders, Denmark, Finland, Germany, Netherlands, Norway and Sweden.

References

Austroads. (2019. Data Standard for Road Management and Investment in Australia and New Zealand (DSRMI), version 3.0. Retrieved from https://austroads.com.au/publications/asset-management/ap-r597-19

buildingSMART. (2020). Industry Foundation Classes (IFC), IfcROAD. Retrieved from http://technical.buildingsmart.org/standards/ifc/ifc-schema-specifications/

Construction Specifications Institute. (2017). OmniClass Construction Classification System. Retrieved from https://www.csiresources.org/standards/omniclass

Highways England. (2020). Asset Data Management Manual (ADMM), version 12.0. Retrieved from https://www.standardsforhighways.co.uk/ha/standards/admm/index.htm

INTERLINK project. (2018). Final Project Report – INTERLINK Approach with European Road Object Type Library. Retrieved from https://roadotl.eu/static/ireport/index.html

International Organization for Standardization. (2001). ISO 12006-2:2001 Building construction — Organization of information about construction works — Part 2: Framework for classification of information

Marcovaldi, E., Biccellari, M. (2018). Asset Data Dictionary, Deliverable D3.1 of the AM4INFRA project. Retrieved from https://am4infra.eu/

Musen, M.A. (2015). The Protégé project: A look back and a look forward. AI Matters. *Association of Computing Machinery Specific Interest Group in Artificial Intelligence, 1(4), June 2015*. DOI: 10.1145/2557001.25757003.

NBS. (2015). Uniclass 2015. Retrieved from https://www.thenbs.com/our-tools/uniclass-2015

Open Geospatial Consortium. (2020). OGC SensorML: Model and XML Encoding Standard, version 2.1. Retrieved from https://www.ogc.org/standards/sensorml

Pan, J. Z., Horrocks, I. (2007). RDFS(FA): Connecting RDF(S) and OWL DL. *IEEE Transactions on Knowledge and Data Engineering, 19(2), 192-206*. https://doi.org/10.1109/TKDE.2007.37

Studer, R., Benjamins, R., Fensel, D. (1998). Knowledge engineering: Principles and methods. *Data & Knowledge Engineering, 25(1–2):161–198,* 1998.

IMPLEMENTATION OF ROAD ASSET MANAGEMENT IN POLAND

Professor Adam ZOFKA[1], Andrzej MACIEJEWSKI, MBA[2]

[1] Institute of Roads and Bridges Research, Warsaw, Poland
[2]General Directorate for National Roads and Motorways, Warsaw, Poland

Abstract. Road asset management (RAM) is a crucial aspect in the development of transportation network. RAM framework should be understood as a rational approach to business model for road authority, which dictates its business processes in a systematic and objective-based manner to ensure that strategic goals of an agency are reached. In a case of road network, such objectives are typically related to travel safety, time and comfort that should be maintained throughout the life cycle of the road assets at optimal costs. As such RAM enables road agencies to meet expectations of its customers to provide safe and reliable road network in effective and efficient way. Nowadays it becomes even more important to adopt performance-based and data-driven approach for road agencies considering current trends in transportation like connected, cooperative and automated mobility.

There are many aspects of RAM including legal, economic, technical and managing aspects. These aspects are inter-connected and nowadays agencies build and incorporate RAM system to establish a common denominator for all agency activities at all organizational levels. This paper discusses selected aspects of RAM in Poland. Since national road network in Poland is relatively new, it is particularly important to go over well designed and detailed implementation process. It should comprise numerous items including self-assessment and gap analysis, change strategy and transportation asset management plan (TAMP). This paper present update on these activities as well as presents future plans in order to establish modern, effective and sustainable RAM framework in Poland.

Keywords: public services, accountability, value, asset management, asset management system, performance, risk, cost, optimisation

Introduction

In the last 15 years Poland has invested tens of billions of Euros in the road infrastructure development, resulting in thousands of kilometres of new motorways and expressways. This huge investment programme is constantly carried out and in the recent years it has been enlarged into the separate programmes regarding construction of at least hundred bypasses, reconstruction of existing network aimed at improving of the road safety and rehabilitation and reconstruction of roads targeted to improvement of current network bearing capacity.

These efforts lead to defining of the possible future challenge in respect to the periodic maintenance regarding funding constraints recognised worldwide and already faced in the countries with extensive road network such as United States or western EU Member States.

For that reason Ministry of Infrastructure, General Directorate of National Roads and Motorways (GDDKiA) and Road and Bridge Research Institute (IBDiM) initiated the special project for defining of the method of the optimisation of the decision-making process regarding the road network rehabilitation and reconstruction.

Ultimately the project has been inaugurated at the end of 2018 by the consortium of the three aforementioned institutions and additionally Wrocław University of Technology.

Works within the project have been foreseen for 3 years, divided into two phases and seven assignments.

Phase A (which has been finalised in May 2021) was focused on the elaboration of the method for the planning of the periodic maintenance works which includes inter alia:

- definition of public services provided by the road network,
- establishment of the performance measurement framework
- root cause distress analysis and associated analytic hierarchy process
- deterioration forecasting models
- relation to the demand forecasting model
- approaches to incorporation of economic efficiency and risk management models
- optimisation model allowing for the multi-objective decision analysis.

Apart from the method Phase A was focused on the creation of the prototype of the set of tools supporting the method and the decision-making processes.

Phase B, which was initiated in June 2021 is focused on the piloting of the defined method and provided prototype and allows all involved parties for the revision of proposed computing models and modules with aim to formally approve the new approach by the end of the project (November 2021).

Concurrently to these the Ministry of Infrastructure carries out works on the revision of the legislation and regulations for the life cycle management of the road infrastructure assets, thus to have comprehensive asset management system for the road infrastructure including not only IT tools but first of all dedicated method, appropriately embedded within the internal business processes of the National Road Authority and aligned with the technical regulations.

1. Understanding of Asset Management

1.1. Accountability of public organisations for service delivery

To address the challenges of road sub-sector governance and to design solutions for better road network performance, these should be considered in the wider perspective of governance of the state. According to the United Nations glossary for Public Administration this may be defined as *"the exercise of economic, political and administrative authority to manage a country's affairs at all levels. It comprises the mechanisms, processes and institutions through which citizens and groups articulate their interests, exercise their legal rights, meet their obligations and mediate their differences"*[1].

According to the United Nations report on Responsive and Accountable Public Governance[2], to enhance public sector responsiveness, it is crucial to focus on satisfying people's expectations of quality, quantity and promptness of the public services delivered within limited resources available. Achievement of responsive governance depends on how policies, strategies, programs, activities and resources are anchored to people's real needs.

In parallel, establishing strong governance and accountability is essential to delivery of expected goals. An accountable organisational culture deters waste and mismanagement of resources. Accountability for performance serves to guide, monitor and evaluate public institutions and programs, informing needed improvements. Therefore, building the capacity for financial and performance accountability builds trust for leveraging resources and safeguarding funds.

For instance, the Office of the Auditor General of Canada defined five principles for effective accountability[3] that have been identified by as being a key to accountable governance:

1. Clear roles and responsibilities: The decision-making roles and responsibilities of the parties in the accountability relationship should be well understood and agreed upon.
2. Clear performance expectations: The objectives pursued, the accomplishments expected and the operating constraints to action, which include means, operating principles and procedures, human resource management issues and adequate financial control should be explicit, understood and agreed upon.
3. Balanced expectations and capacities: Performance expectations should be clearly linked to and balanced with each party's capacities (authorities, skills and resources) to deliver.
4. Credible reporting: Credible and timely information should be reported to demonstrate what has been achieved, whether the means were appropriate and what has been learned (including reporting requirements, modalities, sufficient information for Parliament, etc.).
5. Reasonable review and adjustment: Fair and informed review and feedback on performance should be carried out by the parties, achievements and difficulties recognized, appropriate corrections made with appropriate consequences for the concerned individuals.

In addition, other institutions (such as Australian National Audit Office) emphasize the importance of shared risk management, both in terms of delivery of services and the management of any contract.

Therefore accountability denotes responsibility for results and outcomes, and not only processes (with their related inputs and outputs). When operating effectively, it serves to ensure that public governance can flourish, related institutions perform well, and services are delivered to citizens effectively and efficiently[4].

Adoption of responsive and accountable of governance requires an organisation to re-asses its role in public services delivery and to address a number of challenges for the public sector, like:

1. Change of demographic profiles
2. Increasing customers' expectations
3. Awareness of public services' users and demand for greater transparency
4. Demand for greater transparency
5. Budgetary constraints
6. Global competition to attract investments

Achievement of these requirements for the good governance needs sound understanding of what are the public services and how they may be provided. Delivering public services may be based upon six key strategic enablers, as given below:

---

[1] http://www.unpan.org/Directories/Glossary/tabid/1398/language/en-US/Default.aspx
[2] World Public Sector Report, Responsive and Accountable Public Governance, 2015
[3] Accountability Audit Guide, Office of the Auditor General of Canada, August 2004
[4] Accountability Audit Guide, Office of the Auditor General of Canada, August 2004

1. Understanding the customer
2. Transparency
3. Removing silos between governmental/ public organisations
4. Building capacity
5. Improving the service delivery model
6. Continual improvement

The challenges involved in achieving these requirements include those listed in the following table.

Table 1. Challenges for the strategic enablers for the public services delivery (Source: Maciejewski 2020)

Strategic enabler	Challenges
1. Understanding of the customer	Customer focus is often challenged by public sector culture, hierarchical organizational structures and differing public sector priorities, whereas public agencies priorities need to be aligned to customer requirements.
2. Transparency	To ensure that customers may access information held internally by the organisation. Such information concerns the overall decision making processes - its basis, data taken into consideration, positions and opinions of citizens and both governmental and non-governmental organisations. Transparency therefore promotes good governance through holding government and key decision-makers to account what improves public policy, its effectiveness and efficiency.
3. Removing silos between governmental/ public organisations	This needs "connected government" expressed in common vision, supported by integrated objectives, outcomes or process flows. Key elements for "connected government" should include – *inter alia* – common service standards. It is vital to break down intra-agency silos before starting to break down cross-agency silos.
4. Building capacity	Capacity building main pillars are: • long-term planning, • organisational and process design, • use of technology, • support for people and for the organisation • culture change in the organisation
5. Improving service delivery model	To define the roles organisations should play: • Policy maker, • Regulator, • Service provider. To articulate: • quality of service, • cost of service, • suitability of different service delivery channels for different customers segments. To define: • how technology may support to achieve the organization's goals, • how public-private partnership can deliver targeted outcomes • how to manage the risks.
6. Continual improvement	To benchmark carried out activities, finding the answers for following three questions: • what to innovate? • where to learn from? • how to adopt?

2. Asset Management to facilitate public services delivery

The increase of transport volumes, bringing new investment needs in high-capacity transport networks and ageing infrastructure require improved road construction and maintenance. The traditional way of doing business, which postpones repair activities until major deterioration occurs, is not sustainable. It is too expensive over the long term and it drains the value of road network assets.

As the costs of operating, repairing or constructing are increasing and – at the same time – available funding decreases, it has become more challenging for governments to meet the demands of an ageing infrastructure whilst meeting public expectations.

The challenge is to provide the same or even better value for less money. There is also strong demand for transparency and accountability from road sub-sector organisations, requiring justification for decisions and responsibility taken for results.

As stated above, public sector organizations (irrespective of their legal form) are being increasingly subjected to both legislative and competitive pressures forcing them to reconsider their relationships with users and customers in order to develop a more overt customer orientation (as the primary driver of organisational performance).

The creation of "value" supports the development of a customer orientation, and is a requirement, to which more public sector organisations nowadays subscribe. This applies to all sectors of the economy, also to the road sub-sector.

In modern society, road infrastructure has become an essential part of daily life. Individual road users, a wide variety of commercial enterprises, logistics companies and public transportation agencies expect reliable and safe road infrastructure to carry out their transportation or wider mobility operations, moving goods and people. "Just in Time" supply chains are at the heart of modern manufacturing and retail enterprise and they rely totally on predictable and stable journey time.

Reliable, accessible and safe infrastructure is a cornerstone for socioeconomic progress. It enables productivity growth, shortens travel times and costs, creates jobs, and connects different parts of society.

Recent research[5] indicates that a proactive roads periodic maintenance strategy would only cost 65% when compared with a reactive management strategy. In other words, better management of the network, up to date condition data and planning of the optimum time for an intervention would lead to substantial financial and economic benefits

This is supported by NCHRP Research Report 866 (2018) "*Return on Investment in Transportation Asset Management Systems and Practices*" where a similar level of savings (30-40%) were found to arise from the implementation of a Transportation Asset Management (TAM) system. Such a system is designed to:

- help road authorities make the best use of resources available for maintaining, rehabilitating, and replacing existing physical assets such as pavements, bridges, traffic and safety devices, and facilities; and
- help road authorities make better use of limited resources to maximize asset life, manage risk, and provide safe and efficient travel for passengers and goods.

Asset management has become a popular approach for the asset-heavy industries affecting the organization and management of a number of companies and institutions from multiple sectors of the economy.

This approach is also increasingly popular in the transport infrastructure sector, where planning and scheduling of works or services requires not only the delivery of multifaceted value for road network users, but also for other stakeholders from the general public to particular interest groups (i.e., transport and logistics).

Unfortunately, quite often people confuse the terms asset management and asset management systems, and tend to limit their understanding of asset management systems to a more or less sophisticated set of IT tools.

This understanding is not sufficient and misleads public organisations, which instead of carrying out comprehensive and complex revision of their business models concentrate their efforts on the technical challenges associated with the systems implementation without sound understanding how to redesign their internal processes and procedures, align different level of organisations management and only having these in place adjust IT tools.

Different approach and starting with purchasing of IT system usually leads to the vendor lock-in risk and timely and costly implementation processes. Asset management should therefore be understood as a value creation process which aligns the strategic, tactical and operational levels of an organization's management through technical, engineering, and business principles and practice driven by economic rationale. On the other hand, an appropriate combination of management levels, principles and practices is made possible through the asset management system.

3. Asset Management System

According to the international standard ISO 55000, an asset management system is a set of interrelated and interacting elements of an organization, the function of which is to establish the asset management policy and asset management objectives, as well as the processes needed to achieve those objectives.

The selected processes should therefore enable an organization to understand customer needs and expectations, create and deliver a product and/or service, and collect enough data and information to measure the achieved performance while simultaneously serving as a basis for further improvements.

The general framework for designing the appropriate sequence of processes can be reflected by the popular approach of the Deming Cycle, also known as the Plan-Do-Check-Act (Adjust) cycle (PDCA).

---

[5] Zofka, A. "*Proactive Strategy for Road Infrastructure Management*" p.147, Roads and Bridges Research Institute, Warsaw 2019

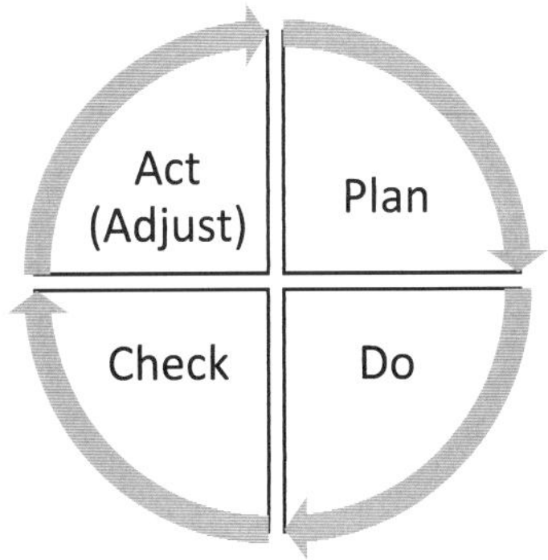

Figure 1. Deming's Circle. *Source*: W. Edwards Deming, *The New Economics for Industry, Government and Education* (Boston, MIT Press, 1993).

The outcome of the planning phase ("Plan") is to establish objectives and furthermore operational processes to deliver desired results.
The phase "Do" is focused on the execution of the planned activities to achieve the given objectives.
During the "Check" phase, the data and results gathered from the previous phases are evaluated. Data is compared to the expected outcomes to identify any similarities or differences between what was planned and what has been achieved. This phase may be called also as "S – study" as it helps to determine the reasons of any observed deviations from the assumed outcomes.
Finally, the phase "Act" (or, even better, "Adjust") is focused on the processes and overall system improvements, building on the results of the performance evaluation done in the preceding step.
PDCA should, however, only be an inspiration for the more detailed design and selection of the system of internal business processes.
As asset management is a value creation process, an asset management system can be described as an organization's value chain and the necessary procedures, systems and competencies enabling the organization to carry out its activities in a logical sequence to ensure operational activities satisfy the strategic objectives of the organization and its customers' needs.
The generic concept of the value chain[6] requires some adjustments for public sector organizations like road authorities, which tend more toward managing road networks through subcontractors than producing services or works by themselves.
Understanding the essence of the value chain for road authorities requires understanding the needs and expectations of communities, tax payers, road users and other stakeholders, translating those needs and expectations into more technical language that makes it possible to plan and programme relevant activities (capital, maintenance and operational interventions in the road network), contracting these activities through the supply chain, supervising the works and services delivery process and, finally, assessing if the quality of the provided works and services is sufficient to achieve satisfactory performance both throughout the whole network and for each individual asset when compared to the expected outcomes.

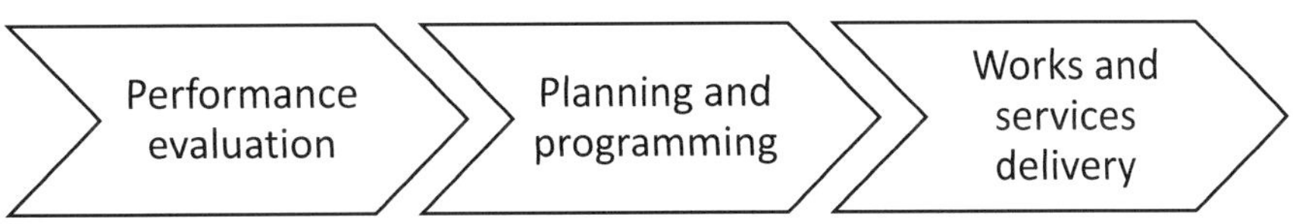

Figure 2. High level value chain for the road authority (Source: Maciejewski 2020)

While the third phase of the above approach seems to be well established among road authorities – essentially their most exercised competency – usually planning and performance evaluation processes require the application of more

[6] Michael E Porter, *Competitive advantage: creating and sustaining superior performance* (New York, Free Press, 1985).

sophisticated tools. This is because road authorities are responsible not only for the delivery of the technical standards of particular assets defined in formal documents but also for providing more diverse outcomes from the whole network and its life cycle perspective, including both economic and customer experience factors.

4. Total Value Management

Adopting an asset management approach and implementing a comprehensive asset management system within the organization (e.g., road authority) requires both top-down and bottom-up activities.
Top-down means a sound understanding of internal and external strategic contexts of an organization, while bottom-up capabilities refer to:

- Operational processes and procedures
- Systems supporting and automating the above procedures
- Competencies reflected within the structure of an organization, job descriptions and the actual capacities of employees

It is not enough to simply have these two approaches implemented or reflected in the organization structure. It is necessary to ensure that they are aligned, meaning that all necessary operational activities, procedures, systems and competencies are designed to fulfil strategic goals and objectives valued by the customers. What is required is a comprehensive management system that identifies and satisfies the needs and expectations of consumers better than the competitors (in this case perhaps seen as other government bodies competing for funding)[7].
Technical quality – as in total quality management (TQM) – addresses aspects of quality with reference to the functions a product must perform, though this is only one of the many value characteristics that need to be considered by an asset manager (including road asset managers)[8].
With conventional TQM processes, it is difficult to address all aspects of the value expected by customers and stakeholders, such as (in the road sector):

- Cost-efficiency
- Effectiveness and performance
- Safety and security
- Network and assets resilience
- Accessibility, connectivity and availability

The response to these challenges could be total value management (TVM). TVM efforts can be achieved through implementation of the asset management approach including all necessary processes, procedures and tools being part of an asset management system (an organization's value chain) to ensure that all detailed activities are aligned and lead to the creation of the expected value (the essence of asset management), as depicted in Figure 3 below.[9]

[7] Competition, as observed in the private sector, may not seem to be applicable for road authorities or more broadly public sector organizations, but taking into account competition in terms of access to taxpayer funds or state budget funds, the relevance of this term should be accepted.

[8] Biren Prasad, "Total Value Management: a knowledge management concept for integrating TQM into concurrent product and process development", *Knowledge and Process Management*, vol. 8, No. 2 (April/June 2001).

[9] Alternatively, instead of implementation of TVM through comprehensive asset management, organizations may wish to work on value creation and management through other approaches like 5P, recently presented by McKinsey: P – portfolio strategy and products (assets in the case of roads), P – people and culture, P – processes and systems, P – performance metrics, P – positions and engagements. The distinction between different approaches may, however, be misleading as the essence is still to deliver value to the customer. Sebastian Leape and others, "More than a mission statement: how the 5Ps embed purpose to deliver value", McKinsey Quarterly, 5 November 2020.

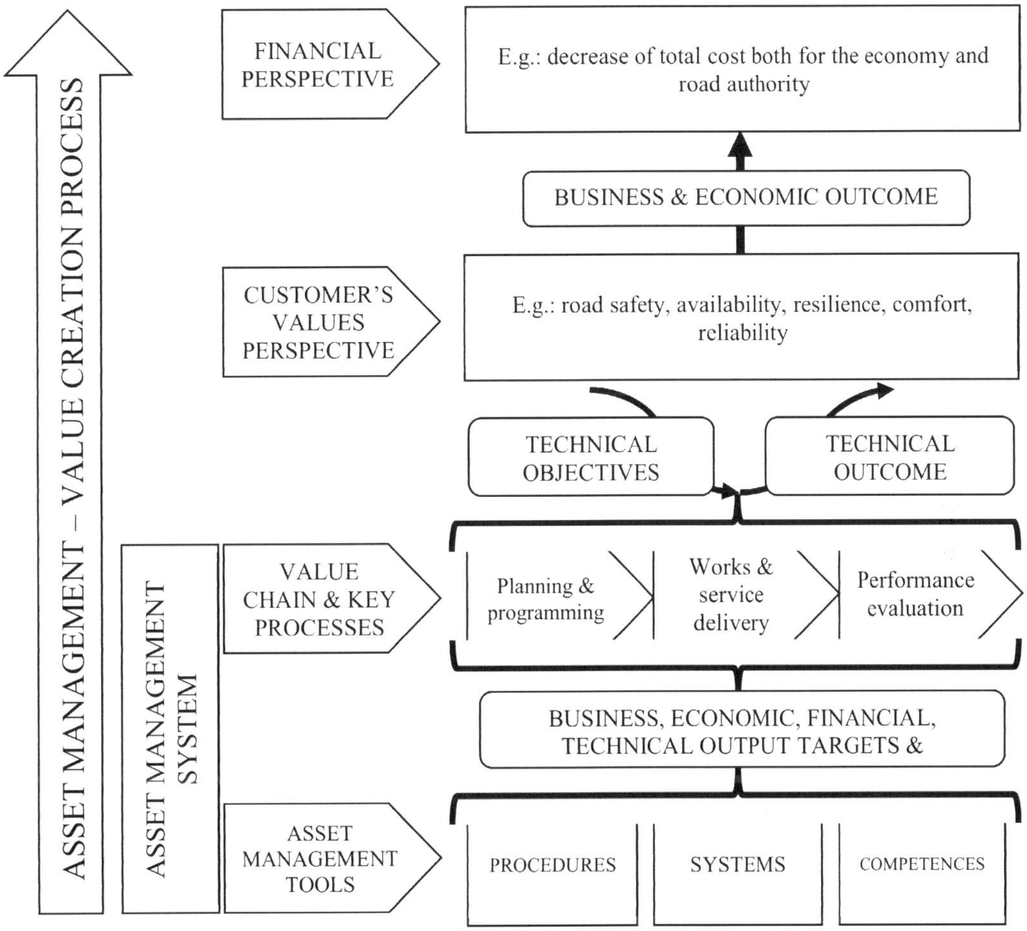

Presented above considerations and understanding of the Asset Management and Asset Management System has been translated within the Project's Team into the proposal of the more implementation-oriented Asset Management Framework, which should support efforts of road authorities in initiation of such processes.

Elaborated framework consists of 6 areas which are as follows:

1. Performance measurement framework
2. Methods and tools for planning and programming
3. Methods and tools for services and works delivery
4. Methods and tools for continuous improvement
5. Organisation
6. IT architecture

5. Performance measurement framework

Performance measurement framework allows an organisation to assess level of its performance regarding both level of service (compare to the customers' value perspective above) and financial outputs and outcomes (see: financial perspective above).

Apart from the customer and financial perspectives performance measurement framework addresses also internal

procedures, methods, systems and competences of an organisation (asset management tools as above) to support both assessment of their performance and to allow for the improvement planning.
The Project's team created an useful tool for creation of such a performance measurement framework as presented below.

Table 2. Metrix for setting of the performance measures (Source: Zofka, Maciejewski, 2021)

Categories and types of measures		**OUTCOMES**	**OUTPUTS**	**INPUTS**
RESILIENCE	FUNCTIONALITY	e.g.: % of gained/lost resilience (resilience sustainability)	e.g.: km of repairs	e.g.: level of funding
SAFETY				
COMFORT				
RELIABILITY				
SERVICE X				
COSTS		e.g.: vehicle operating costs (VOC)		
RISKS		e.g. objectives not achieved	e.g. long lasting procedures	e.g. not sufficient funding limits

6. Method for the periodic maintenance planning

The second area of proposed Asset Management Framework which was elaborated by the Project's team was a method for the periodic maintenance planning.

Proposed method encapsulates both engineering knowledge, standards and practice and then connects them with the economic efficiency model and risk management model. All of these elements are aimed at the preparation of appropriate set of data for the optimisation model and module, thus necessary for the objectives and constraints derived from the above-presented table.

This approach allows for being aware enough when planning the objectives and selecting whether the goal should be more associated to the cost, the functionality or to the risk dimensions. If for example an organisation will decide that its planning process and sub-sequent delivery plans and programmes should provide defined level of reliability, optimisation module will help to determine what costs and risks will this scenario bring.

Achieving of such an analysis is possible due to the maintenance scenarios generator envisaged within the proposed planning method, which derives from the deterioration forecasting model, demand model as well as from the work effect model.

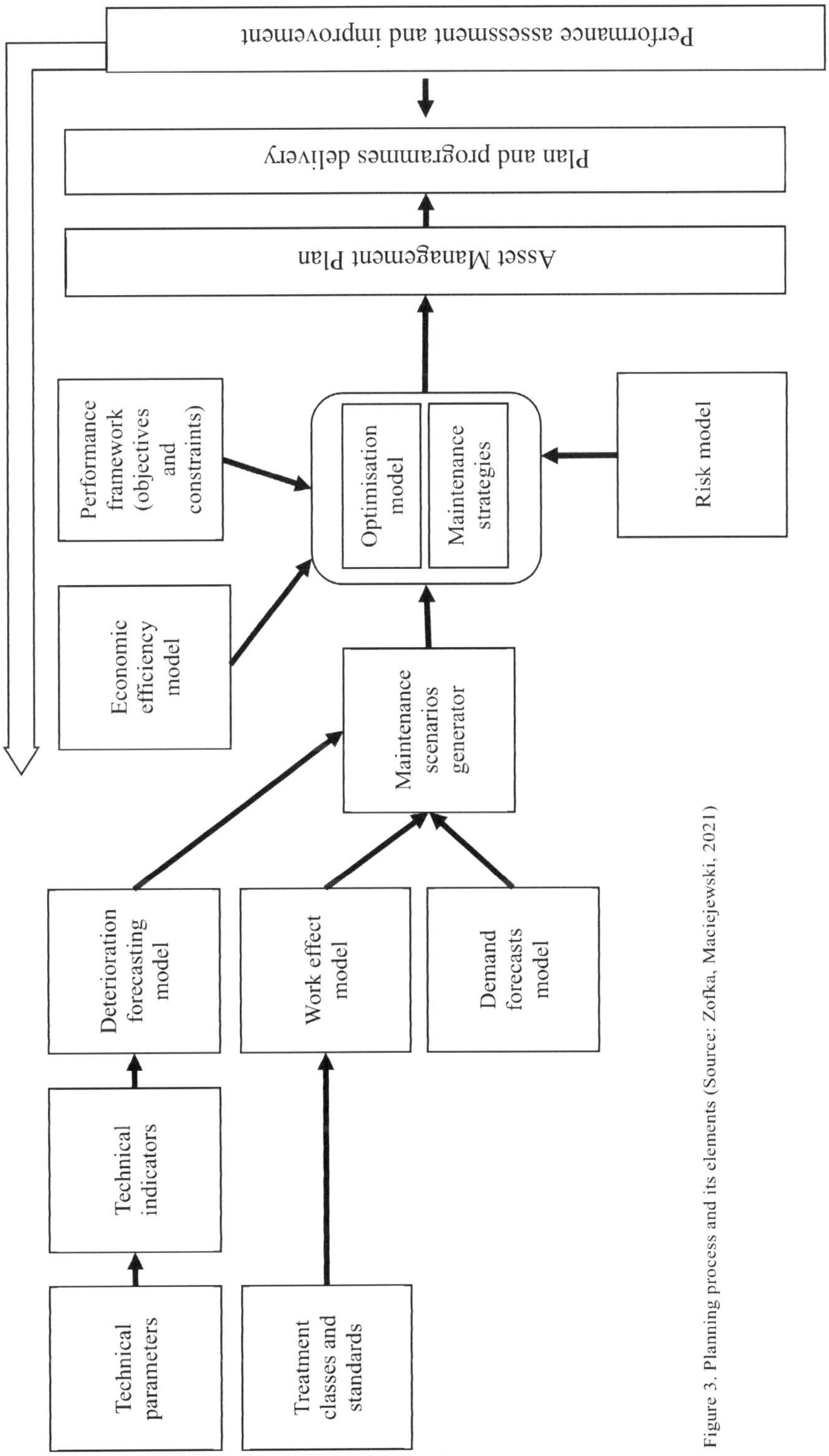

Figure 3. Planning process and its elements (Source: Zofka, Maciejewski, 2021)

7. Supporting tools

To make the method operational it was supplemented within the project's scope with the prototype of the supporting IT system.
Although such a systems are in general available on the market it was decided to assess the real needs regarding particular modules of such a system and adjust it to the overall IT architecture. The IT architecture including both asset and traffic management elements has been already created during the preparatory works for another project – National Traffic Management System, being concurrently under implementation.
This approach allows to ensure appropriate level of interrelations between both processes (of asset and traffic management), envisage one data warehouse and make data collecting and usage processes consistent.
This leads to the increased efficiency of road network operations as data are in reality collected once and used repeatedly.
The main elements of the created tool are:

1. Data warehouse
2. Computing modules for particular models as in the method
3. Business Intelligence platform for the visual analysis of the data.
4. The business intelligence platform is the interface for the end users.

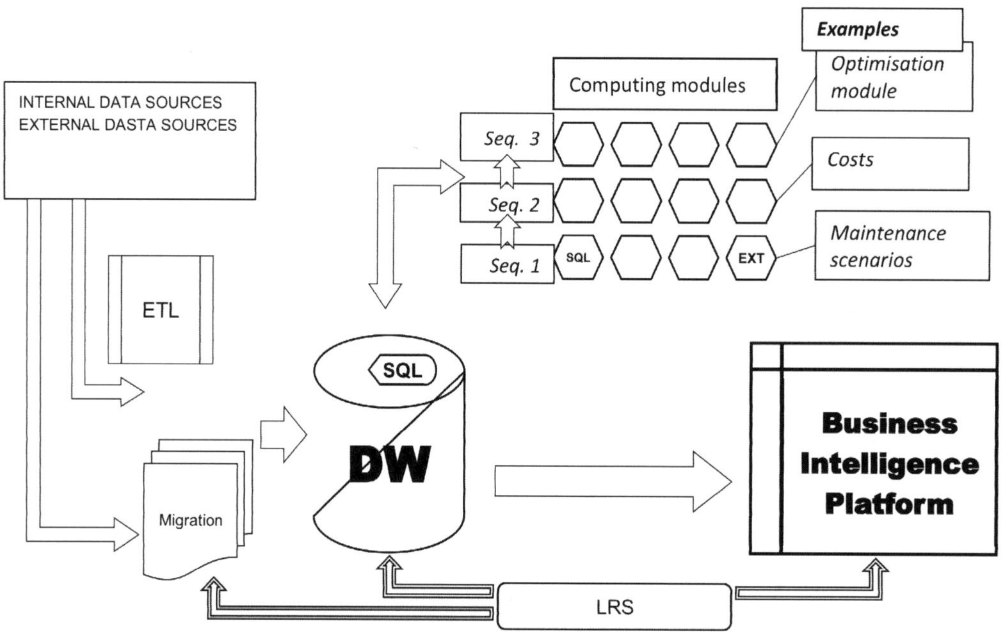

Figure 4. High-level IT architecture. (Source: Zofka 2020).

Conclusions

Implementation of the Road Asset Management System in Poland seems to be scheduled in the appropriate timing regarding the still ongoing process of construction of new network of high speed roads.
As elsewhere such an effort faces some challenges regarding especially the need for moving from the perspective of separate activities to the more asset management mindset what requires seeing of the bigger picture and understanding that the role of the public organisations it to deliver their services in the most effective and efficient wat at once.
Adopted approach is unique taking into consideration elaborated method and also due to independent creation of the supporting IT tools.

Acknowledgments

This paper was created as a part of the research project SIEC11.5t funded by the National Centre for Research and Development (NCBiR) in Poland

References

Alexopoulos, Konstantinos and Wyrowski, Lukasz. *Innovative ways for financing transport infrastructure* (New York and Geneva, United Nations Economic Commission, 2017).
American Association of State Highway and Transportation Officials. *AASHTO Transportation Asset Management Guide* (Washington, D.C., 2011).
American Association of State Highway and Transportation Officials. *AASHTO Transportation Asset Management Guide*, 2nd ed. (Washington, D.C., 2020).
Asian Development Bank. *Reforming the financing system for the road sector in the People's Republic of China* (Mandaluyong, 2015).
Austroads. *Guide to Asset Management*, 3rd ed., July 2018.
Briessinck, Margo and others. *CEDR technical report 2017/06 – TG Asset Management Final Report 2017*, Conference of European Directors of Roads (Brussels, CEDR Secretariat General, 2017).
Brocklebank, Peter. *Private sector involvement in road financing*, Sub-Saharan Africa Transport Policy Program (SSATP), discussion paper no. 102 (Washington, D.C., World Bank, 2014).
Diamond, Jack. *Establishing a performance management framework for government*, IMF Working Paper WP/05/50 (International Monetary Fund, 2005).
Farquharson, Edward and others. *How to Engage with the Private Sector in Public-Private Partnerships in Emerging Markets* (Washington, D.C., World Bank, 2011).
Garemo, Nicklas; Halleman, Brendan; and Hjerpe, Martin. *A better road to the future: Improving the delivery of road infrastructure across the world* (n.p., McKinsey & Company, 2018).
Institute of Asset Management. *An Anatomy of Asset Management*, ver. 3, December 2015.
Institute of Public Works Engineering Australasia. *International Infrastructure Management Manual,* 5th ed. (n.p., 2015).
ISO 55000:2014, 55001:2014, 55002:2014.
Kaplan, Robert S. and Norton, David P. *The Balanced Scorecard: Translating Strategy into Action* (Harvard Business Review Press, 1996).
OECD. *Asset Management for the Roads Sector*, Road Transport and Intermodal Linkages Research Programme (Paris, 2001).
OECD. *Performance Indicators for the Road Sector* (Paris, 2001).
Osborne, David. "Reinventing Government: what a difference strategy makes", 7th Global Forum on Reinventing Government: Building Trust in Government, Vienna, 26-29 June 2007.
Porter, Michael E. *Competitive advantage: creating and sustaining superior performance* (New York, Free Press, 1985).
Prasad, Biren. "Total Value Management: a knowledge management concept for integrating TQM into concurrent product and process development", *Knowledge and Process Management*, vol. 8, No. 2 (April/June 2001).
Public Sector Research Centre PricewaterhouseCoopers. *The road ahead for public service delivery: delivering on the customer promise* (n.p., 2007).
Queiroz, Cesar and Kerali, Henry. *A review of institutional arrangements for Road Asset Management: Lessons for the Developing World*, Transport paper series, no. TP-32 (Washington, D.C., World Bank, 2010).
Talvitie, Antti. *International experience in restructuring road sector* (Washington, D.C., World Bank, 1998).
Task Group S4 Optimizing the financial resources available for roads. *Funding formulas for roads: inventory and assessment* (Brussels, CEDR Secretariat General, 2017).
Transportation Research Board of the National Academies. *NCHRP Report 551: Performance Measures and Targets for Transportation Asset Management*, National Cooperative Highway Research Program (Washington, D.C., 2006).
United Nations Department of Economic and Social Affairs. *World Public Sector Report 2015: Responsive and Accountable Public Governance* (New York, 2015).
World Bank Group. "PPIAF Toolkit for Public–Private Partnerships in Roads and Highways" (*January* 2009).
World Bank Group. Road Financing and Road Funds Knowledge Base. Available at www.worldbank.org/transport
World Economic Forum. *Strategic Infrastructure: Steps to operate and maintain infrastructure efficiently and effectively* (Cologny/Geneva, April 2014).
World Road Association (PIARC). *Asset management manual: a guide for practitioners* (Paris, 2017).
Yeang Yng Ling, Florence and Hartman, Andreas. "Value creation of road infrastructure networks: A structural equation approach", *Journal of Traffic and Transportation Engineering*, vol. 3, No. 1 (February 2016).
Zofka, Adam. "Proactive strategy for road asset management", *Studia i Materiały,* No. 82 (Warsaw, Instytut Badawczy Dróg i Mostów, 2019).

CONSTRUCTION OF EXPRESS ROADS IN LATVIAN STATE MAIN ROAD NETWORK

Dace BĒRZIŅA
"Latvian State Roads", 3 Gogola Street, Riga, LV-1050, Latvia
Dace.Berzina@lvceli.lv

February 18, 2021

Abstract. Latvian State Road Development Strategy for 2020 – 2040 was approved in the year 2020. It is a vision that includes long-term perspectives, strategic goals, tasks and priorities for road network development, and it is created for more efficient planning of state road network development, as well as, attracting additional funding for state owned roads.
The aim of the Strategy 2040 is to create an efficient road network ensuring that the bypass of the Riga City is accessible within two hours from every national and regional development centre in Latvia. Development centres would be accessible within 45 minutes from every Latvian urban settlement along state regional and local roads.
The strategic task is to create sections of express roads in the total length of 1000 kilometres thus improving traffic safety and reducing the emissions of greenhouse gases.
When creating the Strategy 2040 both the changes in traffic and road use and the changes in the location of population were studied. The Strategy foresees that high-speed express road sections would connect the Riga City with the biggest cities. The proposed plan of road network development is created with the aim to cover as large a population as possible.
Special attention is paid to the Riga City, as it serves as the central hub for Latvian and Baltic transportation. After the implementation of the Strategy 2040, the Riga City would be reached within 30 minutes from the nearest development centres.

Keywords: Strategy, state road development, accessibility, express roads.

Introduction

1. Background

The main task of the state road network is to ensure public mobility in all regions of the country and to promote economic development. And this prerequisite sets the vision for the long-term development of state road network by 2040.
The goal of the 2040 strategy is to create an efficient road network, ensuring that the Riga bypass may be reached from any national and regional development centre in Latvia within a maximum of two hours. In turn, development centres along state regional and local roads from any settlement in Latvia could be reached within 45 minutes.

- Road sections of two carriageways have been created in the state main road network.
- Efficient network of state regional roads has been established.
- Efficient network of state local roads has been established.

When developing the Strategy 2040, studies were carried out both how traffic and road user habits have changed and the location of the population has changed in Latvia in the last 20-30 years. It is planned that high-speed road sections would connect Riga with Ventspils, Liepāja, Jelgava, Bauska, Jēkabpils and Daugavpils, Rēzekne, Cēsis and Smiltene, as well as Ainaži. The proposed road network development plan is designed to cover as large a population as possible.
Priority is given to the reconstruction of state main roads by creating two carriageway road sections in the length of 1000 km in the state main road network, thus improving traffic safety and reducing emissions of greenhouse gases.

1.1. Network of state main roads and its quality

The total length of state roads is 20,061 km, of which 46% have asphalt pavement and 54% are paved with gravel. The average density of the national road network is 0.310 km per 1 km^2. 8%, or 1673 km are state main roads.
State main roads connect the national road network with the main road network of other countries and the capital city with other state cities or bypasses of state cities. There are 15 main state roads in Latvia with route indices A1 to A15. Some of state main roads are city bypasses: roads A4 and A5 form the Riga bypass, while roads A14 and A15 are part of the Daugavpils and Rēzekne bypasses.
In recent years, the renovation and reconstruction of the existing pavements on state main roads has been carried out purposefully. 85% of state main roads are in very good, good and satisfactory condition.

1.2. Traffic intensity on state roads

Changes in traffic intensity in the state road network are directly related to changes of the population in the territory of Latvia. Most of the country has experienced rapid depopulation over the past 15 years. Since 1995 the population of Latvia has decreased by ~24 %. Migration has taken place to large cities, especially to the Riga city and Riga city area, which now has up to 33% of the total population in the country (data from 2019). During this period traffic intensity on state main roads has increased 2-6 times.

It may be expected that by 2040 many parishes will lose a significant part of the population, except for the Riga city area. This will have an impact on main roads, in particular in the Riga city area, as the loading data of Latvian state road network shows that two thirds (66.9%) of vehicles for daily trips use only 10% of state roads. The highest traffic intensity is on roads in the Riga planning region (see Figure 1).

Because of the increase in traffic intensity and rapidly developing territories in the Riga city area, the number of dangerous road sections and intersections has rapidly increased. As the traffic in intersections with minor roads increases, it becomes more difficult or almost impossible for a road user to use such roads safely.

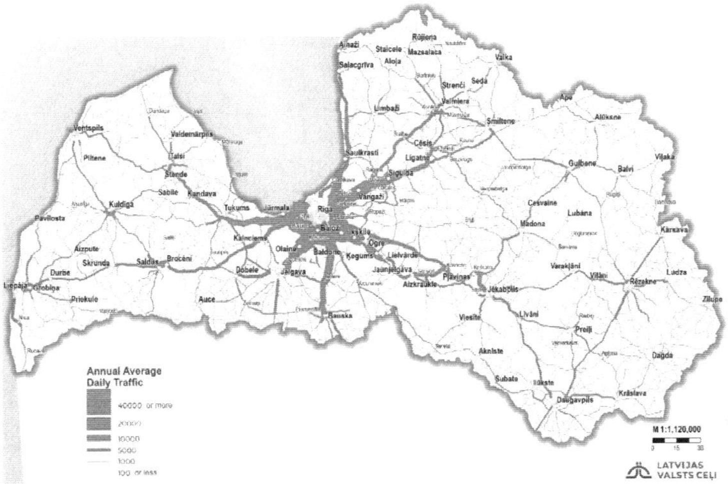

Figure 1. Traffic intensity on state roads

Currently, the maximum traffic intensity permitted on single carriageway two-lane roads is 18,000 - 20,000 vehicles per day. Due to the factors reducing road capacity (road curvature that reduces visibility, poor longitudinal alignment parameters, large number of accesses, etc.), the real value is lower. In some places road capacity is already insufficient, they generally do not meet the needs of economical, safe, comfortable and environmentally friendly traffic. Roads that do not comply with technical parameters cause extensive losses to the economy every year, because road traffic congestions increase the operating costs of road transport, travelling time and fuel consumption.

Latvia is crossed by several road corridors of European significance that are used by transit traffic. The busiest section is the international route E67 from Ainaži to Grenctāle, where truck traffic intensity is on average 3100 trucks per day and the international route E22 in the section Riga - Jēkabpils with an average intensity of 1600 trucks per day.

1.3. Problem areas in the existing network of state

The busiest state main road is the Riga bypass, which includes road A4 Riga bypass (Baltezers - Saulkalne) (hereinafter - road A4) and road A5 Riga bypass (Salaspils - Babite) (hereinafter - road A5). Both of these roads have a high proportion of heavy traffic, and the existing road alignments no longer meet the required traffic capacity. The total traffic intensity is in the range of 8900 - 22200 vehicles per day (on average 13600 vehicles per day). Heavy traffic is in the range of 1800 – 4600 trucks per day (on average 3000 trucks per day). Only 9% of the Riga bypass has two carriageways (4 lanes), and in the rest of the bypass the capacity of the road has reached its limit - the level of traffic comfort is low. It is characterized by relatively low speeds and driving in a dense column with limited opportunities to overtake.

Both roads forming the Riga bypass are not directly connected and the main state road A6 Riga - Daugavpils - Krāslava - Belarussian border (Pātarnieki) (hereinafter - road A6) serves as their connection. The junctions of these roads with the road A6 are considered to be bottlenecks where traffic jams occur.

Road A5 from 0.0 km to 6.0 km is located along the dam of Riga hydro power station. The existing Daugava crossing is the most problematic spot along the entire Riga bypass. This area has low road capacity and it is not technically possible to widen the existing bridge. The traffic flow along the Daugava crossing of the Riga hydro power station should be reduced, as any potential road traffic accident may have a negative impact on the structure of power station itself, which is indeed dangerous.

The existing single carriageway two-lane roads A4 and A5 have significantly reduced road safety and capacity, given that:

- the existing parameters of intersections in separate grades do not comply or only partially comply with the requirements of road design norms, and there are illegal access roads within the boundaries of some of these intersections,
- roads A4 and A5 have several unregulated and several regulated road junctions at grade with permitted left turns, which considering the proximity of Riga are overloaded during morning and evening peak hours. So called "bottle necks" are created and the capacity and road safety of both main and secondary roads is reduced,
- roads A4 and A5 roads have a large number of connections, which reduces the length of road sections with overtaking options. Overtaking in short sections is carried out at significantly increased speed, entering the lane in the opposite direction and often causing the most dangerous frontal collisions,
- one of the most dangerous turns - left turns - is allowed in road connections to houses and companies, because due to insufficient number of two-level road junctions it is not possible to allow only right turns in junctions,
- in the event of road repairs or in the event of accidents or forced stopping of vehicles, the capacity will be significantly reduced considering the existing and increasing traffic volumes,
- pedestrian and cyclist infrastructure that meets traffic safety requirements has not been built on roads A4 and A5.

In total, there are 14 dangerous road sections and intersections on roads A4 and A5 roads, which is 30% of the total number of such sections and intersections. The concentration of dangerous road sections and intersections on these roads is very high, as the total length of these roads is only 3.5% of all state main roads. These sections and intersections are mainly associated with a large number of junctions at grade.

In Latvia there are currently 109.2 km of two carriageway roads in the state main road network. They were built more than 30 years ago and over time many road accesses have been created which may not be allowed if the permitted driving speed is to be increased.

In 2013 the two-lane road P80 Tīnūži - Koknese (hereinafter - the road P80) was rebuilt, and by 2030 it is planned to build new two-lane road section from Koknese to Jēkabpils. In the future it is planned to build a second carriageway next to the existing carriageway in order to provide faster traffic. In addition to that, it is planned to widen the existing regional road P5 Ulbroka - Ogre in the section from road A4 to Tīnūži.

Traffic intensity in the section Koknese - Jēkabpils of the road A6 varies from 3700 to 7000 vehicles per day (on average 6000 vehicles per day). Truck traffic intensity is in the range of 1500 - 1700 vehicles per day (on average 1600 vehicles per day). Although traffic intensity is generally lower than on the busiest roads, road alignment is problematic in this section - many curves with small radii affect overtaking possibilities, driving speed and traffic safety in general.

Via Baltica road included in the E67 route is the E67 section from Tallinn to Warsaw. Via Baltica is an important transit corridor, which facilitates traffic both between the Baltic States and further in the north-south direction. It is also part of one of the core corridors of the Trans-European Transport Network (TEN-T): the North Sea-Baltic Sea corridor, which connects the eastern coastal ports of the Baltic Sea with the North Sea ports and provides intermodality. Via Baltica corridor includes the road A7 Riga - Bauska - Lithuanian border (Grenctāle) (hereinafter – the road A7). The traffic intensity on this road is in the range of 6000 28300 vehicles per day and truck traffic intensity is in the range of 2770 - 4170 trucks per day. Road A7 has no two carriageway road sections. Currently the entire flow of traffic passes through the towns of Iecava and Bauska, significantly deteriorating the quality of life in the immediate vicinity of the road. Therefore, these towns need to have bypasses to relieve them from heavy transport. With the construction of such bypasses 100% of transit traffic, as well as 80-90% of local heavy traffic will move to bypasses if entry bans to towns are introduced. Bypasses will also be used by light transport and it is forecast that 50-70% of the total traffic flow will be transferred to them, thus relieving the city streets.

Road A1 Riga (Baltezers) - Estonian border (Ainaži) (hereinafter - the road A1) is also part of the Via Baltica section of the E67 route. The traffic intensity on this road is in the range of 4600 - 27400 vehicles per day and heavy traffic intensity is in the range of 2000 - 3800 trucks per day. There are no two carriageway road sections on the route. The problem area on this road is the town of Baltezers, where the housing is dense and no high-speed road exists. As the Rail Baltica railway route is planned from Vangaži to the north, it is useful to consider the possibility to provide a road in the railway corridor, which could lead from the road A2 Riga - Sigulda - Estonian border (Veclaicene) somewhere between from Vangaži to Sēnīte to the road A1 motorway near Skulte. In 2021, a feasibility study for the construction of this road section with an environmental impact assessment was started.

1.4. State main roads un the European road network

In 1996 Latvia joined the Agreement on International Motorways of European Countries (hereinafter - AGR), which coordinates the development of the road network in accordance with the requirements of future international traffic. In Latvia 1225 km or 74.3% of state main roads are included in the E-road network and thus have acquired international significance. E-road network in Latvia is shown in Figure 2.

According to the AGR agreement, the E-road network mainly includes roads with a strong north-south and west-east direction. States having signed the Agreement are not required to ensure that the roads included in the network immediately meet the high technical requirements or are built according to such requirements. The Agreement allows for the gradual improvement of roads, improving the technical parameters in accordance with the requirements of increasing traffic.

Figure 2. Network of international European highways in Latvia

The White Paper on European Transport Policy "Roadmap to a Single European Transport Area - Towards a competitive and resource efficient transport system" sets the goal of a competitive and resource efficient transport system - a fully functional and EU-wide TEN-T "core network" by 2030 and high quality and high-performance network by 2050. The core network should provide efficient multimodal connections between the capitals of the EU Member States and other major cities, ports, airports and major land border crossing points, as well as other major economic centres. It should focus on filling in the gaps - mainly cross-border sections and problem areas / detours.

All state main roads in Latvia are part of the TEN-T network (see Figure 3).

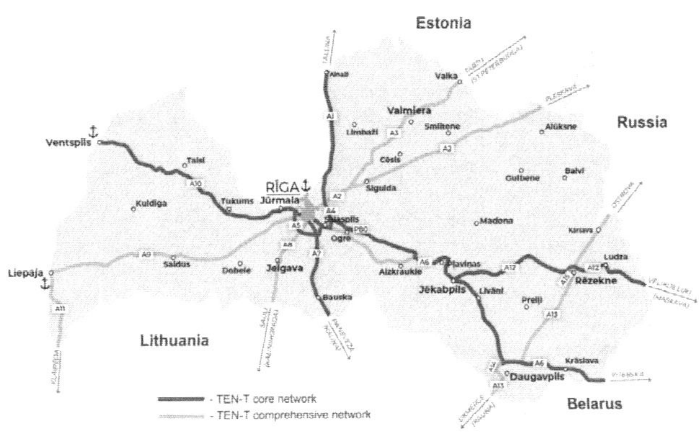

Figure 3. Network of TEN-T roads in Latvia

Regulation (EU) No 1315/2013 of the European Parliament and of the Council of 11 December 2013 on Union guidelines for the development of the trans-European transport network (hereinafter "the EU Regulation"), adopted to ensure the development of the EU transport sector and implementing the TEN-T "comprehensive" network and TEN-T "core network" states that the comprehensive network aims to ensure accessibility and connectivity for all EU regions and should be completed by 2050. The core network, on the other hand, includes those parts of the comprehensive network that are strategically most important for achieving the TEN-T objectives and that have to be completed by 2030. For the core network to be considered complete, roads have to meet the standards of motorways or express roads, but exceptions are allowed where the construction / reconstruction of a road is not economically justified.
There are no requirements set for roads in the TEN-T comprehensive network and they may be planned according to the expected traffic volume, as long as a sufficient level of road safety is ensured. In such cases public roads may become part of the comprehensive network, provided that they are sufficiently safe for traffic.
At the request of a Member State, the Commission may, in duly justified cases, grant an exemption from the requirement to build motorways or express roads in the TEN-T network as long as an adequate level of safety is ensured. In order to justify infrastructure investments, they should also be based on a socio-economic cost-benefit analysis. In road projects the benefits consist of the savings in time spent by road users and vehicle operating costs, so the most important factor in these calculations is the number of road users. Taking into account the relatively low average daily traffic intensity on roads and significant capital investments needed for the construction of high-speed roads or motorways, only the reconstruction of the road A7 Riga - Bauska - Lithuanian border (Grenctāle) in the TEN-T core network by 2030 is economically justified. Also, taking into account the load of the road, the reconstruction of the 59 km long Riga bypass is economically justified. This route is currently being proposed to be included the TEN-T core network.

1.4.1. Economically justified high-speed roads in 2021

1) Riga bypass
The Riga bypass will be a unified system, which will consist of the existing Riga bypasses A4 and A5, as well as the new Daugava bridge and its connections to both bypasses. The road under consideration is 61.7 km long. Currently, 84% are single carriageway two-lane roads and 16% are two carriageway four-lane roads. The traffic intensity ranges from 8416 to 22737 vehicles per day (on average 12704 vehicles per day). 72% of roads are in good, 6% - in satisfactory, 22% - in poor technical condition. Technical - economic calculations show that it is economically justified to rebuild this road into a high-speed highway or motorway already now, because the value of Internal Rate of Return exceeds the discount rate of 5% (IRR = 8.06%).

2) Road A7, section from Ķekava bypass to the Lithuanian border, including Bauska and Iecava bypasses
The road connects Riga with the Lithuanian border (Grenctāle). In the section from the border of the Riga city to the 25th km, the Public-Private Partnership project "E67 / A7 Ķekava bypass" has been started, therefore the economic substantiation has been performed for the remaining section of the A7.
The considered section is 60.1 km long with similar characteristics, including traffic intensity. It is in the range of 4395 - 12669 vehicles per day, but on average 9519 vehicles per day. 95% of road are in good, but 5% - in satisfactory technical condition. Technical - economic calculations show that it is economically justified to rebuild this road into a high-speed highway or motorway already now, because the value of Internal Rate of Return exceeds the discount rate of 5% (IRR = 5.66%).

1.4.2. Economically unjustified high-speed roads in 2021

1) Road A12 Jēkabpils – Rēzekne – Ludza – Russian border (Terehova)
The road connects Jēkabpils with Rēzekne and the Russian border (Terehova). The considered section is 160 km long with similar characteristics, including traffic intensity. It ranges from 861 to 3630 vehicles per day, but on average it is 2414 vehicles per day. 89% of road are in good, 7% - in satisfactory, 4% - in poor technical condition. Technical and economic calculations show that the costs of road reconstruction are 6.7 times higher than the planned benefits for road users. Internal Rate of Return may not be calculated, which indicates that it is not economically justified to rebuild this road into a high-speed road or a motorway. If the condition of the road does not change, the traffic intensity has to increase 7 times in order for these values to equalize. No such increase in traffic is expected in the next 20 years.

2) Road A6 Riga – Daugavpils – Krāslava – Belarusian border (Pātarnieki)
The road connects Jēkabpils with Daugavpils and the Belarusian border (Pātarnieki). The considered section is 170 km long with similar characteristics (the exception is ~22 km long two carriageway road section before the city of Daugavpils), including traffic intensity. It ranges from 1500 to 4211 vehicles per day, but on average it is 2650 vehicles per day. 62% of road are in good, 16% - in satisfactory, 22% - in poor technical condition. Technical and economic calculations show that the costs of road reconstruction are 4.8 times higher than the planned benefits for road users.

Internal Rate of Return is negative (-8.26%), which indicates that it is not economically justified to rebuild this road into a high-speed highway or highway. If the condition of the road does not change, the traffic intensity has to increase 5 times in order for these values to equalize. No such increase in traffic is expected in the next 20 years.

3) Road A10 Riga – Ventspils

The road connects Jūrmala with Ventspils. The section in question is 170 km long with similar characteristics, except for traffic. It ranges from 2635 to13147 vehicles per day. Due to large differences, the road is divided into two sections for calculations. The first section is Jūrmala – turn to Tukums [P121] (traffic intensity ranges from 5249 to 13147 vehicles per day and on average it is 10824 vehicles per day), the second section – turn to Tukums - Ventspils (traffic intensity ranges from 2635 to 6600 vehicles per day and on average it is 3756 vehicles per day).

73% of the 1st section are in good, 27% - in poor technical condition. 88% of the 2nd section are in good, 4% - in satisfactory, 8% - in poor technical condition. Technical-economic calculations show that the costs of the reconstruction of the 1st section 1.53 times exceed the planned benefits for road users. Internal Rate of Return is less than the discount rate of 5% (0.71%), which indicates that it is not economically justified to rebuild this road into a high-speed road or a motorway. If the condition of the road does not change, the traffic intensity has to increase 1.6 times in order for these values to equalize. Such an increase in traffic intensity is projected around 2040.

Technical-economic calculations show that the reconstruction costs of the 2nd section exceed the planned benefits for road users by 6 times. Internal Rate of Return may not be calculated, which indicates that it is not economically justified to rebuild this road into a high-speed road or a motorway. If the condition of the road does not change, the traffic intensity has to increase 6.5 times in order for these values to equalize. No such increase in traffic is expected in the next 20 years.

4) Road A1 Riga (Baltezers) - Estonian border (Ainaži)

The road connects Riga with the Estonian border (Ainaži). The section in question is 101.7 km long with similar characteristics, except for traffic intensity. It ranges from 3778 to 26859 vehicles per day. Due to large differences the road is divided into two sections for calculations. The first section is Riga - Saulkrasti (traffic intensity in the range of 7969 - 26859 vehicles per day, on average 14704 vehicles per day), and the second section is Saulkrasti - Ainaži (traffic intensity in the range of 3778 - 7655 vehicles per day, on average 5187 vehicles per day). The technical condition of the whole road is good.

Technical-economic calculations show that the reconstruction costs of the 1st section road are 1.33 times higher than the planned benefits for road users. Internal Rate of Return is less than the discount rate of 5% (2.04%), which indicates that it is not economically justified to rebuild this road into a high-speed road or a motorway. If the condition of the road does not change, the traffic intensity has to increase 1.4 times in order for these values to equalize. Such an increase in traffic intensity is forecast around 2035.

Technical-economic calculations show that the reconstruction costs of the 2nd section are 2.4 times higher than the planned benefits for road users. Internal Rate of Return is negative (-3.12%), which indicates that it is not economically justified to rebuild this roads into a high-speed road or highway. If the condition of the road does not change, the traffic intensity has to increase 2.5 times in order for these values to equalize. No such increase in traffic is expected in the next 20 years.

5) Section E22 Riga - Jēkabpils

The section of the route E22 consists of several sections of roads P5 Ulbroka - Ogre, P80 Tīnūži - Koknese and A6 Riga - Daugavpils - Krāslava - Belarusian border (Pātarnieki), which connect Riga with Jēkabpils. The considered section is 115.6 km long with similar characteristics, including traffic intensity. It ranges from 4542 to 9124 vehicles per day, but on average it is 6417 vehicles per day. 74% of road are in good and 26% - in satisfactory technical condition. Technical and economic calculations show that the costs of road reconstruction are 1.4 times higher than the planned benefits for road users. Internal Rate of Return is less than the discount rate of 5% (1.80%), which indicates that it is not economically justified to rebuild this road into a high-speed road or a motorway. If the condition of the road does not change, the traffic intensity has to increase 1.7 times in order for these values to equalize. No such increase in traffic is expected in the next 20 years.

1.4.3. Justification of the construction of high-speed roads in 2021

Table 1 . Economically justified and unjustified high-speed roads in 2021

No.	Road	IRR, %	Discounted road user benefits calculated for 20-year period, mill. EUR	Discounted costs calculated for 20-year period, mill. EUR	Real costs, mill. EUR
1	2	3	4	5	6
1	Riga bypass (roads A4 and A5 & the new Daugava bridge)	8.06%	553.3	426.5	516.1
2	Road A7, section from Ķekava bypass to the Lithuanian border, including Bauska and Iecava bypasses	5.66%	310.3	292.5	353.9
1	Road A12 Jēkabpils – Rēzekne – Ludza – Russian border (Terehova)	n/a	118.9	796.4	963.6
2	Road A6 Riga – Daugavpils – Krāslava – Belarusian border (Pātarnieki)	-8.26%	142.4	688.8	833.4
3.1	Road A10 Riga – Ventspils (section 1)	0.71%	139.5	213.4	258.2
3.2	Road A10 Riga – Ventspils (section 2)	n/a	82.9	502.1	607.5
4.1	Road A1 Riga (Baltezers) - Estonian border (Ainaži) (section 1)	2.04%	117.8	156.8	189.7
4.2	Road A1 Riga (Baltezers) - Estonian border (Ainaži) (section 2)	-3.12%	103.1	246.4	298.1
5	Section E22 Riga - Jēkabpils	1.80%	282.5	385.0	465.8

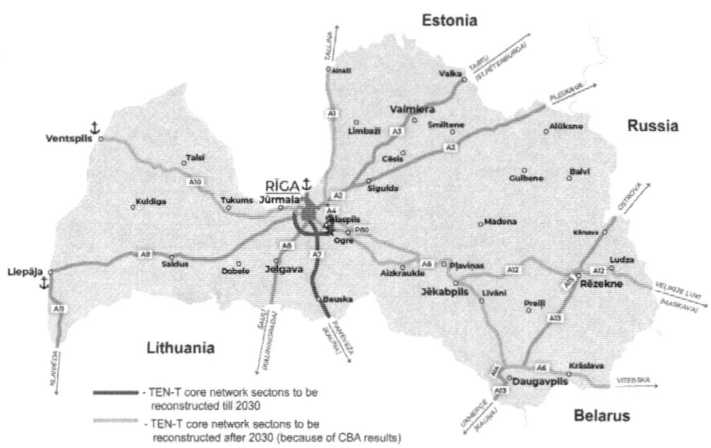

Figure 4. Improvement of TEN-T core network roads until 2030

2. Development of state roads

2.1. Aim

The main task of the state road network is to ensure public mobility in all regions of the country and to promote economic development. And this prerequisite sets the vision for the long-term development of state road network by 2040.
The goal of the 2040 strategy is to create an efficient road network, ensuring that the Riga bypass may be reached from any national and regional development centre in Latvia within a maximum of two hours. In turn, development centres along state regional and local roads from any settlement in Latvia could be reached within 45 minutes.

2.2. Strategic tasks

- Road sections of two carriageways have been created in the state main road network.
- Efficient network of state regional roads has been established.
- Efficient network of state local roads has been established.

When developing the Strategy 2040, studies were carried out both how traffic and road user habits have changed and the location of the population has changed in Latvia in the last 20-30 years. It is planned that high-speed road sections would connect Riga with Ventspils, Liepāja, Jelgava, Bauska, Jēkabpils and Daugavpils, Rēzekne, Cēsis and Smiltene, as well as Ainaži (see Figure 5). The proposed road network development plan is designed to cover as large a population as possible.

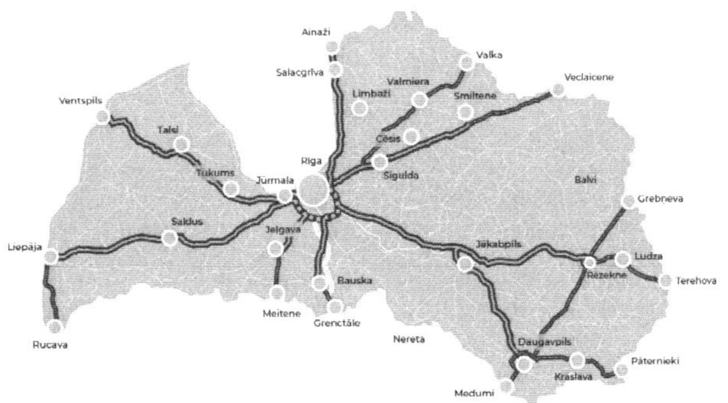

Figure 5. High-speed road sections planned in Strategy 2040

Special attention is paid to the Riga city, as Riga is the central hub of Latvian and Baltic traffic. Almost every main road joins the Riga bypass, however, the traffic intensity on this bypass currently significantly exceeds the capacity for which it was built. As a result of the implementation of the Strategy 2040, Riga has to be reached within 30 minutes from Tukums, Jelgava, Bauska, Ogre, Sigulda and Saulkrasti (see Figure 6).

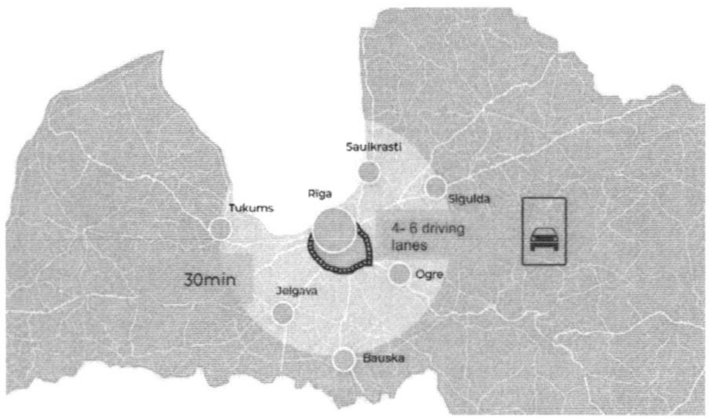

Figure 6. Reachability of the Riga bypass in the Riga city area defined in Strategy 2040

2.3. Priority – reconstruction of state main roads

Priority is given to the reconstruction of state main roads by creating two carriageway road sections in the length of 1000 km in the state main road network, thus improving traffic safety and reducing emissions of greenhouse gases.
Such rebuilding of main roads is necessary to ensure simultaneously fast and safe road traffic. It is necessary to provide a separate or separated carriageway for each direction of traffic, as well as ensure other additional safety solutions: rebuilding the existing one carriageway and two carriageway roads so that they could be accessed only from intersections in separate grades or controlled junctions, prohibiting the stopping of vehicles on carriageways and eliminating intersections with railways or pedestrian ways.
By rebuilding a two-lane road into a multi-lane road with separate carriageways it is possible to eliminate the most dangerous road traffic accidents - frontal collisions, as well as collisions with left-turners, which are often the cause of fatality or serious injury. Road traffic accidents are expected to decrease by 50% in these sections.
Transport sector is one of the main sources of greenhouse gas emissions, accounting for 28.5% of total emissions, and it is important to reduce them. The reconstruction of state main roads into two carriageway roads will improve the quality of driving conditions, the traffic flow will be smoother and the waiting times for certain traffic flows will be significantly reduced or eliminated. Smooth driving is characterized by lower fuel consumption, which is the cause of GHG emissions. It is estimated that improved or rebuilt roads will reduce annual GHG emissions from road transport (carbon dioxide CO_2 and methane CH_4) by around 16%.
Feasible reconstruction option is two carriageways with four lanes, taking into account traffic safety and construction costs (see Figure 8). According to the standard LVS 190:2 “Road design regulations” the optimal traffic intensity for such a technical solution is in the range of 18000 - 65000 vehicles per day and it may be considered that in the foreseeable future the capacity of such roads will be sufficient.

Technical description of a two carriageway road is the following:
1. A road with at least four lanes with separate carriageways;
2. Accessible only from intersections in separate grades or regulated intersections, but in certain justified cases unregulated right entries and exits are allowed. The distance between the nodes is at least 3 km, but it has to be possible to place traffic organisation furniture;
3. Is marked with a special road traffic sign 552;
4. Local traffic in the vicinity of a two carriageway road shall be organized on local roads and may cross such road via intersection in separate grade;
5. No stopping, parking, as well as reversing the car on the carriageway is allowed;
6. Slow traffic, bicycles, as well as pedestrians are not allowed on the road;
7. Bus stops should be separated from the carriageway. The flow of bus passengers across the carriageway has be provided via intersection in separate grade, eliminating the possibilities to cross the road at the same level as the carriageway.
8. Roads shall be equipped with animal restraint fences and animal passages, as well as with noise reduction elements.

In parallel with the road infrastructure, appropriate cycling infrastructure and utility channels which would provide for the development of 5G infrastructure should be established in co-operation with municipalities.

When rebuilding the existing one carriageway roads or constructing new two carriageway roads, the average construction cost per 1 km at 2020 prices is approximately 6.0 million EUR, which consists of:
- reconstruction or construction of the main road - 1.8 million. EUR,
- parallel roads for local traffic - 0.7 million. EUR,
- intersections in separate grades - 1.6 million. EUR,
- engineering structures - 1.1 million. EUR,
- other (expropriation of land, parking lots, additional equipment [animal fences, noise reduction elements, bus stops, etc.]) - 0.8 million. EUR.

When rebuilding the existing two carriageway roads, the reconstruction costs are approximately 1.0 - 3.0 million EUR. The exact costs of each project will be determined after the development of a detailed construction design. They depend on the location of the project, the degree of complexity and other factors.
From the idea of road construction or reconstruction to construction works, several design stages have to be completed – research including an environmental impact assessment, construction design with a minimum composition, land expropriation process, detailed construction design. This whole process takes on average 6-7 years. The duration of construction works depending on the size of the construction site is on average 2 - 4 years.
After the analysis of the situation in the state main road network, the reconstruction or construction of state main road sections is planned in three stages.

First stage of the implementation of Strategy 2040 for the period of 2020 - 2030

Priority 1: reconstruction of Riga bypass.
The following projects are planned for the full reconstruction of the Riga bypass into two carriageway road (see Figure 9):

Project 1: Reconstruction of road A4 Riga bypass (Baltezers – Saulkalne)" - 20.5 km.

Project 2: Reconstruction of road A5 Riga bypass (Salaspils - Babīte), section from state main road A10 to perspective Ķekava bypass - 26.5 km.

Within the framework of these two projects, a new second carriageway, intersections in separate grades, parallel local roads and the necessary equipment (lighting, noise barriers, animal fences, etc.) will be built.

Project 3: Construction of a combined road and railway bridge over the Daugava and related road infrastructure. Within the framework of this project a new combined bridge over the Daugava and related road infrastructure will be built - bridge approaches with two carriageways, reconstructed intersection in separate grades of roads A4 and A6, new connection to the existing national regional road P85, parallel local roads and necessary equipment (lighting, noise barriers). etc.

Project 4: Construction of road A5 Riga bypass (Salaspils-Babīte), section from the new bridge over the Daugava to the junction with Ķekava bypass – 12.0 km.

Priority 2: reconstruction of the existing two carriageway roads.
The reconstruction of the following sections two carriageway roads is planned (see Figure 10):

Project 5: Reconstruction of road A2 Riga – Sigulda – Estonian border (Veclaicene), section from road A4 to the Lorupe ravine (section with two carriageways) - 32.0 km.

Project 6: Reconstruction of road A8 Riga – Jelgava – Lithuanian border (Meitene), section Riga – Jelgava – 36.5 km.

Project 7: Reconstruction of road A10 Riga – Ventspils, section Riga – Jūrmala – 6.7 km.

Priority 3: reconstruction of Via Baltica.

Project 8: Reconstruction of road A7 Riga - Bauska - Lithuanian border (Grenctāle), including Bauska and Iecava bypasses– Lietuvas robeža (Grenctāle) pārbūve, iekļaujot Bauskas un Iecavas apvedceļu izbūvi" – 60.1 km.

In total it is planned to reconstruct 194.3 km of roads during Stage 1 (see Figure 7).

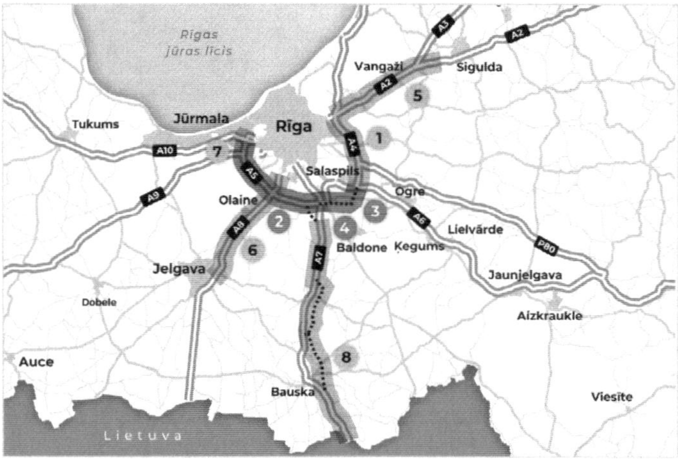

Figure 7. Newly constructed and reconstructed roads during Stage 1 of Strategy 2040

Second stage of the implementation of Strategy 2040 for the period of 2030 - 2035

The reconstruction of the Koknese - Jēkabpils section of the road E22 into a two carriageway road will be carried out gradually. Stage 1 will include the building of one carriageway with 2 lanes, intersections in separate grades and parallel local roads, as well as animal fences). In Stage 2 a second carriageway will be built next to the adjacent carriageway and the works are planned in the next stage from 2035 to 2040.

Project 9: Construction of road E22 section Koknese – Pļaviņas – 17.0 km.
Project 10: Construction of road E22 section Pļaviņas - Jēkabpils - 25.3 km.
These two projects together with road P80 rebuilt in 2013 (stage 1 of the project, 1 carriageway with 2 lanes) will result in a straight, comfortable and safe road in the direction of the eastern border of Latvia (Russia and Belarus).
Project 11: Reconstruction of road A1 Riga (Baltezers) – Estonian border (Ainaži) – 101.7 km.
Project 12: Reconstruction of road A10 Riga – Ventspils, section from Jūrmala to turn to Tukums - 48.4 km.
Project 13: Reconstruction of road A2, section from the Lorupe ravine to turn to Cēsis – 31.6 km.

In total it is planned to reconstruct 224.5 km of roads during Stage 2 (see Figure 8).

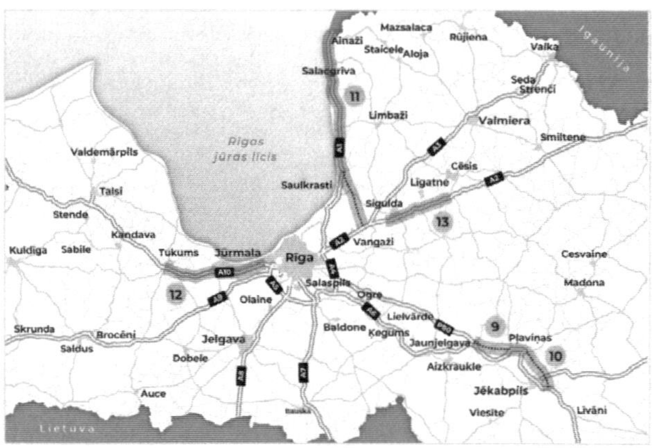

Figure 8. Newly constructed and reconstructed roads during Stage 2 of Strategy 2040

Third stage of the implementation of Strategy 2040 for the period of 2035 - 2040

Project 14: Reconstruction of road E22, section from road A4 to Jēkabpils, stage 2 – 116.8 km.
Project 15: Reconstruction of road A12 Jēkabpils - Rēzekne - Ludza – Russian border (Terehova), section Jēkabpils – Rēzekne - 89.0 km.
Project 16: Reconstruction of road A6 Riga – Daugavpils – Krāslava – Belarussian border (Pātarnieki), section Jēkabpils – Daugavpils – 76.7 km.
Project 17: Reconstruction of road A10 Riga – Ventspils, section Tukums – Ventspils – 114.6 km.
Project 18: Reconstruction of road A9 Riga (Skulte) – Liepāja – 191.7 km.
Project 19: Reconstruction of road A2, section from turn to Cēsis to Smiltene – 48.7 km.

In total it is planned to reconstruct 637.5 km of roads during Stage 3 (see Figure 9).

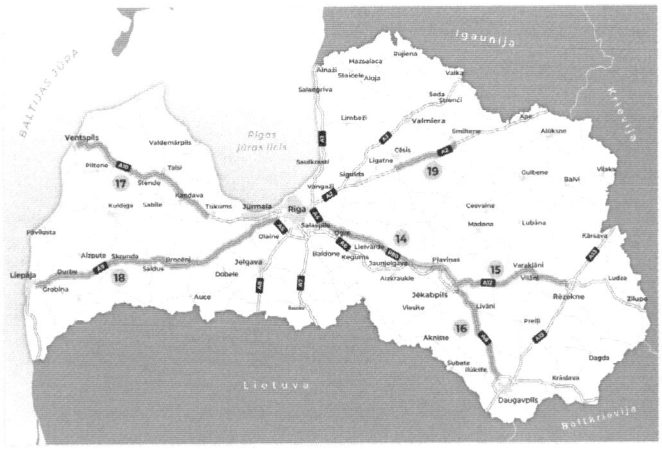

Figure 9. Newly constructed and reconstructed roads during Stage 3 of Strategy 2040

Conclusions

1. In carrying out the development of state roads in accordance with the Strategy 2040, approximately 1000 km of state main roads in the state road network would be rebuilt into high-speed roads, and this implies that:
   1.1. Construction cost in the prices of 2020 is 5.2 billion EUR;
   1.2. The benefits to the Latvian economy from the reduction in travelling time, reduction of road traffic accidents and reduction of CO_2 emissions would reach 169.4 million EUR per year.

2. Project implementation and funding schedule has been prepared for Stage 1 of the reconstruction of state main roads. These projects have to be implemented until 2028 and the necessary funding amounts up to 542.29 million EUR.
   2.1. After the implementation of Stage 1 the benefits to the Latvian economy from the reduction in travelling time, reduction of road traffic accidents and reduction of CO_2 emissions would reach 71.93 million EUR per year.
   2.2. As soon as Stage 1 is implemented interim review of the development of state main roads, as well as Strategy 2040 will be carried out and after that priorities will be adjusted and future action plan and funding schedule will be prepared accordingly.

THE FIRST HIGHWAY INFRASTRUCTURE PPP PROJECT IN THE BALTIC STATES ACCORDING TO DBFM MODEL – CHALLENGES AND OPPORTUNITIES

Verners AKIMOVS, Indra MUIŽNIECE, Ilze Kristine APSALONE, Liesma GRINBERGA

State-Owned Limited Liability Company "Latvian State Roads" (Gogoļa str.3, Riga, LV-1050, Latvia)

E-mail: KekavaBypass@lvceli.lv

Received date 15.06.2021; accepted date: __.__.2021

Abstract.

Kekava Bypass public-private partnership (PPP) project is the first highway infrastructure PPP project in the Baltic States to be implemented according to Design-Build-Finance-Maintain (DBFM) model. Its objective is to solve the "bottleneck" of the TEN-T network road section entering Riga City with only one lane in each direction. The duration of the project is 23 years, including 3 years for design and construction of the infrastructure and 20 years for its maintenance. Kekava Bypass consists of the main road 17.22 km, auxiliary roads 20.66 km, 2 tunnels, 5 two-level road junctions, 1 bridge (all in all more than 100 lane km), 2 pedestrian bridges/tunnels etc.

The Latvian government required that the assets of the Kekava Bypass project are classified off government balance sheet during the whole duration of the PPP contract. Thus, the Kekava Bypass PPP tender documentation was elaborated strictly observing this off-balance sheet treatment frame. Eurostat assessed the project documentation and issued its opinion that the project corresponds to the off-balance sheet criteria.

Latvian State Roads on behalf of the Ministry of Transportation launched Kekava Bypass tender in December 2018. The tender was implemented in four sequent phases, namely, the qualification phase, the submission of initial offers, negotiations with the selected bidders about the initial offers, the submission of the best and final offers. On August 13, 2020, Latvian State Roads announced the winner and the financial due diligence phase to be performed by financial institutions started.

Keywords: road construction, public – private partnership, DBFM, off balance sheet

Introduction

The concept of the public-private partnership (PPP) has become more and more popular in both developed and developing countries in recent decades (Osei-Kyei & Chan, 2015) because it is an opportunity to develop the public infrastructure even if the country does not have financial resources at the moment for such purposes. However, the origins of the PPP can be traced back to ancient times. More detailed evidence about PPP-like cooperation dates to the 17th century in Great Britain (Link, 2006). In the past, it was usually an initiative of individual entrepreneurs rather than an initiative of the government for the development of the public infrastructure. Only since the beginning of the 20th century, public administrations have begun to practice PPP as a targeted activity (Cheney, 2018).

In the EU Member States, the PPP concept became particularly relevant with the 2014 Investment Plan for Europe or the so-called *Juncker Plan* that was designed to reduce the drop of investment with the attraction of private and public resources.

Regarding the very first highway infrastructure PPP project in Latvia, it is important to define the scope: the project includes different phases like pre- preparation, preparation, procurement, PPP contract, hands- off period etc. At the moment of the development of this report, PPP project "Kekava Bypass" implementation is still ongoing and there are different risks to be managed to implement the whole project successfully. The success or failure of the project will create a precedent to apply PPP concept to the other infrastructure projects in Latvia therefore it has been very important from the very beginning to consider each step and to make sure it is the most transparent and the suitable way how it can be implemented.

The first highway infrastructure PPP project in the Baltic States according to DBFM model – challenges and opportunities

1. Public – private partnership as an opportunity

1.1. Different forms of PPP

The public-private partnership does not have one strict definition because it has different forms. The PPP concept is mainly used to identify the type of cooperation between the public and private sectors to achieve a common goal (The World Bank, 2014). For example, the Organisation for Economic Co-operation and Development (OECD) has defined PPP as "an agreement between the government and one or more private partners (which may include the operators and the financers) according to which the private partners deliver the service in such a manner that the service delivery objectives are aligned with the profit objectives of the private partners and where the effectiveness of the alignment depends on a sufficient transfer of risk to the private partner." (OECD, 2008). In turn, according to the Reference Guide of The World Bank PPP is a long-term contract between a private party and a government entity, for providing a public asset or service, in which the private party bears significant risk and management responsibility, and remuneration is linked to performance (The World Bank, 2014). Garvin and Bosoo (2008) explains the meaning of PPP as "a long-term contractual arrangement between the public and private sectors where mutual benefits are sought and where ultimately (a) the private sector provides management and operating services and/or (b) puts private finance at risk" (Garvin & Bosso, 2016). Essentially, it is a procurement method providing "value for money" for the development of public infrastructure taking advantage of competitive tenders, flexible negotiations and risk-sharing between the parties concerned (Akintoye et al., 2003).

In their study, Hodge and Greve (2016) acknowledge that, regardless of how different PPPs are defined, they share several common aspects. In most of the PPP definitions appears the "risk" aspect to be shared between the parties involved in the conclusion of the contract. Another important aspect is "innovation", as PPP implementation is expected to go beyond the usual cooperation. The public and private sectors need to offer new solutions and work together to achieve results. This gives hope for a long-term and successful cooperation, which is not usually the case with a traditional service contract. The successful implementation of projects is also encouraged by the fact that in this case, unlike other procurement methods, the parties have carefully identified and shared risks before concluding a PPP agreement (Li et al., 2005). The main feature of PPP is that they indicate the assets or services provided as a result, rather than how they are provided. In general – it is determined what is needed, not how to achieve it. Therefore, PPP contracts are "combined" with several project phases or functions. Depending on the specifics of the project, features may vary. For example, typical features that could be included in a private partnership agreement, is project development, construction, financing, maintenance, operation (The World Bank, 2014).

1.2. The main challenges

Although the PPP approach is becoming widely used for infrastructure development worldwide, it is still being considered controversial. Some authors emphasize that this is due to the high costs of implementing a PPP, its long procurement process, the lack of experience, the unattractive financial market, the high degree of risk, and, in many cases, the higher costs for end-users (Osei-Kyei & Chan, 2015). Other authors emphasize that the main benefits of using the PPP concept are improved public administration capacity through integrated procurement solutions, promotion of innovative approaches, reduction of project implementation costs and time, transfer of risks to a private partner, acquisition of new skills and technologies (Akintoye et al, 2003).

Bayliss and Waeyenbege conclude at the end of their study that PPP would be more important for developing countries, but they tend to have weak project implementation and regulation capacity, and potential investors have a small interest in investing in lower income countries. The authors of this study doubt the usefulness of PPPs in poor countries and believe that it is not an appropriate tool to indirectly reduce poverty and promote the country's economic growth and prosperity (Bayliss & Waeyenbege, 2017). Other researchers, on the other hand, are daring to argue that PPPs will become a powerful policy tool for development in the future (Hodge & Greve, 2016). Through the PPP scheme, the government has an opportunity to focus on the development of several sectors of the economy by promoting infrastructure development (Cumming, 2007). Sergi et al. (2019) mention India and China as examples of the appropriateness of using the PPP concept to promote the sustainability of the developing countries, at the same time recognizing that the primary obstacle to the use of PPP by this type of developing country is the lack of initiative at the national level, as a private investor cannot initiate this process without a proper regulatory and legal basis.

This disagreement is exacerbated by the fact that there is still no common approach to determining whether a particular project is performing well. As a possible solution for the evaluation of PPPs, researchers have proposed the determination

The first highway infrastructure PPP project in the Baltic States according to DBFM model – challenges and opportunities

of the success factors of a PPP and believe that it could be used to make PPPs more successful in the future. Success factors of PPP is one of the key areas that attract the attention of PPP researchers (Osei-Kyei & Chan, 2015). For example, Osei-Kyei et al. (2017) identified in their study the most important criteria determining the successful implementation of a PPP project are effective risk management, meeting output specifications, reliable and quality service operations, adherence to time, satisfying the need for public facility/service, profitability and long-term relationship and partnership. In the field of PPP research, increased attention is also paid to such important issues as PPP governance and regulation, performance management, economic viability, risk management (Cui et al, 2017).

It is now hard to predict the practice of using the PPP concept in the future and whether it will continue to increase. In the World Bank's annual report on private partner investment in the first half of 2020 is concluded that restrictions imposed to reduce the spread of the Covid-19 virus have had an impact on PPP funding (The Worlds Bank, 2020). Many ongoing projects are delayed or even completely stopped. This is mainly due to the decline of the macroeconomic indicators and the fear of private investors to invest in sectors whose development in the future and return on investment are uncertain. For example, according to the World Bank's private partnership database, private investment in the road sector fell by 79% in the first half of 2020 compared to 2019.

1.3. The local experience so far

So far, a small number of PPP projects have been implemented in Latvia, none of which have been for the development of transport infrastructure, even though the legal basis for this type of co-operation in Latvia was established already in 2009 with the adoption of the Law on Public-Private Partnerships. The first attempt to introduce PPP for the development of transport infrastructure was for the road section Riga - Senite but that was not led to implementation, although all the necessary documentation had already been prepared. The main reason for this outcome was a change of priorities for the state-related institutions that caused this object to lose its importance. The next attempt to implement PPP in the transport sector in Latvia is the construction of the Kekava bypass, which will be described in more detail in this publication.

2. Pre-preparation phase of the Kekava Bypass PPP project

2.1. Description of the necessity

In Latvia, the highest traffic volume is around Riga City, therefore all roads entering Riga, namely, from Ventspils (A10), Daugavpils (A6), Tallinn (A1), Jelgava (A8), have several lanes in each direction. The existing road E67 / A4 in its section Riga – Ķekava has one of the highest volumes – capacity ratio in Latvia and the traffic volume has been multiplied several times during the last 15 years exceeding 17 000 vehicles per day. Currently, the planned traffic volume for the existing road construction has already been exceeded approximately three times. Moreover, Bauska highway (A7) is the only road entering Riga City with one lane in each direction.
In compliance with the Final Report on North Sea-Baltic core network corridors (European Commission, 2014), Kekava Bypass has been acknowledged as the "bottleneck" of the TEN-T network and it is recommended to make actions to improve cross-border connections and increase road capacity.

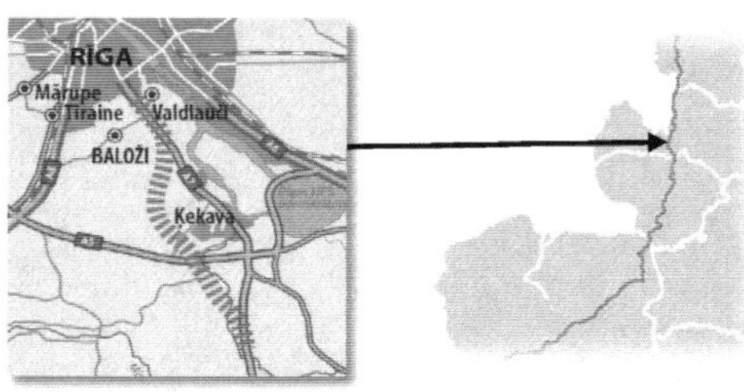

Figure 1. Planned location of the Kekava Bypass.

The first highway infrastructure PPP project in the Baltic States according to DBFM model – challenges and opportunities

Kekava Bypass public-private partnership (PPP) project is the first highway infrastructure PPP project in the Baltic States to be implemented according to Design-Build-Finance-Maintain (DBFM) model. Its objective is to solve the "bottleneck" of the TEN-T network road section entering Riga City with only one lane in each direction.
The Kekava Bypass after its construction will be the new section of the road A7 Riga – Bauska – the border of Lithuania (Grenctale) from km 7.78 to km 25.0 and it will be a part of the international road E67 Via Baltica (Helsinki – Tallinn – Riga – Panevezys – Kaunas – Warsaw – Prague). It will be also a new section of the TEN-T network ensuring the most efficient road connection between the capital of Latvia – Riga and the neighbor country Lithuania as well as the rest of Europe. The road E67 provides traffic stream through Riga city to the bypass of Riga (A4 Baltezers - Saulkalne), the bypass of Riga (A5 Salaspils - Babite), the road A6 Riga-Daugavpils – Kraslava -the border of Belorussia (Patarnieki), the road A8 Riga – Jelgava - the border of Lithuania (Meitene) as well as the road A9 Riga (Skulte) - Liepaja.

2.2. Decision making

As a high priority project, Kekava Bypass was included in the Declarations of several Governments. An activity 1.2.1.2. Preparation of the E67/A7 Kekava Bypass construction project for the commencement of its implementation according to the public-private partnership model and commencement of its implementation, if a relevant decision of the Cabinet of Ministers will be adopted within the measure 1.2.1 Rehabilitation and development of roads in the TEN-T network was included in the Transport Development Guidelines for 2014-2020 (approved by the Cabinet of Ministers with an Order No.683 of 27 December 2013 "On Transport Development Guidelines for 2014–2020". However, the initial decision of the Government to start preparation works of the Kekava Bypass project was taken only on October 7, 2014 (protocol No. 53, § 35) with following development of the first financial and economic calculations.

Analyzing data on availability of the funds for the development of the road infrastructure in Latvia (state budget, EU structural funds) it was concluded that investments into the road infrastructure were insufficient continuously for many years. Thus, the implementation of a new road construction project of such a scale would mean postponing renovation projects of the existing roads for several years. On the other hand, application of the PPP model provides advantages comparing with other funding methods including an alternative funding in comparison to the state budget and EU structural funds and an optimal risk division achieved by transferring to the private partner risks which the private partner will be able to manage most effectively. Considering this, the private – public partnership was assumed as one of the potentially most effective methods for implementing the project.

On the base of the results, the Government decided to start preparation works for the procurement under the PPP model considering also terms set by the EC decision about the European Fund of Strategic Investments (the so-called *Juncker Plan*). This decision can be considered as the official start of the project.

Considering all the above and the scope of the project, the necessity of international financiers was identified at the very beginning of the project. Therefore, Kekava Bypass as a potential PPP project was presented to one of the main potential financiers – the Europe Investment Bank (EIB). The presentation was successful - on December 15, 2015, by sending the Letter of Intent on possibility to fund the project, EIB expressed readiness to support the project. This fact served as a basis for the next step towards the project.

2.3. Selection of the PPP model

Before starting the construction of any infrastructure object, it is necessary to evaluate whether the obtained benefit from the construction of the object is higher than the construction costs. Public infrastructure objects cannot be recouped directly, for example, by setting tolls, however, they have socio-economic benefits.

In March 2014, the first Financial Economic Analysis (FEA) was developed by KPMG Baltics SIA. FEA was developed on the base of the guidelines issued by the Ministry of Finances for the PPP projects. The document was also assessed and approved by the Central Finance and Contracting Agency (CFCA). In December 2015, FEA was updated.

The assessment has been made on the assumptions that:
- the first 5 years of the project implementation will be needed for the purchase of land for the project,
- road construction works will be carried on for 3 years,
- road operation will last 20 years.

The following socio-economic benefits of road construction can be identified for road construction projects and during the period of subsequent use:
- accident cost savings,
- cost savings for road users in terms of time spent on the road,
- depreciation savings on vehicles,
- tax revenue.

The analysis of socio-economic factors showed a positive result - the socio-economic benefits exceeded the total costs of the project - an assessment was made of whether it is more advantageous for the state to implement the project itself through the classical public procurement mechanism, or also transfer these rights to the private partner through a public-private partnership structure. The performed analysis showed that the PPP model is more advantageous for the state, because it has significantly reduced number of risks and amount of costs related to them, which were mainly due to higher prices in the case of public procurement. In PPP model many of the risks related to the public procurement process will be transferred from the public partner to the private partner. The next step was to choose exactly which PPP model to implement.

Transport infrastructure PPP projects are usually based on the principle of subsidies, i.e., they are based on demand, not so much on quality. In such model, the compliance with quality or accessibility standards are not the main issues to watch for; the private partner primarily seeks profitability. In developed countries the risk of demand for yet unbuilt roads are taken over by the private sector, as historical traffic data is stable and reliable. However, countries such as Poland and Hungary had unsuccessful experiences where private operators were unable to bear the traffic risk. In case of Kekava Bypass, the length of the road section, current and projected bypass traffic volume is insufficient to generate sufficient revenue and ensure successful implementation of the Road Subsidy (*Shadow Toll*) mechanism.

Considering that it was concluded that the accessibility payment scheme is the most appropriate mechanism for the Kekava Bypass project due to the following factors:
- the accessibility payment is directly aimed at meeting the established infrastructure standards which are fully compliant to the objectives of the public partner regarding the rehabilitation and maintenance of the road,
- there is no link between the request (traffic) and the compensate to the private partner; traffic risk is replaced by "availability risk",
- the availability fee is usually a fixed amount - in real terms - throughout the project implementation period. In contrast, the *Shadow Toll* varies annually according to traffic volume changes. The public partner knows in advance what amount will have to be paid each year.

Table 1. Summary of the chosen PPP model

Type of the contract	DBFM (Design, Build, Finance, Maintain)
Duration of the PPP contract	23 years (3 [designing and building] + 20 [maintenance])
Type of payment	Availability payment
Customer / Public partner	Ministry of Transport
Balance sheet treatment	Off government balance sheet

Conclusions of the FEA were used to issue on March 10, 2016 an order No. 172 by the Cabinet of Ministers thus launching preparation of the design, construction, financing, and maintenance procurement of the Kekava Bypass under the structure and principles of the public – private partnership.

Considering complexity of the project and the fact that there is no local PPP experience to be used as an example, already in June 2016, SJSC "Latvian State Roads" and EIB agreed to involve the European PPP Expertise Centre in the development process. Additionally, legal, and financial advisory services were purchased on February 16, 2017 to involve more high-level local and international experts in the project implementation.

The first highway infrastructure PPP project in the Baltic States according to DBFM model – challenges and opportunities

3. Preparing and structuring of a benchmarkable PPP model

3.1. Technical research stage

The planned Kekava bypass route includes the sections of the existing road A7 and a section to be built, which crosses mostly the areas of Kekava municipality and Kekava rural territory. The section of the existing road to be reconstructed passes through Krustkalni, Ramava, Valdlauci, Lapenieki and Katlakalns, but the newly built section mainly passes through the sparsely populated areas (the nearest populated place is Skujenieki village). Only a small part of the route in the beginning of the road A7 bypass is in Balozi town – part of Kekava municipality.

Following the regulations of the Republic of Latvia, in September 2015, development of the Environmental Impact Assessment (EIA) was launched to investigate the territory, potential impact of the highway from different aspects as well as to define potential technical solutions – location of the main road, auxiliary roads, traffic management solutions etc.

Table 2. Technical parameters of the Kekava Bypass

Total length	17.58 km
Construction of a new express road	14.40 km
Reconstruction of the existing road	3.18 km
4 lane road	12,22 km
2 lane road	5,36 km
Auxiliary roads	20,66
2-level road junctions	5
Tunnels (…<100 m)	2
Bridges (…<100 m)	1
Roundabouts	10
Pedestrians' bridges and tunnels	2

The project falls under Annex I of the EIA Directive (2011/92/EC). Compliance analysis with the EIA Directive (2014/52/EU), Habitats Directive (92/43/EEC, as amended) and Birds Directive (2009/147/EC) has been part of the EIA process and decision. Initial EIA for Kekava bypass was prepared and Environmental Decision was issued in 2008, but, as the validity date of the decision has expired and certain technical solutions have been changed since then, a repeated EIA was required. The EIA procedure for the Intended activity was applied by the Decision No. 129 of the State Environment Bureau, dated 20 May 2015 "On the application of the environmental impact assessment procedure." The program for the EIA was issued by the Competent Authority on 17 February 2016. EIA report was submitted on 19th January 2017. After comments and clarification, a positive Environmental Decision was issued on 3rd March 2017 stating that the foreseen project will not cause significant negative impacts on environment and is acceptable once indicated mitigation measures (mainly addressing noise and construction calendar) are considered in the project design and implementation. Soon after – on 9th March 2017, Kekava Municipal Council made a Decision accepting the project.

The project does not pass-through Natura 2000 sites. The closest one, Habitats Directive site "Dolessala" (LV0301900), is located 1.6 km from the route on the other side of Daugava river. Compliance with the Habitats Directive (92/43/EEC) and Birds Directive (2009/147/EC) has been addressed in the EIA study and EIA Decision where it is stated: "Taking into account the nature park location, according to the estimates it cannot be expected that the Kekava bypass construction would affect the ecological functions of integrity, development and conservation objectives of the area Natura 2000."

The main residual negative impacts of the project are:

- conversion and permanent loss of about 115 hectares of largely agricultural and forest land,
- additional noise,
- vibration and visual intrusion for those properties close to the road,
- degraded local air quality next to the new road,
- separation of territories which changes the pattern of access.

The project is expected to have some positive environmental impacts due to displacement of local emissions away from built up areas, as well as road safety improvements. The project does not have a particular exposure to climate change.

The first highway infrastructure PPP project in the Baltic States according to DBFM model – challenges and opportunities

The EIA process was carried out ensuring thorough public consultation and stakeholder engagement. In line with regulatory requirements, public consultation took place during the EIA proceedings prior to granting the environmental consent. Several suggestions were received from the individuals and entities. After proper analysis these were considered as much as feasible.

3.2. Specific requirements

Between April and May 2017, the EIB assessed the initial readiness of the PPP project to be financed. The technical readiness, financial economic calculations, impact to the environment of the project were evaluated. As a result of the appraisal mission, on July 18, 2017, the EIB Board of Directors decided on the readiness to finance the PPP project, if the private partner selected in the procurement would express a wish to attract financing from the EIB. At the same time, EIB informed that financial, technical, and legal due diligence of the Private partner will be carried out a before granting funding.

The draft of the PPP contract was build considering the precondition set by the Minister Cabinet: the risks between the public partner and the private partner should be distributed in such a way that the assets of the PPP project are accounted in the balance sheet of the Private partner without adversely affecting the general government budget balance and debt. The maintenance of the off-balance sheet records has been made a mandatory requirement throughout the term of the PPP agreement (unless a decision of the Cabinet of Ministers on the abolition of this requirement is adopted in the future). Therefore, prior to launching the tender, a positive EUROSTAT opinion was received that the project assets are treated off-balance sheet (EUROSTAT, 2018). The Latvian State Roads were the first to receive a positive Eurostat opinion in the Baltic states.

At different project implementation stages, the trends of the construction market of the Baltic States have been analyzed on daily base. One of the tendencies in the area of the large-scale construction having potential to jeopardize the implementation of the project was a complicate structure of the contractor and sub-contractors. To avoid extension of the chain of contractors creating risks for project implementation in time and quality so called *critical works* were defined to be performed by the private partner itself, its member or that subcontractor, on whose abilities and experience in construction the private Partner relays to perform all technical qualification criteria, respectively, works that cannot be outsourced:

- construction of the cold enduring layer (-s),
- construction of all layers of the broken stone bedding,
- construction of the upper layers of the ardent bituminous concrete (asphalt),
- construction of the bond layers of the ardent bituminous concrete (asphalt),
- construction of the sub-layers of the ardent bituminous concrete (asphalt).

To inform the potential market players about the upcoming public-private partnership tender, the Latvian State Roads put much effort in organizing and hosting Kekava Bypass Open Days. These events were widely attended and acknowledged by potential tenderers. In July 2017, the principles of the chosen procurement procedure, PPP model and the main requirements for the applicants and technical solutions were presented to inform potential tenderers and to get a feedback on details needing to be developed. In 2018, Open Days were organized already as individual sessions with the potential tenderers short before procurement announcement to inform them about the more detailed requirements of the upcoming public-private partnership tender to facilitate the timely establishment of the necessary consortia and to give more time to get prepared for the participation in the tender.

3.3. Procurement procedure

In order to determine the private partner who would be granted the right to enter into the PPP contract, the Latvian State Roads launched a public procurement procedure – competitive procedure with negotiations. This procedure was chosen as the most suitable one meeting the specific needs of the contracting authority, namely, to give the selected tenderers, who have complied with the qualification criteria, a possibility to improve the content of their bids during the negotiations held separately with each tenderer.

The procurement procedure was implemented in two stages – the qualification stage and the selection stage. The selection stage also included negotiations with the selected tenderers.

On December 6, 2018, the Latvian State Roads announced the qualification stage. On January 30, 2019, the contracting authority received 5 qualification applications, out of which one was disqualified. Hence, on May 17, 2019, the four qualified candidates were invited to participate in the selection stage by submitting their initial bids comprising both a financial offer and a technical offer.

On August 9, 2019, initial bids were received from two bidders, namely, an association of persons "Cointer and Igate" and an association of persons "Kekava ABT". Both bidders have previous PPP project experience in Europe and in other parts of the world. Cointer Concesiones, S.L. is a Spanish company – a member of the Azvi group specializing in infrastructure and public service projects in Europe and other countries (for more information www.cointer.eu) and Igate is a Latvian road construction company. As much as 80% of the total investment in the association of persons "Kekava ABT" belong to a Luxembourg investment fund "Transport Infrastructure Investment Company 2 S.C.A., SICAR" (TIIC 2 S.C.A., SICAR) specializing in transport and public infrastructure projects with subsidiaries in Portugal, France and Luxembourg, 10% belong to a Latvian road construction joint-stock company "A.C.B." and the remaining 10% - to a Latvian road construction limited liability company "Binders" (for more information www.tiic.pt).

Initial bids comprised technical solutions (can be compared to a technical design though in a limited amount) and a financial model. Evaluation of both initial bids was completed in October 2019 by starting individual negotiations with each tenderer. Latvian State Roads negotiated with each tenderer for 30 hours within 3 months. In frame of these negotiations, both technical solutions and financial models were discussed with the aim to improve the quality of the bids and to decrease the price. Along with the technical solutions and financial models, the draft PPP contract was also negotiated though in a limited amount to preserve the off-balance sheet treatment condition. To keep focused on the off-balance sheet treatment, 5 out of 23 chapters of the PPP contract were closed for any negotiations, because they contain provisions which are crucial for preserving the off-balance sheet treatment in the PPP contract both at the time of signing the PPP contract and during its implementation.

On February 28, 2020, both tenderers were invited to draft and submit their best and final offers (BAFO). On April 21, 2020, the contracting authority received both bidders' BAFOs.

Economically the most advantageous bid was determined basing on the following criteria:
- Financial Model - 80%; and
- Design and other requirements listed in the Award Procurement Document - all jointly 20%.

The latter 20% were further divided as follows:
- Design (15% (fifteen per cent)),
- Scheduled Availability Date (25% (twenty-five per cent)),
- Quality and Risk Management (15% (fifteen per cent)),
- Maintenance Plan (15% (fifteen per cent)),
- Traffic Handling During Construction Works (15% (fifteen per cent)); and
- Traffic Safety Description (15% (fifteen per cent)).

When analyzing the received bids, it was concluded that:
- the fact that there were two bids received was a good achievement if compared to a range of other PPP projects in other countries,
- the 11% difference in price between the two tenders corresponds to the existing market and competition practice,
- if compared to the initial bid, during the negotiations with the tenderer who submitted the BAFO with the lowest price, the total price was reduced by 7% which means a reduction of 19.22 mln EUR. Such a price decrease after the negotiations corresponds to the best practice in the EU in similar PPP projects which had been implemented by using a similar procurement procedure and approach.

Prior to announcing the results of the procurement, the Latvian State Roads submitted a report to the Minister Cabinet and received its approval to proceed with the project. After receiving the Minister Cabinet approval about granting the funding needed for the project, the association of persons "Kekava ABT" was granted the right to sign the PPP contract as its BAFO was recognized as the most economically advantageous bid, which, inter alia, was with the lowest price – gross availability payment of EUR 265 729 046.65 excluding VAT as well as with the shortest construction period (in compliance with the Scheduled Availability Date - less than 2,3 years).

After announcement of the procurement results, in October 2020, the due diligence procedure was started by the financiers chosen by the potential private partner. It was completed in April 2021. The process included both technical, financial, and legal due diligence.

After completion of the due diligence, the contracting authority addressed the EUROSTAT for a repeated opinion on the off-balances sheet treatment of the Kekava Bypass PPP contract as there were amendments in the contract introduced after negotiations with the tenderer during the selection phase and the financiers after the due diligence.

The first highway infrastructure PPP project in the Baltic States according to DBFM model – challenges and opportunities

Latvian State Roads will enter the PPP contract when it receives a positive EUROSTAT opinion on the amendments. Along with the PPP contract, the contracting authority will sign a direct agreement with the financiers chosen by the private partner and the private partner himself. Accordingly, the private partner will sign a financial contract with the financiers.

Conclusions

PPP is a good tool to satisfy public needs for infrastructure while there is limited financing available on the state side. At the same time, one must bear in mind that it is a long way to go from the start of the project till signing a PPP contract. As a public authority, the customer will need deep expertise while handling not only the several- stage public procurement, including negotiations with the bidders, but also conducting negotiations with financiers during the due diligence phase. This whole procurement and due diligence process in case of Kekava Bypass took about two and a half years. If the government decides to implement the project as one being off-balance, the implementing public authority will have to ensure off-balance sheet treatment throughout the negotiations with the bidders and also with the financiers, which is a challenge. The public authority as a customer might also need a repetitive Eurostat decision for the government to be on the safe side that the project is still off-balance after the amendments introduced during the negotiations with the bidders and financiers.

Funding

On the base of the procurement results, an information report was prepared and submitted to the Cabinet of Ministers of the Republic of Latvia. On August 13, 2020, an order No. 442 "On the long-term commitment of the Ministry of Transport to implement the state's main highway "E67 / A7 Ķekava bypass" public-private partnership project" was issued thus giving permission to continue project implementation and to provide a necessary resource within the state budget in period 2021-2043.

It is planned that the PPP contract will be concluded in July 2021.

Author Contributions

Verners Akimovs conceived the study, Ilze Kristīne Apsalone, Indra Muižniece, and Liesma Grīnberga were responsible for the collection, interpretation, design, and development of the data analysis.

Disclosure Statement

There are no competing financial, professional, or personal interest from other parties.

References

Akintoye, A., Beck, M., Hardcastle, C. (2003). Public-private partnerships. Managing risks and opportunities. Blackwell. ISBN 0-632-26465-X

Akintoye, A., Hardcastle, C., Beck, M., Chinyio, E., Asenova, D. (2003). Achieving best value in private finance initiative project procurement. *Construction Management and Economics*, 21 (5), 461–470. https://doi.org/10.1080/0144619032000087285

Bayliss, K. & Waeyenberge, E. (2017). Unpacking the Public Private Partnership Revival. *The Journal of Development Studies*, DOI:10.1080/00220388.2017.1303671

Cheney, D.V. (2018). A Brief History of Public Private Partnerships. Available: https://www.lorman.com/resources/a-brief-history-of-public-private-partnerships-16968

Cumming, D. (2007). Government policy towards entrepreneurial finance. Innovation investment funds. *Journal of Business Venturing*, 22(2), 193–235. https://doi.org/10.1016/j.jbusvent.2005.12.002

Cui, C., Liu, Y., Hope, A., Wang, J. (2018). Review of studies on the public–private partnerships (PPP) for infrastructure projects. *International Journal of Project Management*, 36(5), 773-794. https://doi.org/0.1016/j.ijproman.2018.03.004

Eurostat (09/07/2018). Ref. Ares (2018)3645147. Ex-ante advice on the statistical treatment of the Public Private Partnership (PPP) Project E67/A7 Kekava bypass

Garvin, M. J., Bosso, D. (2008). Assessing the effectiveness of infrastructure public private partnership programs and projects. *Public Works Management & Policy*, 13, 162-178.

Hodge, G.A. & Greve, C. (2016). On Public-Private Partnership Performance. A Contemporary Review. *Public Works Management & Policy*, 1-24. https://doi.org/0.1177/1087724X16657830

Li, B., Akintoye, A., Edwards, P.J., Hardcastle, C. (2005). The allocation of risk in PPP/PFI construction projects in the UK. *International Journal of Project Management*, 23 (1), 25–35. https://doi.org/10.1016/j.ijproman.2004.04.006

Link, A.N. (2006). Public/Private Partnerships. Innovation Strategies and Policy Alternatives. Springer, Boston. https://doi.org/10.1007/0-387-29775-8

The first highway infrastructure PPP project in the Baltic States according to DBFM model – challenges and opportunities

OECD (2008). Public-private partnerships: In pursuit of risk sharing and value for money. Paris, France. Available: https://www.oecd.org/gov/budgeting/public-privatepartnershipsinpursuitofrisksharingandvalueformoney.htm
OECD (2012). Recommendation of the Council on Principles for Public Governance of Puplic-Private Partnerships. Paris, France. Available: https://www.oecd.org/governance/budgeting/PPP-Recommendation.pdf
Osei-Kyei, R. & Chan, A.P.C. (2015). Review of studies on the Critical Success Factors for Public–Private Partnership (PPP) projects from 1990 to 2013. *International Journal of Project Management*, 33, 1335-1346. https://doi.org/10.1016/j.ijproman.2015.02.008
Osei-Kyei, R., Chan, A.P.C., Javed, A.A., Ameyaw, E.E. (2017). Critical success criteria for public-private partnership projects: international experts' opinion. *International Journal of Strategic Property Management*, 21:1, 87-100, https://doi.org/10.3846/1648715X.2016.1246388
PROXIMARE Consortium to the European Commission (2014). North Sea – Baltic Core Network Corridor Study. Final Report. Available: https://ec.europa.eu/transport/sites/transport/files/north_sea-baltic_study_0.pdf
The World Bank (2014). Public-Private Partnerships Reference Guide. Version 2.0. Available: http://documents1.worldbank.org/curated/en/600511468336720455/pdf/903840PPP0Refe0Box385311B000PUBLIC0.pdf
The World Bank (2020). Private Participation in Infrastructure (PPI) 2020. Half year report. Available: https://ppi.worldbank.org/content/dam/PPI/documents/PPI_2020_Half-Year_Update.pdf

Notations

Abbreviations
BAFO – the best and final offer
CFCA - Central Finance and Contracting Agency
DBFM - design, build, finance and maintain
DB - design and build
EIA – Environment impact assessment
EIB – European Investment bank
EUR - euro
FEA – Financial Economic Analysis
mln - million
PPP – public – private partnership
VAT – value added tax

PLANNING OF STATE ROAD RENEWAL IN THE CONTEXT OF TERRITORIAL REFORM

Salvis Skabs
"Latvian State Roads", 3 Gogola Street, Riga, LV-1050, Latvia
Salvis.Skabs@lvceli.lv

February 15, 2021

Abstract. In 2021 a new territorial reform will be carried out in Latvia, and the number of existing 110 local governments will be reduced to 42 regions (previous territorial reform in 2009 resulted in the decrease of the number of local governments from 522 parishes down to 110 local governments).

Each territorial reform has always regarded the importance of the network of state owned roads in ensuring successful operation of local governments, management of their territories and provision of municipal services to local inhabitants.

After the previous territorial reform carried out in 2009 there were 104 centres of local governments that were not connected with paved roads. At present only 59 such centres without paved road connections have remained, and improvements on roads serving these local governments is continued in the scope of annual improvement programmes financed from the state budget.

The Ministry of Environment and Regional Development within the scope of the new territorial reform in 2021 is preparing a new Investment Programme for Road Development in the Context of Territorial Reform. Respective road sections are identified by the administrations of planning regions in co-operation with local governments, the programme itself is compiled by the Ministry of Environment and Regional Development, but State Limited Liability Company "Latvian State Roads" is providing consulting on the choice of most efficient rehabilitation methods, preparation of technical documentation and potential construction costs.

Keywords: territorial reform, local governments, investment programme, road development

Introduction

To enable local governments to perform their functions efficiently, one of the most important factors is high-quality and well-developed road network. Therefore roads are one of the most important issues during the implementation of each territorial reform.

This report gives an overview of the national road network and travelling habits of the population in order to receive the services needed in everyday life. Territorial reforms of local governments carried out in the Republic of Latvia and related state road improvement programmes, which are intended to improve the accessibility of administrative centres are reviewed. Improvement of roads needed to reach administrative centres within the framework of annual state road improvement programmes, are reviewed, as well. The experience of repairing roads important for local governments in 2004-2009 has been summarized. Finally, the State Road Improvement Programme developed by the Ministry of Environment and Regional Development is reviewed in the context of the territorial reform of local governments planned for 2021.

1. Historical classification of road network

The development of road network and the territorial division are closely interlinked, and it is often not clear whether the territorial division serves the road network or the road network serves the territorial division. In any case, changing the territorial division changes the travelling habits of people between settlements and administrative centres, as well as, between adjacent administrative centres. If the lifetime of territorial division is constantly longer than the lifecycle of roads (25-35 years), the road network fully adapts to the territorial division. Roads that are needed to reach the new administrative centres will develop and roads that have led to previous administrative centres will gradually deteriorate and become insignificant. Such a stable situation in the territory of Latvia began to develop in 1960-ies, and it existed until 2004. During this time, most of the roads with bound pavements were newly constructed or rebuilt. Depending on travelling habits and the intensity of economic activity, the roads were widened, and their technical category was increased.

Administrative management model was organized in two levels during this period. Local governments of the lowest level were parishes, and the second level municipalities were districts. In accordance with this administrative model a road network was developed, that with slight changes still exists today. Travelling habits of the population developed accordingly. The most necessary services (shops, schools, etc.) were available in parish centres, and daily travel patterns were developed by the people living in parish territories. This type of mobility was mainly provided by roads, which are now municipal roads. The range of services available in district centres (hospitals, specialized shops, secondary schools, etc.) was wider and job opportunities were greater. This influenced travelling patterns from the parish centres and other settlements to district centres. This type of travel took place along roads that are now classified as state local and regional roads.

Roads between district centres and larger cities, as well as, from parish centres to district centres in this administrative model were of much higher quality but connections between parish centres usually had unbound pavements.

2. Territorial reform in 2004 - 2009

As the population of Latvia was decreasing, a decision was made in 2004 to implement the territorial reform with the aim to transfer from a two-level administrative model to one-level model. District councils were abolished, and parishes were allowed to unite voluntarily and form regions. In many cases the territories of these regions were created ignoring the location of the existing roads. In the result, if there was a need to travel from many settlements to the centre of a region, it was necessary to travel by roads through the territory of another region or even through the centre of the former district. With the establishment of the new regions, the issue of the quality of roads from the settlements in the region to regional centres became especially topical. To solve this problem, a "Programme for the Improvement of State Second Class Roads for Regional Support" was developed.

2.1. Programme for the Improvement of State Second Class Roads for Regional Support

The commencement of this Programme is marked by the Order No. 152 of the Cabinet of Ministers "On the State Support Programme for the Development of Regional Infrastructure", which envisaged the adoption of new policy initiatives and the development of a programme for the implementation of these initiatives.

The goal of the Programme was to stop further collapse of state local roads and to promote their improvement in accordance with economic and social interests, thus promoting the development of regions and raise the living standards of the population.

Before the implementation of the Programme the basic principles were determined, which had to be considered when creating implementation mechanisms, as well as, selecting and prioritizing the road sections. The involvement of local governments and regions in the development of this Programme was the main provision, but the management of the Programme was ensured by the Ministry of Transport. After that secondary principles were determined to serve the development of the Programme and co-operation with local governments:

- Openness and volunteering - the development of programmes takes place by informing the population and local governments; expression of opinion is not imposed.
- Stakeholder participation - officials undertake to ensure the participation of stakeholders in the design and implementation of the programme.
- Principle of co-ordination - programme projects are discussed and co-ordinated mutually.
- Focus on the common good - stakeholders prefer measures that benefit larger groups.
- Rational use of resources - the use of local materials, while ensuring the longevity of the renovated structures is promoted.

The introduction of such principles at the very beginning of Programme development determined success of the programme, reduced the risk of arbitrary interpretations by individual stakeholders, and thirdly determined the open coverage of the implementation of the Programme to the stakeholders and the public.

During the implementation of the Programme, problems initially arose with the co-ordination of individual projects. When the principle of common good was initially introduced and the improvement of road sections that were in the worst condition and with the highest traffic intensity was planned, some mayors refused to co-ordinate such a draft programme and insisted on the inclusion of road sections of specific local governments and the changes in project priority list. As a result, deviations from the original principles had to be made in order to harmonize the draft Programme.

The following two alternatives were initially proposed for Programme implementation:

- Alternative 1: with limited funding, low-cost measures should be planned and at the same time maximum length of improved road kilometres should be reached. In this Alternative funding would be allocated to the renovation of asphalt concrete pavements by laying asphalt concrete levelling layers, as well as, to the renovation of gravel pavements by repairing water drainage systems and gravel pavements. Alternative 1 did not envisage any paving of gravel roads.
- Alternative 2: more complex and resource-intensive road reconstruction measures should be planned that would ensure long-term efficiency of financial investments but reduce the total length of improved roads. This Alternative envisaged both the paving of gravel roads and full or partial reconstruction of asphalt concrete pavements.

When both Alternatives were analysed and when limited state funding was considered the Alternative 1 was chosen to ensure the most intensive improvement of roads in the routes most important for local governments in the shortest possible time.

Programme implementation envisaged the following two options for the selection of road sections to be improved:

- Option 1. According to specific criteria the Ministry of Transport would calculate the share of funds from the state basic budget to be allocated to specific districts or regions thus enabling local governments themselves to assess the importance and condition of state local roads in their territory and select priority road sections and improvement works within the scope of available funding. An important advantage of this option is the possibility for local governments to evaluate and decide on the development of the most important local roads for their needs and the use of the allocated funding for road improvements. On the other hand, the disadvantage of this option is the risk of setting useless priorities, as there is a possibility to subjectively select road sections to be reconstructed and volumes of improvement works, without considering technical condition of roads, traffic intensity or recommendations of experts.
- Option 2. The state basic budget would include specific funding for the implementation of the Programme and the development of the Programme would be delegated to the Ministry of Transport. The Ministry of Transport in co-operation with the Latvian Association of Local Governments would develop criteria for project evaluation and selection. The Programme would include road sections that are most important for the performance of municipal functions and the social needs of population, which urgently need improvements. The development of the Programme would take into account the priority lists prepared by local governments to show the importance of the selected road sections for regional development as well as for the promotion of interregional communications. The advantage of this option is the established project selection procedure, respect to approved criteria, national road development strategy and recommendations of experts, as well as ensuring that priority projects are selected and co-ordinated between neighbouring regions. However, the disadvantage is the centralisation of the selection of local projects.

When the risks for programme implementation were considered, it was decided to choose Option 2 and to develop the Programme centrally, but in co-ordination with local governments.

As recommended by experts, about 30% of state local roads or about 4,000 km were recognized as important for regional development, of which 800 km were paved with asphalt and 3,200 km were paved with gravel. The improvement of these roads was determined as the key indicator to be achieved by the Programme for the Improvement of State Second Class Roads for Regional Support.

The total amount of funding required for the Programme from the state basic budget in 2004 was calculated at 180 million Latvian Lats (~255 million EUR).

Initially, 3 different schedules for the improvement of the mentioned 4000 km of roads were offered with the following calculated funding and the total length of improved roads per year:

- Schedule 1: 130 km of roads with asphalt pavement and 550 km gravel roads would be improved annually. With this schedule the whole Programme for the Improvement of State Second Class Roads for Regional Support would be implemented within 6 years and the calculated annually needed funds would reach 30 million Lats (~42 million EUR).
- Schedule 2 envisaged the improvement of 87 km of roads with asphalt pavement and 362 km of gravel roads per year. With this schedule the whole Programme for the Improvement of State Second Class Roads for Regional Support would be implemented within 9 years and the calculated annually needed funds would reach 20 million Lats (~28 million EUR).
- Finally, Schedule 3 envisaged the improvement of 54 km of roads with asphalt pavement and 212 km of gravel roads per year. With this schedule the whole Programme for the Improvement of State Second Class Roads for Regional Support would be implemented within 15 years and the calculated annually needed funds would reach 12 million Lats (~17 million EUR).

Considering the availability of funds from the state budget, Schedule 3 was chosen. At the same time the Programme specified that the normative lifetime until periodic maintenance works are needed should be 12 to 15 years for asphalt pavements and 5 to 8 years for gravel roads. This indicates that the Programme would not be fully completed when periodic maintenance of asphalt pavements would be already needed, and for approximately half of improved gravel pavements periodic maintenance would have to be performed twice during the Programme period. However, the Programme did not provide any funding for periodic maintenance works.

In 2007 the basic provisions of the Programme for the Improvement of State Second Class Roads for Regional Support were revised. The biggest changes were the allocation of greater priority to the reconstruction of roads with bigger traffic load and roads in urban areas and followingly the reduction of the amount of gravel road improvement works.

It has to be pointed out that the Programme for the Improvement of State Second Class Roads for Regional Support was essentially of a social nature and in financial terms the capital investments invested in this Programme could not provide quick return. Given the low intensity of traffic on state local roads, their maintenance and development in most cases would not be economically justified by cost-benefit analysis programmes applied in the road sector.

An important remark in the guidelines of this Programme stated that the improvement of state local roads may not be a campaign-like activity for one or a few years. Therefore, the guidelines were developed for 7 years, envisaging a sharp increase in funding, and the Programme would mark only the starting point in the improvement of all state local roads. Unfortunately this did not become true and already in 2009 the funding within the scope of Programme for the Improvement of State Second Class Roads for Regional Support was allocated only for the completion of road sections where construction contracts had already been concluded.

March 10, 2010 may be regarded as the official termination date of this Programme when the Cabinet of Ministers of the Republic of Latvia adopted its Order No. 140 "Amendments to the Guidelines for the Development of the Transport System for 2007 - 2013" which stated that the Order of the Cabinet of Ministers No. 64 "Guidelines for the Improvement of State Second Class Roads for Regional Support" was declared invalid.

As a result, this Programme instead of initially planned 15 years was carried out only for 4 years and the funding that was initially planned was available only for 2 years. Summary of Programme achievements is shown in Table 1.

In per cent, the amount of funding allocated to the Programme was 16% of the initially planned amount. The total length of improved roads with asphalt concrete pavement was 181.84 km which was 23% of the initially planned, but the length of improved gravel roads was only 317.34 km instead of the initially planned 3200 km (in per cent it was only 10% of the originally planned amount).

Table 1. Implementation of the Programme for the Improvement of State Second Class Roads for Regional Support

Year	Funding, thous. EUR	Improved roads, km		
		Paved	Gravel	Total
2006	4544	24.95	68.48	93.43
2007	18327	63.1	162.77	225.87
2008	16728	92.05	86.09	178.14
2009	1099	1.74	0	1.74
Total:	**40697**	**181.84**	**317.34**	**499.18**

As soon as the Programme for the Improvement of State Second Class Roads for Regional Support was prematurely terminated due to the lack of funding, the improvement of state local roads was completely suspended.

3. Improvement of regionally important roads without specific programmes

After the completion of activities of the Programme for the Improvement of State Second Class Roads for Regional Support in 2009, this Programme was not resumed. Therefore regionally important roads had to be renewed and maintained with the funding available within the scope of routine road maintenance programmes. At that time renewal of pavements on state local roads was practically not performed. The only exception was the year 2013 when gravel pavements were strengthened and double surface treatment was performed on 53.3 km of state local roads within the scope of routine maintenance works.

In the period when budget funds for state road improvement were lacking, it became clear that it was necessary to develop and implement selection and prioritisation of road sections to be improved so that it would be possible to commence road works immediately after the funds for road improvement become available. The main criteria for road selection were two: traffic intensity and pavement condition. In addition to these two criteria and opportunity to apply a special promotion factor was provided. Such promotion factor allowed the advancement of specific road section in the priority list. The weight of this promotion factor in setting the priority of a road section was 20% of the total road section assessment. This promotion factor facilitated the involvement of local governments in assessing specific reasons for the improvement of specific roads. Mostly these reasons were of social nature and were related to the accessibility of regional centre, pupil transportation routes, public transport, etc. The most important reason for using the promotion factor was ensuring the connection of parish centres to the state road network with a road with asphalt pavement and ensuring the connection of parish centres with regional centre with a road with bound pavement.

In 2012 the connections of all previous parish centres to the network of roads with asphalt pavement were assessed and 104 parish centres were identified that still were lacking the connection to a regional centre or to the network of roads with asphalt pavement by a road with bound pavement. To connect these centres to the network of paved roads, optimum sections of gravel roads were identified that could be reconstructed or strengthened with further application of double surface dressing. The total length of these identified roads reached 859 km.

3.1. Triannual year programmes for works in the state road network

In 2014 the funding from the state basic budget for the improvement of state local roads became available and on the basis of prepared programmes for works to be performed in the state road network road improvements are carried out. Although programmes for works in the state road network are not specifically created to support local governments, almost all works on state local roads and majority of works on regional roads are directly needed to ensure the functions of local governments.

State Limited Liability Company "Latvian State Roads" (further in the text - LSR) develops programmes for works in the state road network in accordance with its quality management procedures. The preparation of these programmes commences with specific LSR order that identifies specific programmes to be developed and persons responsible for each programme. Terms of Reference are prepared for developing programmes for works to be performed in the state road network for the period of three years. Terms of Reference define basic principles for programme development, provide short description of each programme and identify annual funding allocated for each programme for the period of 3 years. Basing on these Terms of Reference methodological guidelines are prepared for the preparation of priority lists of works and selection of specific projects. The aim of these guidelines is to provide guidance for a uniform approach of specialists both from LSR Road Parametre Measurement Department (further - RPMD) and from local units in identifying the priority road sections and selecting the programme projects.

Actual process in programme preparation is commenced with the inspection of all state roads and assessment of their condition. Inspections and assessments of state main and regional roads with asphalt concrete pavements are carried out by RPMD engineers (one engineer per state planning region). State regional roads with gravel pavement and state local roads are inspected and assessed by the engineers of LSR local units. During or directly after road inspection engineers identify road sections that need road renewal or reconstruction works, prepare the summaries of these sections (special sheet appended to the methodological guidelines) and submit them to LSR Programme Planning Department for compilation and programme preparation. Prepared programmes are approved in the meeting of LSR directors.

Most significant advantages of such programme planning and developing process are the following:

- all roads are inspected and assessed,
- road sections are selected by road engineers that are trained for such tasks,
- priority of individual road section is calculated automatically thus excluding any influence of subjective factors,
- any corrections in road section priority list with the application of promotion factor are clearly explained,
- political influence in road selection and prioritising process is reduced to a minimum.

Principles of such a planning system originated about ten years ago, the system itself has been in operation for seven years, and now the system has been improved and shortcomings have been remedied. The process is routine and annual, with each participant aware of their responsibilities, resulting in a lower risk of errors.

From time to time local governments object to this planning process as they consider that they are not sufficiently involved in the planning of roads that are important to them. There is no direct involvement of local governments in this planning process, but there are several opportunities for them to be indirectly involved in road selection and prioritization process. Local governments may communicate with LSR local units directly and explain the importance of specific state local roads and regional gravel roads and the need to improve them. Based on such communication, LSR engineers may apply promotion factor to particular road section, explaining that it is a road important to the local government. Another option how local government may be involved in this planning process is through state planning regions established by local governments. A planning region that consists of all local governments in this region may voluntarily summarise proposals from local governments for road improvements and may submit them to LSR for further review. Such a summary is regarded only as a recommendation and it is not mandatory for programme preparation.

Programmes for works in the state road network are annual and every year new projects are added to each programme. At the same time completed projects are removed from the programmes. Table 2 provides a summary of improvements on local roads within the scope of State Road Improvement Programme. In 2014 – 2020 connections from 47 parish centres to the network of state paved roads were created. In most cases the existing gravel pavements were strengthened and double surface dressing was carried out.

Table 2. Implementation of State Road Improvement Programme, local road component

Year	Funding, thous. EUR	Improved roads, km			
		Paved	Gravel	Bound pavement	Total
2014	2836	60.7	12.1	0	72.8
2015	3753	52.81	0	7.59	60.4
2016	8767	90.21	150.36	9.28	249.85
2017	8342	98.63	8.37	16.45	123.45
2018	30654	57.56	197.15	132.21	386.92
2019	18181	26.4	33.45	38.74	98.59
2020	13598	162.6	122.58	25.71	310.89
Total:	**86131**	**548.91**	**524.01**	**229.98**	**1302.87**

4. Territorial reform in 2021

Before the territorial reform it was concluded that in the future significant decrease in the population of Latvia may be expected mainly in those local governments that are not located close to the capital city Riga. As the population decreases, revenues of local governments would decrease and, as a result, local governments would face increasing difficulties in providing quality services to their population at reasonable costs. One of the basic principles for the creation of new administrative territories states that an administrative centre should be located at a centre of development of regional or national significance. Area close to the Riga city is an exception as the population density is several times higher than in the rest of Latvia. Over time adjustments were made to the original plan to allow the creation of separate local governments in large cities but at the same time cities also will have to create joint institutions with the surrounding region. Therefore the actual centre of a region near a large city will be the in the city. Comparing the new administrative division with the division that existed in Latvia before the reform of 2004 it may be said that the newly established regions will be almost identical to the previously existing districts. Significant differences may be seen only in Riga city area and two regions have been established around development centres, which were not district centres previously (Smiltene and Līvāni). Riga city area will be administratively divided into 8 regions and 2 large cities. Taking into account the historical travelling habits of the population, the road network is fully suitable for such a territorial model.

The distribution of the population in Latvia is monocentric. Capital city Riga is the centre, more than 1/3 of the Latvian population lives there, and more than half of the Latvian population lives in the Riga city area. An integral part of such population distribution is the concentration of economic activity around the capital, which to a large extent also determines the specifics of population and transportation of goods between development centres (new regional centres) and Riga. These needs are met by the network of state main roads, which has improved significantly in recent years and is likely to be good in the near future. Most of the newly established regional centres are located at or in the immediate vicinity of state main roads except for 8 new regional centres.

The connection of these centres to the capital city requires the use of state regional roads. The connection of the new regional centres with the previous regional centres is also ensured mainly by state regional roads, and in the result regional roads play a significantly greater role in the implementation of this territorial reform and provision of local government functions. Local roads will not lose their importance in providing local government functions either, but their role in significantly larger municipalities is mostly to provide access to regional roads than to create full route connections between settlements and administrative centres. Nevertheless, the issue of reaching the former parish centres by roads with bound pavements is still relevant, and it is mainly possible to solve it by constructing bound pavements on all state local roads. In general, the location of roads is already optimal for reaching the new regional centres from parish centres, former regional centres or other settlements in a region, but unfortunately in many cases the quality of road pavements is not good enough to travel to the new regional centres sufficiently quickly. In order to address this situation, in parallel with the new territorial reform, it is planned to implement an Investment Programme for Road Development in the Context of Territorial Reform.

4.1. Investment Programme for Road Development in the Context of Territorial Reform

Investment Programme for Road Development in the Context of Territorial Reform is based on the initiative of the Ministry of Environmental Protection and Regional Development (further MoEPRD) and is developed as an investment programme that would foster territorial reform. Considering that a large number of local governments are sceptical about the merger of local governments, an immediate benefit is needed to be shown in order to reduce this scepticism. In this case it is improved roads.

This investment programme is intended for the reconstruction and renovation of state regional and local roads. The initially defined approximate volume of funding for this programme is set at 300 million EUR.

At the beginning of programme development the following road selection criteria were defined, according to which sections of state regional and local roads to be improved should be identified and included in the programme:

- road pavement condition (poor, very poor),
- annual average daily traffic,
- connections between the present administrative centres of local governments with the potential administrative centres of planned regions after the territorial reform and connections of the administrative centres in regions the territories of which will not be changed,
- the role of functional use in the context of sectoral reforms.

Investment Programme for Road Development in the Context of Territorial Reform was initiated by MoEPRD which gave the task to the administrations of planning regions formed by local governments to identify and summarize the road sections important for the establishment of new regions, which need to be improved basing on the above-mentioned criteria. The administrations of planning regions, when determining the amount of funding available to each newly formed local government, transferred this task to the administrations of the existing regions. Local governments of the existing regions, which will be merged into the new regions, by mutual agreement selected the road sections to be improved and submitted the list of selected road sections to the administrations of the planning regions.

The administrations of the planning regions compiled the submitted priority lists and submitted them to the MoEPRD for the development of the programme. MoEPRD compiled the submitted lists and submitted them to LSR in order to specify the addresses of road sections and costs. This shows that a lot of organizations are involved in the development of the programme and multiple compilation of prepared road lists takes place repeatedly.

As the investments are planned in state-owned road sections, it is provided that when the funding for the new programme is available the beneficiary will be the Ministry of Transport. In its turn, LSR will administer the allocated funding, organize procurement, manage the construction programme and supervise the construction in accordance with the procedures specified by the Cabinet of Ministers on the basis of the current list of priority regional and local road sections prepared by the MoEPRD.

During the development of the programme ~800 km of state regional and local roads were identified, the reconstruction and renovation of which is a priority for the implementation of territorial reform. However, the programme stipulates that after receiving funding, the MoEPRD in co-operation with the Ministry of Transport, State Limited Liability Company "Latvian State Roads", planning regions and the Latvian Association of Local Governments will prepare an updated list of priority regional and local road sections. This option allows the programme to be adjusted according to the available funding, but in practice it is used by local governments, which are unable to agree on the sections of roads to be improved in mutual discussions and quite often change their priorities.

If this programme will receive funding from the state basic budget or any other source without any redistribution of funds from the state road fund programme it will be a substantial investment in state road infrastructure.

Positive aspects in the process of developing and implementation of the Investment Programme for Road Development in the Context of Territorial Reform are the following:

- all stakeholders interested in road improvements may take part in the selection of road sections directly or through local governments,
- two ministries with joint efforts may have greater opportunities to attract funding to the programme.

There are also the following negative aspects in the process of developing and implementation of the Investment Programme for Road Development in the Context of Territorial Reform:

- selection of road sections to be improved is done by local governments that in most cases are lacking qualified personnel for such task,
- priority road sections are demanded in territories of the new regions that would consist of several existing regions which at the moment are not able to reach agreement,
- long bureaucratic process creates substantial risks of accidental errors,
- frequent changes in the programme will not enable timely drafting of documents needed for reconstruction or renovation works.

Despite several drawbacks in developing the Investment Programme for Road Development in the Context of Territorial Reform, such a process may be regarded very positively as it emphasizes the issue of road improvements and involves local governments of the new regions in the process.

Conclusions

Regardless of any programme titles, progress of their development or organizations involved, the most important aspect is the improvement of road network condition. If the creation of a separate programme for a specific purpose is successful in attracting additional funds for the improvement of state roads, it has to be supported, but if the funding for the implementation of such programme is drawn from already existing road improvement programmes, then such a programme creates additional risks for less transparent and less optimal use of funds.

Regardless of any identified goals or specifics, the process of developing any road improvement programme has to be transparent and open, the decisions taken have to be traceable and the road sections included in the programmes have to be well-grounded.

As our experience shows, the implementation of the road improvement programme developed in the context of the previous territorial reform will lose its priority after the implementation of the new territorial reform, and the improvement of roads important for local governments could be effectively performed within the existing framework of LSR work procedures.

When larger local governments are created, they will have the opportunity to attract more qualified specialists including road engineers, and this will significantly improve the competence of local governments in road related issues. Consequently, in the future it would be possible to involve local governments in the selection of state road sections to be improved in a wider and better degree.

References

2004. gada 9. marta Ministru kabineta rīkojums Nr. 152 „Par Valsts atbalsta programmu novadu infrastruktūras attīstībai".
„Programma valsts otrās šķiras autoceļu sakārtošanai novadu atbalstam2006.gadam", VAS „Latvijas Valsts ceļi", 2005.
„Programma valsts otrās šķiras autoceļu sakārtošanai novadu atbalstam2008. - 2009.gadam", VAS „Latvijas Valsts ceļi", 2007.
Darba uzdevums valsts autoceļu tīklā veicamo darbu programmu izstrādāšanai 2021. – 2023. gadam, VAS „Latvijas Valsts ceļi", 2020.
Metodiskie norādījumi valsts galveno, reģionālo un vietējo autoceļu tīklā veicamo darbu programmu un prioritāšu sarakstu izstrādāšanai 2021. – 2023. gadam. VAS „Latvijas Valsts ceļi" Reģionālo programmu daļa, 2020.
Administratīvi teritoriālās reformas iespējamā ietekme uz publiskās infrastruktūras tīklu, Latvijas Republikas Saeima, 2019.
Informatīvais ziņojums "Par investīciju programmu autoceļu attīstībai administratīvi teritoriālās reformas kontekstā", Vides aizsardzības un reģionālās attīstības ministrija, 2019.

Notations

Abbreviations

LSR - State Limited Liability Company "Latvian State Roads"

MoEPRD – the Ministry of Environmental Protection and Regional Development

RPMD - Road Parametre Measurement Department of State Limited Liability Company "Latvian State Roads"

DEVELOPMENT OF VIA BALTICA IN LITHUANIA

Justas Norbutas[1], Jolanta Nemaniūtė-Gužienė[2]

[1,2] State Enterprise Lithuanian Road Administration, J. Basanavičiaus str. 36, 03109 Vilnius, Lithuania
E-mails: [1]justas.norbutas@lakd.lt; [2]jolanta.nemaniute@lakd.lt

Abstract. The total length of Via Baltica corridor, which consists of five roads (A5 *Kaunas–Marijampolė–Suwałki*, A1 *Vilnius–Kaunas–Klaipėda*; A8 *Panevėžys–Aristava–Sitkūnai*, A17 Panevėžys bypass and A10 *Panevėžys–Pasvalys–Riga*) makes up 268 km on the Lithuanian territory. In 2019 the highest traffic volumes on this corridor were 55,942 veh/day. It is the highest-volume heavy vehicle road carrying the greatest loads. Via Baltica is a transit road; therefore, it shall comply to the requirements set to high quality roads. In 2020-2030 it is planned to implement 9 projects, the total value of which is ca 704 million EUR (194.18 km to be reconstructed). Until 2018 having implemented 156 million EUR projects, Via Baltica road A5 *Kaunas–Marijampolė–Suwałki* 17.24–56.83 km section (from Kaunas to Marijampolė) was reconstructed into a motorway until the end of 2018. Currently, preparations are made for Via Baltica reconstruction from Marijampolė to the Lithuanian-Polish border (A5 road from 56.83 km to 97.06 km). Strategic Environmental Assessment has already been completed and a special territorial planning document has been approved by the Lithuanian Government. According to this document, the road section shall be widened up to four traffic lanes by constructing two safe grade-separated intersections (viaducts) and roundabouts, connecting roads, a new heavy vehicle parking lot, by widening the existing bridges, by building new bridges and by implementing various environmental protection measures. At present design works of the above-mentioned 40.23 km-long-road section are underway. The works are due to be completed by the end of 2025.

Keywords: Via Baltica, development, road network, traffic volumes, traffic safety, environment, public health.

Introduction

The development of Trans-European Transport Network road E67 is of utmost importance for the infrastructure of north-south transport corridor. TEN-T road E67 *Helsinki–Tallinn–Riga–Panevėžys–Kaunas–Warsaw*–Wrocław–*Prague* connects North-European transport system with the Central and Western Europe via Poland and the Kaliningrad region. This road carries a lot of international traffic; therefore, it is a very important logistics chain connecting Northern European countries with Southwestern Europe countries. Road E67 is on the route *Helsinki–Tallinn–Riga–Panevėžys–Kaunas–Warsaw*–Wrocław–*Prague*, and Via Baltica is a road section from Warsaw to Tallinn. The total length of Via Baltica makes up 649 km, out of which 268 km on the territory of Lithuania.

On the territory of Lithuania Via Baltica extends from the state border with Poland bypassing Kalvarija, via Marijampolė to Kaunas, from there to Panevėžys, and later to Pasvalys and from there until Saločiai at the Latvian border. In Lithuania, Via Baltica includes the following roads: A10 (*Panevėžys–Pasvalys–Riga*), A17 (Panevėžys bypass), A8 (*Panevėžys–Aristava–Sitkūnai*), part of A1 (*Vilnius–Kaunas–Klaipėda*) and A5 (*Kaunas–Marijampolė–Suwałki*) (Fig. 1). This road is significant for the development of logistics and freight carriage both in Lithuania and other neighbouring countries. Via Baltica today is one of the key transport arteries. Unfortunately, due to the current road infrastructure, complicated traffic conditions and poor driving habits, the road does not meet high traffic quality requirements set to a transit road.

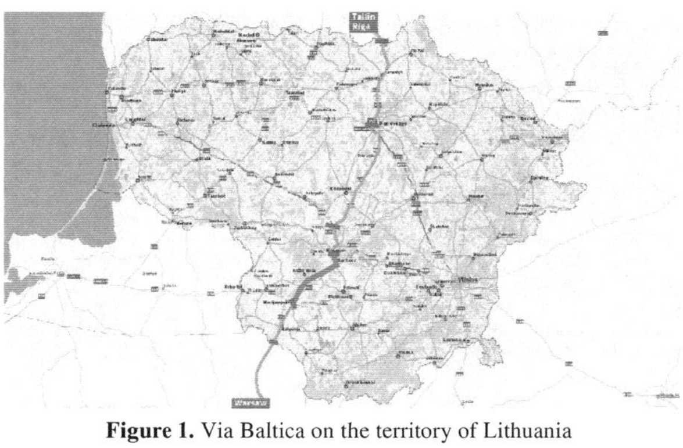

Figure 1. Via Baltica on the territory of Lithuania

Key problems of Via Baltica roads:

- Cross-section does not meet the standard of motorway;
- Cross-section does not meet high quality of traffic;
- Too many intersections;
- Bus stops at the carriageway;
- Pedestrian traffic on carriageway;
- Pedestrian and bicycle traffic on shoulders;
- Dangerous left-turns to petrol stations and rest areas;
- One-level turns on a 4-lane road;
- High density of exits;
- Insufficient length of acceleration and deceleration lanes;
- Lack of environmental measures and their insufficient effectiveness, poor quality of human life due to too much noise caused by vehicles (esp. heavy goods vehicles).

1. Traffic Volumes and Road Accidents

In 2016-2020 traffic volumes on Via Baltica remained the same: ca 11,600 veh./day (yet, there was insignificant decline in traffic volumes in 2020 due to COVID-19) (Fig. 2). Heavy goods vehicles account for 33 per cent of the total traffic volumes. Traffic volumes of heavy goods vehicles have increased by up to 53 per cent on the road section between Marijampolė and the Polish border.

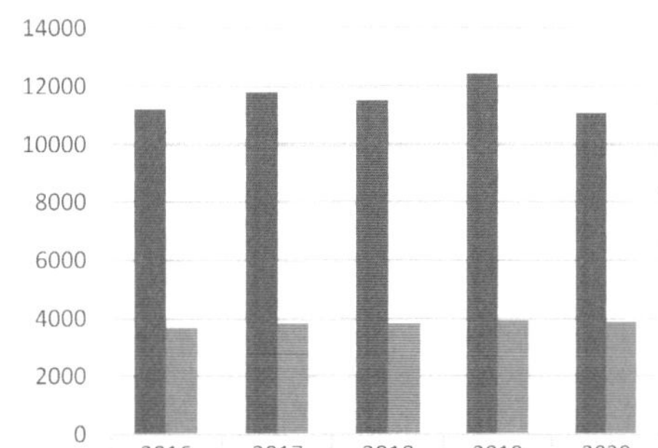

	2016	2017	2018	2019	2020
■ Total vehicles, veh./day	11220	11766	11520	12438	11067
■ Heavy vehicles, veh./day	3670	3839	3818	3923	3872

Figure 2. Via Baltica traffic volumes in 2016-2020

The most intense Via Baltica sections are A5 *Kaunas–Marijampolė– Suwałki* from Kaunas to Marijampolė (Fig. 3) and A1 *Vilnius–Kaunas–Klaipėda* from 102.0 to 114.5 km (Fig. 4), where in 2019 AADT was 20,983 veh./day and 36,762 veh./day respectively.

IOP Conf. Series: Materials Science and Engineering 1202 (2021) 012007 doi:10.1088/1757-899X/1202/1/012007

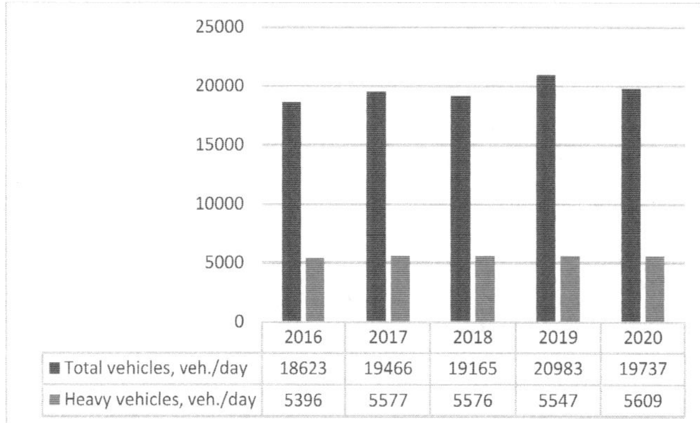

Figure 3. Traffic volumes on road A5 *Kaunas–Marijampolė– Suwałki* section from Kaunas to Marijampolė in 2016-2020

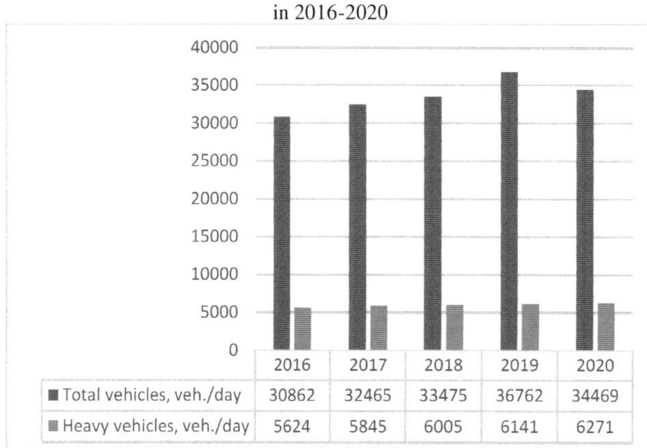

Figure 4. Traffic volumes on road A1 *Vilnius–Kaunas–Klaipėda* section from 102.0 to 114.5 km in 2016-2020

In 2016-2020, 197 road accidents were registered on Via Baltica, with 41 fatalities and 243 injured (Fig. 5). Due to high traffic volumes and especially high number of heavy goods vehicles, this road is dangerous to manoeuvre, drivers overtake other vehicles dangerously. The causes of such traffic accidents vary; however, the most frequent ones mentioned by the Road Police are: 'when making left-turn the drivers do not yield the right of way to the vehicles that are travelling in a straight line; when driving in the opposite traffic direction drivers manoeuver, brake suddenly'. The main road traffic accident types are head-on collisions, collisions with a pedestrian and vehicle rollover. Head-on collisions account for 56.35 per cent, collisions with a pedestrian account for 11.65 per cent and vehicle rollover account for 8.12 per cent of the total number of road accidents.

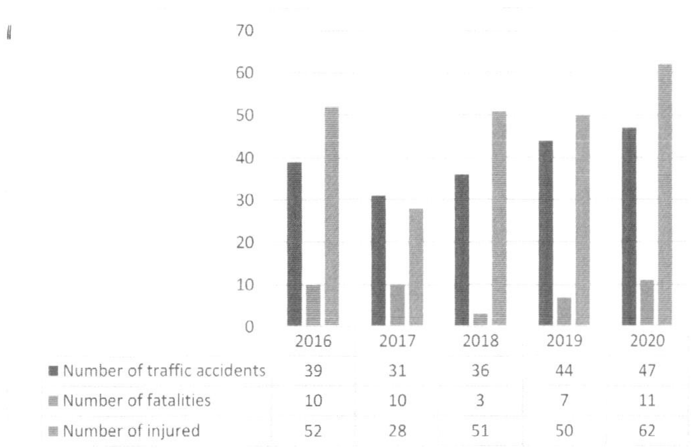

Figure 5. Via Baltica traffic accident statistics in 2016-2020

The greatest number of estimated traffic accidents and fatalities have been registered on the following Via Baltica sections: A8 *Panevėžys–Aristava–Sitkūnai* from 7.50 to 87.86 km (Table 1) and A10 *Panevėžys–Pasvalys–Riga* from 9.17 to 66.09 km (Table 2).

Table 1. Traffic accidents on road A8 *Panevėžys–Aristava–Sitkūnai* section from 7.50 to 87.86 km

Year	Number of traffic accidents	Number of fatalities	Number of injured
2016	9	1	11
2017	6	3	4
2018	9	1	11
2019	12	4	13
2020	8	3	12
In total	44	12	51

Table 2. Traffic accidents on road A10 *Panevėžys–Pasvalys–Riga* section from 9.17 to 66.09 km

Year	Number of traffic accidents	Number of fatalities	Number of injured
2016	13	6	23
2017	11	3	8
2018	10	2	17
2019	21	3	23
2020	17	5	25
In total	72	19	96

The greatest number of road accidents has been registered on unreconstructed road sections, where 1+1 traffic lane and junctions prevail.

2. Implementation Stages of Via Baltica Roads' Development

2.1. Reconstruction of Panevėžys bypass section from 0 to 22.25 km

In 2017-2019 the road was reconstructed from 1+1 to 2+1 traffic lane road. **Value of works: 45 million EUR.** Five bridges and viaducts were reconstructed, one tunnel passage was constructed. The existing reinforced structures for water removal were reconstructed; lighting was installed on roundabouts. The following environmental measures were implemented: 5 new passes for amphibian populations, wildlife guards blocking access to the road, reflectors (52 pcs) deterring fauna from passing the road, net fencing for wildlife (30.09 km), two turb*o* roundabouts were constructed.

2.2. Reconstruction of A1 *Vilnius–Kaunas–Klaipėda* section from 102.9 to 107.0 km

In 2020-2021 reconstruction works are underway. **Value of works: 18 million EUR.** During reconstruction one of the most dangerous junctions in the country will be replaced by a grade-separated intersection. Connecting roads on both sides of the road as well as exit roads with acceleration and deceleration lanes will be constructed. Pedestrian lanes leading from bus stop areas towards adjacent population centres will be constructed; bus stop areas will be reconstructed; roundabouts with entries/exits to/from connecting roads will be constructed. It is planned to construct a pedestrian viaduct and a transport viaduct on this road section, which will remove the existing left-turn. It is also planned to install noise barriers near residential areas, where maximum permitted noise level is exceeded. It is plannned to install lighting on the main and connecting roads as well as to reconstruct the existing engineering networks.

2.3. Reconstruction of A1 *Vilnius–Kaunas–Klaipėda* section from 101.79 to 102.03 km (Sargėnai transport centre)

In 2019-2020, a viaduct was reconstructed. **Value of works: 1.0 million EUR**. The viaduct was widened by constructing an additional traffic lane for transport taking the main road A5 from Marijampolė and turning to the main road A1 in the direction of Klaipėda. Moreover, an additional left-turn was set up on the connecting road at the entry to Kaunas city in the direction of *Klaipėda–Sargėnai.*

2.4. Reconstruction of A5 *Kaunas–Marijampolė–Suwalki* junction at km 6.67 (Akademijos transport centre)

In 2018-2020, the Akademijos transport centre was reconstructed. **Value of works: 3.15 million EUR.** The existing northern four-leg traffic-light intersection was reconstructed into a turbo roundabout, and the southern junction was reconstructed into one traffic lane roundabout. Exit to road A5 *Kaunas–Marijampolė– Suwałki* with a 190-metre-long acceleration lane was constructed. Almost 220 metre-long noise barrier was installed and the existing viaduct was repaired.

2.5. Reconstruction of A5 *Kaunas–Marijampolė– Suwałki* section from 17.34 to 23.40 km

In 2015-2017 the road section was reconstructed according to motorway technical category requirements. **Value of works: 38.0 million EUR.** A new connecting road and a pedestrian-bicycle lane connecting Stanaičiai and Juragiai population centres, pedestrian viaducts across the main road on the mentioned population centres were constructed. The reconstruction of the road section from 22.03 to 23.40 km (on the right side) into a motorway category road was completed when reconstructing connecting roads of the right-side of Mauručiai grade-separated intersection. A three-leg intersection within Mauručiai grade-separated intersection, a roundabout, connecting roads were constructed and lighting was installed. A new railway viaduct and a tunnel passage were reconstructed at km 23.1. The reconstruction of the road section from 21.84 to 23.40 km (on the left side) into a motorway category road was completed when reconstructing the left side connecting roads of the grade-separated Mauručiai intersection. Pavement of the section was strengthened. The existing railway viaduct was reconstructed and the tunnel passage at km 21.3 km was constructed. Noise barriers at Mauručiai (right side of A5 road) and Jurginiškės (left side of A5 road) population centres were installed.

2.6. Reconstruction of A5 *Kaunas–Marijampolė– Suwałki* section from 23.40 to 35.40 km

In 2017-2019 a 12 km-long road section on the territory of Prienai municipality from 23.40 to 35.40 km was widened from two to four traffic lanes. **Value of works: 37.2 million EUR.** To improve traffic safety on this road section, junctions were liquidated, connecting roads, 4 viaducts, one of which tunnel, were constructed. At Skriaudžiai population centre a grade-separated intersection was constructed at km 31.68. To reduce the risk of collision with wildlife and to improve environmental protection, net fencing, ramps, barriers and one underground passage were constructed.

2.7. Reconstruction of *A5 Kaunas–Marijampolė– Suwałki* section from 35.40 to 45.15 km

In 2016-2018, the reconstruction of the road section from 35.40 to 45.15 km from two to four traffic lanes was completed. Former junctions with regional roads No 2606 *Gudeliai–Smilgiai–Būdviečiai* at 35.68 km and No 2617 *Selema–Grigaliūnai* at km 39.96 km were replaced by a grade-separated intersection, which improves traffic safety and separates transit and local traffic flows. **Value of works: 28.1 million EUR.** The junction with local significance road *Kvietkiškis–Utalina* at km 42.22 was replaced by a tunnel passage. At km 38.51 another tunnel passage was constructed for the movement of agricultural machinery and livestock depasture to another side of A5. To ensure access to the existing land plots, connecting roads were constructed on both sides of the main road. Moreover, to maximize environmental protection, noise barriers, net fencing and an underground wildlife passage were constructed.

2.8. Reconstruction of A5 *Kaunas–Marijampolė– Suwałki* section from 45.15 to 56.83 km

In 2017-2019, the road section was upgraded to four lanes, connecting roads, six viaducts, two roundabouts and two grade-separated intersections at Sasnava (52.23 km) and Puskelniai (56.66 km) population centres were constructed. **Value of works: 52.7 million EUR.** To avoid collisions with wildlife, net fencing and barriers were installed

2.9. Reconstruction of A5 *Kaunas–Marijampolė– Suwałki* section from 56.83 to 97.06 km

The existing road section complies with II technical category requirements. Currently, most of it is only 1+1 road. Increased traffic volumes, especially those of heavy goods vehicles cause a high number of road accidents. The road has become dangerous for manoeuvring, drivers think that they can time and beat the oncoming traffic. This road section is specific because heavy goods vehicles account for even 53 per cent of the whole transport flow.

The Government of the Lithuanian Republic by its 20 November 2019 Resolution No 1177 "On Approval of the Special Plan and Starting the Procedure of Land Acquisition of the Main Road A5 *Kaunas–Marijampolė– Suwałki* Section from 56.83 to 97.06 km Reconstruction", approved the special plan of the main road A5 *Kaunas–Marijampolė–Suwałki* section from 56.83 to 97.06 km and allowed to start the procedure of land acquisition for public needs. It shall be pointed out that in 2020 the procedures of land acquisition for public needs were carried out successfully.

When continuing the project implementation, the section to be reconstructed was divided into 4 stages; in **2020** design contracts were signed and the following project preparation works were performed:

- preparation of project of works on road A5 *Kaunas–Marijampolė– Suwałki* section from 56.83 to 72.50 km (**Stage I**), section from 72.50 to 79.00 km (**Stage II**), section from 79.00 to 85.00 km (**Stage III**). Designing period: July 2020 – November 2021;
- preparation of project of works on road A5 *Kaunas–Marijampolė– Suwałki* section from 56.83 to 97.06 km (**Stage IV**). Designing period: January 2021 – April 2022.

The road section of the total length 40.23 km will be reconstructed into a motorway category, and the maximum speed limit of 130 km/h will be applied. It is planned to construct 5 junctions, to reconstruct 6 existing and to construct new grade-separated intersections, to construct 6 tunnel passages, 2 green bridges, 17 roundabouts on connecting roads, deceleration and acceleration lanes, 3 tunnel passages for wildlife, connecting roads, the total length of which is 79.1 km. It is planned to construct 2 new and reconstruct 3 existing bridges, to liquidate junctions, to carry out reconstruction at the border according to the solutions coordinated with Poland. The completion of works on the whole road section is planned by the end of 2025. **Value of works: ca 300 million EUR.**

2.10. Reconstruction of A8 *Panevėžys–Aristava–Sitkūnai* from 7.5 to 87.86 km

It is planned to reconstruct this section in **2026-2030**. Currently, a feasibility study is underway, which aims to evaluate the existing infrastructure, to select principal solutions (2+1 or 2+2 traffic lanes). Having selected principal solutions, projects of works will be prepared. If needed, a special territorial planning document on the procedures of land acquisition for public needs will be prepared. **Value of works: ca 250.0 million EUR.** It is planned to implement the project in three stages: sections from 7.50 to 37.78 km, from 37.78 to 53.54 km, from 53.54 to 8.86 km.

2.11. Reconstruction of A10 *Panevėžys–Pasvalys–Riga* from 9.17 to 66.09 km

It is planned to reconstruct this road section in **2026-2030**. Currently, a feasibility study is underway, which aims to evaluate the existing infrastructure, to select principal solutions (2+1 or 2+2 traffic lanes). Having selected principal solutions, projects of works will be prepared. If needed, a special territorial planning document on the procedures of land acquisition for public needs will be prepared as well. **Value of works: 147.0 million EUR.** It is planned to implement the project in three stages: sections from 9.17 to 38.83 km, from 38.83 to 47.83 km, 47.83 to 66.09 km.

3. Environment and Public Health

Planning and detailed design of the reconstructed sections of Via Baltica aim to contribute in achieving sustainable development goals, to ensure (to maintain where appropriate and to achieve where necessary) environmental and public health protection norms and requirements. Strategic environmental assessment (SEA) procedure is followed by Environmental Impact Assessment (EIA) procedure. Both procedures ensure consultation with state and municipal institutions, the public.

From the perspective of public health the implementation of the reconstruction projects is positive. Significant decrease in traffic accidents would result in significant decrease of injuries and death cases. Healthier environment due to decrease of environmental noise after implementing recommended measures will lead to better public health. The road is adapted for all users, passengers, including those with reduced mobility. Pedestrian and cycling infrastructure will contribute to making mobility more healthy and sustainable.

Although Via Baltica bypasses urban centres, some villages, small towns and detached homesteads remain in the vicinity of the road. Detailed computer noise modelling reveals conflict areas and necessary noise mitigation measures, noise barriers are being designed and will be implemented. Three types of principles need to be considered in the detailed design and in later installation of noise barriers: general non-acoustical principles of the construction, principles of acoustical design, landscape and architectural design principles. First of all, barriers must be acoustically effective. Acoustical design considerations include barrier material, barrier locations, dimensions and shapes. Non-acoustical design considerations are also important. The solution of one problem (noise) may cause other problems such as unsafe conditions, visual blight, maintenance difficulties, lack of maintenance access due to improper barrier design and other problems. With proper attention to maintainability, structural integrity, safety, aesthetics, and other non-acoustical factors potential negative effects of noise barriers can be avoided or reduced. The detailed design of noise barriers, as well as other environmental mitigation measures, will be based on scientific evidence, good practice and innovation.

In order to prevent and reduce animal-vehicle collisions risk, to preserve ecological connectivity and to reduce disturbances caused by traffic to adjacent ecosystems, fauna protection systems including fences, wildlife guards, crossings and related necessary for safe migration infrastructure are designed. The barrier effect for wildlife will be at most compensated. Two green bridges will provide animals with safe passage over the road. Water and soil protection measures will contribute to protecting natural environment.

Up to date reconstruction of the existing road will contribute to making infrastructure more climate-resilient (e.g., renewal of run-off water treatment systems, renewal of road pavement). This will contribute to actions ensuring a well-functioning transport system.

Via Baltica is one of the priority roads for implementation of charging stations for electric vehicles.

The projects respect the precautionary principle, the principle of preventive action. The follow-up of implementation of environmental impact mitigating measures will be carried out.

According to durability design principles, the reconstructed road should last and be reliable for 20-30 years. The planned green public procurements will secure proper quality of construction works, resource efficiency.

4 Implementation Risks of Via Baltica Development

To develop Via Baltica successfully, the following risks shall be managed:

- Insufficient financing;
- Public procurement procedures/ legal disputes;
- Lengthy procedures of territorial planning and land acquisition for public needs and design works;
- Meeting environmental requirements;
- Insufficient quality of project implementation activities;
- Defects of contract works;
- Market capability to implement two large-scale projects *Rail Baltica* and *Via Baltica*.

Conclusions

In 2016-2020, 197 traffic accidents were registered on Via Baltica, with 41 fatalities and 243 injured. The main types of traffic accidents registered on Via Baltica are head-on collisions, collisions with a pedestrian and vehicle rollover. Out of the total number of road traffic accidents head-on collisions account for 56.35 per cent, collisions with a pedestrian account for 11.65 per cent and vehicle rollover account for 8.12 per cent.

It is planned to complete all nine Via Baltica projects with the total value of ca 704 million EUR by 2030.

Having implemented the planned projects, Via Baltica will comply with the technical requirements set to TEN-T roads. The implementation of the above-mentioned projects will improve the infrastructure of TEN-T road on the territory of Lithuania and will have a long-term impact. It will create more favourable conditions for transit and local traffic; it will reduce travel time, fuel consumption. The upgraded road pavement and applied environmental measures will improve the living conditions in the residential areas.

IOP Conf. Series: Materials Science and Engineering 1202 (2021) 012008 doi:10.1088/1757-899X/1202/1/012008

TRANSFORMATION OF LITHUANIAN ROAD CHARGING

Darius RAŽINSKAS, Aidas SALADŽIUS
[2] *Intelligent Transport Systems and Road Charging Department, Vilnius, Lithuania*
Corresponding author: Darius Ražinskas, darius.razinskas@lakd.lt

2021

Abstract. The current road user charge in Lithuania is based on a temporary system and is not connected to the actually covered mileage. The time-based charge is not considered to be socially fair; therefore, the principles 'user pays' and 'polluter pays' shall be employed. The charge to be introduced will take into account the interests of local road infrastructure users who are currently paying unreasonably high charge for a short distance covered.

If the current system is retained and e-tolling system is not established, heavy vehicle drivers would continue paying aroad user charge by purchasing e-vignette. It would mean that road user charges paid by vehicle owners/ holders would not cover the damages directly made by the user and the current road charging system would not guarantee sufficient financing for the maintenance and development of the road infrastructure.

Legal decisions on electronic road charging system have not been taken yet; however, the project implementation preparation is underway. The latest Parliamentary approvals were received in September 2020, and it is believed that all necessary legal acts will be amended by the end of 2020.

The road charging technology has not been selected yet. However, the conducted project's implementation analysis, the assessment of necessary investments and road charging system maintenance costs revealed that a GPS-based technology would be the most economical one.

It is planned to implement the electronic road charging system by the beginning of 2023. Upon the project's implementation, additional financing to ensure high quality and safe main roads should be allocated.

The Current Situation

Currently, road users shall carry a valid e-vignette for the use of toll main road A1–A18 sections (a full list is available at: https://lakd.lrv.lt/lt/keliu-mokesciai-ir-rinkliavos/naudotojo-mokestis, Chapter *Apmokestinti keliai* (Toll Roads)) by buses (vehicle category M2–M3), heavy goods vehicles (vehicle categories N1–N3) and their combinations and special purpose road vehicles. Since 21 August 2018, only e-vignettes are sold.

E-vignette is an electronic record in the Road User Charge Payment Register Modulus of the State Significance Road Traffic Information System, confirming the payment fact and entitling to use toll main roads for the set period.

E-vignettes are distributed at www.keliumokestis.lt and selling points.

The Lithuanian Transport Safety Administration, the Customs of the Republic of Lithuania, territorial and specialized Police institutions check if vehicle owners/ holders carry necessary documents confirming the payment of the user charge of the vehicles registered in the Republic of Lithuania, foreign countries, including EU member countries, on the road or at the border crossing points. In case a vignette is invalid and the user charge is not considered to be paid on time or improperly paid, sanctions laid out in the Code of Administrative Offences of the Republic of Lithuania are applied. Vehicle owners/ holders are imposed a fine, which shall be paid into the State Tax Inspectorate account, and the collected funds are transferred to the state budget.

The generated income (revenues from e-vignette selling and paid fines), as defined in the Law on the Financing of the Road Maintenance and Development Programme (later – RMDP), are collected by the Ministry of Finance in the state budget and are used according to the annual estimate of RMDP fund allocation approved by the Government of the Republic of Lithuania, which is implemented by SE Lithuanian Road Administration.

Target and Technologies

The new e-tolling system to be implemented that is used to collect the charge for the use of toll main roads and to finance the costs of relevant road network infrastructure should, first of all, meet the following key requirements:

- the information about the road users (owners and holders of heavy goods vehicles) shall be identifiable and protected from possible forgery;
- the system shall be based on modern technologies, yet shall be road user-friendly;

- the equipment used shall be easily accessible, installable and replaceable;
- confidentiality of legal and physical entities shall be ensured;
- road charging shall be non-discriminative, and its collection shall not cause any additional obstacles at the state border crossing points;
- the system's infrastructure on the roads shall be sufficient in order to ensure effective toll collection.

As specified in the Directive (EU) 2019/520 of the European Parliament and of the Council of 19 March 2019 on the interoperability of electronic road toll systems and facilitating cross-border exchange of information on the failure to pay road fees in the Union, all new electronic road toll systems shall use one or more of the following technologies:

(a) satellite positioning;
(b) mobile communications;
(c) 5.8 GHz microwave technology.

Currently, the European countries that have implemented e-tolling systems and are introducing the directives mentioned above, apply the following three technological schemes:

- GNSS-based (Global Navigation Satellite System) based system;
- DSRC-based (Dedicated Short-Range Communication) system;
- ANPR-based (Automatic Number Plate Recognition) system.

Based on the best practices of foreign countries and technological systems used in the country, Lithuania could also employ GNSS, DSRC and ANPR technologies or their combinations when charging heavy goods vehicles. In most European countries that employ GNSS-based technology, the combination of GNSS and DSRC technological systems is used. In that case Global Positioning System (GPS) and mobile communication are used as the base of the system that is controlled via DSRC-based technology.

Comparison of E-vignette and E-tolling

Time-based vignette	Distance-based e-tolling
– socially unfair and disproportional charge since it does not ensure the principles 'user pays' and 'polluter pays'; – does not enable to compensate for the real damage made to roads TP (> 3.5 t); – does not enable to collect the funds for the maintenance of proper road condition. Funds collected from vignettes are considerably lower than the damage made by vehicles (> 3.5 t) to the road infrastructure; – discriminates road users, when unreasonably high charge is paid for a short distance covered; – distorts competition and creates better conditions for carriers whose vehicles cover longer distances on the Lithuanian roads (> 200 km per day).	– socially fair and proportional charge for the use of the road infrastructure; – promotes the use of environmentally-friendly vehicles, which reduces the impact on climate change and the environment; – ensures fair distribution of funds for road infrastructure development and maintenance; – enables to collect funds for the maintenance and development of infrastructure; – creates conditions to harmonize technical and legal conditions applied in EU member countries by developing a uniform system and fair competition.

Comparision of Technological Systems

Technology	Advantages	Disadvantages	Accuracy, in %	General Capex and Opex in 15 years, excluding VAT
GNSS (satellite positioning)	Flexible amendment of toll road sections; Investments required only for the enforcement of road infrastructure; Easily applicable on other roads.	High initial costs (OBU[1]) compared to other technologies.	99	147.038.286.76
DSRC (Dedicated Short Range Communication (5.8 GHz microwave technology)	High dependability, low signal interference; Inexpensive OBU (compared to GNSS); Gradually integrated system.	High initial investments required when installing necessary road infrastructure; Expensive and complicated application on other roads; Costly on the roads with numerous intersections.	99.5-99.9	239.762.486.94
ANPR (Automatic Number Plate Recognition)	Does not require OBU device; Does not require enforcement infrastructure; Gradually integrated system.	High quality vehicle number plates required; Quality depends on lighting and weather conditions; Manual control may increase operational costs; Non-standardized EETS.	90-98	237.135.215.12
SmartPhone technology	Flexible amendment of toll road sections; Investments required only for the road enforcement infrastructure; Easily applicable on other roads. Does not require OBU device	Data protection shall be guaranteed when collecting data; Non-standardized EETS.	99	117.734.567.36

[1] On-board unit

Conclusions

The key factor in selecting the best technology has been its compliance with the aims and tasks of the investment project. Having evaluated the needs and emerging problems of the project, the focus was laid on the activities related to the road infrastructure. Technological progress was another important factor in the selection process. Having investigated technical, socioeconomic and financial advantages and disadvantages of the alternatives, SmartPhone Technology was selected as the most advantageous and beneficial one.

Lithuania would be the first country in the European Union to employ SmartPhone technology in implementing distance-based e-tolling, yet not the first one in the world. SmartPhone mobile applications have been employed in Australia and the United States. To our mind, taking into account the progress of technologies and mobile devices, SmartPhone mobile application is a future technology in e-tolling.

UTILIZING EFFECTIVE TECHNOLOGY IN HIGH-STANDARD ROAD DEFECTS INVENTORY SYSTEM IN ESTONIA

Marek TRUU (TEEDE TEHNOKESKUS AS)[1], Romet RAUN (TEEDE TEHNOKESKUS AS) [1], Maret JENTSON (TEEDE TEHNOKESKUS AS) [1]

[1]*Research and Development Department, Tallinn, Estonia*
Corresponding author: Marek Truu, marek.truu@teed.ee

Received 15 March 2021

Abstract. Road pavement is expected to withstand enormous traffic loads for long time but sooner or later the deterioration reaches levels when its optimal to apply treatment. While easy to measure roughness or rutting in normal traffic speed, defects are in most countries still collected by means of time-consuming visual inspection in low traffic speeds or expensive and difficult to-use equipment. Also, most visual inspection systems only operate with aggregated inspection data. That makes data-collection expensive and defects-based decision-making inefficient. In Estonia, defects inventory system utilizes high quality panoramic and orthogonal images to enable data collection in traffic speeds and detailed mapping of pavement defects in 10 classes. Defects mapped in full detail means that location, shape and size of each defect is known and classified data can be effectively used twice in pavement maintenance planning: for section selection planning in road network level when aggregated and for work method selection in design process when analyzed in detail. Combined with measured lidar-based point-cloud data, detailed 3d-basemap saves both road-owner's and road designer's valuable time in design phase. In period of 2016-2020, around 35000km of state roads were analyzed with one of the most efficient road defects inventory systems in the world. Also, around 25000 km of municipal and forest roads have been captured with same technology covering several pavement types from bicycle paths to multilane streets and motorways. Current presentation discusses outcomes of Estonian defects inventory study in 2020.

Keywords: road network, asset management, pavement, defects inventory, mobile mapping, lidar, point cloud, panoramic images, orthoimages, maintenance planning

Introduction

Road pavement is expected to withstand enormous traffic loads for long time but sooner or later the deterioration reaches levels when its optimal to apply treatment. While easy to measure roughness or rutting in normal traffic speed, defects are in most countries still collected by means of time-consuming visual inspection in low traffic speeds or expensive and difficult-to-use equipment. Also, most visual inspection systems only operate with aggregated inspection data. That makes data-collection expensive and defects-based decision-making inefficient. Current presentation discusses outcomes of Estonian defects inventory study in 2020 and benefits of system.

1. Traditional pavement defects mapping method

Although laser-based road measurement high-quality crack measurement systems are available for many years, these systems are expensive for many users and still need significant amount of manual work. Even more – these systems are not effective in detecting certain type of defects, eg. typical ravelling in Estonia.

Due to the above, many countries and roadowners continue defects mapping through visual inspection in actual traffic. Such systems are very slow affecting traffic safety and lack of possibilities to ensure inspection quality for no raw data documentation will remain if there are claims later.

2. Pavement defects mapping methods in Estonia

Each year defects inventory is carried out in about 7000-8000 km of national roads in Estonia. Defects inventory will start immediately after snow melting and shall be completed by mid-May. That is required because of visibility of cracks due to moisture condition and crack width during spring period. Mapping is described in following figure.

Utilizing effective technology in high-standard road defects inventory system in Estonia – Marek Truu & Romet Raun

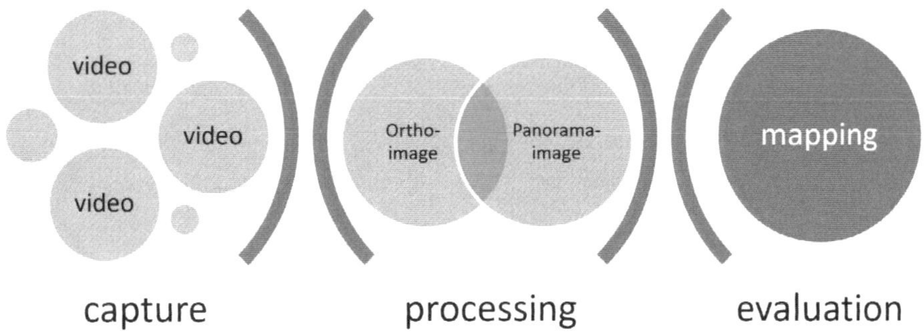

Figure 1. Defects mapping process in three steps

Data is captured so that roads with higher traffic volumes AADT ≥ 2000 need to be finished by end of April and roads with lower traffic volumes AADT < 2000 shall be finished by middle of May. Mobile mapping system is equipped with panorama camera, GNSS+INS device, lidar and data capture hardware and software. Data processing is carried out in fast server and images are published in web application taking. Special web application is used for defects mapping.

Figure 2. Data capture equipment

In Estonia, defects inventory system utilizes high quality panoramic and ortho-images. Data collection is carried out in traffic speed. Pavement defects are then mapped in full detail and classified into 10 defect classes:

- Joint cracks (narrow, wide)
- Longitudinal cracks (narrow, wide)
- Transversal cracks
- Alligator cracking
- Ravelling
- Edge break
- Patching
- Potholes

Proper classification is essential to make the adequate linking of mapped defects with possible causes. Each part of the process is fully digital, which makes the raw data reusable and creates the opportunity for different processing and analysis. Defects mapped in full detail means, that location, shape and size of each defect is known and classified data can be effectively used repeatedly in pavement maintenance planning throughout the road life-cycle.

Mapping is carried out in ortho-images while valuable information about surrounding from panorama images has significant importance when determining actual reason for pavement collapse.

Figure 3. Classified pavement defects data. Alligator cracking in red, longitudinal cracking in yellow, edge break in violet and transversal cracking in green can be seen in ortho-image view.

Some of the most common uses for defects data are:

1. evaluation of structural properties of pavement based on progression of cracking for each individual defect separately or in network level, especially when analyzed with structural data from falling weight deflectometer and rutting;
2. maintenance section selection planning in road network level when aggregated;
3. work method selection in design process when analyzed in detail based on defect types;
4. identification of reflective cracking after maintenance works (mapping of "first out" defects);
5. identification of defects caused by poor backfill compaction for utilities.

Figure 4. Panorama image makes it possible to adequately evaluate influence of surrounding area for defects' causes

Figure 5. Measurement tools allow to locate defects or other subjects for more detailed planning of maintenance works

3. Defects inventory on national roads 2020

During the defects inventory project of national roads 7570,063 km of paved roads were mapped. Mapped sections can be seen in the map below.

Utilizing effective technology in high-standard road defects inventory system in Estonia – Marek Truu & Romet Raun

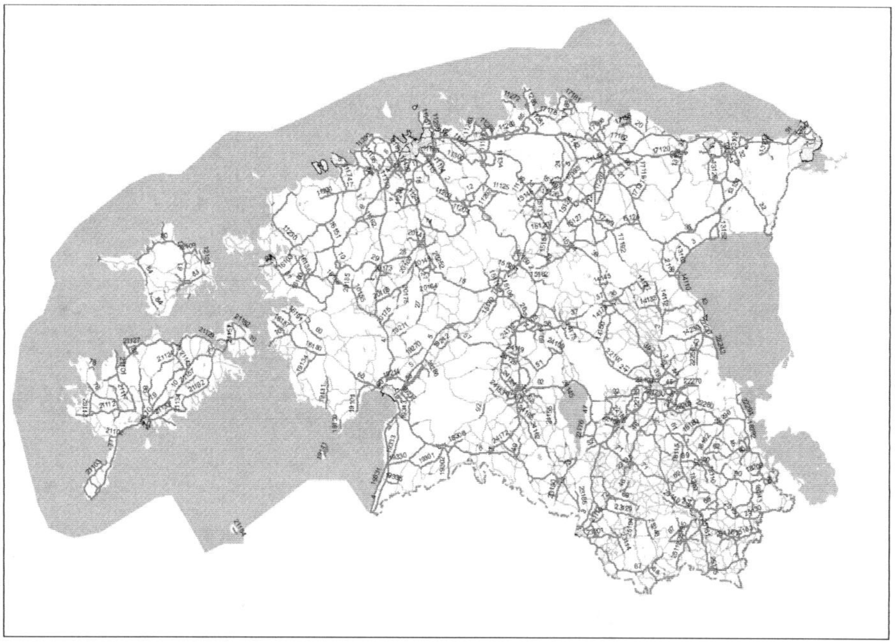

Figure 6. Map of roads with defects inventory in 2020

Below few panorama views from some sample road sections in poor condition.

Figure 7. Panorama view on road no 25 Mäeküla-Koeru-Kapu km 16,7

Figure 8. Panorama view on road no 16136 Taebla-Kullamaa km 14,0

Figure 9. Panorama view on road no 22290 Rasina-Meeksi km 5,6

Figures below represent average values of classified cracks for different road categories – main roads, connectors, basic roads and secondary roads. One can see that there are virtually no wide longitudinal cracks or joint cracks in Estonian national road network. Also potholes are very rear.

Utilizing effective technology in high-standard road defects inventory system in Estonia – Marek Truu & Romet Raun

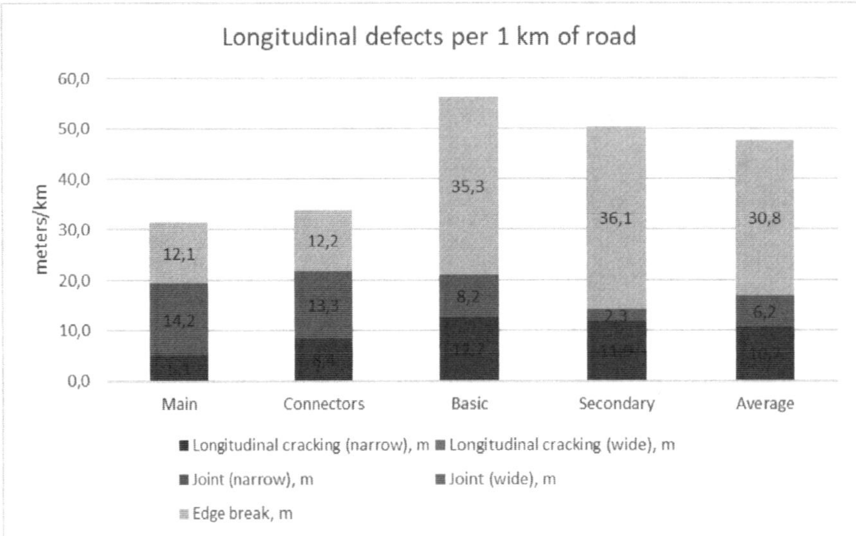

Figure 10. Longitudinal defects per 1 km of road

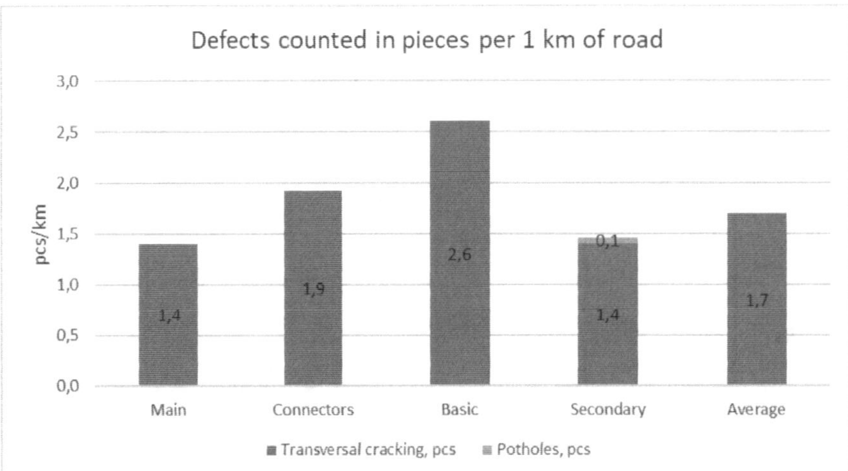

Figure 11. Defects counted in pieces per 1 km of road

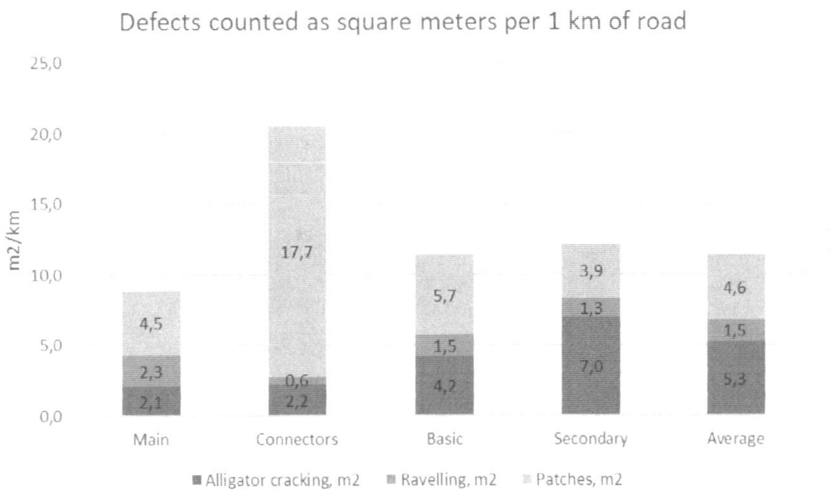

Figure 12. Defects counted as square meters per 1 km of road

4. Pavement defects in City environment

In City environments pavement defects often reflect poor backfill compaction of utilities trenches or backfill material with different water permeability. In these cases it is useful to see utility line data together with pavement defects, giving clear linkage with reason.

Figure 13. Utilities data can effectively explain reason for defects or patching

5. Point cloud adds important aspect for defects evaluation

Point clouds are often considered only as a tool for design purposes and less utilized for as-built data and understanding the problems of existing roadway. Pont clouds make it possible to quickly evaluate longitudinal ditch slopes or kerb locations for water movement, or traffic safety issues from visibility point of view. Driverless cars need far more detailed pavement maps to compare observed date with ground truth, so lidar data adjusted with data from image is ideal to quickly prepare large areas of network to the level needed for autonomous vehicles.

Figure 14. Point cloud makes accurate measurements possible in all directions – x, y and z.

6. Pavement defects mapping widely used

Combined with measured lidar-based point-cloud data, detailed 3d-basemap saves both road-owner's and road designer's valuable time in design phase. In period of 2016-2020, around 35000km of state roads were analyzed with one of the most efficient road defects inventory systems in the world. Also, around 25000 km of municipal and forest roads have been captured with same technology covering several pavement types from bicycle paths to multilane streets and motorways.

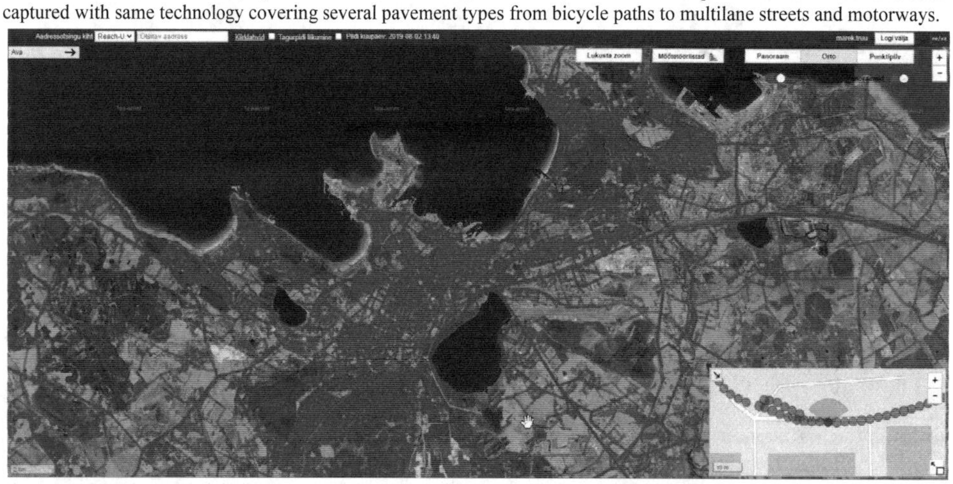

Figure 15. Estonian capital Tallinn covered with high quality image and lidar data

7. Conclusions

Although used for many applications (eg. locating traffic marking, preparing digital road twin for autonomous vehicles), currently the most widely used application is pavement defects monitoring, covering 7000-8000 km of national roads per year and mapping defects into 10 classes. Combined with measured lidar-based point-cloud data, detailed 3d-basemap saves both road-owner's and road designer's valuable time in design phase. In period of 2016-2020, around 35000km of state roads were analyzed with one of the most efficient road defects inventory systems in the world. Also, around 25000 km of municipal and forest roads have been captured with same technology covering several pavement types from bicycle paths to multilane streets and motorways.

Authors hereby confirm that we do not have any competing financial, professional, or personal interests from other parties.

Template: Anonymized Document

ROLLER COMPACTED CONCRETE – BEST PRACTICE OF LITHUANIA

Audrius VAITKUS ORCID iD (0000-0001-5103-9747)[1], Rafal MICKEVIČ ORCID iD (0000-0003-0832-3536)[1], Ovidijus ŠERNAS ORCID iD (0000-0002-0019-4242)[1]

[1]*Road Research Institute, Vilnius Gediminas Technical University, Vilnius, Lithuania*
Corresponding author: Rafal Mickevič rafal.mickevic@vilniustech.lt,

Abstract. In these days constantly being looking for solution to reduce construction costs, the amount of materials used and the negative impact on the environment. Designing pavement structures with top layer of traditional concrete, the structures become very massive. An alternative to traditional concrete is roller-compacted concrete, the concrete with significantly larger fine aggregates which lead concrete mix to be non-slip. The roller-compacted mix can also achieve high concrete density and consolidation by rolling. Roller-compacted concrete is also an economical and fast-construction alternative for many pavement applications. In Lithuania, roller-compacted concrete as the top layer is an innovation and first application of it was few years ago. However, in recent years application of roller-compacted concrete increased in industrial areas and low-volume rural roads. The best practice of application of roller-compacted concrete was in Klaipeda Free Economic Zone, where roller-compacted concrete was used as the top layer of industrial area for heavy load traffic. Fresh roller-compacted concrete workability decreases with time, to reduce transportation time and avoid excessive moisture loss, which can cause problems in placement of roller-compacted concrete, mobile concrete batching plant used. Mobile concrete batching plant was located near construction site.

Keywords: concrete, roller-compacted concrete, mobile concrete batching plant, paver, mixture design, cement and special additives stabilized base, fly ash

Introduction

Roller-compacted concrete (RCC) is an economical, fast-construction candidate for many pavement applications (Chhorn et al., 2017). RCC has similar strength properties and consists of the same basic ingredients as conventional concrete—well-graded aggregates, cementitious materials, and water—but has different mixture proportions. The largest difference between RCC mixtures and conventional concrete mixtures is that RCC has a higher percentage of fine aggregates, which allows for tight packing and consolidation. Typical material comparisons of conventional concrete and RCC are represented in Figure 1 (National Concrete Pavement Technology Center, 2010). RCC is typically placed with an asphalt-type paver equipped with a standard or high-density screed, followed by a combination of passes with rollers for compaction (National Concrete Pavement Technology Center, 2010).

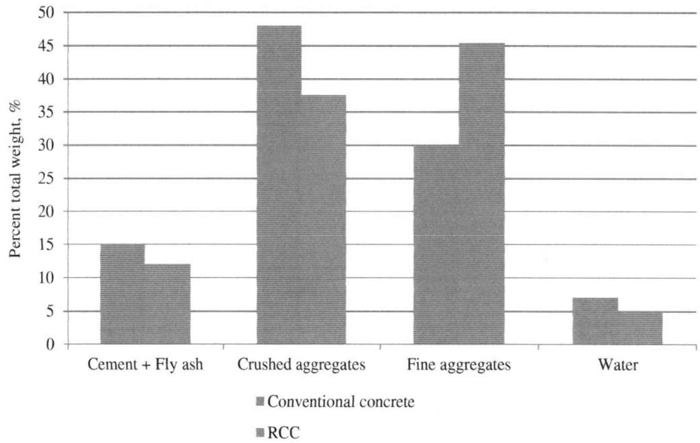

Figure 1. Typical material comparisons of conventional concrete and RCC

Template: Anonymized Document

RCC typically was used for pavements to carry heavy loads. However, in recent years its use in commercial areas and for local streets and highways has been increasing (National Concrete Pavement Technology Center, 2010). Advantages of RCC:

- Fast construction;
- Economical;
- Longevity;
- Ability to withstand high stresses;
- High bearing capacity.

In these days in Lithuania there is a big interest in RCC pavement structures and the number of locations where RCC was used is growing every year. Lithuanian scientists, contractors and other institutions are all working together to design the most economical RCC pavement structure and RCC mixture that could outstand design traffic loads. In Lithuania there are over 4 locations where RCC was used. All the locations are different according to RCC pavement structure and RCC mixture. In these locations few experimental tests were made measuring surface deflection with falling weight deflectometer (FWD). The main aim of these experimental tests is to collect deflection values from different RCC structures with different RCC mixture.

1. Typical concrete pavement structures

Concrete pavement structures should meet "Rules for the design of road standard pavement structures" KPT SDK 19 requirements. KPT SDK 19 is a document which regulates all types of pavement structures used in Lithuania. Traditional concrete pavement structures are represented in KPT SDK 19 according to traffic loads, which are expressed in 10 t equivalent standard axle load passages number. These pavement structures sometimes are irrational because they will satisfy a certain 10 t equivalent standard axle load range. When the design load falls to the lower limit, the standard pavement structures may have huge bearing capacity reserve in the individual case. Traditional concrete pavement structures regulated by KPT SDK 19 are represented in Table 1.

Table 1. Traditional concrete pavement structures used in Lithuania

Pavement structure class		DK 100	DK 32	DK 10	DK 3	DK 2	DK 1	DK 0,3	DK 0,1
Design load A (ESAs), mln.	A	> 32	> 10–32	> 3,0–10	> 2,0–3,0	> 1,0–2,0	> 0,3–1,0	> 0,1–0,3	≤ 0,1
		Layer thickness, cm							
Concrete surface course		27	26	25	24	23	–	–	–
Nonwoven geotextile		+	+	+	+	+	–	–	–
Concrete base course		15	15	15	15	15	–	–	–
Frost-resistant layer		*	*	*	*	*	–	–	–
Concrete surface course		27	26	25	24	23	–	–	–
Nonwoven geotextile		+	+	+	+	+	–	–	–
Stabilized layer of frost no-sensitive materials		20	15	15	15	15	–	–	–
Frost no-sensitive materials (well graded soil)		*	*	*	*	*	–	–	–
Concrete surface course		27	26	25	24	23	20	20	20
Nonwoven geotextile		+	+	+	+	+	+	+	+
Stabilized layer of frost no-sensitive materials (well graded soil)		25	20	20	20	20	15	15	15
Frost no-sensitive materials (poorly graded soil)		*	*	*	*	*	*	*	*
Concrete surface course		26	25	24	23	22	*	*	*
Asphalt base course		10	10	10	10	8	*	*	*
Frost-resistant layer		*	*	*	*	*	*	*	*
Concrete surface course		29	28	27	26	24	–	–	–
Unbound base course		30	30	30	30	30	–	–	–

Template: Anonymized Document

Pavement structure class		DK 100	DK 32	DK 10	DK 3	DK 2	DK 1	DK 0,3	DK 0,1
Design load A (ESAs), mln.	A	> 32	> 10–32	> 3,0–10	> 2,0–3,0	> 1,0–2,0	> 0,3–1,0	> 0,1–0,3	≤ 0,1
		Layer thickness, cm							
Frost no-sensitive materials		*	*	*	*	*	–	–	–
Concrete surface course		29	28	27	26	24	–	–	–
Unbound base course		20	20	20	20	20	–	–	–
Frost-resistant layer		*	*	*	*	*	–	–	–
Concrete surface course		–	–	–	–	–	21	21	21
Frost-resistant layer		–	–	–	–	–	*	*	*
* – the thickness is adjusted according to KPT SDK 19 depending on the maximum freezing depth.									

2. Road pavement structures design practice

RCC pavement structures and mixtures are designed individually. Typically for the production for concrete pavement are used C30/37-C40/50 class concrete (Vaitkus et al., 2019). Typical specification of fresh concrete mixture: water-cement ratio from 0.37 to 0.45; minimum air voids in fresh concrete with a maximum 8 mm aggregate size – 5.5%, and with 32 mm or 22 – 4.0% (Vaitkus et al., 2019). Hardened concrete typical specification: flexural strength after 28 days from 4.5 MPa to 5.2 MPa; compressive strength after 28 days from 27.5 MPa to 40.0 MPa (Vaitkus et al., 2019). The most important mechanical properties of concrete are compressive strength, flexural strength and tensile splitting strength (Vaitkus et al., 2020). Depending on design methodolody different mechanical properties of concrete should be used. Designing concrete pavement structures according to Richtlinien für die rechnerische Dimensionierung von Betondecken im Oberbau von Verkehrsflächen RDO Beton 09, one of the most important mechanical properties is tensile splitting strength, but when using design programs such like Streetpave or FAARFIELD, the most important mechanical properties is flexural strength. Designing RCC pavement structures and mixtures all this properties were evaluated. In Lithuania there is a patent for RCC mixture and there are four pavement structures designed and installed according to that patent and atleast one location is going to be instaled in 2021. Different compositions of RCC mixtures used in Lithuania are represented in Table 2. Important to admit that precise composition should be each time realized performing type conformity test in the lab. RCC pavement structures used in Lithuania consists of subgrade, cement and special additives stabilized base and RCC course. Cement and special additives stabilized base is one of the best solution in areas with weak soil. Use of supplementary cementing materials has become an integral part of the high strength and high-performance stabilisation methods (Ghorbani et al., 2018).Some of the commonly used supplementary cementing materials are fly ash, silica fume, ground granulated blast furnace slag and rice husk ash (Silitonga et al., 2010). RCC pavement structure used in one of the Lithuanian project are represented in Figure 2.

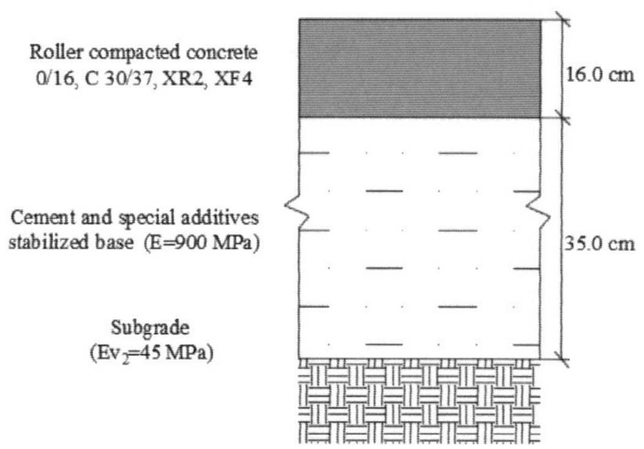

Figure 2. Typical RCC pavement structure used in Lithuania

Template: Anonymized Document

Table 2. Practice of using RCC mixture compositions in Lithuania

Type	Component	Proportion, %
1	4/16 fraction crushed aggregates	40.4
	0/4 fraction fine aggregates	36.2
	0/0.063 fraction fly ash	4.0
	Cement	14.0
	Concrete plasticizer	0.06
	Water	5.3
2	4/16 fraction crushed aggregates	39.1
	0/4 fraction fine aggregates	37.2
	0/0.063 fraction fly ash	4.4
	Cement	14.1
	Concrete plasticizer	0.07
	Water	5.1
3	5/16 fraction crushed aggregates	22.5
	5/8 fraction crushed aggregates	22.7
	0/4 fraction fine aggregates	31.2
	0/0.063 fraction fly ash	3.8
	Cement	14.2
	Concrete plasticizer	0.05
	Water	5.6
4	5/16 fraction crushed aggregates	23.5
	5/8 fraction crushed aggregates	20.6
	0/4 fraction fine aggregates	32.4
	0/0.063 fraction granite powder	3.5
	Cement	14.5
	Concrete plasticizer	0.06
	Water	5.4

RCC mixtures typically have a lower volume of cementitious materials, coarse aggregates, and water than conventional concrete mixes and a higher volume of fine aggregates (National Concrete Pavement Technology Center, 2010). RCC mixture can be produced in central mix concrete plant and then transported to the site or can be produced in mobile concrete batching plant, which can be constructed at site. Typically RCC mixture consists of fine aggregates, fraction 5/8 and fraction 4/16 crushed aggregates. Only in one Lithuanian project was used mobile concrete mixer plant. Used mobile concrete mixer plant could mix two aggregates, so to fit mobile concrete batching plant, in RCC mixture were used fine and one type of fraction crushed aggregates. Mobile concrete batching plant used in one of the projects is represented in Figure 3. Practice of using RCC structures and mixtures in Lithuania are represented in Table 3.

Template: Anonymized Document

Figure 3. Mobile concrete batching plant used in one of the Lithuanian RCC projects

Table 3. Practice of using RCC structures in Lithuania

Location	Year of construction	Mixture		Structure			
		Components	Mixured in	Layer	Thickness, cm	Flexural strength, MPa	E modulus, MPa
Local road No. 130	2021	RCC 0/16 C30/37 XR2 XF4	Central mix concrete plant	RCC	16	5.5	–
				Cement and special additives stabilized base	40	–	500
				Frost-resistant layer	34	–	100
Klaipeda FEZ	2020	RCC 0/16 C30/37 XR2 XF4	Mobile concrete batching plant	RCC	16	4.2	–
				Cement and special additives stabilized base	35	–	900
Klaipeda city	2019	RCC 0/16 C30/37 XR2 XF4	Central mix concrete plant	RCC	16	No data	
				Cement and special additives stabilized base	40		

Template: Anonymized Document

Location	Year of construction	Mixture		Structure			
		Components	Mixured in	Layer	Thickness, cm	Flexural strength, MPa	E modulus, MPa
Pier berth in Kaunas city	2018	RCC 0/16 C30/37 XR2 XF4	Central mix concrete plant	RCC	16	No data	
				Cement and special additives stabilized base	20		
Private road	2016	RCC 0/16 C30/37 XR2 XF4	Central mix concrete plant	RCC	varied from 10 cm to 16 cm	No data	
				Cement and special additives stabilized base	30		

3. Surface deflection measurement on RCC pavement structure

RCC pavement surface deflections were measured with falling weight deflectometer (FWD) in Klaipeda free economic zone (FEZ). FWD measurements were carried out in November 2020. Measurements were taken in the middle of the RCC slab, in the corner and at the joint. Moreover measurements were taken in two different locations on site which called lane 1 and lane 2. Due to varying loads for each measurement, the deflections cannot be directly compared to each other without their normalisation to a standard load 50 kN. According to Motiejunas et al. (2010) surface deflections were normalized to a standard load 50 kN.

As seen from Table 4 and Figure 4 surface deflection in lane 1 in the middle of the RCC slab varied from 128 μm to 244 μm, average – 190 μm, standard deviation – 48 μm. In lane 1 in the corner of RCC slab surface deflection varied from 178 μm to 256 μm, average – 225 μm, standard deviation – 36 μm. In lane 1 at the joint of RCC slab surface deflection varied from 142 μm to 247 μm, average – 193 μm, standard deviation – 36 μm. Surface deflection in lane 2 in the middle of the RCC slab varied from 184 μm to 337 m, average – 252 μm, standard deviation – 64 μm. In lane 2 in the corner of RCC slab surface deflection varied from 210 μm to 381 μm, average – 277 μm, standard deviation – 71 μm. In lane 2 at the joint of RCC slab surface deflection varied from 261 μm to 449 μm, average – 346 μm, standard deviation – 65 μm.

Table 4. Statistical indicators of RCC surface deflection measured in Klaipeda FEZ

Statistical indicator	Lane 1			Lane 2		
	Middle	Corner	Joint	Middle	Corner	Joint
Minimum	128	178	142	184	210	261
Maximum	244	256	247	337	381	449
Average	190	225	193	252	277	346
Standard deviation	48	36	36	64	71	65

Template: Anonymized Document

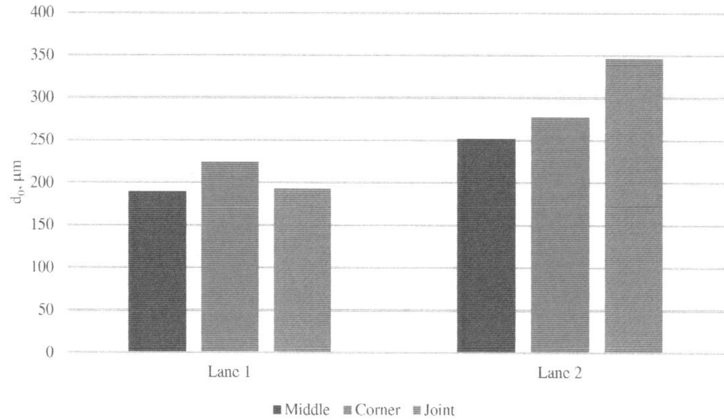

Figure 4. Averages of RCC surface deflection measured in Klaipeda FEZ

Conclusions

1. Starting from 2016 Lithuania gain experience of RCC pavement structures with cement and special additives stabilized base. Cement and special additives stabilized base thickness varied from 20 cm to 40 cm, E modulus varied from 500 MPa to 900 MPa. RCC course thickness varied from 10 cm to 16 cm, flexural strength of RCC mixture varied from 4.2 MPa to 5.5 MPa.
2. Average surface deflection measured with falling weight deflectometer (FWD) on pavement structure with 16 cm thickness RCC course (flexural strength 4.2 MPa) and 35 cm thickness of cement and special additives stabilized base (E modulus 900 MPa) in lane 1 in the middle of the RCC slab 190 μm, in the corner of RCC slab – 225 μm, at the joint of RCC slab – 193. In lane 2 average surface deflection in the middle of the RCC slab 252 μm, in the corner of RCC slab – 277 μm, at the joint of RCC slab – 346 μm.
3. Looking forward is important to evaluate RCC pavement structure with cement and special additives stabilized base sensitivity to frost and change of pavement structure bearing capacity dependent on subgrade hydrothermal conditions.

Disclosure Statement

This project has not any competing financial, professional, or personal interests from other parties.

References

Vaitkus A, Gražulytė J, Vorobjovas V and Šernas O 2020 Cement Concrete Modular Pavement Implementation for Pedestrian and Bicycle Path, *IOP Conf. Series: Materials Science and Engineering* **987.**

Vaitkus A, Kleizienė R, Vorobjovas V and Čygas D 2019 Mixture strength class and slab dimensions' effect on the precast concrete pavement structural performance Baltic Journal of Road and Bridge Engineering 14(3) pp 443–71

Ghorbani A, Salimzadehshooiili M, Medzvieckas J, Kliukas R 2018 Strength characteristics of cement-hice husk ash stabilised sand-clay mixture reinforced with polypropylene fibers Baltic Journal of Road and Bridge Engineering 13(4)

Chhorn, C., Hong, S. J., & Lee, S.-W. (2017). A study on performance of roller-compacted concrete for pavement. Construction and Building Materials, 153, 535–543. https://doi.org/10.1016/j.conbuildmat.2017.07.135

Owino, J. P., Fomunung, I., & Khalafalla, M. (2014). The recent development in the roller compacted concrete. BIE Journal of Engineering and Applied Sciences, 5(2), 42–49.

Silitonga, E., Levacher, D., & Mezazigh, S. (2010). Utilization of fly ash for stabilization of marine dredged sediments. European Journal of Environmental and Civil Engineering, 14(2), 253-265. https://doi.org/10.1080/19648189.2010.9693216

Motiejunas, Algirdas et al. 2010. "Research on the Dependence of Asphalt Pavement Stiffness upon the Temperature of Pavement Layers." Baltic Journal of Road and Bridge Engineering 5(1): 50–54.

National Concrete Pavement Technology Center. (2010). Guide for roller-compacted concrete pavements. Iowa.

Template: Anonymized Document

KPT SDK 19 Automobilių kelių standartizuotų dangų konstrukcijų projektavimo taisyklės. Lietuvos automobilių kelių direkcija prie Susisiekimo Ministerijos (LAKD) (in Lithuanian)
Richtlinien für die rechnerische Dimensionierung von Betondecken im Oberbau von Verkehrsf lächen RDO Beton 09 (in German)

ADVANTAGES OF FILLER AND SURFACTANT ADDITIVE TO IMPROVE MOISTURE SENSITIVITY OF HOT MIX ASPHALT MIXTURES.

Asres Simeneh, ALAMREW (https://orcid.org/0000-0002-5517-5380)[1], Konrad MOLLENHAUER (https://orcid.org/0000-0002-9091-7335)[2]

[1]*Civil and Environmental Engineering, University of Kassel, Kassel, Germany (asres.simeneh@student.uni-kassel.de)*
[2]*Civil and Environmental Engineering, University of Kassel, Kassel, Germany (k.mollenhauer@uni-kassel.de)*

Corresponding author: Asres Simeneh, Alamrew, asres.simeneh@student.uni-kassel.de

Received 15th March 2021, accepted date

Abstract.

This research investigated the effect of mineral composition of aggregate on moisture sensitivity of bituminous mixtures and explored the benefits of hydrated lime filler and Wetfix BE surfactant additive to improve the resistance of the mix against moisture sensitivity. Basalt, quartzite, and limestone aggregates were selected based on their different mineralogy and 70 -100 penetration graded bitumen binders used during the study. Four laboratory tests the rolling bottle, shaking abrasion, pull-off tensile strength and indirect tensile strength tests were applied to study the effects of aggregate minerals and benefits of hydrated lime and Wetfix BE. Statistical analysis using Two-way ANOVA test conducted for each test to check the outcome significance. Results from each test revealed that mineral composition of aggregate have significant effects on the moisture resistance performance of bituminous mixtures and hydrated lime filler and Wetfix BE surfactant additives have advantages to improve the performance of bituminous mixture against moisture sensitivity and improves the long-term performance of asphalt mix.

Keywords: Aggregate mineralogy, Moisture susceptibility, Hydrated Lime, Wetfix BE, Hot Mix Asphalt.

Introduction

The exposure of bituminous mixtures to water is an inevitable problem. Even though it has been centuries to studying the effect of moisture in bituminous mixtures, the problem leading to significant costs for pavement maintenance and reconstruction. The properties of bituminous mixtures on moisture sensitivity depend on internal factors that arise from the properties of the constitute materials such as aggregate, bitumen, fillers, asphalt mixtures, and the external factors including temperature, moisture, loading, adhesion promoters, etc.

Many researchers attempted to express the mechanism of moisture damage through different approaches to measuring its effect in bituminous mixtures. The mechanism of moisture damage is very complex, and it is a combined effect of physical and chemical effects of the component materials. Extensive researches have been done to study the mechanism of moisture damage in bituminous mixtures and reached different conclusions. However, the conflicting implicates the problem of stripping is still not fully understood and no single theory finds that effectively measures the damage due to moisture exposure. (Bagampadde et al. 2006; Rahim et al. 2019)

Moisture damage is a combination result of the loss of cohesion and adhesion. The water in the bituminous mixture ultimately enters between aggregate and bitumen binder and weakening the bond and separate the adherence. Loss of cohesion refers to the reduction of material integrity within the bitumen binder and within the mastic asphalt and loss of adhesion is the removal of the asphalt film from the aggregate surface. The characteristic of adhesion and cohesion is significantly different in terms of their damage and contributions to lose the strength of the mix

The effect of moisture in bituminous mixtures studied considering the physical and chemical properties of the mixture components separately. Though each study approaches showing the effect of moisture in bituminous mixtures, the single method or procedure can't entirely explain the overall effects of the complex influences of moisture in the bituminous mixture. The bituminous mixtures composed of coarse fraction aggregate, fine aggregate, and filler. These components of materials have significant implications in the bituminous mixtures to resist the effect of moisture damage.

This research tries to investigate the effect of moisture in bituminous mixtures through four different laboratory tests. Selected laboratory tests considered loose and compacted mixtures, effects on small and larger fractions of aggregates, and to study the cohesive and adhesive effects. The shaking abrasion test used for a small fraction of the aggregates and the Rolling Bottle test used for large size of aggregate. Moreover, these effects also are studied using the Pull-off strength test and Indirect tensile strength test.

This paper aims to study the combined effects of cohesive and adhesive failures in the bituminous mixture and the advantages of using adhesion promoters. The goal of this research is to investigate the moisture resistance of the Hot Mix Asphalt Mixtures prepared from three different sources of aggregates having different mineral compositions. The specific objectives of this research include

1. Explore properties of different aggregate sources on moisture resistance capabilities.
2. Study the advantages of Hydrated lime filler and Wetfix BE liquid adhesion promoter to resisting the moisture susceptibility of bituminous mixtures.
3. Study the effect of Aggregate mineralogy and adhesion promotors on the adhesion properties of Asphalt mixture on the different material scales: Asphalt mix, Asphalt mortar and bitumen-aggregate interface.

Methodology

This research has undertaken a laboratory investigation on moisture susceptibility of a bituminous mixture by using three different aggregate and one bitumen binder. Four laboratory tests employed to study the effect of moisture in the loose mix and compacted mixtures. Therefore, the bitumen-aggregate interaction was studies in different scales:

- Adhesion between binder and aggregate surface by Pull-off tensile strength tests and rolling-bottle tests,
- Cohesion of asphalt mortar by shaking abrasion tests including the bitumen, and fine aggregates in compacted specimens,
- Cohesion of the compacted asphalt mixture by indirect tensile strength tests were used.

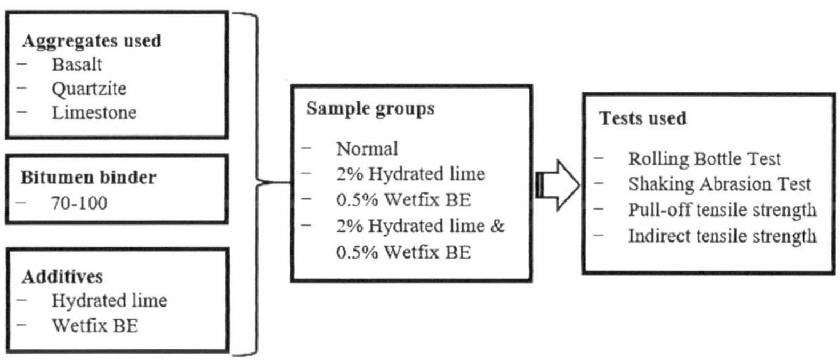

Figure 1. Materials and Tests used in the test

Sampling of Material
Bitumen

The 70-100 penetration graded bitumen binder used in all tests to study the effects of moisture in bituminous mixtures.

Aggregate

The aggregate properties such as texture, angularity and air void are influencing the adhesion properties of the mixture. The effect of aggregate discussed in terms of physical and chemical effects. The chemical bonding theory explains the possible adhesions properties of aggregate and bitumen binder through the chemical reaction between reactive minerals of aggregate and carboxylic acid components in asphalt binder.

Minerals in the aggregate crystallized by the geological process and having different mineral compositions. However, some rocks such as limestone and quartzite are dominated by one type of mineral. Aggregates from different types and sources show different adhesion performances on bituminous mixtures and their chemical composition affects the physical and chemical properties of the asphalt mixtures. (Cho and Kim 2010; Chaturabong and Bahia 2016; Guo et al. 2020)

The effect of water in bituminous mixtures described with an interaction of water, bitumen, aggregate, and the interfaces between the three. The aggregates are influenced more by the presence of water comparing with the bitumen binder. This is because aggregates composed of different minerals and these minerals have distinct relations with water. (Zhang et al. 2015)

The water entrance within interface of bitumen and aggregate leading to a chemical reaction between the material components. The interface between the aggregate and bitumen binder is vulnerable to moisture damage due to polarity difference of the bitumen and the aggregate, and results in a complex physical and chemical reaction between the two components. Literature shows that aggregate has a more dominant factor than bitumen for the effect of moisture susceptibility. (Bagampadde et al. 2006; Guo et al. 2020)

Literature shows the specific elements in the aggregate influence the adhesion properties of bitumen and aggregate. Minerals in the aggregate composed of an alkali metal such as sodium and potassium are highly moisture susceptible. Alkali metals in the aggregate form water-soluble salts on the interface of bitumen binder and aggregate. Similarly, aggregates which have a large siliceous component has poor adherence. The acidic aggregates such as quartzite are siliceous aggregates that contain SiO2 which is hydrolytically unstable in water. Silica releases H+ ions and reduces pH which results in weaker bonds between bitumen and siliceous aggregates. The compounds include iron, magnesium, and calcium have good resistance to moisture. The aggregates surface rich in calcium improving the stripping resistance by reacting with bitumen and forming hydrophobic salts. (Lyne et al. 2013; Bagampadde et al. 2006 Zhang et al. 2015; Asif et al. 2018)

Hence it is very important to study the effect of aggregate chemical composition in the mixtures using three different aggregate sources with different mineral composition Limestone, Quartzite and Basalt aggregate selected to study the effects of moisture in bituminous mixtures.

Dark-colored Basalt rocks are igneous rock commonly forms from a volcanic lava flow that has a low silica content and is relatively rich in minerals formed from iron and magnesium. Whereas Quartzite is a geologically metamorphic rock made from quartz sandstone, it is highly composed of silicate minerals. Similarly, Limestone is a sedimentary rock that has high calcium contained carbonate minerals.

Igneous rock basalt has a higher density (3.038 g/cm^3) compared to the limestone (2.711 g/cm^3) and quartzite (2.648 g/cm^3) aggregate. The limestone contains high calcium and other alkaline metals that help to create a strong adhesion bond with bitumen binder in comparison with basalt and quartzite. Quartzite is containing a high amount of siliceous material and has a poor bond with bitumen binder with the presence of water due to its high polarity and hydrophilic nature.

Adhesion promoters

Adhesion promoters work by modifying the interface properties of asphalt binder and aggregate by improving the chemical affinity and chemical-physical interaction of these two materials. Passive and active adhesions promoters (PAP / AAP) are commonly used to improve the adhesions between the aggregate and bitumen binder. The PAPs improving the moisture resistance of the mix by preventing the water from entering the interface of the two materials. Whereas the AAPs promote the bonding between asphalt binder and aggregate even with the presence of water between the interface. The AAPs have two parts the positive charge head and hydrocarbon tails. The positive parts are in contact tightly with the aggregate and the hydrocarbon tails are well compatible with asphalt binder. Amines, poly-amines, and amido-amine adhesion promoters are some of the most common types of AAPs (Guo et al. 2020)

The APA Wetfix BE is well known surfactant adhesion promoter used to increase the performance of bituminous mixture to resist the effect of moisture and reducing the antistripping effect. Wetfix BE is manufacturing by *Akzo Nobel Surface Chemistry* and it is heat resistant adhesive used in Hot Mix Asphalt mixtures. It contains both hydrophobic (non-polar) or hydrophilic (polar) groups in one molecule. The application of Wetfix BE in bitumen binder can influence by changing the polarity of the bitumen surface, which helps as abridge to create a good attachment with the aggregate having poor adhesions with the bitumen binder. The Wetfix BE also helps to create a chemical bond between the bitumen binder and an aggregate. 0.5% of Wetfix BE by weight of bitumen binder used to study its effectiveness in moisture resistance.

The mineralogy of fillers used in bituminous mixtures also has a vital role to resist moisture damage. Moreover, the increased surface area of the filler has advantages to absorb the bitumen binder successfully. Mineral fillers affect the bonding between asphalt and aggregates and improve the performance of the mix by increasing the cohesive performance of the mix under moisture exposure. Fillers are also improving resistance to deformation at high temperatures by stiffening the binder. (Chaturabong and Bahia 2016)

The active filler hydrated lime is one of the best-known additives mainly used to decrease the moisture susceptibility of bituminous mixtures. Some of the advantages of hydrated lime are reducing the moisture sensitivity of the mix, minimizes low-temperature toughness, and increases rut resistance. Hydrated lime applied as a filler during mixing with a bitumen binder, on the wet aggregate, or mix it with a bitumen binder and then with aggregate. Earlier research shows the advantages of hydrated lime to resist the moisture in terms of mechanical properties of the mixtures without showing the physical and chemical effects of hydrated lime in the aggregate and bitumen. (Dallas N. Little et al. 2006; Zaidi et al. 2019)

The researches show that the application of 1 - 3 % of hydrated lime improves the stiffness (rutting deformation) and moisture resistance of bituminous mixtures. In this research 2% of Hydrated lime by weight of aggregate is used to study the benefits of moisture resistance in bituminous mixtures. The chemicals in the bitumen binder interact with hydrated lime and with the aggregate minerals to form the strong adhesion. In this research four sample groups, the normal bitumen, modified 2 % of hydrated lime, bitumen modified with 0.5% Wetfix BE and a combination of 0.5% Wetfix BE and 2 % of hydrated lime are formed to study the effect of adhesion promoters on moisture sensitivity.

Test procedures

Different test procedures are using to investigate the moisture damage in bituminous mixtures. However, limitations are predictable to properly quantifying the results or to explains the moisture damage process. Hence, selecting a proper method of laboratory adhesion testing is crucial to predict the long-term performance of bituminous mixtures under moisture exposure. Each test has drawbacks due to its complexity, sample conditioning, uneven test results, and/or difficulties to separate interrelated factors influencing the test results. (Grönniger et al. 2010) In this research, laboratory tests were selected based on their relevance to see the cohesive and adhesive properties of the mixtures as discussed below.

Rolling Bottle tests

The Rolling Bottle Test according to DIN EN 12697-11 was applied to assess the adhesion behavior between aggregate and bitumen by visually estimating the percent bitumen coated after rolling in a bottle with water.

An aggregate passing 11.2 mm sieve and retained on 8 mm (8/11) is prepared for the test. Then 510 g from the dried aggregate is weighed and heated to the mixing temperature in the heating chamber for three hours. Similarly, 70 - 100 bitumen used and heated to the mixing temperature of 170°C in sealed metal cups for 3hrs.

2% hydrated lime by mass of aggregate was applied on the moist aggregate prior to drying and kept for 6 h before putting it in the oven. The duration of keeping at room temperature is important because to facilitate the chemical reaction between the lime and aggregate surface in the presence of water. After oven drying, the hydrated lime resulted in a total cover of the aggregate surface, compare figure 2.

Likewise, 0.5% Wetfix BE by the weight of bitumen binder mixed at 150 °C using a mechanical mixer with a duration of 10 min at a rotating speed of 50 rpm. Then the bitumen is kept in the oven for a minimum of 3hrs before mixed with the aggregate.

Figure 2. Quartzite and limestone aggregates without and with application of hydrated lime

The Rolling Bottle Test results highly depending on the coated thickness of the bitumen binder on the aggregate surface. With the same volume of aggregates, the higher density aggregates have higher mass comparing with low-density aggregate. Therefore, to keep the thickness of bitumen constant between different aggregates, the amount of bitumen binder mixed with the aggregate determined using an equation of $\alpha \cdot (16.0 \pm 0.2)$ g where $\alpha = 2.65/\rho_m$, where ρ_m is the aggregate density. Three aggregates with different density used in this test and based on their densities different amounts of bitumen binder used.

The aggregate and bitumen mixed in the mixing pan with a spatula up to the entire surface of aggregates have covered with bitumen. Then the mixed sample spread on baking paper and stored for 24 h at (20 ± 5) °C protected from sun rays and dust. After 24 h on the air, the mixed sample divided into three equally sized (150 ± 2) g and fill in the bottles which are half-filled with demineralized water at a temperature of (5 ± 2) °C. For each test sample category, three bottles are filled. Then the bottles with coated aggregate sample and filled water placed on the rolling device as shown in figure 3. The bitumen used for this test is 70-100, so the bottle rotation was ste to 60 rpm. The water temperature initially is at (5 ± 2) °C and changed to the room temperature during the experiment. After 24 h of rotation, the samples are changed from bottles to glass trays for visual inspection as shown in figure 3. The remaining degree of bitumen coating on the rock surface is estimated visually by two inexperienced individuals. The test result is the average of the two reported proportion of binder coated surfaces.

a) Rotated bottles b) Samples for visual assessment

Figure 3. Rotated bottles and ready samples for visual assessment

Shaking abrasion test

To assess the effect of water to the cohesion properties of an asphalt mortar and therefore the adhesive interaction of bitumen and fine aggregates, the shaking abrasion test according to TP Gestein-StB Teil 6.6.3. was applied. The test is adopted from EN 12774-7, originally applied on slurry surfacing and determines the water sensitivity of asphalt mixes made from 0/2fine aggregates with the gradation as shown in table 1. The test measures the abrasion of standardized specimens produced when cylindrical specimens of compacted asphalt mix placed in water-filled shaking cylinders, which in turn rotated about an axis in a suitable device with overhead movement. The cylindrical test specimens are also tested for swelling and water absorption. In each test series, three cylindrical test specimens with a diameter of 30 mm prepared and tested.

Table 1. Mixture gradation for shaking abrasion test

Aggregate Size, mm	Weigted Aggregate [g]
0,71 -2,0	48 ± 0,2
0,25 - 0,71	84 ± 0,2
0,125 - 0,25	48 ±02
0 - 0,125	60 ± 0,2

The quantity of binder mixed with the fine aggregate is determined based on the density of the aggregate 0/2using

$$B = \frac{15.5 * 2700}{\rho_g} [g]$$

B is the quantity of binder, rounded to 0.1g and ρ_g is bulk density of the aggregate 0/2 mm

Note: For a bulk density of aggregate 0/2 of 2.700 g/cm^3 the weighed-in quantity is 15.5 g (corresponding to about 6.5 wt.%).

The fine aggregate sample is mixed with the binder at a temperature of (150±5) °C A mass of 40±0.3 g of the fine asphalt mix is filled to the cylindrical mould and compacted by applying a compaction force of 10 kN. After cooling to room temperature, the specimens are extracted from the molds and stored 24 h and their dry mass is determined. For the assessment of water absorption, swelling properties and the moisture-effect to the cohesion, the specimens are stored at subsequent conditions and durations according to table 2.

Table 2. Water storing conditions for assessment of absorption, swelling properties and the moisture-effect to the cohesion

storing step	temperature (°C)	pressure	duration	weighing
1	20 ± 2	normal	90 min	under water and surface dried
2	20 ± 5	6.7 kPa	30 min	-
3	20 ± 2	normal	90 min	under water and surface dried
4	20 ± 2	normal	72 h	under water and surface dried

From the weighing, the absorption volume and swelling properties are assessed.

Directly after the moisture conditioning, each specimen is transferred to a shaking cylinder previously filled with (750±5) ml of fresh portable water. The shaking cylinders are rotated over head with a speed of (20±2) revolutions per minute at room temperature until a (3600±10) revolutions are reached. Finally, the specimens are removed from the cylinders and surfaced-dried with a damp leather cloth and their mass is determined for the assessment of the shaking abrasion value.

Pull off strength test

Pull-off test helps to investigate the adhesion potential between the aggregate and bitumen binder. It measures the required limit force to cause adhesion failure of bitumen on the aggregate surface (Guo et al. 2020; Zaidi et al. 2019). Further, the type of failure, whether it is adhesive or cohesive, can be determined. Many factors including the film thickness, nature of bitumen and aggregate, and the temperature affect the cohesion of a bitumen film and adhesion of a bitumen-aggregate interface(Zhang et al. 2016; Bagampadde et al. 2006; Asif et al. 2018; Bagampadde et al. 2006).

The asphalt film thickness within the mixture is one of the important factors that affect the bituminous mixtures resistance to moisture damage. It has also been reported that the bond strength is directly related to film thickness. Samples with thicker bitumen film tend to have cohesive failure and with thinner bitumen film have an adhesive failure. In addition to the thickness, the testing temperature has a significant role for the type of failure whether to be cohesive or adhesive. (Zhang et al. 2016; Bagampadde et al. 2006; Chaturabong and Bahia 2016). However, the failures depend on different factors and the authors who worked on this research believe it is difficult to judge the failures by categorizing them into adhesion and cohesion. Some of the most important factors that change the result of failure are the testing temperature and viscosity of the bitumen binder. As the testing temperature increases, because of the viscoelastic properties of the bitumen binder, the failure becomes more cohesive. However, by selecting a single temperature result between sample groups compared each other.

In this research new design of dollies prepared as shown in figure 4b was applied. The dollies were designed to consider the asphalt thickness to be 0.5 mm and the rough surface of aggregate. It has circular hills and valleys narrowed from the outer edge towards the center to represent the rough surface of aggregate in the bituminous mixtures. It has a diameter of 20 mm with a maximum depth of the stub 0.5 mm and the hills touched the surface of stone directly when it attached with the stone plate.

The PosiTest AT-A pull-off adhesion tester used in this test and it automatically measures the force required to pull a specified test diameter of coating away from its substrate using hydraulic pressure. All tests run at a rate of 0.7 MPa/s.

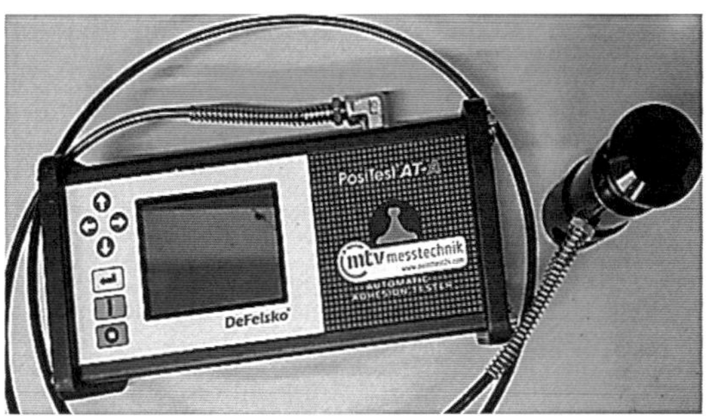

b) Pull off adhestion tester

b) Dollies used to attach

Figure 4. Posi Test AT-A pull-off adhesion tester & modified dollies

Large size rock blocks (side length ~ 30 cm) were sampled from the aggregate production site. From the blocks, stone plates were cut in the laboratory by using a diamond-edged saw cutter. Ultrasonic cleaning was applied to remove contaminants from the plate surfaces. Therefore, distilled water was added to the ultrasonic cleaner at a temperature of 60°C and the stones kept in the cleaner for 60 min. The ultrasonic cleaning helps to produce high-frequency pressure (sound) waves to agitate a liquid. The agitation produces high forces on contaminants adhering to stone plates and it helps to clean the stone surface. After removing the plates from the ultrasonic cleaner, their surface was cleaned with acetone to remove remaining impurities from the aggregate surface and dried in the oven at a temperature of 65°C at least for 1 h. Similarly, dollies attached with the stone are also cleaned with acetone and put at the same temperature into the oven.

At one prepared aggregate plate, twelve dollies could be fixed for one test. Four sample groups the normal bitumen, bitumen mixed with hydrated lime, bitumen mixed with 0.5% Wetfix BE and combined mix of hydrated lime and 0.5% Wetfix BE used to study the change in the strength of dollies detachment from the stone plate before and after water conditioning. Considering 2% hydrated lime and 6% of bitumen binder by mass in the bituminous mixtures the hydrated lime and bitumen binder mixed with a proportion of 1:3. For each test, three dollies were attached with each prepared binder on one aggregate plate. .

The bitumen binder used to attach the dollies with the aggregate keep in the oven at a temperature of 150°C. The bitumen sample poured onto the surface of the dollies and placed on the stone plate by applied a vertical force manually for 10 s. The excessive binder flows out through the dolly diameter was trimmed using a smaller spatula.

After the dollies were attached, the stone plates samples were kept at room temperature for 24 h. Afterward, the plates were conditioned in three variations: dry conditioned, water conditioned for 24 h and 48 h at 40 °C. After conditioning, the stone plates were stored dry for 2 h at a temperature of 15 °C before conducting the and the pull-off tests. The conditioning was organized in a way, that the pull-off tests were applied 4 days after the gluing of the dollies for all conditioning variations.

Indirect tensile strength tests

The cohesion of the asphalt mix specimens were assessed by water sensitivity tests according to DIN EN12697-12. The water sensitivity measured by the ITSR (Indirect Tensile Strength Ratio) of the indirect tensile strength of samples stored in a water bath and dry samples.

For the tests, a dense-graded asphalt mix with a nominal aggregate size of 16 mm was applied, usually designed for heavy-trafficked asphalt binder courses, referred as asphalt concrete AC 16. The applied aggregate mix grading is in the figure 5. The aggregate gradation included 5% limestone filler for basalt and limestone aggregates and 6.63% for quartzite aggregate.

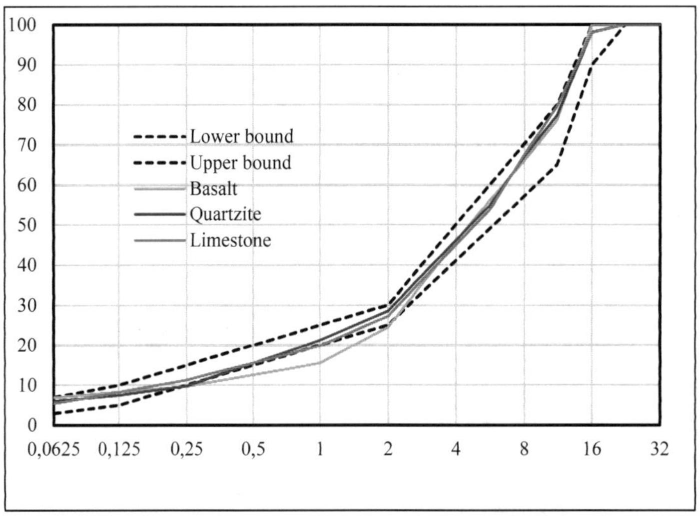

Figure 5. Mix gradation for Indirect tensile strength

For each aggregate source, four sample binder groups (normal bitumen, 2% hydrated lime additive used as a filler, bitumen with 0.5% Wetfix BE, and combined effect of 2% hydrated lime and 0.5% Wetfix BE) created. For each sample groups, six cylindrical test specimens prepared for testing water sensitivity. Hence a total of 72 samples prepared 24 from each aggregate source.

The amount of bitumen binder used varying according to the aggregate density. The higher density of aggregate refers to small surface areas comparing with aggregates which have the same volume but low density. Hence the bitumen estimated for each aggregate based on density of aggregate. The average density of basalt is 3.03gm/cm3. Similarly, the limestone and quartzite aggregates have 2.72 and 2.66 gm/cm^3. Considering the bitumen amount which has equal coat thickness between the aggregates, 5%, 5.58%, and 5.69% are used for each of Basalt, Limestone and Quartzite aggregate.

Specimens prepared by the Marshall compacting procedure with 35 hits on each side of the sample. The samples must have a diameter of (100 ± 3) mm and a thickness of between 35 to 75 mm. Specimens with the same sample groups prepared on the same day, and the heights and densities of the specimens determined. After checking the mean heights and mean densities of the specimens do not exceed 5 mm and 0.030 g/cm3, the samples divided into two groups with each group have three specimens. After production, the specimens stored at room temperature for 24 hours. Then specimens for determination of indirect tensile strength in the dry state stored at a temperature of (20 ± 5) °C on a flat surface. Whereas conditioned specimens store in a vacuum filled with distilled water at a temperature of (20 ± 5). Applying the vacuum with a residual pressure of (67 ± 3) KPa within (10 ± 1) min and holding the vacuum for (30 ± 5) min. Then samples transferred into a water bath at a temperature of (40 ± 1) °C and store a period of 68 to 72 hours.

Result and discussion

Rolling Bottle Test

The effect of water to the adhesion between binder an aggregate was assessed by using the Rolling Bottle Test for four test samples: Normal (without any additives), Hydrated lime, Wetfix BE, and combined effect of Hydrated lime and Wetfix BE. The remaining surface area covered with bitumen after 24 h of sample conditioning in rotating bottles is shown in figure 6. In figure 7, the actual aggregate surfaces after rolling-bottle tests are shown. The actual visible difference in covered surface may be interfered by the photography and is usually affected by aggregate color and light reflections.

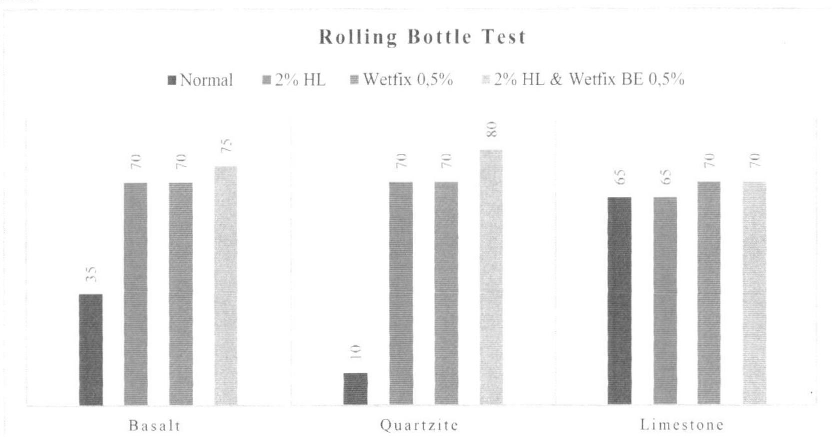

Figure 6. Rolling Bottle Test result

For the neat bitumen, the loading within the rotating bottles result in a low proportion of covered aggregate surface for the basalt (35 %) and quartzite (10 %) sample, indicating low adhesion properties. Wefix BE liquid additive and hydrated lime have significant benefit in improving adhesion of bitumen with the basalt and quartzite aggregate. With addition of adhesion promoters, the binder coverage proportions exceed 70 % for all same and result in values which were similarly obtained for the limestone aggregate. Limestone aggregates have a high amount of calcium minerals and have better adhesion with bitumen binder under normal conditions.

Figure 7. Aggregate surfaces after Rolling Bottle Test for basalt (top), quartzite (middle) and limestone (bottom).

Shaking abrasion Test

The shaking abrasion test was applied for the assessment of the water effects to cohesion of asphalt mortar. The results of the test procedure are water absorption, volume swelling and the shaking abrasion.

In figure 8, the water absorption is shown. Here especially the individual grading within the fine aggregates results in varied void content of the resulting specimens, where the basalt reaches highest and limestone lowest absorption value. Application of adhesions promoters of the Wetfix BE liquid additive and hydrated lime filler shows insignificant changes to the absorption properties of fine fractions.

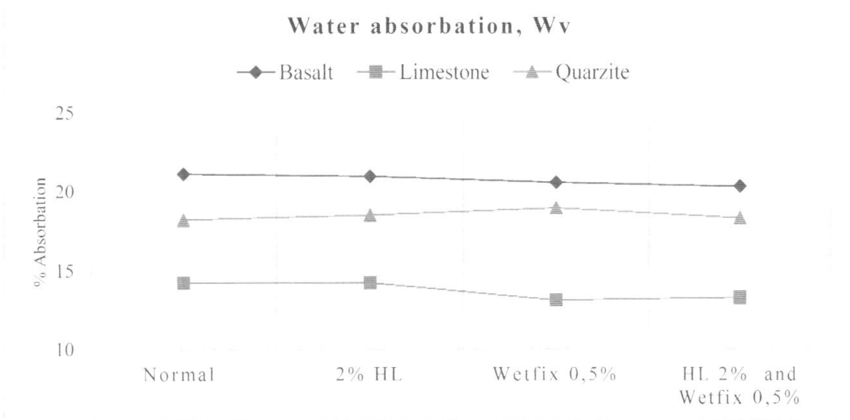

Figure 8 Water absorption of prepared samples

The swelling effect is explained by an increase in the volume of the sample because of water after submergence in water during vacuum and prolonged storing time. An increase in the volume of the bituminous mixture increases the void in the mix and reduces its durability. As shown in the Figure 9, basalt and limestone have higher swelling index under normal conditions, but application of 2% of hydrated lime in basalt and limestone aggregate benefits to reduces the swelling effect by 46% and 68% respectively. On the other hand, the addition of Wetfix BE doesn't affect the swelling values

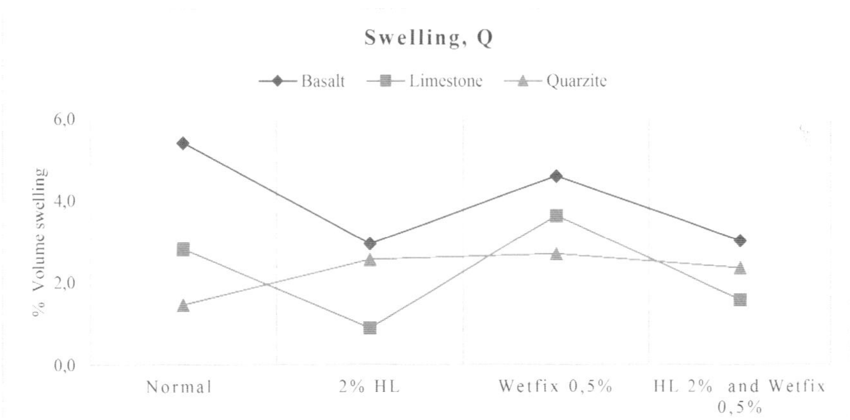

Figure 9: Swelling properties of prepared samples

The shaking abrasion value is the mean test result parameter indicating the effect of water to the asphalt mortar cohesion and/or the adhesion between binder and fine aggregates. The initial samples used have a diameter of 30mm and a height of 25mm. After 3600 rotations, the combined mechanical and physical impact of the water flow within the tubes will result in abrasion of the specimens which is identified by the mass loss (in %). Hence, the smaller value of shaking abrasion results shows more resistance against moisture susceptibility. Figure 10 shows one specimen for each tested sample variation after the abrasion test. The measured mass loses are plotted in figure 11.

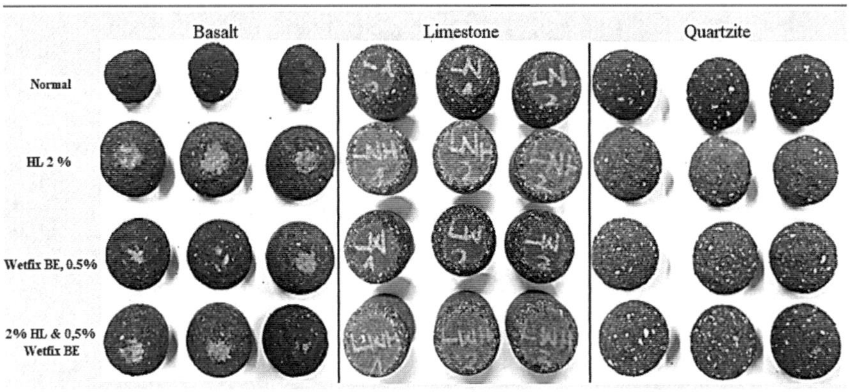

Figure 10: Sample photos after shaking abrasion test

The hydrated lime shows significant advantages on Basalt and Limestone aggregates to lower the shaking abrasion value of the mixtures and its implication is that it has a significant advantage to improve the performance and durability of the mixture under water exposure. The positive effects hydrated lime and Wetfix BE is especially observed for the basalt aggregate.

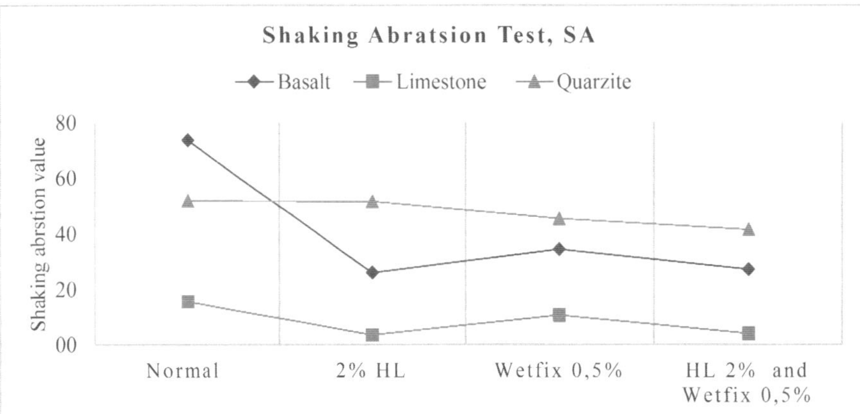

Figure 11. Shaking abrasion value

The swelling effect is directly related to the shaking abrasion values. As the swelling decreases with presence of Hydrated lime and Wetfix BE adhesion the samples perform better, and the shaking abrasion value becomes smaller.

Table 3: Two-way ANOVA of shaking abrasion result; α=0.05

Source of Variation	SS	df	MS	F	P-value	F crit
Sample groups	2828,57	3	942,86	431,74	5,25E-21	3,01
Aggregate type	10426,68	2	5213,34	2387,20	2,45E-28	3,40
Interaction	2284,18	6	380,70	174,32	1,41E-18	2,51
Within	52,41	24	2,18			
Total	15591,85	35				

The statistical analysis using two-way ANOVA conducted with hypothesis of H_{01}: All sample groups have equal mean shaking abrasion value, H_{02}: All aggregate sources have same mean shaking abrasion value and, H_{03}: there is no interaction between the sample groups and aggregate types. The results in table 3 show that for all three hypothesis F > F crit. Hence, application of hydrated lime and Wetfix BE have significant improvements and the aggregate type (mineralogy) also significantly affect the shaking abrasion value.

Pull off test

By pull-off tests, the adhesion between aggregate and bitumen is assessed with a controlled surface area. Figure 12 shows the pull-off strength for the basalt aggregates. The Wetfix BE has the advantage to increase the pull-off strength of dry samples and samples under moisture exposure.

It was expected that samples that exposed to water decrease their strength comparing to the dry condition. However, in the third and fourth categories samples which exposed for 48hrs in water and samples exposed for 24hrs in the fourth sample group show an increase in detachment strength comparing to the dry conditions, and that is not expected in the real world. This might be a limitation of the pull-off strength test.

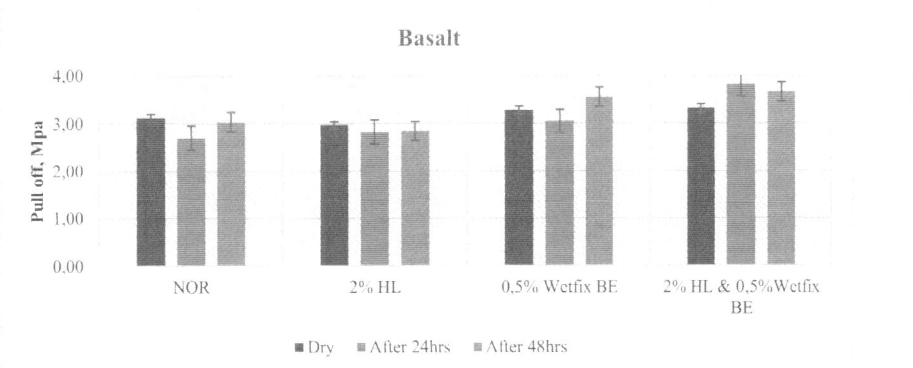

Figure 12: Pull off test strength for basalt stone

The test results obtained for the quartzite stone plate are plotted in figure 13. Application of hydrated lime increases detaching strength of the dry and moisture conditioned samples compared to a normal bitumen binder. For the samples with Wetfix BE additive, an obvious reduction of pull-off strength is measured.

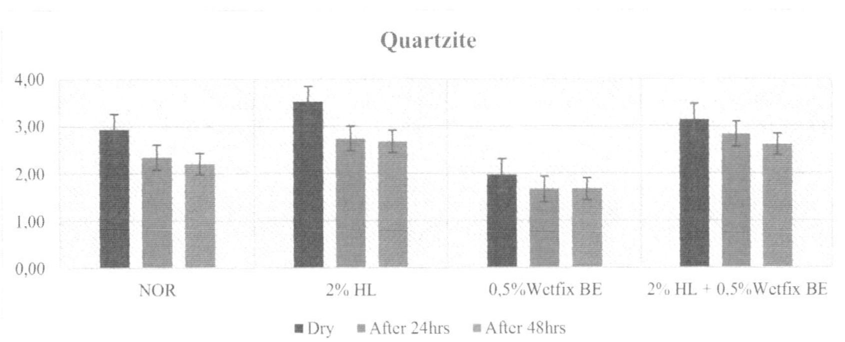

Figure 13. Pull off test strength for quartzite stone

As shown in the photos of detached dollies in Figure 14, for the samples with Wetfix BE, the detachment happens not between the bitumen and the aggregate surface but between binder and dolly. This was not observed for the basalt and limestone aggregates. Instead of the dollies surface, the aim of the pull-off test is to measure the detachment strength between the stone plate and bitumen binder attached to the dollies

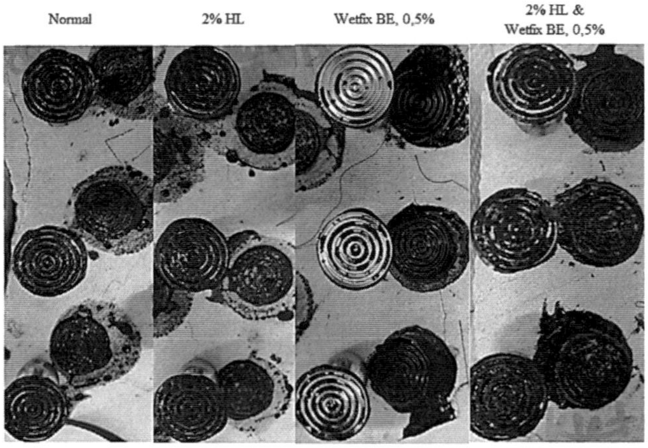

Figure 14: Detaching of dollies from quartzite after 48hrs in water bath

For the limestone aggregate, the water conditioning results in a decrease of pull-off strength when neat bitumen is used. By adding hydrated lime, the adhesion loss due to water is not observed any more. For the samples with Wetfix BE, the reduced adhesion between binder and metal of the dollies results in decreased strength readings, see figure 15. The results can't be used for the assessment of adhesion parameters between binder and aggregate.

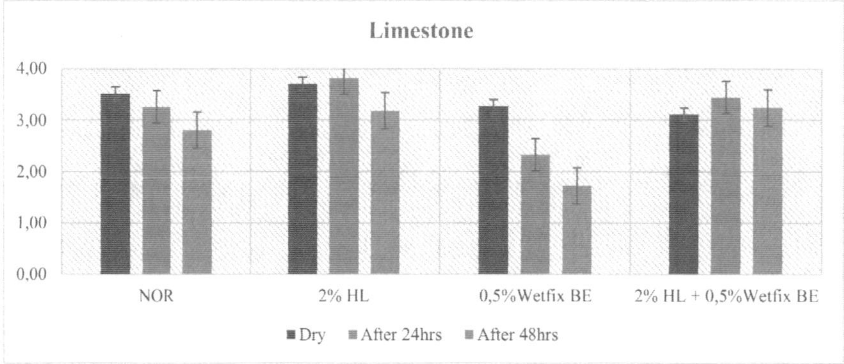

Figure 15. Pull off test strength for quartzite stone

Even though it is not expected that the moisture-conditioned samples have less pull-off strength than the dry, it is possible to show that the hydrated lime has a benefit to improve the adhesion strength between the stone and dollies.

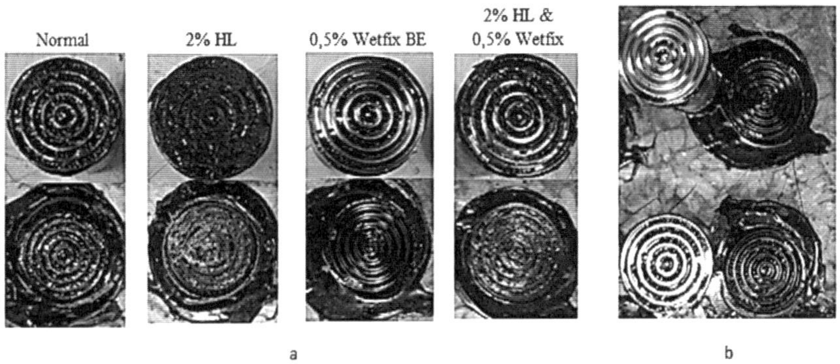

Figure 16: a) Detaching dollies from limestone after 48 hrs water conditioning b) Detaching from Wetfix BE

Limestone has better adhesion strength comparing with basalt and quartzite stones. In dry, 24hrs and 48rs moisture conditioned samples, the limestone at normal bitumen and with an application of hydrated lime shows higher strength comparing with the quartzite and basalt aggregate.

Two-way ANOVA Statistical analysis for basalt, quartzite and limestone plates was conducted whether hydrated lime and Wetfix BE adhesion promoters, and the moisture conditioning has a significant effect on the test values. The analysis conducted with $\alpha = 5\%$ and the null hypothesis were:

- H_{01}: the normal sample and samples with hydrated lime and wetfix BE have equal mean pull off tensile strength,
- H_{02}: The dry and water conditioned samples have equal mean stress, and
- H_{03}: there is no interaction between the sample groups and water conditioning.

The result in table 4 shows that the application of hydrated lime and/or Wetfix BE (indicated as sample group) have a value of $F > F_{crit}$ or p-value < 0.05 for all three aggregate types. Therefore, the null hypothesis is rejected, and it can be concluded, that the application of hydrated lime and Wetfix BE significantly affect the mean values of the pull-off adhesion test.

Similarly, for the moisture conditioning results in $F > F_{crit}$ or p-value < 0.05 for the Quartzite and Limestone plates but not for the basalt plates. Hence, for Quartzite and Limestone, the moisture conditioning will reduce the pull-off strength, whereas for the basalt aggregate there is no significant effect to be observed

Table 4: Two-way ANOVA for pull-off test; α=0.05

Source of Variation	SS	df	MS	F	P-value	F crit
			Basalt			
Sample groups	3,89	3	1,3	13,01	0.000	2,87
Moisture condition	0,1	2	0,05	0,5	0,6125	3,26
Interaction	0,93	6	0,15	1,55	0,1895	2,36
Within	3,59	36	0,1			
Total	8,51	47				
			Quartzite			
Sample groups	8,12	3	2,71	41,36	0.0000	3,01
Moisture condition	2,48	2	1,24	18,96	0.0000	3,4
Interaction	0,34	6	0,06	0,87	0,5313	2,51
Within	1,57	24	0,07			
Total	12,51	35				
			Limestone			
Sample groups	6,13	3	2,04	25,33	0.0000	3,01
Moisture condition	2,77	2	1,39	17,16	0.0000	3,4
Interaction	2,45	6	0,41	5,05	0,0018	2,51
Within	1,94	24	0,08			
Total	13,29	35				

The result shows that an application of adhesion promoters has a significant difference between the strength of the pull-off test. Similarly, the dry and moisture conditioning samples have statically significant differences. There is also an interaction between the moisture conditioning of samples and adhesion promoters.

Indirect tensile strength

The indirect tensile strength tests were conducted at 15°C. The remaining indirect tensile strength for the tested combinations of aggregates and binders is plotted in figure 17. For basalt and limestone, the water condition results in a small strength reduction. However, for the quartzite aggregates, the low ITSR value obtained for the neat bitumen can be increased by adding of 2% hydrated lime and/or 0.5% Wetfix BE.

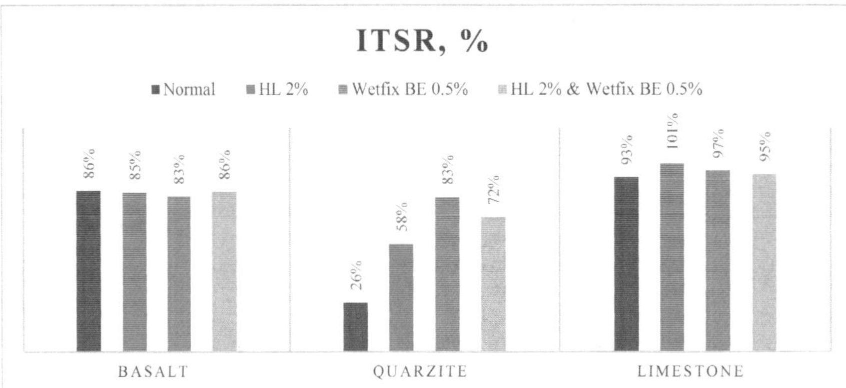

Figure 17: indirect tensile strength ratio between sample groups of each aggregates

To check the significance of sample groups and water conditioning, A statistical analysis with $\alpha = 0.05$ is conducting using two-way ANOVA test. The result summarized as shown in table 5.

Table 5: Two-way ANOVA for ITS test

Source of Variation	F	P-value	F crit
		Basalt	
Sample Groups	41,13	0,0000	3,24
Moisture conditioning	149,76	0,0000	4,49
Interaction	0,01	0,9984	3,24
		Quartzite	
Sample Groups	34,28	0,0000	3,24
Moisture conditioning	351,56	0,0000	4,49
Interaction	32,35	0,0000	3,24
		Limestone	
Sample Groups	3,77	0,0319	3,24
Moisture conditioning	3,37	0,0851	4,49
Interaction	0,78	0,5198	3,24

The checked null-hypotheses (H_{01}: All sample groups have equal mean stress, H_{02}: The dry and water conditioned samples have equal mean stress, and H_{03}: The sample groups and water conditioning are independent or there is no interaction effect) result in following conclusions:

H_{01} is rejected for all the aggregates, indicating that the application of hydrated lime and Wetfix BE have a significant effect. Similarly, the moisture conditioning values for Basalt and Quartzite show that F > Fcrit & P-value > 0.05 and the dry and moisture conditioned samples have significant mean differences. However, limestone have a contrary result that shows F < Fcrit & P-value < 0.05 that indicates there is no significant mean differences between the dry and moisture conditioned samples and conclude that limestone have better adhesion with bitumen binder comparing with basalt and quartzite aggregates.

Findings and conclusions

Three aggregate sources were selected to study the effects of aggregate mineral composition on the performance of bituminous mixtures. Rolling Bottle Test, Shaking Abrasion Test, pull off strength test and Indirect tensile strength tests were used to investigate the adhesion and cohesion performance of mixtures produces from each aggregate. Hydrated lime and Wetfix BE adhesions promoters were used to study their effect to moisture susceptibility of bituminous mixtures. In addition to the normal bitumen binder three sample groups were created which are the hydrated lime and Wetfix BE. Based on the result found from this research, the following conclusions can be drawn.

1. For neat bitumen the highly silicious quartzite aggregate shows poor performance comparing with basalt and limestone within the tests on coarse aggregate (rolling bottle tests), asphalt mixture (ITSR) and the pure adhesion between binder and aggregate (pull-off test). This is because the high proportion of hydrophilic silicious minerals in quartzite aggregates results in low adhesion performance in the presence of moisture. Comparing with the other two aggregates, Limestone show better initial performance in all tests. This is because the high proportion of calcium mineral in limestone aggregate gives an advantage of resistance against moisture. Hence, we conclude that the mineral composition of aggregates affects the moisture resistance of bituminous mixtures.
2. For the results of shaking abrasion test, this negative effect of the quartzite aggregate is not shown. A reason for this could be positive effects from the individual aggregate composition in the fines fraction which results in a low void content (and water absorption) and a low swelling.
3. The application of Hydrated lime and Wetfix BE show an advantage on improving the resistance of moisture damage in all aggregates. The application of hydrated lime and Wetfix BE improves the performance of basalt aggregate in Rolling Bottle test, shaking abrasion test and pull off tensile strength. However, it shows no improvement in the ITSR of basalt aggregate. In quartzite aggregate the application of hydrated lime and Wetfix BE shows an improving in all tests. However, there is a problem in pull off tensile strength test with the application of Wetfix BE because of failing from the surface of dollies. For limestone aggregates the application of adhesions promoters hydrated lime and Wetfix BE shows a better performance in shaking abrasion and pull off tensile strength tests. However, there is no significant changes in the Rolling Bottle test and ITSR values. Hence, it can be concluded that the application of hydrated lime and Wetfix BE show benefits to increase sample performance of the dry and moisture susceptible conditions. However, the benefit ranging between the aggregate mineralogy.

Reference

Asif, Syeda Aamara; Ahmed, Naveed; Hayat, Aneeqa; Hussan, Sabahat; Shabbir, Faisal; Mehmood, Khalid (2018): Study of adhesion characteristics of different bitumen–aggregate combinations using bitumen bond strength test. In *Journal of the Chinese Institute of Engineers* 41 (5), pp. 430–440. DOI: 10.1080/02533839.2018.1490205.

Bagampadde, U.; Isacsson, U.; Kiggundu, B. M. (2006): Impact of bitumen and aggregate composition on stripping in bituminous mixtures. In *Mater Struct 39*, pp. 303–315. Available online at https://doi.org/10.1007/s11527-005-9040-5.

Chaturabong, Preeda; Bahia, Hussain U. (2016): Effect of moisture on the cohesion of asphalt mastics and bonding with surface of aggregates. In *Road Materials and Pavement Design* 19 (3), pp. 741–753. DOI: 10.1080/14680629.2016.1267659.

Cho, Dong-Woo; Kim, Kyoungchul (2010): The mechanisms of moisture damage in asphalt pavement by applying chemistry aspects. In *KSCE J Civ Eng* 14 (3), pp. 333–341. DOI: 10.1007/s12205-010-0333-z.

Dallas N. Little; Jon A. Epps; Peter E. Sebaaly (2006): The Benefits of Hydrated Lime in Hot Mix Asphalt.

Grönniger, Jens; Wistuba, Michael P.; Renken, Peter (2010): Adhesion in Bitumen-Aggregate-Systems. In *Road Materials and Pavement Design* 11 (4), pp. 881–898. DOI: 10.1080/14680629.2010.9690311.

Guo, Fucheng; Pei, Jianzhong; Zhang, Jiupeng; Xue, Bin; Sun, Guoqing; Li, Rui (2020): Study on the adhesion property between asphalt binder and aggregate: A state-of-the-art review. In *Construction and Building Materials* 256, p. 119474. DOI: 10.1016/j.conbuildmat.2020.119474.

Lyne, Åsa Laurell; Redelius, Per; Collin, Måns; Birgisson, Björn (2013): Characterization of stripping properties of stone material in asphalt. In *Mater Struct* 46 (1-2), pp. 47–61. DOI: 10.1617/s11527-012-9882-6.

Rahim, Abdur; Thom, Nick; Airey, Gordon (2019): Development of compression pull-off test (CPOT) to assess bond strength of bitumen. In *Construction and Building Materials* 207, pp. 412–421. DOI: 10.1016/j.conbuildmat.2019.02.093.

Zaidi, Syed Bilal Ahmed; Airey, Gordon D.; Grenfell, James; Alfaqawi, Rami M.; Ahmed, Imtiaz; Ahmad, Naveed; Haynes, Mike (2019): Moisture susceptibility of hydrated lime modified mastics using adhesion test methods and surface free energy techniques. In *International Journal of Pavement Engineering*, pp. 1–13. DOI: 10.1080/10298436.2019.1648811.

Zhang, Jizhe; Airey, Gordon D.; Grenfell, James R. A. (2016): Experimental evaluation of cohesive and adhesive bond strength and fracture energy of bitumen-aggregate systems. In *Mater Struct* 49 (7), pp. 2653–2667. DOI: 10.1617/s11527-015-0674-7.

Zhang, Jizhe; Apeagyei, Alex K.; Airey, Gordon D.; Grenfell, James R.A. (2015): Influence of aggregate mineralogical composition on water resistance of aggregate–bitumen adhesion. In *International Journal of Adhesion and Adhesives* 62, pp. 45–54. DOI: 10.1016/j.ijadhadh.2015.06.012.

EVALUATION OF THE RESIDUAL LOAD-BEARING CAPACITY OF THE EXISTING ROAD USING PLATE LOADING TEST

Endijs VIRSIS[1]* (ORCID iD 0000-0001-6565-7239), Ainars PAEGLITIS[2] (ORCID iD 0000-0002-1916-7682), Atis ZARINS[3] (ORCID iD 0000-0003-1731-2206)

[1-3]*Department of Roads and Bridges, Faculty of Civil Engineering, Riga Technical University, Riga, Latvia*
*Corresponding author. E-mail: endijs.virsis@rtu.lv

Received day month year; accepted date (use style Received dates, or Alt + Ctrl + R)

Abstract. In the last few years as the road construction budget has been decreasing in Latvia, the number of road construction reinforcement design and construction objects has been increasing. At the beginning of the project development of the existing road condition is assessed, taking into account various pavement evaluation criteria and it is determined on which road sections it is possible to reinforce the pavement and where full construction is required. The road pavement structure in Latvia is developed using "Recommendations for road design. Pavement" and inaccurately defining the bearing capacity of the existing foundation can significantly affect the service life of the designed structure. During the construction of the road, establishing that the bearing capacity of the existing foundation is lower than specified in the project incurs additional costs for the customer. Project changes are made, and special solutions are provided in order to achieve the bearing capacity on the mineral material layers defined in the project. One of the most accurate ways to determine the bearing capacity of existing road structural layers is the static plate test. However, the results of this test are also not 100% accurate and any of them may give unreasonable results due to various influencing factors. The aim of this work is to analyze the results of static plate test by determining the most important factors that affect the obtained load-bearing capacity values, identify biased / erroneous test results, and determine which results reflect the residual load-bearing capacity of the existing road structure.

Keywords: Road bearing capacity; Roads; Road construction; Soil reinforcement; Static plate load test;

Introduction

In Latvia significant funding for the road sector came from the European Union structural funds, but from 2018 it is decreased from 124 million euros to 63.2 million euros in 2019. It means that in the coming years, road construction and renovation will become increasingly dependent on the state budget. Europe has defined the goals for the next period, which do not actually include road construction and similar projects, however, the European Commission points out it is possible that Latvia could obtain a small amount of funding for roads. However, we cannot rely only on the European Union structural funds, and we need to think of another way to increase funding for Latvian roads. To keep the existing national road network in its current state by providing daily maintenance and road construction repairs around 600 million funding for the road sector every year would be needed (Latvian State Roads, December 2019).

According to statistics, 28% of main roads in Latvia are still in poor and very poor condition. Comparing to year 2012, proportion of regional roads in poor and very poor condition has decreased only by 7%, in 2020 a reduction of 50% was planned. Comparing year 2018 to 2012, proportion of regional roads with black pavement has increased only by 5.2%, but 80% were planned. In order to significantly improve the overall condition of the Latvian road network, it is necessary to increase funding for road reconstruction or to look for innovative and cheaper methods to reduce construction costs (Latvian State Roads, December 2019).

In recent years because of decreasing road sector funding, the number of pavement reinforcement projects has increased significantly. By not envisaging a full pavement construction, it is possible to significantly reduce the project costs, thus it is possible to optimize the available funds for road reconstruction. Not realizing a full road construction, it is possible to significantly reduce the project costs, thus it is possible to optimize the available funds for road reconstruction. In order to perform a high-quality road construction reinforcement project, it is very important to analyze the load-bearing capacity and structural condition of the existing road structure and foundations with the help of geotechnical or another research. Incorrect assessment of the existing foundation structure properties and residual load-bearing capacity can create significant risks of structural deformations in the new road structure. As a result, the road may lose load-bearing capacity and incur unforeseen costs.

The static plate loading test is used to assess the deformation properties of the soil, the load-bearing capacity of the existing foundation, as well as the compaction of structural layers. This test can be performed on all types of dispersed (loose) soils, embankments and rocky soils, but is not normally used on very soft and fine-grained soils. If in the road reconstruction or reinforcement project are specified requirements for the load-bearing capacity or compaction of the prepared soil layers, then the static plate test is considered to be one of the most accurate methods to estimate load-bearing capacity and degree of compaction. Due to the accuracy of the test results and the fast data processing, it is very often used to estimate the residual load-bearing capacity of the existing road base. However, we must take into account the fact that the static slab test must be carried out on a perfectly flat and unobstructed surface, either on the ground surface or at a certain depth at the bottom of the trench, thus significantly determining whether the results obtained are adequate and true.

1. Objectives

Inaccurately determining the residual load-bearing capacity of the foundations of the existing road structure can significantly affect the service life of the reconstructed road, increase construction costs by eliminating unforeseen problems, thus reducing the overall quality of the road network.

The aim of this work is to analyze the results of static plate test by determining the most important factors that affect the obtained load-bearing capacity values, identify biased / erroneous test results and determine which results reflect the residual load-bearing capacity of the existing road structure. As part of this work, the data of geotechnical research and construction work quality (load-bearing capacity indicators and compaction of the prepared layers) of the motorway A10 Riga-Ventspils 13.30 - 19.20 km section were analyzed.

2. Plate load test (PLT)

Plate load test is a field test which is commonly adopted to determine the bearing capacity and settlement of soil under a given condition of loading as well as the quality of the compaction works performed on shallow foundations. This test can be performed on all types of dispersed (loose) soils, embankments, and crushed stone layers, but is generally not used in very soft and fine-grained soils. Testing of shallow foundation with a plate must be performed on a flat and smooth surface - either on the surface of the ground or at the foundation of the excavation at a certain depth. If the construction design specifies requirements for the bearing capacity or compaction of the prepared soil layer (ledge), the static plate method is one of the most accurate methods for determining this load-bearing capacity and compaction degree. It should be noted that with a standard stamp of Ø 300 mm the soil is tested at a depth of about 60 cm (x2 stamp diameter). Therefore, in cases where a compaction test of a thicker embankment layer is required, the tests should be carried out gradually when the corresponding embankment layer is heaped up and compacted (Ģeo eksperts, 5 March 2021).

Plate load test is performed based on German standard DIN 18134 (Testing procedures and testing equipment – Plate-loading test). The loading of the stamp must be carried out with a gradual increase in pressure. The pressure recommended by DIN 18134 for the standard stamp of Ø 300 mm, which must be achieved by loading, is 0.5 MN/m2 (or 0.5 MPa). The load shall be increased in not less than six stages or steps approximately of the same degree until the aforementioned maximum pressure is reached. Each change of load (from step to step) must be completed within one minute. The loading test consists of three stages – the first-time loading when a pressure of 0.5 MN/m2 is reached with at least 6 loading steps; unload (3 pressure reduction stages (50%, 25% and ~ 2% of maximum load)); followed by a second time loading, during which the load must be increased to the pre-final load level of the first cycle (so that a full load of 0.5 MPa would not be achieved).

After performing the test at the site, the data obtained are processed and the results are evaluated in the office. Based on the recommendations given in DIN 18134, calculations can be done manually based on the calculation formulas provided by this standard but can also be done using special computer software designed for this purpose. As a result of the static plate-loading test, the following main parameters are obtained: EV1 (after first-time loading results) and EV2 (after second time loading results) deformation modules as well as the ratio of these modules EV2/EV1. In the construction practice, the value EV2 and the ratio EV2/EV1 are mainly used. The value EV2 shows the bearing capacity and deformation properties of the tested soil, but the ratio EV2/EV1 indicates the degree of compaction of the particular soil type (Ģeo eksperts, 5 March 2021).

3. Geotechnical research

During road reinforcement and reconstruction projects, geotechnical research of the existing soil and road surface is always performed. Depending on the road category and the traffic intensity of the vehicles, the design task defines the minimum requirements for geotechnical research, which includes soil drilling, various soil in situ tests, static loading plate, laboratory tests of soil samples and other studies.

Justification of geotechnical research - to provide the customer with the necessary information to ensure the development of a full-fledged road pavement reinforcement (reconstruction) project. To provide the information necessary for road engineers about the existing roadway pavement structure, road embankment and natural subsoils including the structural layer of the road surface the composition, properties and distribution of the soil layers, both in section and in direction. The task of geotechnical research is to obtain information on the structure of the road surface, the properties of the road pavement structural layers and the characteristic values of the geotechnical parameters of the natural subsoil by installing exploration wells and removing soil samples, as well as performing in-situ tests and laboratory tests. (Ceļuprojekts, 2018).

Geotechnical research on the state main highway A10 Riga-Ventspils 13.30 - 19.20 km section was performed by the Geology Department of AS "Ceļuprojekts" (see Figure 1). During the geotechnical research, a total amount of 170 survey points was made on the road section: 34 boreholes next to the road slope foot (borehole depth 1.0… 3.0 m) and 136 boreholes on the roadway part, as well as on the roadside (borehole depth average 1.0 m). The total volume of drilling works was 228.6 m. In order to determine the physical and mechanical properties of the road structural layer and natural soils, dynamic probing was performed at a depth of 3.0 m at 15 research points. The total volume of dynamic probing was 45.0 m. In order to determine the deformation and strength properties of the existing base layer of the road, static plate loading test were performed at 34 points in the carriageway part. During the drilling works, 68 samples of pavement asphalt concrete were taken from the exploration boreholes to determine the binder content, granulometric composition, penetration and softening temperature; 34 samples of structural layers of the road with disturbed structure for determination of granulometric composition; and 25 samples of lying natural soils with disturbed structure for soil type determination. (Ceļuprojekts, 2018).

Figure 1. Geotechnical research on the state main highway A10 Riga-Ventspils 13.30 - 19.20 km

The base of unbound mineral materials in the exploration site of the existing road base is mostly variable in composition and thickness, it consists of dolomite crushed stone with gravel or gravel with pebbles. But the base sub-layer consists of gravel with pebbles or gravel with a mixture of pebbles and dolomite chips. The total thickness of the road base layer varies from 0.22 to 0.55 m, average 0.35 m. The deformation modulus/ bearing capacity (Ev2) of the existing base layer in the research phase was significant and indicates a high degree of compaction of the base layer. The modulus of deformation Ev2 varies from 161 to 374 MPa. In road kilometer 15,8km (SKT-51 / PLT-15) Ev2 was relatively lower and reached 121 MPa. The parameter ratio Ev2 / Ev1 characterizing the degree of compaction varies from 1.07 to 5.00. (Ceļuprojekts, 2018).

The frost-resistant layer was found during the field work at all research points, just below the base layer. The frost-resistant layer consists of fine-grained, dense to medium-dense sand. The thickness of the layer varies from 0.20 to 0.60 m, the average thickness is 0.30 m. In some places, the frost-resistant sandy mixture contains pebbles or crushed stone. The content of clay particles (particle size <0.063 mm) varies in the range from 2.0 to 5.6%. The organic matter content in the round is <2%.

Using the results of the geotechnical research of the static plate test, a summary of the basic load-bearing capacities of the existing structure has been developed (see Figure 2).

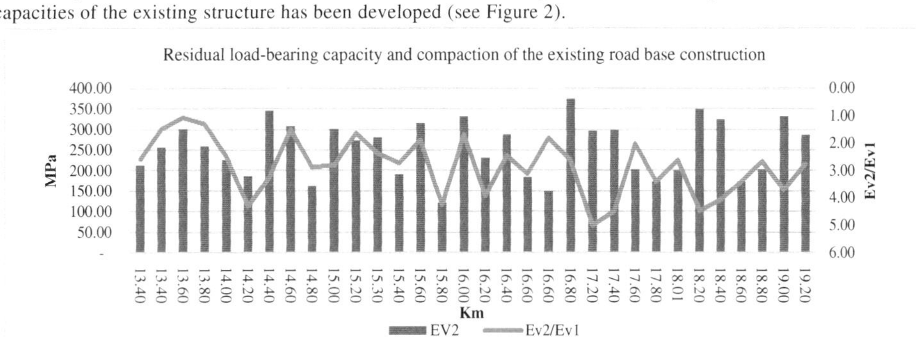

Figure 2. Residual load-bearing capacity and compaction of the existing road base construction

The Figure 2. shows the load-bearing capacity and the compaction of the existing base. Obtained result analyze proved, that it is possible to observe the correlation between compaction and load-bearing capacity. As the compaction of the base layer is increasing, its bearing capacity also increases. In the obtained results there are some sections in which this correlation is not observed, for example, in the section from 16.8 -17.4km, 18.2-18.4km and 19.0-19.2km. To make sure, why these sections, have such a high load-bearing capacity, considering low degree of compaction, the relationship between the load-bearing capacity and the existing crushed stone / gravel base thickness must be compared (see Figure 3).

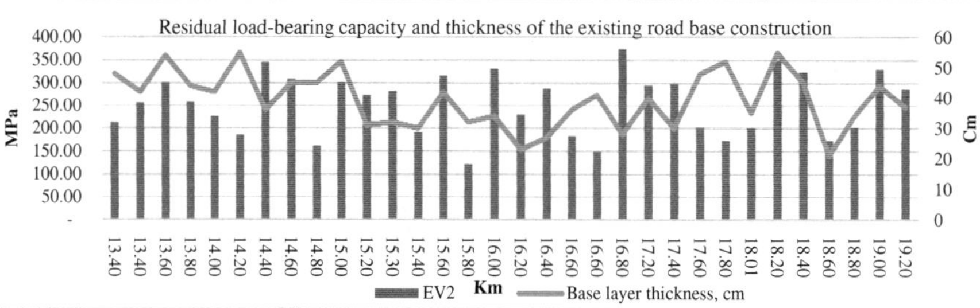

Figure 3. Residual load-bearing capacity and thickness of the existing road base construction

The Figure 3. shows the load-bearing capacity of the existing foundation depending on the thickness of the base layer. There is no unambiguous correlation in the obtained results and it is possible to conclude that the compaction of the structure has a much more significant effect on the load-bearing capacity of the existing foundation. In the previously determined sections from 16.8 -17.4 km, 18.2-18.4 km and 19.0-19.2 km, it can be observed that the thickness of the basic structure varies in the range from 30 to 55 cm. In the previously determined sections from 16.8 -17.4 km, 18.2-18.4 km and 19.0-19.2 km, it can be observed that the thickness of the basic structure varies in the range from 30 to 55 cm. In the section from 18.2-18.4 km and 19.0-19.2 km the thickness of the existing base is larger than in the adjacent sections and it is possible that a high load-bearing capacity of the existing base structure was achieved here despite the low degree of layer compaction. The high load-bearing capacity of the section 16.8-17.4 km with the base thickness and compaction cannot be justified.

4. Road reconstruction

According to the data of geotechnical research, during the design the load-bearing capacity of the existing foundation is analyzed and the calculation of the road pavement structure is developed. There are no common assumptions about how to accurately determine the remaining load-bearing capacity of an existing foundation, so it is the engineer's responsibility and duty to carefully analyze the available information to develop a quality project. Inaccurately determining the remaining load-bearing capacity of the existing foundation may affect the course of construction, and during the construction of the real situation, it may be necessary to develop special solutions for strengthening the existing foundation or significantly change the design of the intended pavement structure.

In the project was designed the construction of a pavement reinforcement solution consisting of an existing foundation, a 20 cm thick layer of recycled material and three layers of asphalt (see Figure 4). On the existing base layer, was defined that it is necessary to achieve a load capacity of 90MPa. Such load-bearing capacity indicators were determined by performing calculations using the "Road Design Recommendations Road Pavement" methodology. Using this calculation methodology, we obtain the theoretical values of the bearing capacity of the layers. Based on the data of the realized road projects, it can be stated that there is a correlation between the theoretical bearing capacity calculation values Eekv and the static plate test values Ev2, therefore the value obtained in the calculation of the ICP methodology was determined as the remaining load-bearing capacity of the existing base.

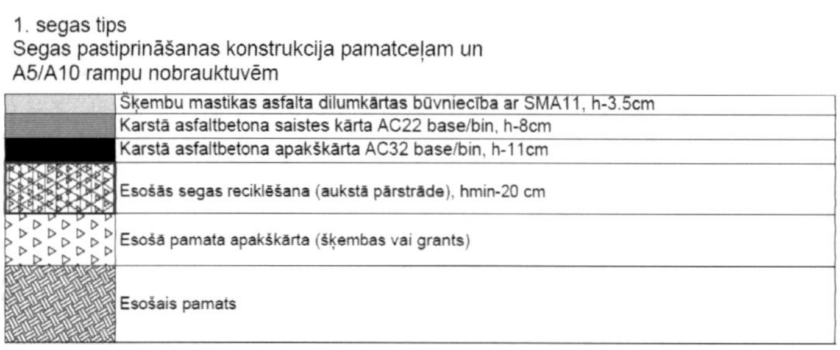

Figure 4. Designed road construction

The static plate test measurements were made on the existing road base layer during the reconstruction to make sure about the possible implementation of the developed solution and about the estimated load-bearing capacity of the existing road base. Initially, the old asphalt concrete structure was milled and the test was performed on the existing foundation under it.

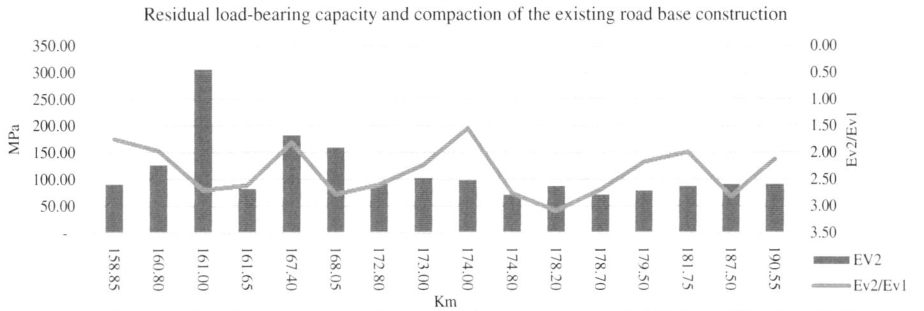

Figure 5. Residual load-bearing capacity and compaction of the existing road base construction

The Figure 5. summarizes the load-bearing capacity indicators of the existing foundation determined during construction. According to the obtained results, it can be seen that the average load-bearing capacity of the existing base is about 90MPa, therefore it can be concluded that the load-bearing capacity of the existing base has been determined accurately. Compared to the results obtained during the geotechnical survey, there is no clear correlation between the basic compaction and the changes in bearing capacity.

5. Residual load-bearing capacity analysis

During the development and implementation of each road project, a geotechnical research is carried out. With the help of geotechnical research engineers try to determine the properties of the existing soil layers and road construction parameters. During construction, the quality of the constructed layers is monitored by determining the load-bearing capacity and compaction of the layer. It is safe to say that the static plate test is performed during the design and construction of each road, so it is important to understand the interrelationships that could be used to determine the load-bearing capacity of an existing foundation. The Figure 6. compares the load-bearing capacity of the existing foundation found during geotechnical research and construction.

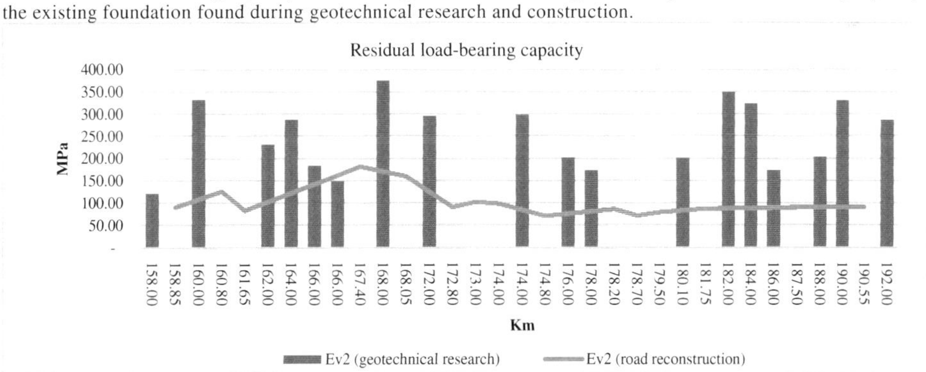

Figure 6. Residual load-bearing capacity

Comparing the load-bearing capacity determined during the project development and construction, it can be concluded that the results obtained in the geotechnical research are about 1.8 times, or by 80% higher than those determined during construction. Such a reduction in load-bearing capacity may create risks that an inexperienced engineer in the project will define the load-bearing capacity of the existing foundation based only on the results of the static plate test results obtained during the geotechnical research. The actual basic load-bearing capacity are influenced by other parameters of soil properties and, of course, there are other factors (heterogeneity of the geological situation, self-weight of asphalt concrete and geotechnical research results and others) that can significantly affect the results achieved.

One of the most important factors influencing the results could be the different static plate test loading conditions. During the geotechnical research, a small area of asphalt concrete is cut out to form a trial site pit in order to be able to get to the existing foundation structure, perform a static plate test and determine which soil layers are in the road structure and ground. During construction, the static plate test is performed when there is no other material on top of the tested layer that could affect the static plate test results. In order to ascertain how much the load-bearing capacity values of the existing base change taking into account different test conditions, a theoretical finite element calculation was performed (see Figure 7) by simulating a static plate test based on geotechnical research interpretations of soil layers and layer thicknesses. The calculation was based on the test procedure described in DIN 18134 (Testing procedures and testing equipment - Plate-loading test).

Name	Colour	Initial Element Loading	Unit Weight (kN/m3)	Peak Cohesion (kPa)	Peak Friction Angle (°)	Peak Tensile Strength (kPa)	Residual Cohesion (kPa)	Residual Friction Angle (°)	Residual Tensile Strength (kPa)	Dilation Angle (°)	Young's Modulus (kPa)
Skembas		Field Stress & Body Force	19	0	35	0	0	35	0	0	260000
SMILTS, smalka, bilva		Field Stress & Body Force	21	0	36.45	0	0	32	0	4	34690
SMILTS, smalka, vidėji bilva,		Field Stress & Body Force	19.7	0	33.85	0	0	31	0	3	25160
metals		Field Stress & Body Force	78	0	90	350000					200000000
Asfaltbetons		Field Stress & Body Force	22	400	45	0	200	40	0	0	2000000

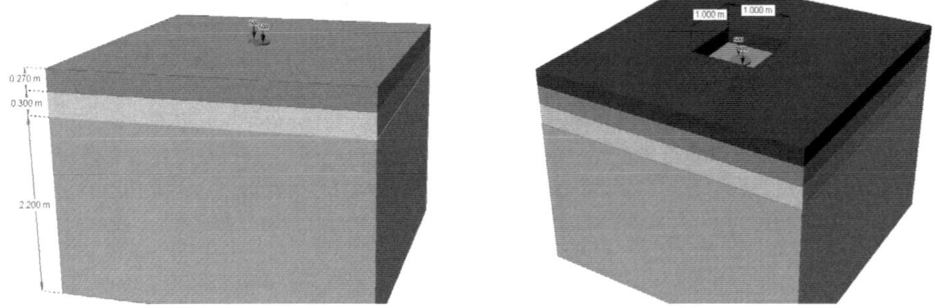

Figure 7. Finite element calculation model

The developed finite element calculation was based on the dynamic probe test DPM-62 (16.4 km). The left side of the Figure 7 shows the static plate loading scheme during construction, and the right side shows the approximate loading conditions during the geotechnical survey. In limited traffic conditions, much smaller trial site pits are often made during geotechnical research, but a 1x1m viewport is assumed for analytical calculation. The diameter of the static slab is assumed to be 0.3 m and the first load cycle is performed in 7 steps until the maximum load is reached, but the second cycle is performed in 6 steps (see Figure 8). At each loading stage, the achieved deformation is recorded and, similarly to the static plate test equipment, deformation curves have been developed.

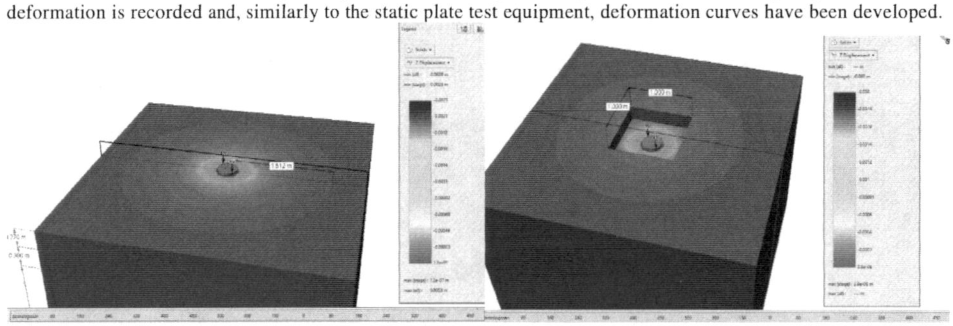

Figure 8. Finite element calculation results

According to the finite element calculation model, it was determined that the applied load spreads within a radius of approximately 1.5 m from the loading center. If the old asphalt concrete is above the existing base layer structure, the soil layer deformations during static plate test is slightly smaller (maximum deformation without asphalt layer - 2.8 mm, but with asphalt layer - 2.3 mm). It's because the base layer cannot deform sideways and up, because the asphalt layer does not allow it.

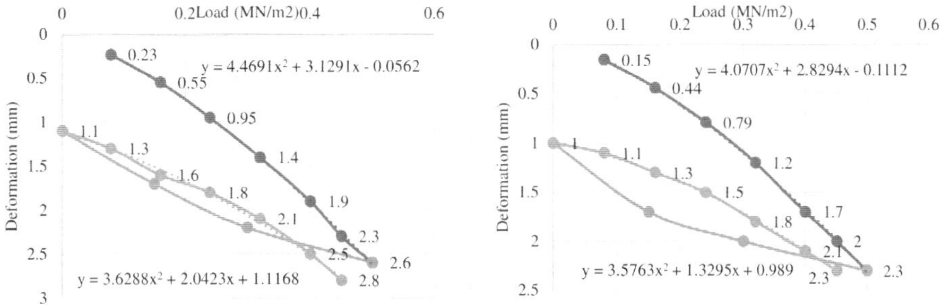

Figure 9. Finite element model load-bearing capacity calculation

Significant differences in the results were obtained in the calculated theoretical load-bearing capacity of the existing base (see Figure 9). The diagram on the left shows the load curve for a foundation without an asphalt concrete structure, but on the right considering the construction of the trial test pit. The following results were obtained:

- Construction without asphalt concrete surface (road construction) – Ev1=105,97MPa; Ev2=152,58MPa; Ev2/Ev1=1,43.
- Construction with asphalt concrete surface (geotechnical research) – Ev1=117,26MPa; Ev2=203,39MPa; Ev2/Ev1=1,73.

According to the finite element calculation models, the load-bearing capacity of the existing basic structure shows a difference of 50MPa, or 30%. In fact, by reducing the trial test pit area, this difference only increases, and to make sure of this, a third calculation model was developed, the trial test pit area was formed in a circular shape with a diameter of 34 cm. The shape and size of such a trial site pit correspond to the specific geotechnical survey point of the A10 motorway (see Figure 10).

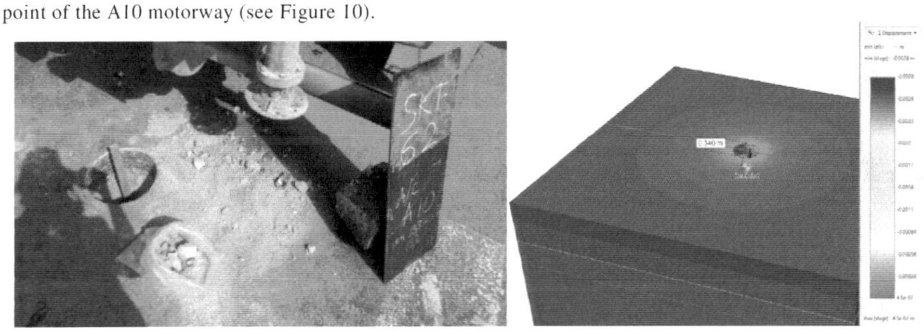

Figure 10. Geotechnical research trial test pit at section 16.4km

The developed third finite element calculation model was based on the same existing soil properties from the dynamic probe test DPM-62. By creating such a small trial test pit, the conformity of the obtained results is significantly reduced in comparison with the real situation during construction, because in addition to the load caused by the static plate test, the self-weight load of the asphalt layer also works on existing foundation structure and asphalt layer limits the deformations of the existing base layer.

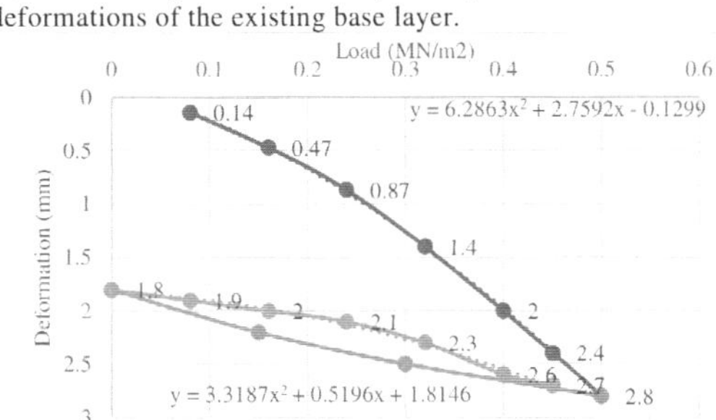

Figure 11. Finite element model load-bearing capacity calculation

The following results were obtained for circular trial test pit (see Figure 11) – Ev1=103,91MPa; Ev2=333,70MPa; Ev2/Ev1=3,21. Comparing the theoretically obtained load-bearing capacities of the existing base with the data of the static plate test results, we can see that during the geotechnical research on section 16.4 km load-bearing capacity determined Ev2=287,1 MPa, but the finite element calculation results showed Ev2 = 333.70 MPa load capacity. The difference of the obtained results is 16% and we can assume that the developed finite element calculation model provides relatively accurate results from which we can conclude the most significant correlations.

All three developed finite element calculation models are based on the same theoretical material properties, but each model has different loading conditions. After analyzing the data, we can conclude that the greatest influence on the test results of the static plate test is provided by the loading conditions, or the fact whether another layer is built above the test layer, which could affect the accuracy of the results. By reducing the trial test pit area from 1x1m to 0.34m in diameter, the value of Ev2 increased by 64%, but comparing to the structure without asphalt, the change in the value of Ev2 is 119%. These indicators give a clear idea that the geotechnical research data do not give accurate results if a sufficiently large trial test pit area is not prepared, but taking into account the fact that it is not possible to create at least 1.5x1.5 m asphalt section during the research, then the engineer must estimate the approximate reduction in load-bearing capacity depending on the size of the trial test pit created during the geotechnical research. Therefore, during the geotechnical research, it is important to perform photo fixations when creating the trial test pits and performing the static plate tests, because with the help of photos it is possible to assess the test conditions.

Conclusions

The actual basic load-bearing capacity are influenced by various parameters of soil properties and there are other factors (heterogeneity of the geological situation, impact of asphalt concrete self-weight on the results of geotechnical research and others) that can significantly affect the achieved results.

The basic bearing capacity determined during geotechnical research can be up to 3 times higher than that found during construction works, therefore it is very important to analyze not only static plate test results, but also soil sample laboratory data, field research data.

Different test conditions have the greatest impact on the load-bearing capacity of the existing base. The existing asphalt concrete or other type of structure located above the test layer prevents obtaining appropriate results, because the created trial site pit has too small area and the base layer cannot deform sideways and upwards, thus achieving higher load-bearing capacity of the existing base than they actually are.

According to the static plate test results, it is possible to observe the correlation that as the compaction of the base layer increases, its bearing capacity also increases. Therefore, in case the existing foundation is not sufficiently dense or the existing foundation layers are made of loose or soft ground the load-bearing capacity will be significantly lower and solutions should be sought during the project development to compact the loose or soft ground so that full road construction does not have to be rebuilt.

During the geotechnical research, it is important to perform photo fixations when creating the trial test pits and performing the static plate tests, because with the help of photos it is possible to assess the test conditions.

Reducing the trial test pit size increases difference between load bearing capacity results comparing the geotechnical research and road construction work conditions.

Acknowledgements

This work was supported by the Riga Technical University's Doctoral Grant programme.

References

Ceļuprojekts. (2018). Construction Project of pavement reinforcement (reconstruction) of the state main highway A10 Riga – Ventspils section km 13.3 - 19.2. Project documentation. Latvia.

DIN 18134:2012-O4 Soil -Testing procedures and testing equipment - Plate load test

Ģeo eksperts (5 March 2021) Plate load test [Article about plate load test procedure]. Retrieved from https://www.geoeksperts.lv/services/plate-load-test/

Latvian State Roads (December 2019) Road newspaper [Article about road sector funding]. Retrieved from https://lvceli.lv/aktualitates/finansejums-bus-jamekle-radosak/

BALTIC COUNTRIES MINERAL RESOURCES AND AGGREGATES USED IN TRANSPORT INFRASTRUCTURE EXTRACTION ANALYSIS

Mantas MAKULAVIČIUS 0000-0001-5666-2298[1], Henrikas SIVILEVIČIUS[2]

[1, 2]*Department of Mobile Machinery and Railway Transport, Transport Engineering, Vilnius Gediminas Technical University, Vilnius, Lithuania*
E-mail: m.makulavicius@milsa.lt; henrikas.sivilevicius@vgtu.lt
(corresponding author Mantas Makulavičius)

Received 15 March 2021; accepted XX

Abstract. Mineral resources and aggregates used in transport infrastructure or any other purposes importance is undeniable. Its usage and extraction not only benefits economics and social environment, but also should be sustainable. Regarding to this, analysis of Baltic countries mineral resources and aggregates extraction was conducted. The main purpose of the article is to analyze mineral resources situation in each of the country and provide extraction data for aggregates used in transport infrastructure. Mining and quarrying sector is evaluated in whole country economic picture and period 2008-2020 was analyzed by combining data from all the Baltic countries for aggregates extraction. The article is valuable by putting all Baltic countries in one perspective and sharing newest data. Further investigations could include analysis of factors for aggregates demand and its forecast.

Keywords: mineral resources, aggregates, extraction, Baltic countries, Lithuania, Latvia, Estonia, analysis.

Introduction

Socio-economic activities such as increasing housing, infrastructure demand and manufacturing sector has guided the growing demand for non-energy mineral resources (Auci & Vignani, 2020). There are over 50,000 mines around the world in which approximately 200 types of mineral resources are in use (Petrić, 2011) Based by European Aggregates Industry data the European aggregates sector represents 15.000 companies (mostly SMEs) operating through 26.000 sites, with over 200.000 employees. Aggregates are the second most used natural materials worldwide, only after water, and consequently are the biggest mining branch by production volume (Escavy et al., 2020). The European average aggregates demand is 6 tons per capita per year, meaning the use of 1 Kg of aggregates per hour.
Aggregates are inert material, while primary aggregates are produced from natural sources extracted from quarries and gravel pits and in some countries from sea-dredged materials (marine aggregates). Secondary aggregates are recycled aggregates derive from reprocessing materials previously used in construction, including construction, demolition residues and by-products from other industrial processes, like blast or electric furnace slags or china clay residues.
There are currently no viable alternatives to the large scale use of aggregates in construction (Brown et al., 2011). Because of this mineral resources and aggregates used in transport infrastructure and other purposes importance is undeniable.

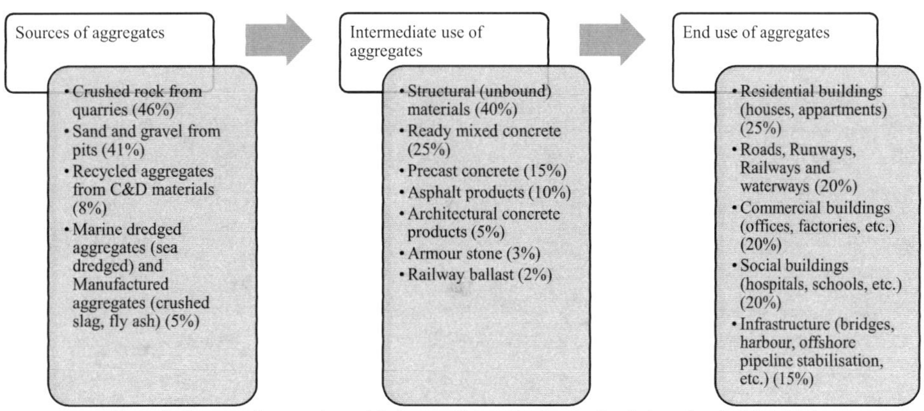

Figure 1. Sources of aggregates and its intermediate and end usage (made by author by EAA data)

Aggregates are the core building material in all our homes, offices, social buildings and infrastructures: without aggregates these would literally fall apart. The construction of a typical new home uses up to 400 tons of aggregates (both end product and concrete) - from the foundations through to the roof tiles. Aggregates are used to build schools, hospitals, museums and other social or public buildings. The construction of a school uses up to 3,000 tons of aggregates.
Figure 1 shows sources of aggregates and its intermediate or end use. 15% of produced aggregates are used for large infrastructures including bridges, harbors, offshore pipeline stabilization, and many other. Aggregates feature at all levels of the road construction up to the surface, which includes aggregates resistant to polishing, ensuring skid-resistance. The construction of 1 km of motorway uses up to 30,000 tons of aggregates. Aggregates are essential as track ballast for Europe's rail network. The construction of 1 meter of railway for a High Speed train (TGV) uses up to 9 tons of aggregates.
Usage of aggregates is a function of the state of a national economy: Together with the growth of the economy, the demand for sand, gravel and crushed stone increases, as they are essential for infrastructural development and commercial and domestic building activities (Tiess & Kriz, 2011). These sources of demand are in turn driven by macroeconomic and demographic factors like population growth and economic growth (Savoy, 1996).
The main aim of this article is to give a brief overview of Baltic countries economic situation and drivers for mineral resource and aggregates extraction while providing an outlook of current mineral resource situation in each of the country and analysis of aggregates used in transport infrastructure extraction for 2008-2020 period.

1. Outlook of the Baltic countries economic situation and the importance of mining and quarrying industry

There are a number of methods for measuring the economic contributions of an industry. Gross output, or turnover, represents the total value of sales produced by an industry within a period of time. However, economic benefit is more accurately measured in terms of 'Gross Value Added' (GVA), which is defined as gross output minus the value of goods and services used to produce that output. There is a very close link between GVA and 'Gross Domestic Product' (GDP) and the GVA of an industry can be thought of as its contribution to GDP. There is also keen public interest in the employment an industry sustains (Brown et al., 2011).

Table 1. Outlook of Baltic countries economies (source Eurostat)

	Units	Estonia	Latvia	Lithuania	Data
GDP	million euro	28,112.4	30,463.3	48,797.4	2019, current prices
Population	people	1,328,976	1,907,675	2,794,090	2020.01.01
Unemployment	Percentage of active population	4.4	6.3	6.3	2019, from 15 to 74 years
GDP per capita	Euro per capita	21,220	15,920	17,460	2019, current prices
Doing business index 2020	Ease of business rank / score	18 / 80.6	19 / 80.3	11 / 81.6	2019.05.01

While analyzing data provided by Eurostat, although Baltic countries could be named as one market, there are slight differences between each one viewed separately. Lithuania is the largest in population and GDP creation in total, however lowest unemployment rate and GDP per capita is biggest in Estonia (higher by 21.5% and 33.3% compared to Lithuania and Latvia accordingly). Since the legislative basis and ease of operation are important in any business, the same rule applies with even more attitude to environmental regulation for mineral resources extraction. The Doing business index shows that all of the Baltic countries are in Top 20, with Lithuania in the lead at 11th position.

Table 2. Road structure in the Baltics by significance (2018 data, source Eurostat)

	Estonia	Latvia	Lithuania
Total	59,008	58,443	85,249
State roads	16,608	19,990	20,915
Provincial roads	24,002	30,147	64,334
Communal roads	18,237	8,306	-

To understand the demand for aggregates both transport infrastructure and construction sectors should be analyzed. Tables 2 and 3 provide information on road structure by significance and pavement in each country.

Lithuania has the biggest network of the Baltic countries with more than 85 thousand kilometers length of roads, while Latvia and Estonia despite the territory size differences have practically the same network around 59 thousand kilometers of roads. While looking just to national significance roads, it must be stated that Lithuania improved its situation by changing road pavement type from gravel roads to asphalt pavement over the last years.

Table 3. National significance road pavement types (2019 data)

	Estonia	Latvia	Lithuania
Total	16,609	20,015	21,238
Length of roads with improved pavement (asphalt, concrete, etc.)	9,447	9,283	15,121
Length of crushed stone and gravel roads	7,162	10,732	6,117

While looking just to national significance roads, it must be stated that Lithuania improved its situation by changing road pavement type from gravel roads to asphalt pavement over the last years. That's why situation in Latvia (having more than 50% of crushed stone and gravel national significance roads) could lead to bigger demand in aggregates in the future when improving the pavement.

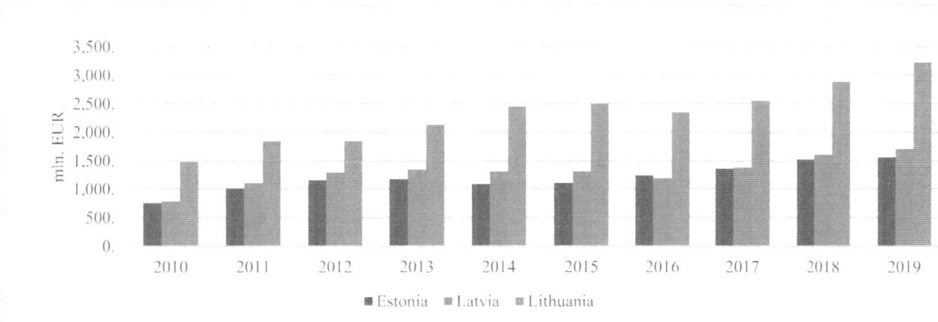

Figure 2. Gross value added in construction among Baltic countries in 2010-2019

The GDP from construction is also a safe indicator of aggregates production (Menegaki & Kaliampakos, 2010). Lithuania GVA from construction is higher by 80% than Estonia or Latvia. But we can see the same trend of construction GVA increasement from 2010 to 2019 period. For example all the countries have doubled the gross value added in construction. All of this combined leads to demand of mineral resources and aggregates extraction. Figure 3 shows gross value added in mining and quarrying industry.

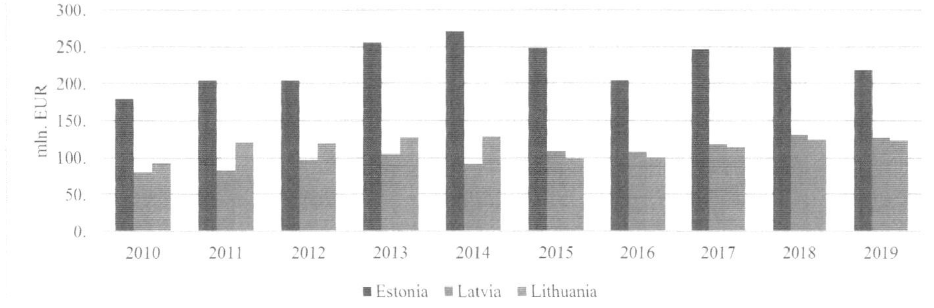

Figure 3. Gross value added in mining and quarrying among Baltic countries in 2010-2019

Despite the fact that construction is a huge part of Lithuania economy, Estonia leads all of the Baltic countries by far with almost 220 million Eur in the mining and quarrying sector, while Lithuania and Latvia 122 and 128 mln. Eur accordingly. It can be explained by countries geology and mineral resources available for extraction, which is analyzed in the sections below.

2. Mineral resources analysis in Lithuania

The legal base of Lithuania provides for that the underground and its mineral resources is an exclusive property of the state or the state has an exclusive right to the underground – this emphasizes that the objects important to national security must be owned by state. Lithuania territory is covered with natural resources, located almost evenly through whole country (fig. 3).

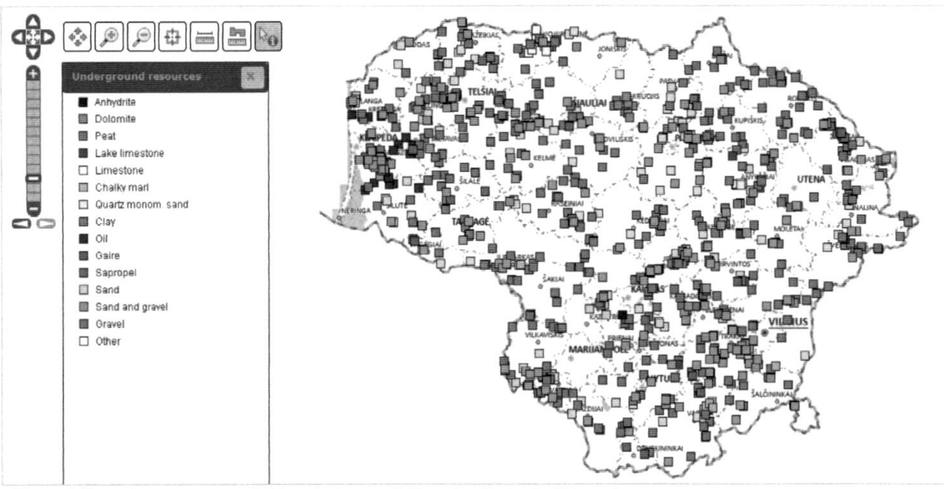

Figure 4. Natural resource map of Lithuania (source: Lithuanian Geological Survey)

There are common geographical situated resources, for example limestone is mainly located in the north of the country at the border of Latvia. Dolomite resources are also located in the north of the country, mainly in Pakruojis district. sand and gravel resources can be found all over, with bigger density of resources in the far west and south east of the country. Dolomite beds in Lithuania are detected in many geological systems, but only dolomite deposits occurring near the land surface in the northern part of Lithuania in the Upper Devonian Pliaviniai, Įstras, Stipinai, Kruoja or Žagarė formations are considered to be of a practical significance. There are some inland oil drills, but the production and resources are quite small compared to oil rich countries. There is valuable resources of Anhydrite as well, situated close to Kaunas, but it is not being extracted so far. Due to geographical placement of Lithuania, there is no opportunity to extract granite aggregates according to economical rationality (Makulavičius & Sivilevičius, 2020). So it must be kept in mind, that igneous rocks are imported to Lithuania.

Table 4. Lithuania natural resources reserves

Resources	Approved reserves	Reserves left at the end of 2020	Percentage Left
Anhydrite	80,694,000	80,694,000	100.0%
Dolomite	167,305,000	127,418,000	76.2%
Peat	1,547,569,900	1,215,554,711	78.5%
Lake limestone	585,000	585,000	100.0%
Gypsum	16,823,000	16,823,000	100.0%
Limestone	290,913,000	223,311,310	76.8%
Chalky marl	8,076,000	8,057,060	99.8%

Quartz monom. sand	6,927,000	3,625,000	52.3%
Clay	170,508,000	146,415,000	85.9%
Gaire	19,773,000	19,063,410	96.4%
Sapropel	15,302,000	15,247,000	99.6%
Sand	419,822,000	383,319,000	91.3%
Gravel	928,154,000	723,595,000	78.0%
TOTAL	3,672,451,900	2,963,707,491	80.7%

Based by Lithuanian Geological survey data, 17 kinds of mineral reserves/resources have been explored to various degree of detail in Lithuania. Nine of these (limestone, dolomite, sand, gravel, clay, chalky marl, peat, sapropel, and oil) are under exploitation, and the exploitation of gaize (opoka) was suspended in the 1990s. According to the legal acts, only the reserves explored in detail can be used. Permission for the utilization of mineral resources and cavities can be given to legal entities by the Government of the Republic of Lithuania or by the Geological Survey, depending on the kind of resources, their quantity and potential impact on the subsurface status of other states. By the end of December 2019, 316 enterprises, 3 natural persons and one group of legal persons acting under the contract of joint activity had permissions to use resources of solid minerals and 107 enterprises and one natural person had permissions to use groundwater resources (J. Čyžienė, 2020)

3. Mineral resources analysis in Latvia

The subterranean depths of Latvia are rich in the resources of various mineral materials that can be used for the production of construction materials: the resources of sand and sand-gravel, dolomite, clay, limestone, gypsum rock, peat, and sapropel are widely available (Sproge et al., 2013). Figure 5 shows the natural resources map of Latvia.

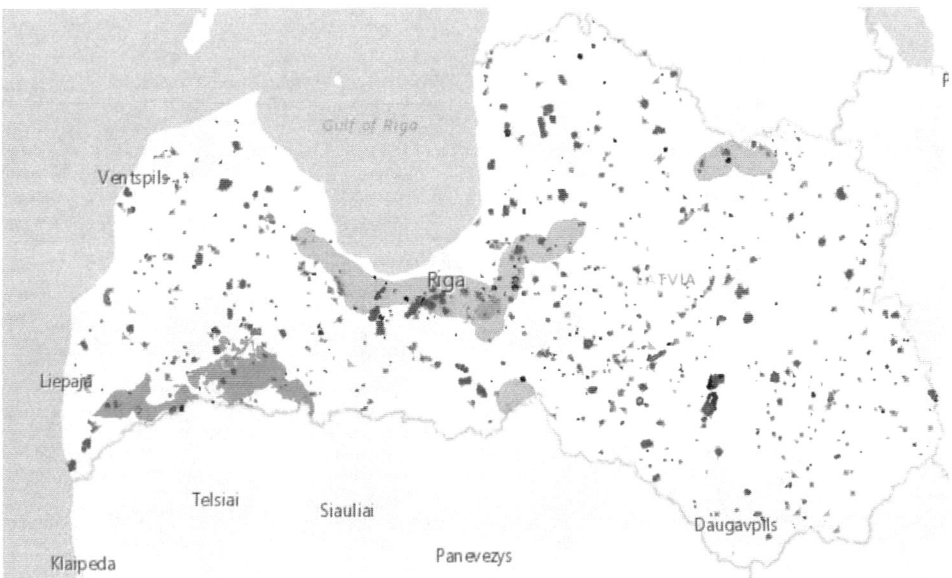

Figure 5. Natural resource map of Latvia (source: Latvian Environment, Geology and Meteorology Center)

Gypsum rock is among the most valuable land resources; Latvia provides this resource to all the Baltic States. Dolomite, in turn, is a widely spread mineral resource and presents a major source of mechanically resistant stone materials in Latvia.

IOP Conf. Series: Materials Science and Engineering 1202 (2021) 012013 doi:10.1088/1757-899X/1202/1/012013

Thirty the most important deposits of gypsum rock, limestone, dolomite, clay, quartz sand, gravel, sand, stone, and sapropel are included in the list of mineral deposits of national importance (Sproge et al., 2013). Gravel is also abundant throughout Latvia territory. However gravel can be an abundant resource at a global scale, however locally it can present a risk of supply constraint (Ioannidou et al., 2017).

Table 5. Latvia natural resources reserves

Mineral resource	Unit	Approved limits	Reserve 01.01.2020	Percentage left
aleuritis	thousand m3	33.92	30.78	90.7%
dolomite	thousand m3	51,016.09	40,983.66	80.3%
healing mud	thousand t	13.43	11.95	89.0%
gypsum	thousand m3	5,536.12	4,062.58	73.4%
limestone	thousand m3	33,726.35	25,274.82	74.9%
peat	thousand t	80,563.84	74,331.33	92.3%
quartz sand	thousand m3	3,042.51	2,873.92	94.5%
boulders	thousand m3	8.1	3.74	46.2%
clay	thousand m3	12,868.82	11,713.03	91.0%
clay sand	thousand m3	242.42	202.54	83.5%
clay sand (covered)	thousand m3	324.2	252.86	78.0%
clay sand, loam	thousand m3	178.25	177.05	99.3%
sapropel	thousand t	7.63	7.63	100.0%
loam	thousand m3	639.48	599.14	93.7%
sand	thousand m3	122,790.60	107,521.17	87.6%
sand (covered)	thousand m3	53.2	36.33	68.3%
sand-gravel	thousand m3	96,318.96	80,628.04	83.7%
sand-gravel (covered)	thousand m3	70.2	68.96	98.2%
sand-gravel and sand *	thousand m3	45	40.15	89.2%
Total		407,479.12	348,819.67	85.6%

The principal mineral found in Latvia is limestone, with an approximated reserve of 6 billion cubic meters. The reserves are sufficient enough to provide 85% of the raw material for its cement industry. The deposits are spread throughout the country and occur at fairly shallow depths. The major field is the Kumu field located in the district of Saldus. Limestone is mainly used as a raw material for cement and concrete.

Peat is one of the greatest resources of Latvia and is of immense significance in the preservation of the county's beauty. Peatland in Latvia covers approximately 6,400 square kilometers or 10% of the total land area with the major deposits located in the eastern plains and near Riga. Peat was first extracted in Latvia in the 18th century but the intensive extraction started after the establishment of the state in 1918. There are over 10,000 peat fields in Latvia capable of producing over 1.7 million tons of peat annually. Approximately 60% of the peat reserves are the high type and the rest are low types. Latvia has increasingly exported its peat resources to other European countries, especially Western Europe because the region's resource has significantly depleted.

4. Mineral resources analysis in Estonia

944 deposits (incl. oil shale, peat, crystalline rocks, gravel, sand, clay, dolostone, limestone, phosphorite, sea mud) were registered as of 4th of January 2021 in Estonia.

The kukersite oil shale is the most important mineral resource in Estonia. The mining of kukersite oil shale began in 1916. At present, of the total 10–12 million tonnes mined per year, slightly over 80% is burnt directly in thermal power plants as pulverised fossil fuel. The chemical industry uses about 15% of the mined oil shale for oil recovery, and about 3% goes to the cement industry. The Estonian deposit is the largest commercially exploited oil shale deposit in the world; its total reserves exceed 7 billion tonnes of oil shale (Raukas & Tavast, 2004).

Figure 6. Natural resource map of Estonia

In Estonia, there are more than 900 sand and gravel deposits for road and industrial building, and unlimited Cambrian, Devonian and Quaternary clay deposits, used mainly as a raw material in the ceramics and cement industries. The unlimited resources of limestone are used as a raw material in producing lime, cement and building stone, but also in glass, chemical, pulp and paper industries. Dolomite is suitable for glass and facing stones, used in the chemical industry and as road and industrial building material. There are also large resources of curative mud, mineral water and different minerals of local importance, as lake chalk, ochre, glauconite, etc. Estonia has big peat resources and nowadays peat is an important export article.

Table 6. Mining mineral deposits in Estonia in 2020, status of resources as of 31.12.2020 Data source: Estonian Land Board.

Mineral resource	Unit	Mineral resources as of 31.12.2020		
		Active		Inactive
		Proved reserves	Probable reserves	Proved+Probable
Oilshale	thousand t	937 683	302 525	3 371 721
Phosphorite	thousand t	–	–	2 935 735
Cement limestone	thousand m3	56 661	76 141	50 994
Technological limestone	thousand m3	23 913	41 024	73 610
Building limestone	thousand m3	128 503	192 037	366 374
High quality building limestone	thousand m3	3 913	–	1 073
Low quality building limestone	thousand m3	4 092	617	1 470
Filling limestone	thousand m3	4 458	–	–
Technological dolostone	thousand m3	12 374	82 582	–
Decorative dolostone	thousand m3	2 520	20 603	2 261
Building dolostone	thousand m3	61 302	97 833	80 908
High quality building dolostone	thousand m3	3 543	–	130

Low quality building dolostone	thousand m3	6070	–	–
Filling dolostone	thousand m3	11 093	–	–
Crystalline building rocks	thousand m3	1 245 062	1 723 932	–
Cement clay	thousand m3	14 902	11 213	489
Hard melting clay	thousand m3	–	–	341
Ceramic clay	thousand m3	9 978	235 893	13 028
Ceramsite clay	thousand m3	8 046	2 248	228
Building gravel	thousand m3	69 674	61 952	17 682
Filling gravel	thousand m3	3 723	–	14
Technological sand	thousand m3	6 065	3 231	2 128
Filling sand	thousand m3	113 383	5 680	8 382
Building sand	thousand m3	272 318	415 513	168 787
Sea mud	thousand t	1 374	–	1 680
Lake mud (field fertilizer)	thousand t	171	1 048	312
Lake mud (therapy)	thousand t	1 129	–	–
Lake lime	thousand t	731	5 120	4 857
Undecomposed peat	thousand t	48 969	77 373	121 177
Well decomposed peat	thousand t	156 808	667 502	517 376

Peatlands occupy 10,091 km2 or 22.3% of the Estonia territory. Within Estonia as a whole there are 165,000 mires and peatlands with an area over 1 hectare, of which 1626 are of commercial importance. At present, total peat reserves are estimated at 2.37 billion tonnes or 15.24 billion cubic metres. Economically exploitable resources constitute 1.52 billion tonnes because 0.85 billion tonnes occur under fields and meadows. 69 mires (156,562 ha by area) or 16% of mires are under conservation. The greatest thickness of peat (16.7 m) was measured in the Võllamäe bog in south-eastern Estonia. As the natural accumulation of peat in mires is slow and natural sites for its accumulation are dwindling rapidly, the Government of the Estonian Republic has adopted the regulations of sustainable use of peat which enacts annual output quotas for every county (Raukas & Tavast, 2004).

5. Aggregates used in transport infrastructure extraction analysis in the Baltic countries

It can be stated, that it is not easy to extract mineral resources not only due to technological reasons. Mineral resources accessibility in the framework of land use planning poses a particularly difficult problem (Nieć et al., 2014). The exploitation of many mineral commodities is often limited by:

- actual or planned land use, precluding mineral exploitation of mineral deposits,
- land property ownership (in case of opencast mining)
- environment protection (landscape protection, , Natura 2000, etc.),
- social opposition and the NIMBY syndrome (Not In My Back Yard),
- costs of extraction depending on natural conditions of mineral deposit occurrence and mineral quality,
- lack of appropriate state policy, incl including fiscal policy, related to mining (Galos et al., 2012).

Vast majority of demand locations for aggregates are satisfied by local producers at distances usually less than 50 km. Therefore, a critical factor for the viability of an aggregates quarry is its location, which is controlled by geology (Bleischwitz & Bahn-Walkowiak, 2007).

All of this effects aggregates extraction. To analyze Baltic countries aggregates used in transport infrastructure extraction only dolomite, limestone, gravel and sand were chosen. The data was gathered from national institutions: Lithuania Geological Survey, Estonian Land Board and Latvian Environment, Geology and Meteorology Center. The period for 2008-2020 is described and it can clearly show full economic cycle with highs before financial crisis and recent year trends. The newest data for 2020 was available for Lithuania and Estonia with slight of grasp for COVID19 effects on aggregate extraction. All the raw data is described in annex table 1. Figure 7 shows each of the aggregate extraction among the countries. It can clearly be seen, that due to mineral resources available, the aggregate differs among the countries as well. For example limestone is widely used in Estonia and Estonia leads limestone extraction (more than 2.3 mln. cubic meters) by 50 percent than Lithuania and Latvia combined (781 thousand and 723 thousand cubic meters accordingly). It also can be stated, that limestone production is quite stable throughout the last years.

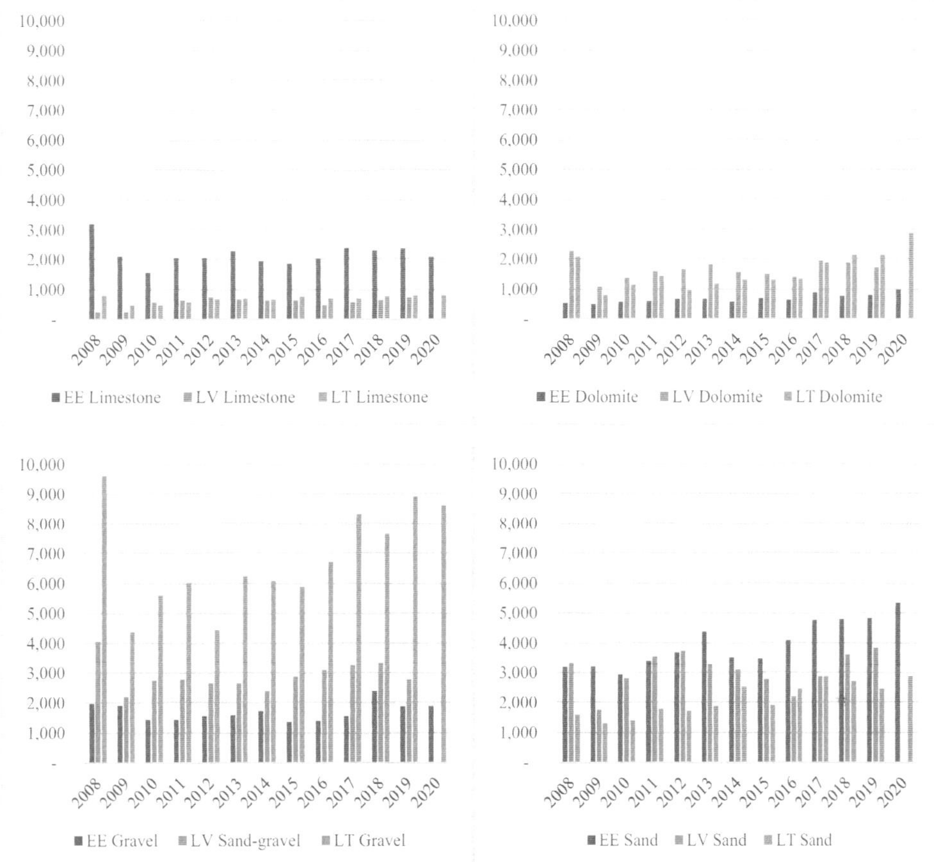

Figure 7. Aggregates extraction among Baltic countries in 2008-2020

However different situation is with dolomite natural resource. Lithuania is leading the Baltic countries with 2.88 mln. cubic meters extraction, Latvia is second with 1.7 mln. cubic meters and Estonia extracts only 983 thousand cubic meters. For the last 5 years huge increase for dolomite demand can be seen. The dolomite extraction from 2015 increase by 220%. Estonian dolomite extraction for the past 5 years is stable. Since gravel is widely used in Lithuania, the numbers are exceptional between Baltic countries. Lithuania gravel extraction in 2019 (8.9 mln. cubic meters) practically doubles Estonian (1.866 mln. cubic meters) and Latvian (2.76 mln. cubic meters) extraction combined. The trend for Latvia and Estonia since 2015 with a peak in 2018 extraction is now downwards for the past two years. Although in Lithuania gravel extraction increased 46%, but still lack behind 2008 extraction which was 9.6 mln. cubic meters. Sand is being extracted most in Estonia with 5.3 mln. cubic meters in 2020 and both Estonia and Latvia has and upward trend since 2015.

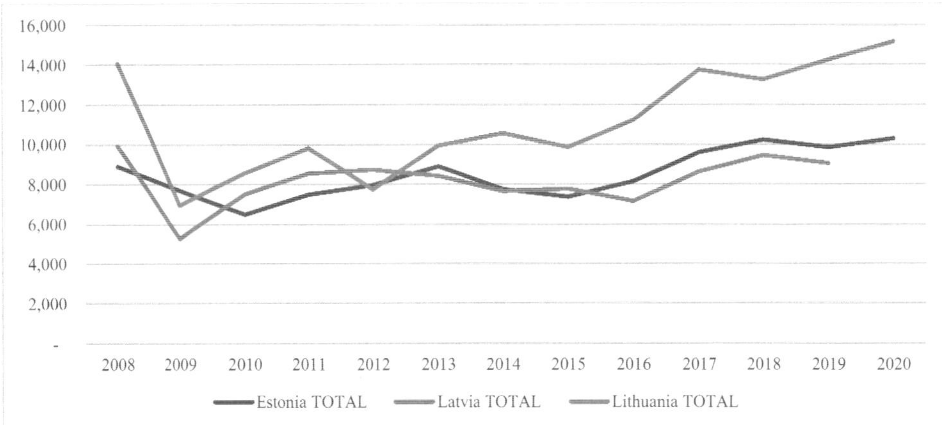

Figure 8. Total aggregates extraction among Baltic countries in 2008-2002

Figure 8 shows total extraction of aggregates (limestone, dolomite, sand and gravel) among Baltic countries for 2008-2020 period. Lithuania leads the Baltic countries with total 15,612 mln. cubic meters extraction in 2020, while Estonia extracts 10,2 mln. cubic meters and Latvia – just above 9 mln. cubic meters. It is obvious, that economic factors hit the aggregate industry in 2009 and all the countries showed decrease from 2008. After that, the industry was on the up and down track since 2010 until 2015. Since 2015 all the countries show upward trend. For example in Lithuania total aggregates extraction rose by 58%, in Latvia – by 16%, in Estonia – by 39%. Latvia still haven't reached 2008 level in 2019. There is no data for 2020 in Latvia, but in Estonia despite COVID19 effects on economy the demand for aggregates even increased compared to 2019 by 4.5%. Meanwhile in Lithuania, the decision for 2020 road sector financing was made in December of 2019, allocating 590 million EUR to the sector. 142 million of them were forbidden to use until other funds were depleted. However then COVID19 crisis strike and Lithuanian government thought of the ways how to boost local economy. Due to this several decisions have been made. Permission to use "frozen" 142 million was issued in 2020 March and it was followed by additional 150 million EUR funding which were assigned in April. These additional funds were explicitly allocated for possible projects which already have final technical design and may be implemented until November 30. That's why overall possible budget for 2020 was higher by 30% of the initial one (Makulavičius & Sivilevičius, 2021). This could have affected the aggregates demand highly and the extraction in 2020 compared to the year before rose by 6.3%.

Conclusions

Baltic countries could be named as one market, but there are slight differences between each one viewed separately. Lithuania is the largest in population and GDP creation in total, however lowest unemployment rate and GDP per capita is biggest in Estonia. Despite the fact that construction is a huge part of Lithuania economy, Estonia leads all of the Baltic countries by far with almost 220 million Eur in the mining and quarrying sector.

While analyzing each countries, it can be stated that geology among the is similar with some exceptions. 17 kinds of mineral reserves/resources have been explored to various degree of detail in Lithuania. Nine of these (limestone, dolomite, sand, gravel, clay, chalky marl, peat, sapropel, and oil) are under exploitation. The most valuable could be called limestone, dolomite, peat and oil with huge potential of anhydrite still waiting for its turn.

In Latvia peat, limestone and gypsum rock are among the most valuable land resources, thirty the most important deposits of gypsum rock, limestone, dolomite, clay, quartz sand, gravel, sand, stone, and sapropel are included in the list of mineral deposits of national importance.

944 deposits (incl. oil shale, peat, crystalline rocks, gravel, sand, clay, dolostone, limestone, phosphorite, sea mud) were registered as of 4th of January 2021 in Estonia. While being alone among the Baltic countries, Estonia extracts oil shale as one of the most important resources with peat and limestone.

While analyzing aggregates extraction for 2008-2020, it can be stated that structure of aggregates differs. For example Lithuania leads with dolomite extraction, while Estonia uses their limestone effectively and by bigger quantities than their neighbors. Lithuania gravel extraction in 2019 practically doubles Estonian and Latvian extraction combined.

Although total aggregates extraction is on upwards trend since 2015, there was a lot of uncertainty in 2020 because of COVID19 effects on economy. For example overall budget for road infrastructure in 2020 was higher by 30% of the initial one and had effect on aggregates demand. It can be stated that Estonia and Lithuania coped extremely well in this industry by even increasing the extraction by 4.5% and 6.3% accordingly.

References

Auci, S., & Vignani, D. (2020). Mines and quarries production: A driver analysis of withdrawals in Italy. Resources Policy, 67. https://doi.org/10.1016/j.resourpol.2020.101657

Bleischwitz, R., & Bahn-Walkowiak, B. (2007). Aggregates and construction markets in Europe: Towards a sectoral action plan on sustainable resource management. Minerals and Energy - Raw Materials Report, 22(3–4), 159–176. https://doi.org/10.1080/14041040701683664

Brown, T., Mcevoy, F., & Ward, J. (2011). Aggregates in England-Economic contribution and environmental cost of indigenous supply. https://doi.org/10.1016/j.resourpol.2011.07.001

Escavy, J. I., Herrero, M. J., Trigos, L., & Sanz-Pérez, E. (2020). Demographic vs economic variables in the modelling and forecasting of the demand of aggregates: The case of the Spanish market (1995–2016). Resources Policy, 65. https://doi.org/10.1016/j.resourpol.2019.101537

Galos, K., Nieć, M., Radwanek-Bąk, B, Smakowski, T., Szamałek, K., 2012. The mineral security of Poland – barriers in meeting the State's needs for non-energy minerals. Biuletyn Państwowego Instytutu Geologicznego 452, 53–58.

Ioannidou, D., Meylan, G., Sonnemann, G., & Habert, G. (2017). Is gravel becoming scarce? Evaluating the local criticality of construction aggregates. Resources, Conservation and Recycling, 126, 25–33. https://doi.org/10.1016/j.resconrec.2017.07.016

J. Čyžienė. (2020). Lietuvos geologijos tarnybos 2019 metų veiklos rezultatai: [metinė ataskaita] = Lithuanian Geological Survey: Annual Report 2019. In Lietuvos geologijos tarnyba. https://www.lgt.lt/images/Metines ataskaitos/LGT ataskaita 2019 visa.pdf

Makulavičius, M., & Sivilevičius, H. (2020). EVALUATION OF GRADATION VARIATION OF THE GRANITE AGGREGATES USED IN ASPHALT MIXTURES AT THEIR MANUFACTURING PLACE. The 11th International Conference ENVIRONMENTAL ENGINEERING 11th ICEE SELECTED PAPERS. https://doi.org/10.3846/enviro.2020.601

Makulavičius, M., & Sivilevičius, H. (2021). EXTRACTION ANALYSIS OF LITHUANIA MINERAL RESOURCES AND AGGREGATES USED IN TRANSPORT INFRASTRUCTURE COMBINING FORECAST AND EVALUATION OF FACTORS FOR ITS DEMAND

Menegaki, M. E., & Kaliampakos, D. C. (2010). European aggregates production: Drivers, correlations and trends. Resources Policy, 35(3), 235–244. https://doi.org/10.1016/j.resourpol.2010.01.003

Nieć, M., Galos, K., & Szamałek, K. (2014). Main challenges of mineral resources policy of Poland. Resources Policy, 42, 93–103. https://doi.org/10.1016/j.resourpol.2014.10.010

Petrić, J. (2011). THE USE OF MINERAL RESOURCES AND ISSUES OF HARMONISATION BETWEEN SPATIAL PLANS FOR THE MINING AREAS IN SERBIA WITH OTHER STRATEGIC DOCUMENTS. 24, 21–26. https://doi.org/10.2298/SPAT1124021P

Raukas, A., & Tavast, E. (2004). SUSTAINABLE MANAGEMENT OF MINERAL RESOURCES, SOIL COVER AND GEOSITES IN ESTONIA. Proceedings of the Conference "Geological Heritage Concept, Conservation and Protection Policy in Central Europe," 191–198.

Savoy, M., 1996. Development of a model for forecasting the consum- ption of aggregates in the Lower Mainland, Coriolis Consulting Corp. Report prepared for: The Ministry of Energy, Mines and Petroleum Resources, B.C., Canada.

Sproge, I., Curkina, I., & Jekabsone, S. (2013). MINERAL RESOURCES AND LONG TERM DEVELOPMENT IN LATVIA. Economic Science for Rural Development Conference Proceedings, 31, 205–214.

Tiess, G., & Kriz, A. (2011). Aggregates Resources Policies in Europe. Development of IT solutions for the Enhancement of Planning & Permitting Procedures. International Journal of Environmental Protection, 1(3), 69–72.

IOP Conf. Series: Materials Science and Engineering 1202 (2021) 012013 doi:10.1088/1757-899X/1202/1/012013

Annexes

Annex Table 1. Baltic countries aggregates extraction in 2008-2020 period

Country	Aggregate	2008	2009	2010	2011	2012	2013	2014	2015	2016	2017	2018	2019	2020
Estonia	Limestone	3,207	2,094	1,542	2,053	2,052	2,272	1,943	1,854	2,020	2,392	2,288	2,342	2,080
	Dolomite	526	481	571	595	672	669	569	686	632	885	767	802	983
	Gravel	1,964	1,907	1,434	1,435	1,552	1,581	1,730	1,360	1,384	1,551	2,383	1,866	1,888
	Sand	3,186	3,201	2,939	3,398	3,680	4,366	3,509	3,469	4,099	4,766	4,783	4,826	5,333
	TOTAL	8,883	7,682	6,487	7,481	7,956	8,887	7,751	7,369	8,135	9,595	10,221	9,835	10,284
Latvia	Dolomite	2,255	1,078	1,376	1,593	1,638	1,828	1,541	1,487	1,384	1,950	1,887	1,717	nd
	Limestone	254	228	570	629	732	651	619	631	457	559	626	723	nd
	Sand	3,334	1,762	2,820	3,558	3,728	3,289	3,090	2,774	2,208	2,872	3,617	3,841	nd
	Sand-gravel	4,073	2,202	2,746	2,765	2,635	2,649	2,399	2,881	3,104	3,249	3,313	2,765	nd
	TOTAL	9,915	5,270	7,513	8,545	8,733	8,417	7,649	7,772	7,153	8,630	9,443	9,046	nd
Lithuania	Dolomite	2,070	799	1,124	1,423	943	1,158	1,291	1,304	1,344	1,873	2,144	2,124	2,880
	Limestone	804	448	454	572	658	681	669	751	686	704	744	781	781
	Sand	1,582	1,314	1,384	1,784	1,700	1,874	2,529	1,901	2,460	2,856	2,709	2,450	2,885
	Gravel	9,606	4,375	5,599	6,021	4,437	6,223	6,070	5,891	6,722	8,325	7,669	8,901	8,616
	TOTAL	14,062	6,936	8,561	9,800	7,738	9,936	10,559	9,847	11,212	13,758	13,266	14,256	15,162

ADVERSE EFFECTS OF FLAMING AS A SURFACE TREATMENT METHOD FOR STONE SLABS USED IN ROAD PAVEMENTS

Bartłomiej GRZESIK 0000-0003-2586-887X[1], Marcin GRYGIEREK 0000-0002-2969-7054[2], Przemysław ROKITOWSKI 0000-0002-5912-0475[3], Zdzisław ADAMCZYK 0000-0002-5925-4676[4], Jacek NOWAK 0000-0003-0307-8295[5]

[1-3]*Department of Geotechnics and Roads, Faculty of Civil Engineering, Silesian University of Technology, Gliwice, Poland*

[4, 5]*Department of Applied Geology, Faculty of Mining, Safety Engineering and Industrial Automation, Silesian University of Technology, Gliwice, Poland*

Corresponding author: Bartlomiej Grzesik, bartlomiej.grzesik@polsl.pl

Received day month year; accepted date (use style Received dates, or Alt + Ctrl + R)

Abstract. This paper presents the results of tests on a damaged pavement made of flamed granite slabs. Due to their architectural value, the use of such pavements made of stone materials is a popular trend in Europe, especially in historic city centres. Faming is a popular method of surface treatment of stone elements, including slabs. The use of flame with a temperature around 1300°C on granite rock leads to allotropic transformations of quartz. The accompanying volume changes lead to flaking of the surface. As a result, the flaming gives the slab a natural texture and improves its anti-slip properties. As it was assessed, most slabs used in pavements exhibited characteristic edge and corner damage. Examination of the mechanical properties of rock taken from a slab revealed different results for samples taken from the high temperature impact zone and from other parts of the slab. The mineralogical tests carried out on samples from the high temperature impact zone showed changes in relation to the original mineralogical composition of granite, including the presence of glaze. These changes resulted in the accumulation of stresses, especially in the areas of slab edges and corners. The analysis of the test results was made in relation to the lack of uniform European standards for stone treatment by flaming and the lack of control procedures for this process. As a result of unrestricted flaming conditions, the originally homogeneous properties of the rock may vary within a single product and lead to its accelerated degradation during exploitation.

Keywords: flaming, granite, quartz, stone slab, damage, road pavement.

Introduction

Despite the enormous progress in the field of road materials, natural rocks remain an important material for the production of paving elements covering road structures. Due to their architectural value, stone road slabs are widely used – primarily in important, representative streets in historic centres of many cities, where this material has been used since the dawn of time. In addition to their undeniable decorative value, properly made stone surfaces are characterised by significant operational durability and resistance to environmental conditions. The functional properties of stone surface elements are determined by both the physical and mechanical properties of the raw material and the production technology used in these construction products. One of the aspects of the production of road stone slabs is giving them an appropriate texture, the purpose of which is to ensure high anti-skid properties. Currently, there are several methods of giving stone slabs appropriate roughness. According to the European standard (EN 1341), the following methods of surface treatment of stone slab products can be distinguished: polishing, grinding (producing a fine texture), and bush hammering, machined with visible tool traces, shot peening or flaming (producing a thick texture).

While the flame treatment technology is widely used in the world, is relatively simple and does not require advanced machinery, the small number of scientific publications on this subject provides mixed and often contradictory information related to it.

Flaming technology is most often mentioned along with other stone processing methods, such as water jet surface treatment (Ozcelik, Ciccu, & Costa, 2011; Ciccu & Costa, 2012), laser surface blasting (Penide, del Val, Riveiro, Soto, Comesaña, Quintero, Boutinguiza, Lusquiños, & Pou, 2019), in the context of flame physics, including numerical analyses (Liakos, Koukou, Founti, & Markatos, 2002), or comparisons of the effects of granite treatment with various other methods (García-del-Cura, Benavente, & Martinez-Martinez, 2008). More broadly, the impact of flame on granite was explained by Donaldson et. al (1988) as consisting of allotropic transformations of quartz at high temperature, leading to a change in its volume and, as a result, peeling of the granite surface, which gives it a characteristic, rough texture resembling a natural rock fracture. It is a classic, well-known series of quartz thermal transformations, in which α-β quartz inversion at 573°C plays a key role. This transformation is accompanied by an increase in volume of about 5% (Heuze, 1983; Grapes, 2006). In fact, the transformations of quartz are accompanied by less significant changes in the feldspars, leading to their vitrification and microcracks (Ciccu & Costa, 2012; Freire-List, Gomez-Villalba, & Fort, 2015) as well as dehydration of the oxyhydroxides from the biotite (Sanmartin, Prieto, & de Melo Silva, 2011). Flame treatment is the only type of thermal heat treatment to produce texture. Flaming can be performed manually or automatically (Fig. 1).

Figure 1. Stone slab flaming with a handheld torch (on the left) and fully automatic machine (on the right)

If accompanied by water cooling, flaming provokes a thermal shock resulting in a breakage or vitrification of the component minerals, revealing the appearance of their crystal structure (Ciccu & Costa, 2012). It should also be noted that of all types of treatments, flaming has the narrowest scope of application. Flaming is used for quartz rich rocks; in practice it is limited to granite and quartzite (Ciccu, 1993; Brite-Euram Report, 1999; Recommended Best Practices, 2010). Flaming is not intended for the treatment of carbonate rocks, where high temperature leads to rock decomposition due to calcination. However, there are known marbles and limestones that can be flamed (García-del-Cura et al., 2008; Ozcelik et al., 2011). It is unclear whether the flame has a strict temperature range for the stone to achieve the desired effect. Some sources report a range between 1280°C and 1360°C (Lorenc & Mazurek, 2007) or around 1600°C (Liakos et al., 2002), while others set it at above 2000°C (Ozcelik et al., 2011). However, temperature is crucial for the phenomena that will occur in the flamed material. In the temperature range from about 573°C to 1711°C phase transformations of quartz, subsequent to α-β quartz inversion, take place, until the quartz changes into liquid form. At 1500°C, cristobalite partially turns into amorphous silica. All quartz transformations are accompanied by significant volume expansion (Casasola, Rincón, & Romero, 2012; Ringdalen, 2015), as is the case with the transformations preceding feldspar vitrification, which takes place at 1000°C (Brown & Parsons, 1989). Rapid cooling of the flamed material stops the reverse phase changes of minerals which occur as the temperature is slowly lowered. In extreme cases, this may lead to the hardening of the glaze in the cracks of the rock. Glaze has a lower density than its mineral components. For example, pure quartz glaze has a density of 2.20 g/cm^3, and quartz 2.65 g/cm^3 (Dai, Yin, Xu, Jin, Li, & Harmuth, 2018). A change in density means a change in volume. Glaze, as a new, uncontrolled, expansive component of granite will have a negative impact on its durability.

High temperatures can also lead to microcracking due to thermal shock. This phenomenon leads to a deterioration of the strength parameters of granite (Mardoukhi, 2017; Rossi, Saar, Kant, & von Rohr, 2018), significantly in case of its bending strength (Donaldson et. Al, 1988; García-del-Cura et al., 2008; Vazquez, Acuña, Benavente, Gibeaux, Navarro, & Gomez-Heras, 2016; Murru, Freire-List, Fort, Varas-Muriel, & Meloni, 2018). Exposure to water and frost in winter season can also weaken the flamed stone surface by advancing the extent of microcracks (Recommended Best Practices, 2010).

The applicable EN 1341 standard does not define flame treatment, and does not set any requirements regarding the method of its performance. In the context of the applicable standards, Factory Production Control does not directly cover techniques, tools, or procedures related to the flaming activity. According to this state of affairs, it should be understood that any defects of the product arising in the course of improper performance of this process are to be revealed on the basis of standard physico-mechanical tests of the flamed stone material (breaking strength, durability of flexural strength against freeze, water absorption, etc.). As the analysed case shows, the consequences of faulty flaming may appear only during the pavement operation.

1. Research material and methods

The subject of the research presented in this paper were flamed road granite slabs with dimensions of 14x50x100 cm, which showed characteristic damage after a short time of operation. A large part of the slabs on the street, about 500 m long, was damaged in the form of chipping of corners and edges, less often cracks. This paper discusses only the material issues related to the slabs, while ignoring the analysis of the behaviour of the entire structure, issues concerning the distribution and method of filling the expansion joints, or the maintenance of the pavement as factors with no clearly confirmed influence on the damage revealed in the examined case.

For the purposes of the planned tests of the pavement (Fig. 2), two granite slabs with the dimensions of 14 x 50 x 100 cm each were obtained. Slab B was characterised by damage the origin of which is the subject of this article.

Figure 2. Section of investigated damaged pavement with marked slab B in the middle of the pictured foreground

Due to the nature of the loss of substance in the slab in the near-edge zone, material tests were focused on these areas. A total of 55 samples were prepared from slab B (Fig. 3), which were used to determine the following physical and mechanical parameters of the rock:

- Determination of uniaxial compressive strength according to EN 1926:2007 standard;
- Determination of frost resistance according to EN 12371:2010 standard (as an effect on compressive strength according to EN 1926:2007 standard);
- Determination of water absorption at atmospheric pressure according to EN 13755:2008 standard;
- Determination of abrasion resistance according to EN 14157:2017-11 standard (Method B – Böhme Abrasion Test);
- Determination of linear thermal expansion coefficient based on EN 14581:2006 standard.

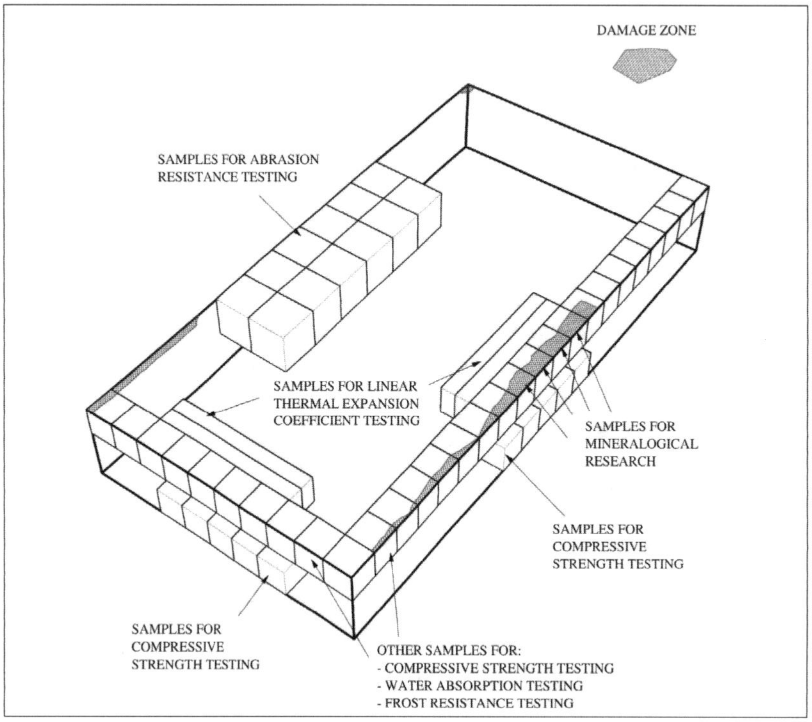

Figure 3. Slab B sampling scheme

The determination of the coefficient of linear thermal expansion constituted a separate issue. In the context of stone materials used in construction, there are currently no requirements concerning this characteristic. The determination was carried out in order to obtain detailed information on the physical properties of the tested granite that could be used in further analyses, e.g., using the finite element method (FEM). The impact of the actual temperatures affecting the granite slabs in the pavement was considered to be of particular importance. The coefficient of linear thermal expansion determined in accordance with the standard applies to the temperature range of +20 to + 80˚C. Therefore, it does not fully correspond to the actual conditions. In order to determine the thermal expansion coefficient in the assumed temperature range from -10 to + 50˚C, the Thin Layer Shrinkage System and Vikasonic devices by Schleibinger Masgerate System (Fig. 4a) were used. The first of the devices enables measurement by means of laser displacement of changes in the length of samples (Fig. 4b), while the second one allows the measurement of temperature changes accompanying changes in sample length. These two properties make it possible to determine the linear coefficient of thermal expansion according to the formula:

$$\alpha = \frac{\Delta L}{L \cdot \Delta T} \tag{1}$$

where:
ΔL – length change due to temperature changes,
ΔT – temperature change,
L – sample length.

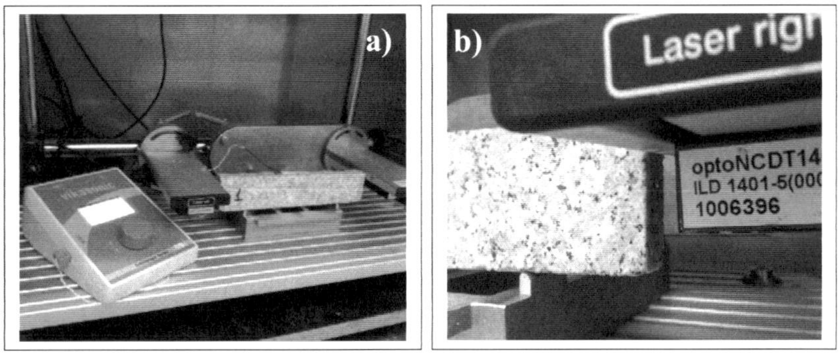

Figure 4. Determination of linear thermal expansion coefficient: a – view of the measuring devices used in the research; b – granite sample with visible reflected light from a laser rangefinder

The test used samples compliant with the PN EN 14581 standard, i.e., rectangular samples with dimensions of 250 (length) x 50 (height) x 25 (width) mm. A total of four samples were tested – two taken so that the largest dimension was parallel to the shorter edge of the slab and two with the largest dimension in line with the longer edge of the slab. Due to the isotropic properties of granite, the same expansion in both directions was to be expected, which was also a part of the verification of the correctness of the applied test method.

Two granite slabs (samples A and B) were used for mineralogical research. One of them was a slab in which no edge damage was observed macroscopically (A), while the other slab showed clear edge damage (B). Cubic samples with a side of 50 mm were cut from both slabs. They were appropriately assigned symbols – constituting subsequent samples. Mineralogical tests were performed for cubic samples with the following symbols: 19 – a sample from a granite slab in which no edge damage was observed, 10, 11 and 12 – samples from a granite slab in which clear edge damage was observed.

At the sampling site of samples 10, 11, 12, the granite slab on the flamed side showed damage at the edges in the form of (Fig. 5):

- presence of cracks perpendicular to the flamed surface,
- rock breaks in the edge area,
- presence of a 2-3 mm thick film running parallel to the flamed surface, showing detachment from the slab (using a slight mechanical force, e.g., fingers, it can be separated).

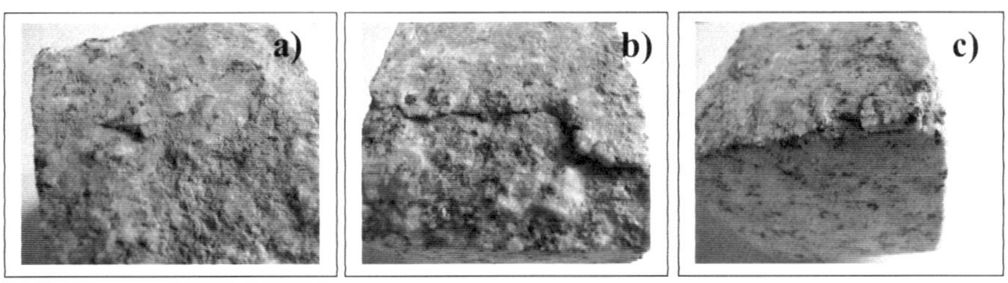

Figure 5. Cubic samples with visible damage in the form of detachment of a thin layer of rock substance: a – no. 10, b – no. 11, c – no. 12

Apart from basic macroscopic observations of rocks, the following research methodology was used in the study:

- transmitted light microscopic observations,
- phase identification by X-ray diffraction.

The transmitted light microscopic observations were made on a Zeiss Axioskop microscope with an image analyser. Microscopic preparations in the form of cuts were made from the collected samples.

Phase identification by X-ray diffraction (XRD) was performed on a Seifert-FPM XRD 7 diffractometer, using a cobalt lamp, Fe filter, 35 kV voltage, 25 mA current.

Complementary mineralogical tests were carried out for samples from granite slabs with clear edge damage, using a scanning microscope, with the possibility of determining the chemical composition in the micro-area. These tests were performed using the Hitachi SEM SU3500 variable vacuum scanning microscope, working with the Thermo Scientific NORAN System 7 EDS UltraDry energy dispersion X-ray spectrometer.

2. Research results

2.1. Selected physical and mechanical properties

In accordance with Figure 3, the determination of the compressive strength was carried out on air-dry samples taken from the upper edge of the slab affected by the flaming process. During the test, successive series of samples from this area were loaded perpendicularly or parallel to the face surface of the slab (subjected to flaming). The comparative series consisted of samples loaded perpendicularly and parallel to the face surface, but taken from outside the flame impact area (lower edge of the slab). The average compressive strengths of the soaked samples taken from the flame impact area (upper edge) were also compared with the samples from the same location previously subjected to 56 freeze-thaw cycles. The averaged results with standard deviation (error bars) are displayed in Figure 6.

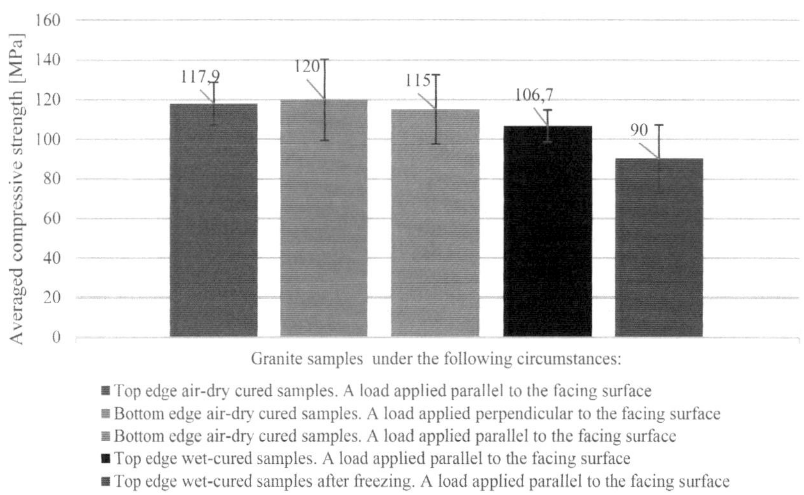

Figure 6. Averaged compressive strength of granite samples from slab B depending on: sampling location, way of cure, way of applying the load

Taking into account the standard deviation of the results of the compressive strength of the samples in the air-dry state, loaded perpendicularly or parallel to the flamed plane, in relation to the strength of the samples outside the high temperature impact area, no clear effect of this process on the deterioration of the marked feature was found. Samples subjected to saturated compression showed a slight decrease in strength in relation to the samples in the dry state from the edge area; slightly greater decrease was observed in the samples subjected to 56 freeze-thaw cycles. Taking into account the slight water absorption determined only in samples obtained from the flame impact area (upper edge) at the level of 0.25% (w/w) (standard deviation $\sigma = 0.08\%$), it can be assumed that under the influence of factors such as water and frost, the compressive strength of granite from the flame impact area is slightly deteriorated.

Determination of abrasion was carried out on four series of samples taken (Figure 3) from the flame impact area (slab B). In the case of two series, the face (flamed) surface of the samples was abraded, while in the case of the next two series, the plane opposite to the face surface (obtained during the sample cutting) was abraded. In both cases, the determination was carried out on air-dry samples and on saturated samples. The results are presented in Figure 7.

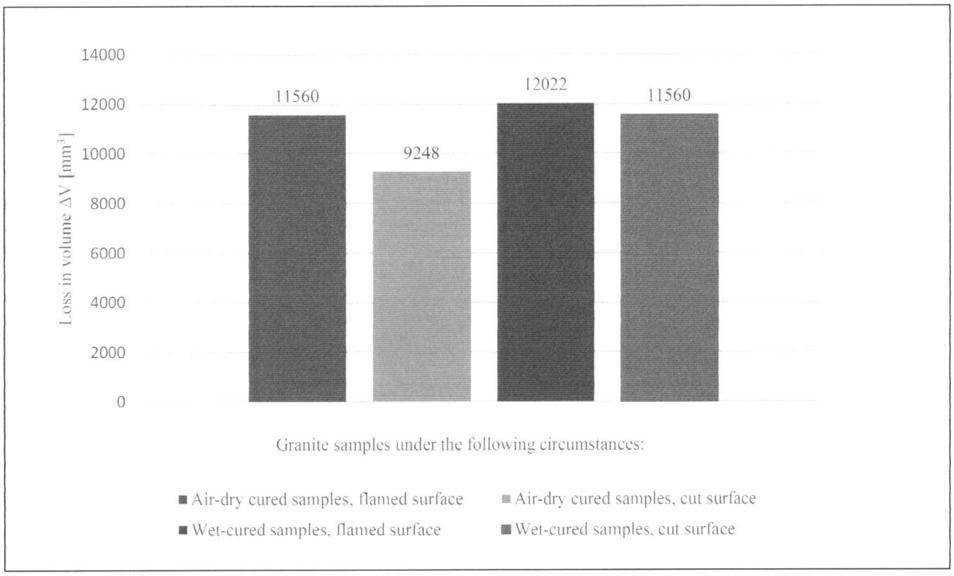

Figure 7. Averaged abrasive wear [mm³] of granite samples from slab B depending on: sampling location, way of cure, surface finish

In case of this feature, it was noticed that the lowest resistance during the Böhme Abrasion Test was shown by saturated samples in which the flamed surface was abraded. In this case, the average material loss was 3.82% of the sample volume. The abrasion of saturated samples with the surface obtained when cutting the slab resulted in the same loss as the air-dry samples abraded with the flamed surface, i.e., 3.68% of the volume. Abrasion of air-dry samples with the cut surface resulted in the smallest average loss of their volume, i.e., 2.94%.

The linear thermal expansion coefficient was determined in two stages. In the first stage, the samples were heated to a temperature of approx. 50°C, and then placed in a climatic chamber where the temperature was kept at 20°C. During the cooling of the sample from 50°C to 20°C, changes in the length of the sample were measured. The second stage consisted of cooling the sample to -10°C and measuring changes in its length during heating to 20°C in a climatic chamber. An example measurement result is shown in Fig. 8. Average thermal expansion coefficient results are shown in Table 1.

Table 1. Average linear thermal expansion coefficient determined over the temperature range -10÷50°C

Average temperature change of samples ΔT [°C]	Average change in sample length ΔL [μm]	Average linear thermal expansion coefficient α [x10^{-6} 1/°C]
57.92	113.80	7.82

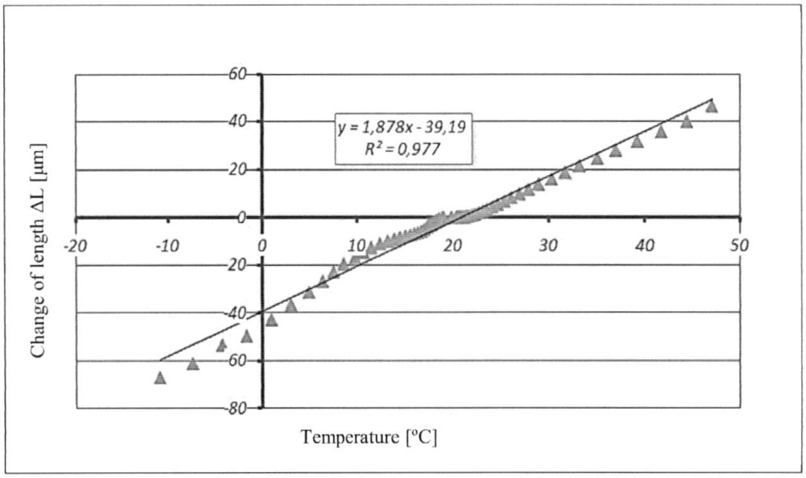

Figure 8. Dependence of the length change on the temperature for the example sample within the range of -10 ÷ 50°C

It should be noted that the obtained linear thermal expansion coefficient was characterised by a value typical for granites in the temperature range -20÷60°C, i.e., 4.8÷8.3 (x10-6 1/°C) (Hockman & Kessler, 1950) and that it was comparable for perpendicular directions, which confirms the isotropic nature of the rock. The obtained results exclude the possibility of the observed damage only occuring due to thermal shrinkage in the pavement operation conditions.

2.2. Mineral composition - microscopy

Macroscopically, the granite sample from slab B was very similar to the samples from slab A (no damage). The structure was crystalline, the grain size was up to several millimetres. The texture of the rock was dense and chaotic. Macroscopically, feldspars (represented mainly by orthoclase), quartz, and biotite have been identified.

The granite slab with damage at the edges on the flamed surface showed a slight discolouration compared to non-flamed surfaces. The flamed surface had a slightly grey tint compared to the non-flamed granite part, which was lighter in shade. This discolouration reached a thickness of 1 to 3 millimetres. Additionally, on the flamed surface, the feldspars acquired a matt gloss, while the surfaces of quartz grains more often showed a more intense, greasy, less glassy shine than on grains observed on non-flamed surfaces. In macroscopic evaluation, the quartz grains and their closest vicinity were vitrified.

The mineral composition of damaged granite slabs at the edges was practically very similar to undamaged granite slabs. In the examined granite, more abundant pseudomorphoses of pennin after biotite and amphibole crystals subject to thermal corrosion were found, compared to the slab without damage.

The microscopic observations carried out on the samples created perpendicularly to the flamed (inner) surface revealed that all the granular components in the outer area constituting the flamed surface were very strongly fractured. The further from this surface and deeper into the slab, the less cracks were observed. This clearly indicates the effect of applying high temperature and then rapidly cooling the material, which results in a typical thermal contraction causing the formation of cracks. The more numerous the cracks, the higher the temperature and the faster the cooling.

Simultaneously, it was established that the main minerals, exposed on the outer surface of the slab, perpendicular to the flamed surface, showed very similar effects of thermal contraction, in the form of numerous cracks, and their amount decreased with the growing distance from this surface (Fig. 9).

In this case, it should be emphasised that the formed cracks usually exhibit a direction that is approximately perpendicular to the cooled surface.

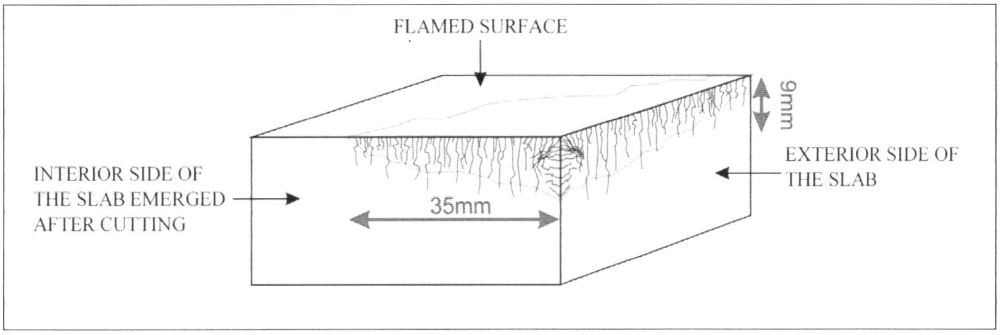

Figure 9. After-flaming cracks located on lateral sides of granite slab B in the edge area – model

Microscopic observations of the samples cut both perpendicularly and parallel to the flamed surface revealed the presence of glaze in the area of the flamed surface, especially in the areas of damage close to the edge. Glaze covered the grains of minerals, mainly alkaline feldspars (Fig. 10). In the outer part, the glaze had a spherical form (Fig. 11) indicating a rapid hardening of the alloy resulting from the melting of minerals.

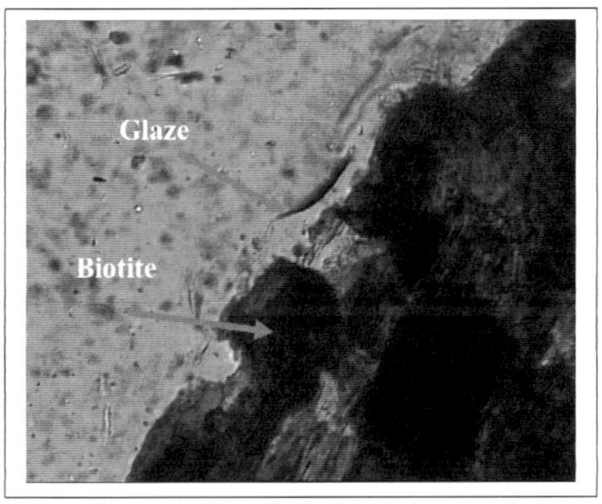

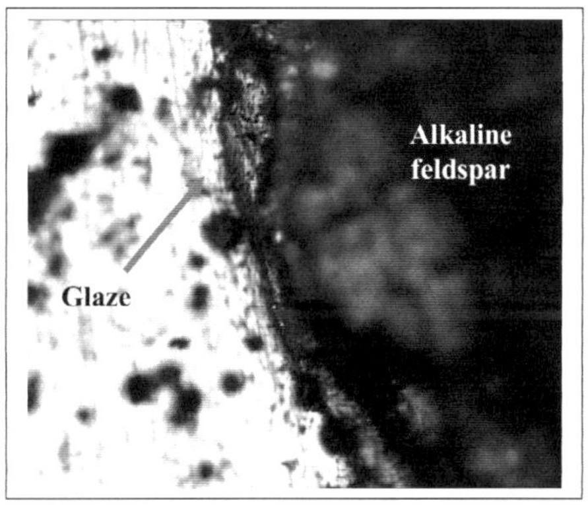

Figure 10. Photomicrograph (transmitted light) of granite from slab B (sample No.11), glaze growing on biotite plates in the center, 1 polar, magn. 200x

Figure 11. Photomicrograph (transmitted light) of granite from slab B (sample No.11), a large grain of strongly fractured alkaline feldspar covered by pherical glaze to the right, 1 polar, magn. 200x

2.3. Mineral composition – X-ray diffraction

The phase identification by X-ray diffraction method allowed the determination of the main minerals present in the examined granite (Fig. 12, Fig. 13). In the diffraction pattern of samples from granite B (Fig. 13), characteristic lines were found for:

- quartz, main lines: 3.34; 4.25, 1.81 Å,
- feldspars (alkaline feldspars and plagioclases), main lines: 3.18-3.19; 3.24; 4.03Å,
- biotite, main lines: 10.03-10.05; 5.03; 2.83 Å,
- chlorite, lines 7.04-7.05; 3.53.

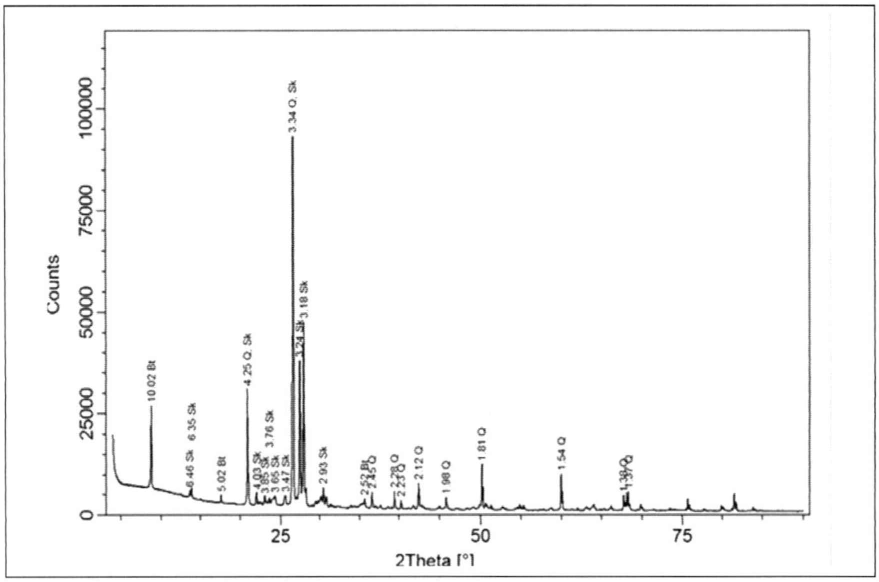

Figure 12. X-ray diffraction pattern of granite from the non-flamed part of slab B. Explanation: Sk – feldspars, Q – quartz, Mgh – maghemite, Pl – plagioclases, Ch – chlorite, Bt – biotite

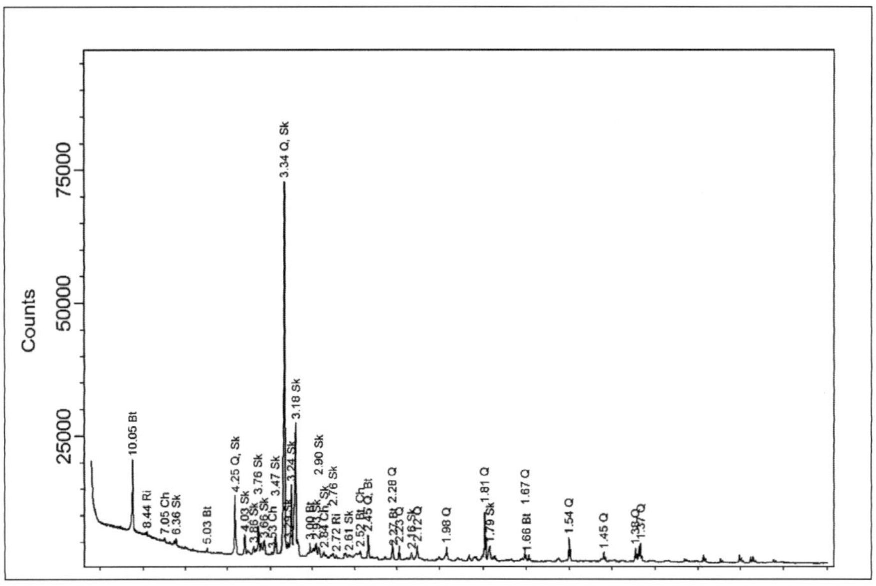

Figure 13. X-ray diffraction pattern of granite from flamed area of slab B. Explanation: Sk – feldspars, Q – quartz, Mgh – maghemite, Pl – plagioclases, Ch – chlorite, Bt – biotite

By comparing the diffraction patterns of samples from flamed surfaces (Fig. 13) and samples that were not exposed to temperature (Fig. 12), it can be seen that although the same major minerals are present (quartz, feldspar, plagioclase, and biotite), the intensities of reflections from individual minerals have changed. This may be the result of high temperature, which has melted some of the minerals into glaze.

2.4. Mineral composition – scanning electron microscopy (SEM)

Due to the fact that macroscopic (optical) observations revealed the presence of vitrified grains and non-transparent minerals in the flamed film, supplementary mineralogical studies were carried out using scanning microscopy (SEM), with the possibility of determining the chemical composition in the micro-area (EDS). The research revealed that non-transparent minerals were represented by: iron and copper sulfides (probably chalcopyrite and/or bornite), iron sulphides (probably pyrite), lead sulphide (galena), titanomagnetite.

This study paid particular attention to the glaze in the immediate vicinity of alkaline feldspars. The feldspar was covered with glaze (Fig. 14) or the glaze filled the gaps in the feldspar grains (Fig. 18). The results of the studies of the chemical composition of the glaze clearly indicate that it was formed in the granite remelting area, in the flamed area. The chemical composition of feldspar is quite typical, because the main chemical components are silicon, aluminium, potassium, and sodium with calcium additions (Fig. 15, Fig. 16). Meanwhile, the chemical composition of the glaze was apparently very similar to feldspar, but always with an addition of: sulphur, iron, and magnesium (Fig. 17, Fig. 19). Thus, the glaze was probably the result of feldspar (alkaline feldspars and plagioclases) melting with sulphide inclusions present in them, which is indicated by the presence of sulphur and iron in the glaze. At the same time, the source of magnesium, as well as silicon, aluminium, and potassium in the glaze, among the minerals present in the granite, could have been biotite. It is also noteworthy that in the EDS spectra of the glaze (Fig. 17, Fig. 19), the Si-derived line was definitely more intense than the Al line compared to the same lines in the feldspar EDS spectra. In turn, the lines derived from potassium, sodium, and calcium also showed lower intensities than the silicon lines in the glaze spectra compared to the feldspar spectra. This clearly indicates that apart from feldspar (the sulphide infiltrates in them) and biotite, quartz was also melted down.

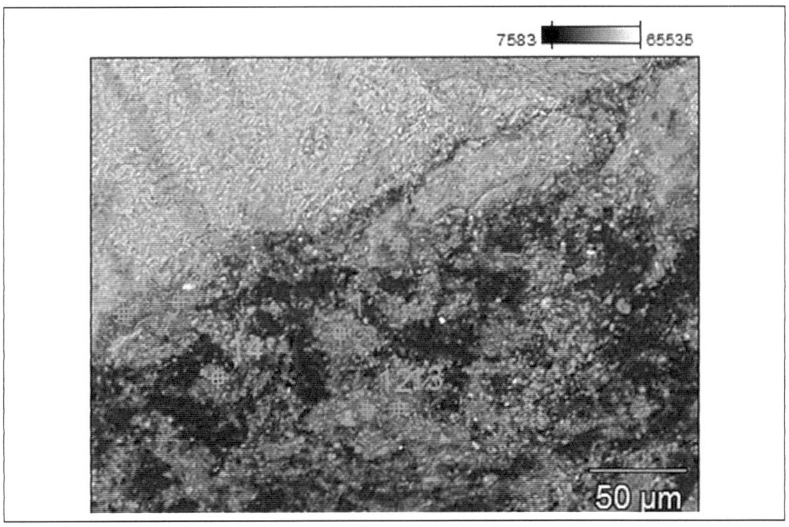

Figure 14. SEM photomicrograph of alkaline feldspar (points 5-11) and glaze (points 1-4, 12-14). Slab B, flamed surface, "granite 15" area

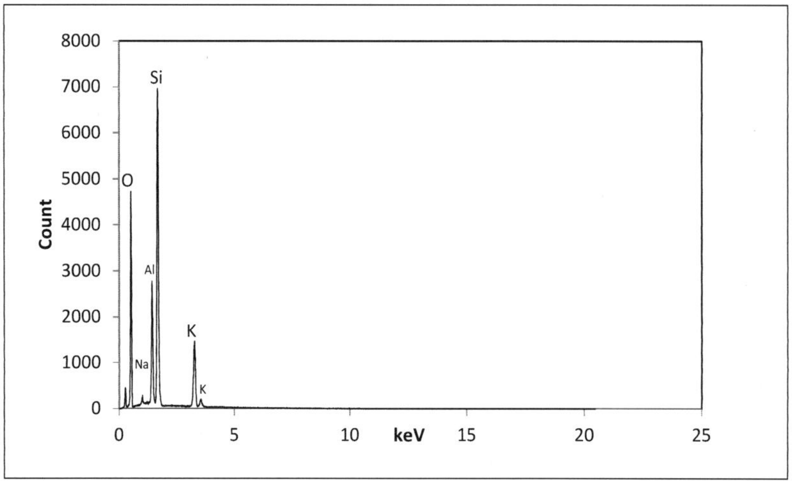

Figure 15. EDS spectrum of alkaline feldspar – measuring point 6, "granite 15" area

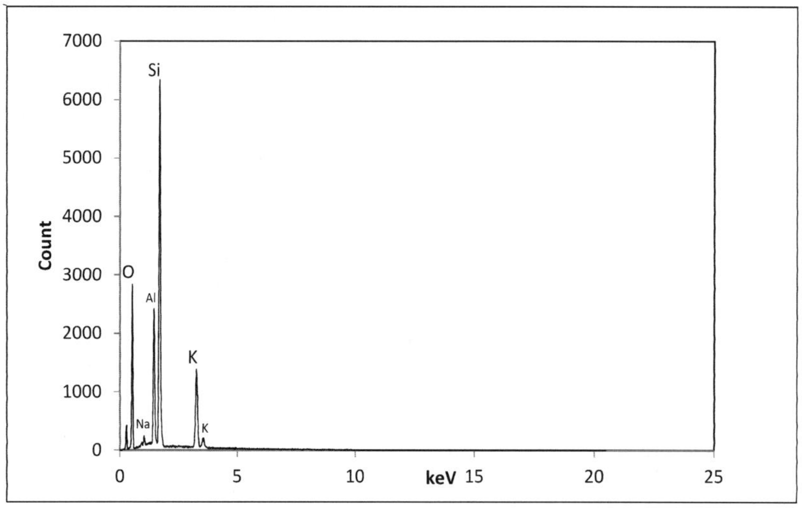

Figure 16. EDS spectrum of alkaline feldspar – measuring point 10, "granite 15" area

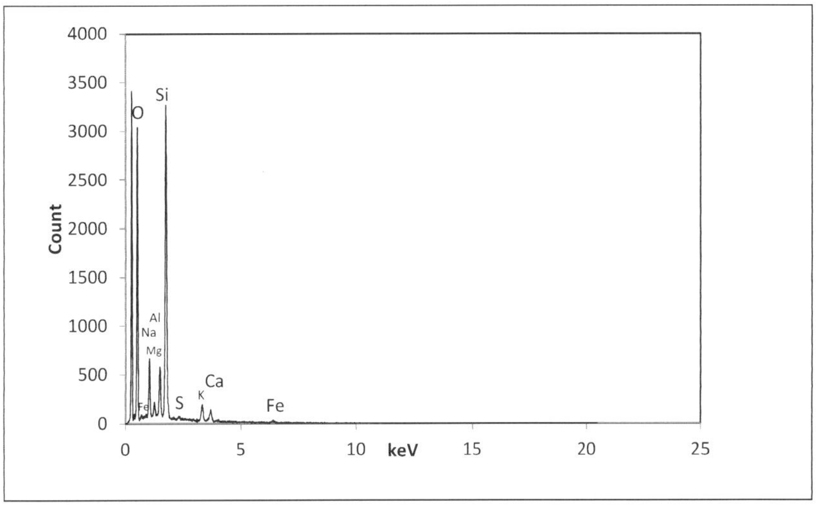

Figure 17. EDS spectrum of glaze – measuring point 14, "granite 15" area

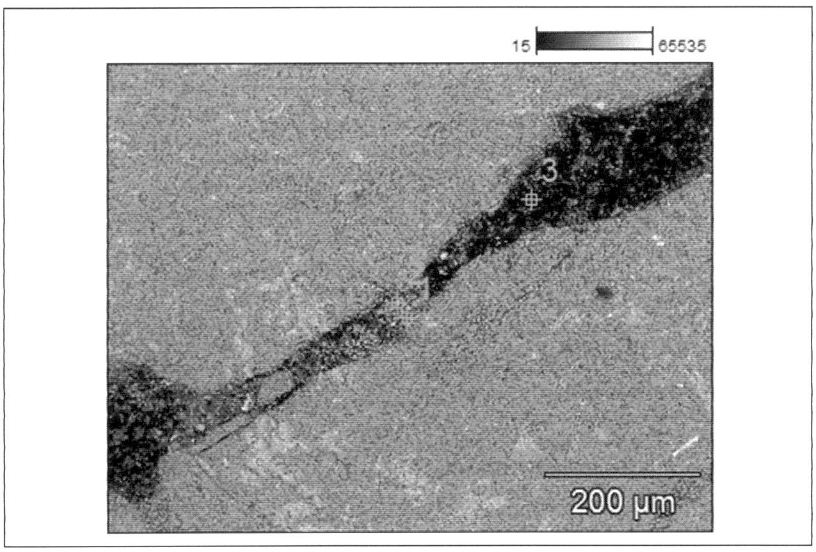

Figure 18. SEM photomicrograph of glaze filling the crack in the alkaline feldspar. Slab B, flamed surface, "granite 13" area

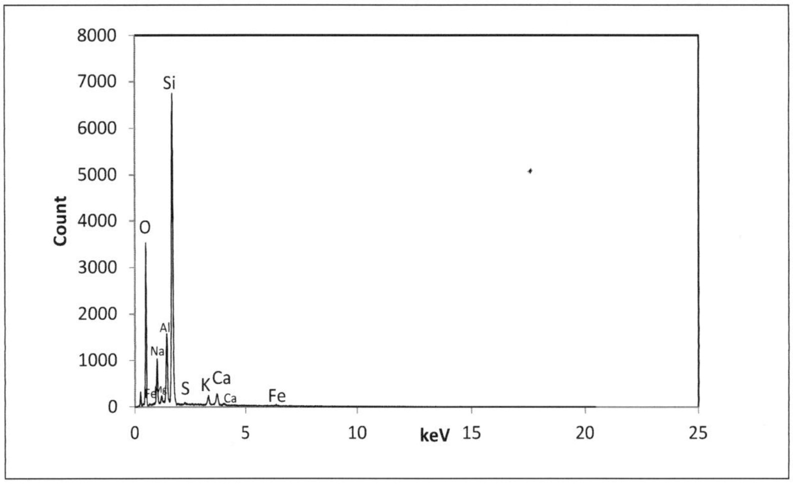

Figure 19. EDS spectrum of glaze – measuring point 1, "granite 13" area

3. Conclusions

As the research has shown, despite the noticeable damage to granite slabs of material origin, standard test methods used to control mechanical properties as part of factory production control in accordance with EN 1341: 2013-05 do not constitute an effective way to detect and eliminate defective products in the case of flamed stone slabs. Apart from the abrasion test, the samples taken from the high temperature impact area did not exhibit any significant differences in relation to the samples taken from outside these areas.

Mineralogical studies should be considered decisive in determining the causes of damage to granite road slabs. Microscopic, XRD and SEM examinations allowed to establish that the causes of damage to granite slabs on the edges were cracks resulting from thermal contraction, as a result of "overheating" of the slabs in the edge areas. This was indicated by the presence of numerous cracks in these areas on the face surface, but also on the outer side, which the flame of the burner used for the flaming should theoretically not cover. In the edge areas, the impact of temperature during flaming may have a much greater spatial range, as it may cover not only the actual flaming surface but also the side surfaces of the slab. The test results indicate that the causes of damage in the edge areas of the tested granite slabs were related to improperly carried out flaming, i.e., too high temperature and/or too long exposure time, which is indicated by the presence of glaze. Glaze is not a natural component of deep-sea volcanic rocks. Its chemical composition and the form of fine veins filling the cracks of cracked minerals clearly indicate that it is the result of the flaming process. Its presence should be considered unfavourable in the context of the durability of stone products for paving. Density changes accompanying the phase transformations of feldspar and quartz preceding the transition of these components into glaze are expansive in nature, causing stresses in the rock, which may lead to damage or destruction of stone products. It should be noted that glaze undergoes devitrification (crystallisation) over time, which in turn causes another change in volume – in this case, shrinkage. Exposure to environmental factors typical for the pavement, such as rainfall, significant temperature fluctuations, and winter maintenance measures typical for road products, will constitute an additional factor magnifying the effect of inadequate flaming. The effects of improperly conducted flaming can thus be significantly spread over time.

Flaming as a cheap, fast, and reliable method of imparting non-skid texture to stone products used in communication construction seems to be carried out largely in a manner based on intuition. However, it requires the development of detailed standards, primarily specifying the duration and temperature of heating, as the local impact of very high temperature on the rock causes much more far-reaching effects than just peeling of the surface observed with the naked eye.

4. References

Ozcelik, Y., Ciccu, R., & Costa, G., (2011). Comparison of the water jet and some traditional stone surface treatment methods in different lithotypes. *Construction and Building Materials, 25,* 678-687. https://doi.org/10.1016/j.conbuildmat.2010.07.021

Ciccu, R., & Costa, G. (2012). Waterjet Rough finishing of stone surface. *Tehnicki Vjesnik, 19(2),* 373-379.

Penide, J., del Val, J., Riveiro, A., Soto, R., Comesaña, R., Quintero, F., Boutinguiza, M., Lusquiños, F., & Pou, J. (2019). Laser Surface Blasting of Granite Stones Using a Laser Scanning System. *Coatings, 9*(131), 1-11. https://doi.org/10.3390/coatings9020131

Liakos, H., Koukou, M., Founti, M., & Markatos, N. (2002). Industrial-scale processing of granite surfaces by natural gas jet flames. *Applied Thermal Engineering, 22,* 393-405. https://doi.org/10.1016/S1359-4311(01)00103-X

García-del-Cura, M., Benavente, D., & Martinez-Martinez, J. (2008). The effect of surface finishes on outdoor granite and limestone pavers. *Materiales de Construcción, 58,* 65-79.

Donaldson, B. et. al (1988). *New Stone Technology, Design, and Construction for Exterior Wall Systems.* American Society for Testing and Materials, Bailtimore, USA, 128 p.

Freire-Lista, D., Gomez-Villalba, L., & Fort, R. (2015). Microcracking of granite feldspar during thermal artificial processes. *Periodico di Mineralogia, 84*(3A), 519-537. https://doi.org/10.2451/2015pm0029

Sanmartín P., Prieto, B., & de Melo Silva, B. (2011). Effect of Surface Finish on Roughness, Color, and Gloss of Ornamental Granites. *Journal of Materials in Civil Engineering, 23*(8), 1239-1248. https://doi.org/10.1061/(ASCE)MT.1943-5533.0000285

Ciccu, R. (1993). Stone Quarrying: The Italian Experience. In *Proceedings of the 13th Mining Congress of Turkey,* İstanbul, Turkey, (pp. 341-354).

Lorenc, M., & Mazurek, S. (2007). *Use a Stone. Genesis - Application – Preparation.* Studio JASA, Wrocław, Poland, 248 p. (in Polish).

Vazquez, P., Acuña, M., Benavente, D., Gibeaux, S., Navarro, I., & Gomez-Heras, M. (2016). Evolution of surface properties of ornamental granitoids exposed to high temperatures. *Construction and Building Materials, 104,* 263-275. http://dx.doi.org/10.1016/j.conbuildmat.2015.12.051

Murru, A., Freire-Lista, D., Fort, R., Varas-Muriel, M., & Meloni, P. (2018). Evaluation of post-thermal shock effects in Carrara marble and Santa Caterina di Pittinuri limestone. *Construction and Building Materials, 186,* 1200-1211. https://doi.org/10.1016/j.conbuildmat.2018.08.034

Grapes, R. (2006). *Pyrometamorphism.* Springer-Verlag, Berlin, Germany, 275.p

Heuze, F. (1983). High-temperature mechanical, physical and thermal properties of granitic rocks – a review. *International Journal of Rock Mechanics and Mining Sciences and Geomechanics Abstracts, 20(1),* 3-10

Brown, W., Parsons, I. (1989). Alkali feldspars: ordering rates, phase transformation and behaviour diagrams for igneous rocks. *Mineralogical Magazin, 53,* 25-42

Dai, Y., Yin, Y., Xu, X., Jin, S., Li, Y., Harmuth, H. (2018). Effect of the phase transformation on fracture behaviour of fused silica refractories. *Journal of the European Ceramic Society, 38 (16),* 5601-5609. https://doi.org/10.1016/j.jeurceramsoc.2018.08.040

Ringdalen, E. (2015). Changes in quartz during heating and the possible effects on Si production. *The Journal of The Minerals, 67,* 484–492. https://doi.org/10.1007/s11837-014-1149-y

Mardoukhi, A. (2017). Effects of microstructural features, thermal shocks and strain rate on the mechanical response of granitic rocks. *Tampere University of Technology. Publication, 493,* 1-58

Hockman, A., Kessler, D. (1950). Thermal and moisture expansion studies of some domestic granites. *Journal of Research of the National Bureau of Standards, Research Paper RP 2087 (44),* 395-410

Casasola, R., Rincón, J., Romero, M. (2012). Glass-ceramics glazes for ceramic tiles – a review. *Journal of Material Science, 47,* 553-582. https://doi.org/10.1007/s10853-011-5981-y

European Committee for Standardization. (2013). EN 1341:2013-05 Slabs of natural stone for external paving – Requirements and test methods.

European Committee for Standardization. (2007). EN 1926:2007 Natural stone test methods – Determination of uniaxial compressive strength

European Committee for Standardization. (2010). EN 12371:2010 Natural stone test method – Determination of frost resistance

European Committee for Standardization. (2008). EN 13755:2008 Natural stone test methods – Determination of water absorption at atmospheric pressure

European Committee for Standardization. (2017). EN 14157:2017-11 Natural stone test methods - Determination of abrasion resistance

European Committee for Standardization.Determination of linear thermal expansion coefficient based on EN 14581:2006 standard.

Brite-Euram 3908 Report. (1999). *Granite and quartzite stone slabs processing by oxy-methane flaming.* Retrieved from https://cordis.europa.eu/project/id/BRPR970441/fr

Building Stone Institute. (2010). *Recommended best practices.* Retrieved from https://www.naturalstoneinstitute.org/

Motivational bonus-system based on pavement installation temperatures measured by thermographic system (TGS Pavement) in Estonia

MOTIVATIONAL BONUS-SYSTEM BASED ON PAVEMENT INSTALLATION TEMPERATURES MEASUREMENT BY THERMOGRAPHIC SYSTEM (TGS PAVEMENT) IN ESTONIA

Romet RAUN (TEEDE TEHNOKESKUS AS)[1], Marek TRUU (TEEDE TEHNOKESKUS AS)[2]

[1]*Research and Development Department, Tallinn, Estonia*
[2] *Research and Development Department, Tallinn, Estonia*
Corresponding author: Romet Raun, romet.raun@teed.ee

Received 15 March 2021

Abstract. Paving is one of the most crucial stage in the matter of road lifespan, since it is the surface layer that has stand up to all the external factors (weather, traffic etc.). Insufficient pavement density caused by thermal segregation during paving works can reduce the lifespan significantly, especially in Estonian climate (freeze-thaw cycles). Modern technology offers different solutions to reduce the risk of low quality in asphalt production and road paving works. Mobile asphalt plant, feeder and thermo-isolated trailers are some piece of equipment, that contractor can use to level up the minimal required quality requirements. The question is, when to use those and which to use? Moreover, is there any possibility to motivate the contractors to put in some extra effort? In Estonia, motivational bonus-system has been established to encourage innovation and reward the extra effort that has been made for quality improvements. The methodology is based on years of experience gained in different researches and pilot-projects. There are no strict rules for the road paving equipment in the methodology – for example contractor can choose himself either the feeder or/and thermo-insulated trailers are used on not. The main requirement is that the temperatures of entire paving process (surface layer) has been measured and analyzed by special thermographic system. Current presentation discusses the symbiosis of bonus-malus system and development of special thermographic system (TGS Pavement) as a multifunctional tool in asphalt paving in Estonia.

Keywords: road, pavement, paving, segregation, homogeneity, thermography, bonus-malus system

Introduction

Paving is one of the most crucial stage in the matter of road lifespan, since it is the surface layer that has to stand up all the external factors (weather, traffic etc.). Insufficient pavement density caused by thermal segregation during paving works can reduce the lifespan significantly, especially in Estonian climate (freeze-thaw cycles). Current presentation discusses the symbiosis of bonus-malus system and development of special thermographic system (TGS Pavement) as a multifunctional tool in asphalt paving in Estonia.

1. Traditional pavement installation measuring method

Asphalt paving temperature measuring is a standard procedure in paving process. The idea and need for this kind of measurement is to assess the quality of pavement homogeneity, segregation and density. In most countries the current practice is that the measurement is done by supervisor using an ordinary spot temperature thermometer. Although nowadays many hand-held systems have been developed and even portable IR cameras are affordable and widely used, this method is really superficial and actually does not give us any valuable information about the pavement quality at all. Why? Measurement made by traditional spot method represents only one single point temperature and is not giving any information about pavement homogeneity in transverse and longitudinal direction. Even more, spot measurements are strongly influenced by measurement distance and angle (different at each measurement), non-systematically stored and not geo-reference, which makes it impossible to link any future defect with the paving temperature of investigated pavement area.

2. Innovative solutions for pavement temperature quality measurement

What roadowner actually needs for pavement quality assessment, is a continuous pavement installation temperature measurement across the entire pavement freshly laid (length and width). Nowadays there are only few technical IR solutions in the world that can provide a data collection, which would be really useful for pavement installation quality monitoring. All these systems are slightly different from the technical parameters and data analyzing software possibilities, but the purpose is the same – irreplaceable tool for pavement quality monitoring and assessment.

Motivational bonus-system based on pavement installation temperatures measured by thermographic system (TGS Pavement) in Estonia

One of the devices on the market is TGS Pavement - innovation by Teede Tehnokeskus, Estonia (see figure below). TGS Pavement's main specifications are:

- Real-video and data monitoring (both on jobsite and in the office)
- Temperature measurement across the entire pavement laid (length and width)
- Temperature measurement for each 10x10 cm
- Geo-referenced data (high precision GNSS)
- 100% pavement quality reporting
- Easy to install and use
- Works with any paver type

Figure 1. TGS Pavement thermographic system attached to paver.

As in most countries, the use of thermographic system while paving asphalt is not required by roadowners, the systems are mostly used by innovative asphalt paving contractors only, to monitor their own quality. Now as we see the benefit for both parties (roadowner and contractor) of the system, how can we popularize the usage of it?

3. Motivational bonus-system in Estonia

Modern technology offers different solutions to reduce the risk of low quality in asphalt production and road paving works. Mobile asphalt plant, feeder and thermo-isolated trailers are some pieces of equipment, that contractor can use to level up the minimal required quality requirements. The question is, when to use those and which to use? Moreover, is there any possibility to motivate the contractors to put in some extra effort?

In Estonia, motivational bonus-system has been established to encourage innovation and reward the extra effort that has been made for quality improvements. The idea of the methodology is that if the pavement installation quality is higher than the minimum requirements, then the contractor is able to earn bonus (up to 5% of the contract fee). The bonus calculation has two components:

Motivational bonus-system based on pavement installation temperatures measured by thermographic system (TGS Pavement) in Estonia

- Bonus for avoiding paver full-stops while paving
- Bonus for avoiding risk-areas

Paver stop is counted when the paver stays in the same location for at least 2 minutes (to earn bonus it is allowed to make 5 stops in the length of 1000 meters of paving). Risk-area is an area with installation temperature lower than the average temperature of the last 100m of pavement installation temperature (the percentage of risk-area has to be lower than 5% to earn bonus).

The methodology is based on years of co-operation between Estonian Transport Administration, Teede Tehnokeskus and multiple asphalt paving contractors in Estonia. The basic criteria was, that there should be no strict rules for the road paving equipment in the methodology – for example contractor can choose wheather the feeder or/and thermo-insulated trailers are used on not. The main requirement is that the temperatures of entire paving process (surface layer) has been measured and analyzed by special thermographic system.

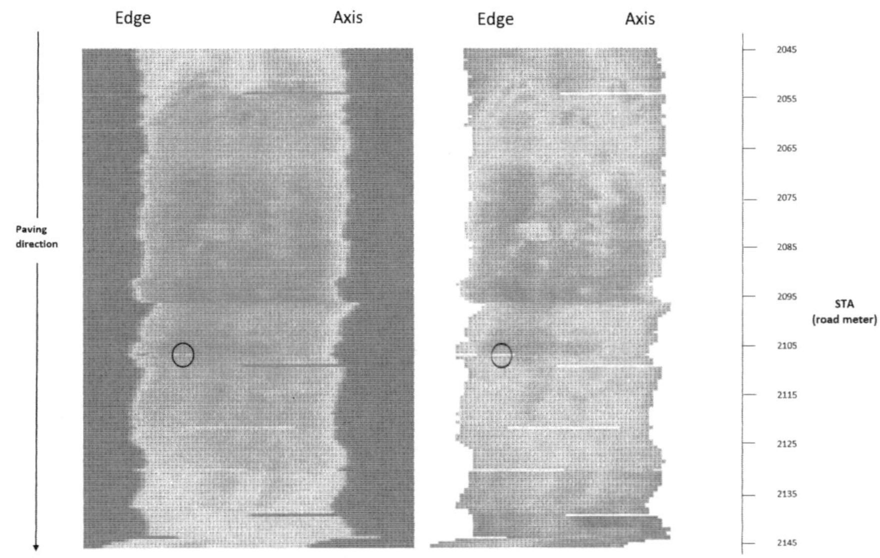

Figure 2. Data analyzing process (raw data on the left side and cleared data on the right side).

The first pilot projects took place in 2016 – three test sites in total length of ca 18 kilometers. In 2020 the numbers had increased to more than 40 jobsites in total length more than 230 kilometers. On that basis it can be stated that the methodology fulfills the purpose.

4. Additional benefits of TGS Pavement

Most clients have seen the benefit of the thermographic system and it has become more and more popular to use the system all the time – even on the sites where it is not able to earn bonus. Why? The system offers feedback and confidence for the contractor, that their quality meets the requirements. If there is any issue with machines (paver, feeder, non-insulated trucks) or really un-homogeneity pavement or the weather (wind, low temperatures) affect to the pavement temperature, then the information is immediately available and contractor can make decisions in the blink of an eye. That saves great amount of money every year for system users.

Motivational bonus-system based on pavement installation temperatures measured by thermographic system (TGS Pavement) in Estonia

Figure 3. Pavement installation real-time monitoring on jobsite (using Wi-Fi or 4G)

The system is user-friendly, as it transforms the results into dynamic images and graphs that are easy to follow. The data is saved and uploaded to cloud, so that it can be monitored on-site or even across the world both in real-time or historically. Some of the most crucial information that can be captured for efficient planning and paving process:

- Weather information (possibility to use data from on-site weather station with forecast)
- Thermal map of full day (immediate feedback of thermal homogeneity)
- Whole cross-section temperature values
- Thermal images for decision making supporting

Motivational bonus-system based on pavement installation temperatures measured by thermographic system (TGS Pavement) in Estonia

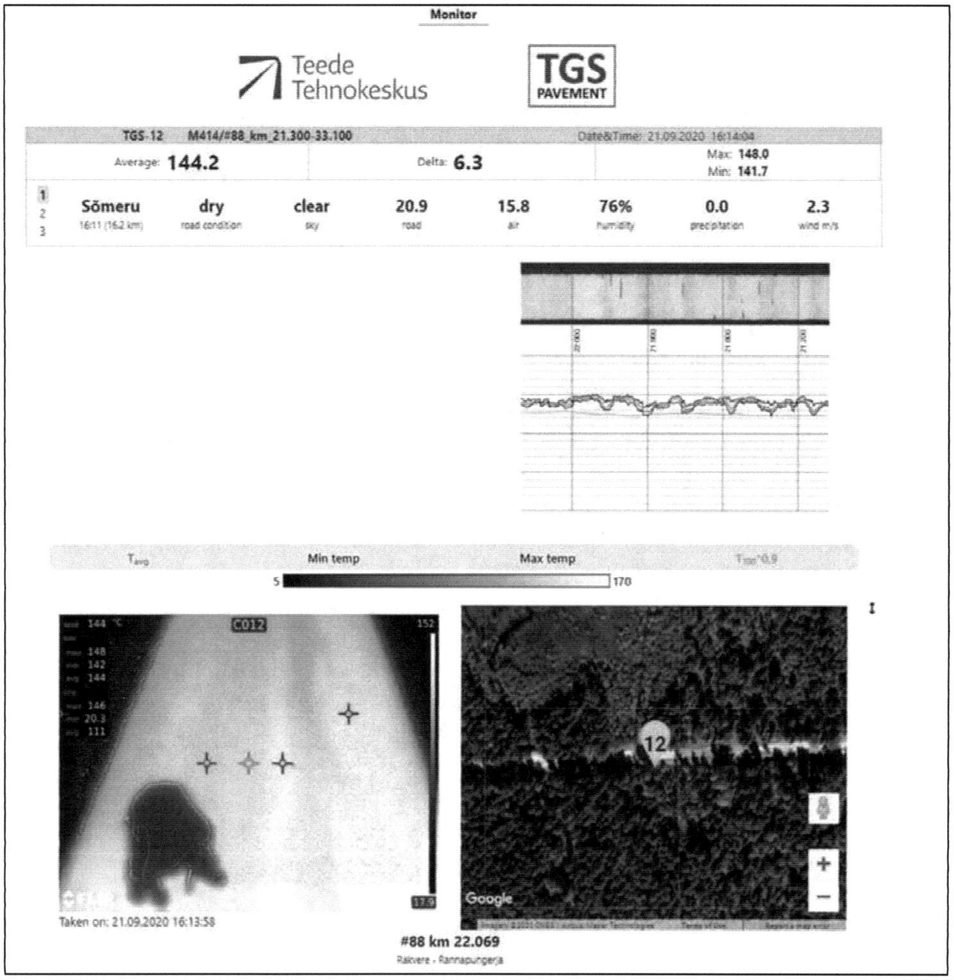

Figure 4. Pavement installation real-time monitoring in backoffice.

5. Conclusions

The most common reason for pavement failure is high-temperature segregation that causes uneven compaction that makes pavement susceptible to damage from moisture and freeze-thaw cycles. TGS Pavement is a thermographic system for pavement quality monitoring that is one of the few solutions in the world that provides real-time pavement quality reporting. Adding a motivational bonus-system to the paving process creates a solid foundation for maximizing the effort from each party.

In period 2019-2020, around 300 kilometers of state roads were measured and analyzed with TGS Pavement system. The pavement quality has become more stable and long-term users have polished their paving process to optimum. Most clients use the system in their everyday job, even if there is no opportunity for bonus.

Template: Anonymized Document

FIELD STUDIES OF MSWI BOTTOM ASH AS AGGREGATE FOR UNBOUND BASE COURSE MIXTURES

Judita GRAŽULYTĖ (ORCID ID 0000-0001-6502-9400)[1], Audrius VAITKUS (ORCID ID 0000-0001-5103-9747)[1], Egidijus Rytas VAIDOGAS (ORCID ID 0000-0002-7187-7972)[2], Ovidijus ŠERNAS (ORCID ID 0000-0002-0019-4242)[1]

[1]*Road Research Institute, Vilnius Gediminas Technical University, Vilnius, Lithuania*
[2]*Department of Building Materials and Fire Safety, Vilnius Gediminas Technical University, Vilnius, Lithuania*
Corresponding author: Judita Gražulytė, judita.grazulyte@vilniustech.lt

Received day month year; accepted date (use style Received dates, or Alt + Ctrl + R)

Abstract. In the European Union, about 30–40 million tonnes of residues known as municipal solid waste incinerator (MSWI) bottom ash is generated and landfilled annually. To address the continuous growth of landfills and to implement zero waste and circular economy policies, researchers are researching ways to turn MSWI bottom ash into a useable resource. The conducted studies show that MSWI bottom ash is suitable for civil engineering, especially for roads, however there is a lack of field studies. As a result, MSWI bottom ash was used to construct unbound base course in heavy vehicles parking lot in 2018 and two pedestrian paths in 2018 and 2020 in Vilnius (Lithuania). This paper focuses on the structures composition and performance of those unbound base courses in terms of stability of particle size distribution, bearing capacity and permeability. The conducted study showed promising results for MSWI bottom ash as aggregate (mixture) to construct unbound base course.

Keywords: bottom ash, municipal solid waste, unbound base course, pavement structure, field study, particle size distribution, bearing capacity, permeability.

Introduction

In the European Union, more than 140 million tons of municipal solid waste is incinerated annually. It generates about 30–40 million tons of residues known as municipal solid waste incinerator (MSWI) bottom ash, which is typically landfilled. To address the continuous growth of landfills and to implement zero waste and circular economy policies, researchers are researching ways to turn MSWI bottom ash into a useable resource.

The main concern of MSWI bottom ash usage as a resource is a leaching of heavy metals and soluble salts to the environment. Seeking to reduce it, MSWI bottom ash is aged (weathered), i.e. stored in uncovered stockpiles with access to water for a specific period. The length of this period differs among countries. For example, in Spain, France, Germany and Lithuania it is three months (del Valle-Zermeño, Chimenos, Giró-Paloma, & Formosa, 2014; ISWA, 2006; Izquiedro et al., 2001; Gražulytė, Vaitkus, Šernas & Žalimienė 2021), whereas in the Netherlands bottom ash can be aged only six weeks, and in Sweden the aging period is extended up to six months (ISWA, 2006). During this aging (weathering), oxides and hydrates that are present in the MSWI bottom ash react with carbon dioxide and water up taken from the atmosphere. These chemical reactions produce stable carbonates and reduce pH of bottom ash up to 8–10 (Chimenos et al. 2000, Dou 2017). MSWI bottom ash can be considered as a resource only if the leaching of heavy metals after aging (weathering) meets the environmental requirements. Otherwise, it has to be landfilled or further aged to improve the quality. MSWI bottom ash without ageing is always landfilled.

MSWI bottom ash consists of ash, ceramics, glass, minerals, ferrous and non-ferrous metals, unburned materials and organic carbon (Chandler et al., 1997; Chimenos, Segarra, Fernández, & Espiell, 1999). Ferrous metals account about 7–15% of MSWI bottom ash mass and non-ferrous metals – 1–2% (Baun, Kamuk, & Avanzi, 2007; Sabbas et al., 2003). It is worth highlighting that ferrous metals constitute more than 60% of all metals and can be easily recovered. MSWI bottom ash with higher amount of metals than 5% cannot be used in civil engineering. Therefore, recovery of metals including both ferrous and non-ferrous metals is crucial in turning MSWI bottom ash into a resource. It is done either directly after the generation of MSWI bottom ash or after ageing (weathering). The recovery of metals also has economic benefits since recovered metals are recycled through the international scrap market.

Thorough investigations that have been made into the physical and mechanical characteristics of MSWI bottom ash have revealed its potential suitability for civil engineering, especially for road construction. The utilization of MSWI bottom ash as material to construct embankment and subgrade as well as unbound base and sub-base are the most promising application areas (Alhassan & Tanko, 2012; An et al., 2014; Becquart, Bernard, Abriak, & Zentar, 2009; Forteza, Far, Seguí, & Cerdá, 2004; Hjelmar, Holm, & Crillesen, 2007; Izquiedro et al., 2001; Xie et al., 2017; Vaitkus,

Template: Anonymized Document

Gražulytė, Vorobjovas, Šernas & Kleizeinė, 2018; Vaitkus, Gražulytė, Vorobjovas, Šernas & Kleizeinė, 2019). However, most studies of MSWI bottom ash as resource to construct unbound base and sub-base layers have only been carried out in a laboratory. Nevertheless, several attempts to test MSWI bottom ash under real traffic and climate conditions have been made.

Sormunen and Kolisoja (2017) analyzed the performance of structural layers (filtration, sub-base and base) of the interim storage field. Those layers were constructed of 100% MSWI bottom ash. The results after half a year of operation showed that MSWI bottom ash is suitable for lower structural layers (filtration and sub-base) of roads and field structures. It was also revealed that the usage of MSWI bottom ash to construct base course is questionable since material is prone to crushing and therefore is not able to resist the higher stresses occurring in the upper parts of structure despite the increase in stiffness over time. In this case, authors recommended constructing an additional base layer of natural aggregate or a thicker asphalt concrete layer on top of base layer from MSWI bottom ash. To better understand the real effect of MSWI bottom ash on environment, some field studies with MSWI bottom ash were dedicated only on the leaching of metals and soluble salts. In those cases, information on the mechanical (structural) performance of MSWI bottom ash was not provided (Hjelmar, Holm & Crillesen, 2007; Izquierdo, Querol, Josa, Vazquez & López-Soler, 2008; Sormunen, Kaartinen & Rantsi, 2018). It has to be noted that leaching of metals in those field studies passed the environmental requirements and was lower or similar to the predicted.

Taking together the conducted studies, there is a lack of field studies. Consequently, the main aim of this paper is determine the performance of base course of heavy vehicles parking lot and two pedestrian paths constructed of MSWI bottom ash with and without natural aggregates in terms of stability of aggregate particle size distribution, bearing capacity and permeability.

1. Experimental research

In Lithuania, until now MSWI bottom ash as a road material has been used in three construction projects: heavy vehicles parking lot (in 2018) and two pedestrian paths (in 2018 and 2020). All of them are involved in this study. In all cases, MSWI bottom ash with and without natural aggregates was used to construct unbound base course. The performance of that layer was evaluated based on the aggregate particle size distribution, bearing capacity and permeability that was determined at the construction phase and within operation.

1.1. MSWI bottom ash

In all cases MSWI bottom ash was generated in a waste-to-energy plant located in Klaipėda (Lithuania) and stored outside more than 3 months in uncovered stockpiles with free access to water and air. Once the aging (weathering) was completed, ferrous and non-ferrous metals were recovered from MSWI bottom ash. It is worth highlighting that because of the applied different technologies to recover metals, five fractions of bottom ash were produced until the autumn of 2018 (0/2, 2/4, 4/8, 5/11 and 11/22) and since then – two fractions (0/5 and 0/16). The characteristics of each fraction are given in Table 1. Their comparison to properties of natural aggregates and requirements for road materials are discussed in details by Vaitkus et al. (2018) and Gražulytė et al (2021). The leaching of metals from MSWI bottom ash passed the environmental requirements for application in civil engineering irrespective of fraction.

Table 1. Characteristics of bottom ash

Characteristic	MSWI bottom ash fraction						
	until autumn of 2018					since autumn of 2018	
	0/2	2/4	4/8	5/11	11/22	0/5	0/16
Water content, %	15.3	7.9	10.1	4.5	4.1	–	
Oven-dried particle density, Mg/m3	2.682	2.093	2.123	2.156	2.152	2.70	2.58
Loose bulk density, Mg/m3	1.141	1.097	1.171	1.105	1.044	–	
Flakiness index (FI), -	–	–	6	10	16	–	11
Shape index (SI), -	–	–	5	12	13	–	10
Percentage of crushed and broken surfaces, %	–	–	96	97	96	–	98
Resistance to fragmentation (LA), -	–	35	37	39	40	–	39
Water absorption, %	–	9.3	8.0	7.3	7.5	27.5	9.0
Resistance to freezing and thawing (loss mass), %	–	12.7	11.5	10.7	10.4	–	10.8

Template: Anonymized Document

1.2. Field studies

1.2.1. Heavy vehicles parking lot: 1st field study

MSWI bottom ash after aging (weathering) and recovery of metals was mixed with natural aggregates and used to construct unbound base course of heavy vehicle parking lot. The parking lot was located in Liepkalnio street (Vilnius). A principal scheme of pavement structure is illustrated in Figure 1 (a).The construction works started at the end of autumn 2017 and was finished in the spring of 2018. Therefore, five fractions of bottom ash (0/2, 2/4, 4/8, 5/11 and 11/22) and crushed gravel (fr. 16/45) were mixed in ratio at 3:1 to produce unbound mixture with nominal particle size of 45 mm. The particle size distribution of produced mixture is given in Figure 2.

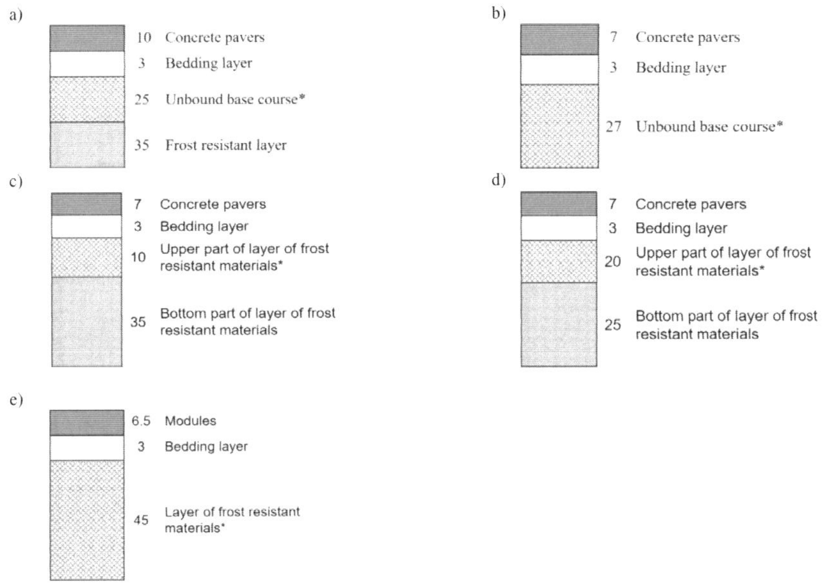

*Note: * – layer, in which MSWI bottom ash was used*

Figure 1. Pavement structures with MSWI bottom ash: a) 1st field study; b) 2nd field study; c–e) 3rd field study

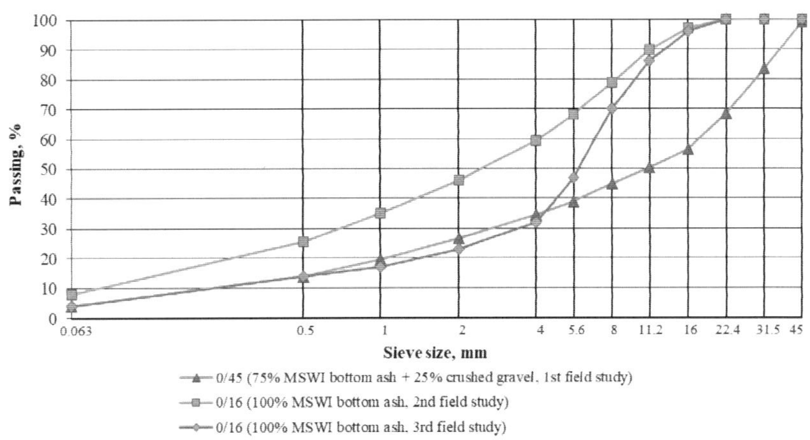

Figure 2. Particle size distribution of produced mixtures with MSWI bottom ash for unbound base course construction

Template: Anonymized Document

1.2.2. Pedestrian path: 2nd field study

MSWI bottom ash after aging (weathering) and recovery of metals was used to construct unbound base course of pedestrian path. The pedestrian path was located in Justiniškių street (Vilnius). A principal scheme of pavement structure is illustrated in Figure 2 (b). Unbound base course with MSWI bottom ash was constructed in 157 m of pedestrian path. The construction works was finished at the beginning of summer 2018. Therefore, five fractions of bottom ash (0/2, 2/4, 4/8, 5/11 and 11/22) were mixed to each other to produce unbound mixture with nominal particle size of 16 mm. The particle size distribution of produced mixture is given in Figure 2.

1.2.3. Pedestrian path: 3rd field study

MSWI bottom ash after aging (weathering) and recovery of metals was used to construct unbound base course of pedestrian path. The pedestrian path was located in Lakštingalų street (Vilnius). A principal scheme of pavement structure is illustrated in Figure 2 (c–e). Unbound base course with MSWI bottom ash was constructed in 63 m of pedestrian path. However, in this case the length of pedestrian path was spitted into three sub-sections and in each of them MSWI bottom ash was used to construct different part of unbound base course (layer of frost resistant materials). For example, 23 m of pedestrian path was with MSWI bottom ash layer in thickness of 10 cm while in other 17 m the thickness of MSWI bottom ash layer was increased up to 20 cm and in the last part of pedestrian path the whole layer was constructed from MSWI bottom ash. The construction works was finished at the end of summer 2020. Therefore, two fractions of bottom ash (0/5 and 0/16) were mixed to each other to produce unbound mixture with nominal particle size of 16 mm. The particle size distribution of produced mixture is given in Figure 2. All tests have been done on the last pavement structure.

1.3. Test methods

In Table 2 is given a summary of the determined properties of unbound base course from MSWI bottom ash at the construction phase and within operation. At least two samples were taken from different site places for testing.

Bearing capacity was determined by plate load test with either static or dynamic load. A bearing plate in a diameter of 300 mm was placed on the constructed base course or in the test pit (if test was carried out within operation) and then was loaded either with hydraulic loading device (static load) or falling weight (dynamic load). In the dynamic plate load testing, two loading cycles with different loads were applied in steps to the loading plate using a hydraulic hand pump. For each loading step, the corresponding settlement (deflection) of the plate was recorded. Based on the measured settlement (deflection) of base course in accordance to bearing pressure, a bearing capacity (modulus E_{v1} and E_{v2}) as well as the compaction level (ratio of E_{v2} to E_{v1}) was determined. In this case, measurements were done according to Lithuanian standard LST 1360-5. If dynamic plate load testing was used instead of static loading, the determined modulus (E_{avg}) was recalculated to the E_{v2}.

Table 2. Summary of determined characteristics of unbound base course from MSWI bottom ash

Characteristics	Test method	Number of field study (construction year)		
		1 (2018)	2 (2018)	3 (2020)
Construction phase				
Particle size distribution	EN 933-1	+	+	+
Optimal water content	EN 13286-2	-	-	+
Proctor density	EN 13286-2	-	-	+
Water permeability	CEN ISO/TS 17892-11	-	-	+
Bearing capacity	LST 1360-5	+	-	+
Within operation				
Particle size distribution	EN 933-1	-	+	-
Optimal water content	EN 13286-2	-	+	-
Proctor density	EN 13286-2	-	+	-
Water permeability	CEN ISO/TS 17892-11	-	+	-
Bearing capacity	LST 1360-5 or specific instruction	-	+	-

Template: Anonymized Document

2. Results and analysis

Particle size distribution of MSWI bottom ash with and without addition of natural aggregates at the construction phase and within operation is given in Figure 3. The results show that it is possible to produce homogenous unbound mixture by mixing different fractions of MSWI bottom ash and adding natural aggregates as needed. In the 1st field study all three samples taken from different site places had almost the same particle size distribution. The highest difference (6.3%) in passing was at sieve size of 31.5 mm and it decreased with lower sieve sizes. At sieve of 1 mm and lower, the difference in passing was lower than 0.6%. The amount of particles smaller than 0.063 mm varied from 3.9% to 4.5%.

In the 2nd field study, mixture with lower nominal aggregate size was used than in the 1st field study and it resulted in finer mixture. It is worth highlighting that in this case mixture was produced purely from MSWI bottom ash. The comparison of particle size distribution among samples taken from different site places showed that they are almost identical. The difference in passing was lower than 1.8% irrespective of sieve size. The test results after 1 year of operation (in 2019) showed that MSWI bottom ash is not prone to crushing under such low loads as pedestrian and bicycles. Contrary to expectations, it seems that some of the particles bounded with each other and formed slightly larger particles. But this phenomenon should be examined further since the difference in passing through sieves was lower than 4%. Regarding the amount of particles smaller than 0.063 mm, it decreased from 8.1–8.2% to 6.3–6.5%.

In the 3rd field study the curve of particle size distribution was not so smooth as in the 1st and 2nd field studies. Nevertheless, the passing through sieves of 11.2–22.4 mm and lower than 4 mm was almost the same as in the 2nd and 3rd field study, respectively. The passing through 4–11.2 mm varied from 32% to 86%. The amount of particles smaller than 0.063 mm was 3.9%. According to Lithuanian normative technical documents, the amount of particles smaller than 0.063 mm in unbound mixtures designed to construct base and sub-base course cannot be higher than 5%.

Optimal water content, Proctor density, water permeability and bearing capacity are given in Table 3. It seems that optimal water content of MSWI bottom ash mixture (fr. 0/16) is about 17–20% irrespective of what fractions of MSWI bottom ash were used to produce that mixture. Such high water is required because MSWI bottom ash is much more porous material than natural aggregates and as a results absorb more water.

Base (sub-base) course constructed of MSWI bottom ash is permeable. According to Lithuanian normative technical documents, coefficient of water permeability has to be higher than 1.0×10^{-5} m/s. Only one sample failed to pass this requirement. The coefficient of water permeability for that sample was 0.41×10^{-5} m/s. Too high amount of particles smaller than 0.063 mm may lead to this since it was more than 6%. Nevertheless, a sample taken in the same field study, but different place had the required permeability. Further study need to be taken to identify why in some places MSWI bottom ash mixture has the required permeability and in others fails. It has to be noted that when amount of particles smaller than 0.063 mm was limited to 3.9%, the coefficient of permeability was higher than 2.2×10^{-5} m/s. Materials with that water permeability can be used to construct sub-base (frost blanket course and the layer of frost resistant material) even in highways.

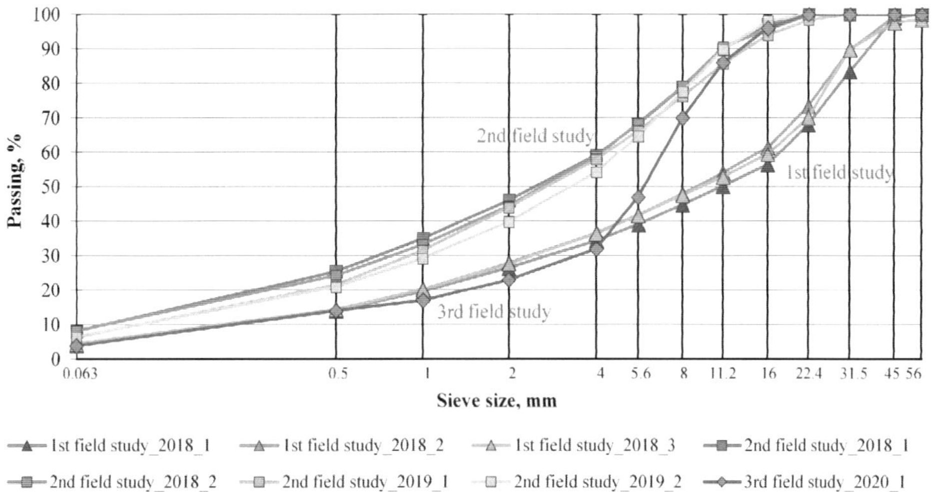

Figure 3. Particle size distribution of produced mixtures with MSWI bottom ash for unbound base course construction

Template: Anonymized Document

Table 3. Optimal water content, Proctor density, water permeability and bearing capacity of MSWI bottom ash mixtures

Characteristics	1st field study (75% MSWI BA + 25% crushed gravel (fr. 0/45))	2nd field study (100% MSWI BA (fr. 0/16))			3rd field study (100% MSWI BA (fr. 0/16))		
	2018	2018	2019		2020		
Optimal water content, %	-	-	21	17	21		
Proctor density, Mg/m^3	-	-	1.65	1.70	1.56		
Coefficient of water permeability, $\times 10^{-5}$ m/s	-	-	0.41	1.20	2.21		
E_{avg}, MPa			45.4	69.9	32.3	33.09	38.9
E_{v2}, MPa	165.9	-	100.7	149.8	64.6	66.2	77.7
E_{v2}/E_{v1}, –	1.82	-	-		-	-	-

It is recommended to achieve at least 100 MPa bearing capacity of unbound base (sub-base) course in pedestrian paths. In all field studies, it was implemented except the third one. There the strain modulus E_{v2} varied from 66.2 MPa to 77.7 MPa. However, the additional compaction or other solutions that increase the bearing capacity (e.g. construction of additional layer from stronger materials) was not applied. Meanwhile, in other field studies (first and the second one) E_{v2} varied from 100.7 MPa to 165.9 MPa. The highest value (165.9 MPa) was achieved in the first field study where 25% was crushed gravel fr. 16/45. Nevertheless, the analysis showed that 120 MPa can be achieved in both cases: with and without addition of natural aggregates into MSWI bottom ash mixture.

Every year all field studies are also visually investigated in order to identify any cracks, settlements, heaves and other defects that may appear due to use of MSWI bottom ash. Until now, all sections perform without premature failure, defects and distresses.

Conclusions

1. The results from conducted field studies support previous laboratory studies in possibility to use MSWI bottom ash as unbound base (sub-base) course material, especially in pedestrian and bicycles paths. For this purpose, mixtures made of either pure MSWI bottom ash (fraction 0/16) or mixed with natural aggregates (fraction 0/45) can be successfully used.
2. Analysis of particle size distribution at the construction phase and after 1 year of operation revealed that MSWI bottom ash does not crush when pavement is affected by equipment maintaining pedestrian and bicycles pathways. Contrary to expectations, it seems that some of the particles bounded with each other and formed slightly larger particles. To better understand this phenomenon, the particle size distribution should be examined further within operation of pedestrian path.
3. This study showed that permeability of MSWI bottom ash mixture is not a concern when amount of particles smaller than 0.063 mm is lower than 4% (in this case, the coefficient of permeability was higher than that required for highways ($\geq 2.0\times 10^{-5}$ m/s)). Otherwise, especially when amount of particles smaller than 0.063 mm is higher than 6%, the results were opposite to each other. Further study need to be taken to identify why in some cases even with higher amount of small particles MSWI bottom ash mixture has the required permeability and in others fails.
4. Tests carried out in the fields showed that mixtures with MSWI bottom ash has the required bearing capacity ($E_{v2} \geq 100$ MPa) for pedestrian and bicycle paths irrespective of mixture fraction and presence of natural aggregates if those layers are properly compacted. Moreover, E_{v2} in some test places was higher than 120 MPa, thus those mixtures could be also used to construct unbound base (sub-base) course in roads. However, a special attention has to be paid on the compaction since MSWI bottom ash has higher optimal water content than natural aggregates.
5. These field studies was a first attempt to use MSWI bottom ash as aggregate (mixture) to construct unbound base (sub-base) course and was limited to the loads generated by the equipment maintaining pedestrian and bicycles pathways. Thus, further comprehensive studies need to be carried out in order to get a full picture of MSWI bottom ash performance under real traffic and climate conditions. For this purpose, MSWI bottom ash as unbound base (sub-base) course mixture has to be tested in real road traffic and its performance has to be compared with that of typical pavement structure.

Template: Anonymized Document

Acknowledgements

This research was funded by Fortum Heat Lietuva, a private limited liability company.

Disclosure Statement

No potential conflict of interest was reported by the authors.

References

Alhassan, H. M., & Tanko, A. M. (2012). Characterization of solid waste incinerator bottom ash and the potential for its use. *International Journal of Engineering Research and Applications, 2*(4), 516-522.

An, J., et al. (2014). *Evaluating the use of waste-to-energy bottom ash as road construction materials*. Report No. BDK78-977- 20. Orlando: University of Central Florida.

Baun, D., Kamuk, B., & Avanzi, P. (2007). Treatment of bottom ash from waste to energy plants: overview and experiences. *11th International Waste Management and Landfill Symposium* (pp. 1-5). October, Cagliari.

Becquart, F., Bernard, F., Abriak, N.-E., & Zentar, R. (2009). Monotonic aspects of the mechanical behaviour of bottom ash from municipal solid waste incineration and its potential use for road construction. *Waste Management, 29*(4), 1320-1329. https://doi.org/10.1016/j.wasman.2008.08.019

Chandler, A. J., et al. (1997). Municipal solid waste incinerator residues. *Studies in Environmental Science, 67*, 1–974.

Chimenos, J., et al. (2000). Short-term natural weathering of MSWI bottom ash. *Journal of Hazardous Materials, 79*(3), 287–299. https://doi.org/10.1016/S0304-3894(00)00270-3

Chimenos, J., Segarra, M., Fernández, M. A., & Espiell, F. (1999). Characterization of the bottom ash in municipal solid waste incinerator. *Journal of Hazardous Materials, 64*(3), 211-222. https://doi.org/10.1016/S0304-3894(98)00246-5

del Valle-Zermeño, R., Chimenos, J. M., Giró-Paloma, J., & Formosa, J. (2014). Use of weathered and fresh bottom ash mix layers as a subbase in road constructions: environmental behavior enhancement by means of a retaining barrier. *Chemosphere, 117*(1), 402-409. https://doi.org/10.1016/j.chemosphere.2014.07.095

Dou, X., et al. (2017). Review of MSWI bottom ash utilization from perspectives of collective characterization, treatment and existing application. *Renewable and Sustainable Energy Reviews, 79*(May), 24–38. https://doi.org/10.1016/j.rser.2017.05.044

Forteza, R., Far, M., Seguí, C., & Cerdá, V. (2004). Characterization of bottom ash in municipal solid waste incinerators for its use in road base. *Waste Management, 24*(9), 899-909. https://doi.org/10.1016/j.wasman.2004.07.004

Gražulytė, J., Vaitkus, A., Šernas, O., & Žalimienė, L. (2021). The impact of MSWI bottom ash as aggregate on concrete mechanical performance. *International Journal of Pavement Engineering*. https://doi.org/10.1080/10298436.2021.1873333

Hjelmar, O., Holm, J., & Crillesen, K. (2007). Utilisation of MSWI bottom ash as sub-base in road construction: first results from a large-scale test site. *Journal of Hazardous Materials, 139*(3), 471-480. https://doi.org/10.1016/j.jhazmat.2006.02.059

ISWA. (2006). *Management of bottom ash from WTE plants*. Copenhagen: ISWA.

Izquiedro, M., et al. (2001). Use of bottom ash from municipal solid waste incineration as a road material. *In: International ash utilization symposium*. Center for Applied Energy Research, University of Kentucky, 1–8.

Izquierdo, M., Querol, X., Josa, A., Vazquez, E., & López-Soler, A. (2008). Comparison between laboratory and field leachability of MSWI bottom ash as a road material. *Science of The Total Environment, 389*(1), 10–19. https://doi.org/10.1016/j.scitotenv.2007.08.020

Sabbas, T., et al. (2003). Management of municipal solid waste incineration residues. *Waste Management, 23*(1), 61-88. https://doi.org/10.1016/S0956-053X(02)00161-7

Sormunen, L. A., & Kolisoja, P. (2017). Construction of an interim storage field using recovered municipal solid waste incineration bottom ash: Field performance study. *Waste Management, 64*, 107–116. https://doi.org/10.1016/j.wasman.2017.03.014

Sormunen, L. A., Kaartinen, T., & Rantsi, R. (2018). MSWI BA treated with Advanced Dry Recovery: a field scale study on materials' leaching properties. *International Journal of Sustainable Engineering, 11*(4), 261–271. https://doi.org/10.1080/19397038.2018.1444679

Vaitkus, A., Gražulytė, J., Šernas, O., Vorobjovas, V., & Kleizienė, R. (2019). An algorithm for the use of MSWI bottom ash as a building material in road pavement structural layers. *Construction and Building Materials*, 212, 456–466. https://doi.org/10.1016/j.conbuildmat.2019.04.014

Vaitkus, A., Gražulytė, J., Vorobjovas, V., Šernas, O., & Kleizienė, R. (2018). Potential of MSWI bottom ash to be used as aggregate in road building materials. *The Baltic Journal of Road and Bridge Engineering, 13*(1), 77–86. https://doi.org/10.3846/bjrbe.2018.401

Xie, R., et al. (2017). Assessment of municipal solid waste incineration bottom ash as a potential road material. *Road Materials and Pavement Design, 18*(4), 992-998. https://doi.org/10.1080/14680629.2016.1206483

Template: Anonymized Document

OPTIMISATION OF NANO-ZnO AND NANO-SiO_2 MIXING TIME FOR BITUMEN MODIFICATION

Judita GRAŽULYTĖ 0000-0001-6502-9400[1], Audrius VAITKUS 0000-0001-5103-9747[1], Simona BITARYTĖ 0000-0002-8331-5870[1]

[1]*Road Research Institute, Vilnius Gediminas Technical University, Vilnius, Lithuania*
Corresponding author: Judita, Gražulytė, judita.grazulyte@vilniustech.lt
Audrius, Vaitkus, audrius.vaitkus@vilniustech.lt
Simona, Bitarytė, simona.bitaryte@vilniustech.lt

Received day month year; accepted date (use style Received dates, or Alt + Ctrl + R)

Abstract. Nanotechnologies have gradually penetrated to the field of bitumen modification especially where durable asphalt mixtures have to be designed. Longer mixing time, higher temperatures or/and higher rotation (shearing) speeds are needed to increase the dispersion of nanoparticles in bitumen. However, this is not necessarily beneficial to the physical and mechanical properties of the final material. As a result, in this study nano-zinc oxide (nano-ZnO) and nano-silica (nano-SiO_2) mixing time for bitumen modification was optimized considering the physical and mechanical properties of the final bitumen. For this purpose bitumen PMB 25/55-60 was modified with nanoparticles at 180 °C using a laboratory high-shear mixer at a rotation speed of 4000 rpm for different modification time selected on the basis of literature review (60 and 90 minutes). Penetration, softening point, viscosity at 135 °C, recovery and non-recoverable creep compliance (multiple stress creep and recovery test) at 60 °C were measured in order to determine the optimal mixing time. The results showed that 60 minutes ensures the dispersion of nano-ZnO and nano-SiO2 in the bitumen PMB 25/55-60 and longer mixing time do not have a significant effect on the properties of nano-ZnO and nano-SiO2 modified bitumen (the difference was less than 7%).

Keywords: bitumen, nano-zinc oxide (nano-ZnO), nano-silica (nano-SiO2), modification time, nanoparticles

Introduction

Bitumen in asphalt pavements is constantly affected by traffic and climate, which inevitably degrades its physical, mechanical and chemical properties. This process is called aging. In recent decades, the problem of bitumen aging has received special attention. To improve the resistance of bitumen to aging, antioxidant additives are incorporated into bitumen. However, they are organic substances and their properties gradually change. It limits their use in the long-term (Du et al., 2015).

Nowadays, nanotechnology is playing an important role in modifying materials to ensure the high quality. Nano particles are defined as a particles with at least one dimension that is less than 100 nm (Yang et al., 2013). Nanomaterials such as nano-zinc oxide, nano-silica, nano-titanium dioxide, nano-clay and carbon nanotubes are suitable for bitumen modification. The conducted studies show that nanomaterials strengthen the binding properties of bitumen, increase the durability of the bitumen and resistance to transport loads. For example, asphalt pavement become more resistance to aging and increases resistance to rutting (Ezzat et al., 2016). Also, nanomaterials give bitumen greater ductility, increase softening temperature and reduce penetration (Rezaei et al., 2016, Ziari et al., 2014). As a result, nanoparticle-modified bitumen becomes more resistant to temperature cracks, permanent deformation and fatigue (Alhamali et al., 2016, Yusoff et al., 2019).

One of the most commonly used nanomaterials in bitumen modification is nano-zinc oxide (nano-ZnO) and nano-silica (nano-SiO_2). Nano-ZnO is an organic compound that looks like a milky white powder and is insoluble in water. Previous studies indicated that bitumen modified with nano-ZnO has lower stiffness, higher softening temperature, ductility, viscosity, resistance to rutting, aging and resistant to temperature cracks (Du et al., 2015, Zhang et al., 2018)

Nano–SiO_2 is an inorganic dioxide that looks like a white powder and is also insoluble in water. Nano-SiO_2 added in bitumen improves pavement performance. For example, pavement becomes more resistant to temperature cracks rutting, aging and fatigue (Alhamali et al., 2016, (Yusoff et al., 2019, Lazzara et al., 2010, Lee et al., 2013).

In order to ensure appropriate dispersion of nanoparticles to bitumen, mixing has to be performed using special equipment and specific mixing time, temperature and rotation (shearing) speed. However, different modification conditions are used among different scientist and there is no universally accepted technique. For example, Khairul et al. (2018) slowly added nano–ZnO in bitumen 60/70 and mixed them with high shear mixer at a rotation speed of 2000 rpm for 30 min at 135–150 °C. While other scientists, the rotation speed and mixing time reduced up to 12000 rpm and 4–5 min, respectively (Azarhoosh et al., 2018). Also, bitumen modification with nano-ZnO particles was conducted at a

Template: Anonymized Document

rotation speed of 2000–5000 rpm for longer period, i.e. 60–90 min (Du et al. 2015, Zhang et al. 2015, Liu et al. 2015). In previous researches, bitumen was also modified with nano–SiO_2. Sezavar et al. (2019) done it with with high shear mixer at 4000 rpm and mixed for 30 min at 163 °C. Other scientists, for modification also used a high shear mixer, but the rotation speed was lower (1500–3000 rpm) and mixing time was longer (for 60 min). The temperature was kept at 145–160 °C (Alhamali et al., 2016, Ezzat et al., 2016, Yusoff et al., 2019). Saltan et al. (2017) nano–SiO2 added at 160 °C and mixed at a rotation speed of 4000 rpm for 120 min.

It is obvious that modification conditions as time, temperature and rotation speed are still not optimized although it has a significant effect. For example, mixing the nanomaterial for too short or at too low temperature does not ensure the dispersion of the nanomaterial, on the other hand, mixing too long or at too high temperature, the bitumen will age and this will affect its performance. Therefore, the purpose of this work is to evaluate the effect of mixing time on the physical and mechanical properties of nano–ZnO and nano–SiO_2 modified bitumen.

1. Experimental research

In this study, polymer modified bitumen PMB 25/55-60 was modified with nanomaterials. Chemical composition and physical properties of bitumen PMB 25/55-60 are shown in Figure 1 and Table 1. The bitumen was modified with 6% nano-ZnO or 1% nano-SiO_2. Table 2 represents the physical and chemical properties of nano-ZnO and nano-SiO_2.
The bitumen heating temperature for bitumen modification was selected in accordance with the requirements of standard EN 12594. Bitumen PMB 25/55-60 was heated at 180 °C in oven until it flowed fully. A laboratory high-shear mixer with rotation speed of 4000 rpm was used to disperse nano–ZnO or nano-SiO_2 into bitumen. Two different mixing times (60 and 90 minutes) were selected based on the literature review. To acquire a homogeneous dispersion, nanoparticles were gradually added into the bitumen.

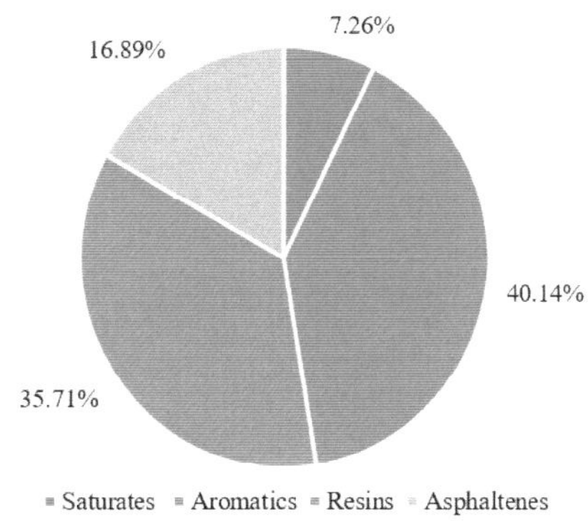

Figure 1. Chemical composition of bitumen PMB 25/55-60

Table 1. Physical properties of bitumen PMB 25/55-60

Characteristic	Test standard	Value
Softening Point (°C)	EN 1427:2015	65.0
Penetration at 25 °C (dmm)	EN 1426:2015	31.2
Viscosity at 135 °C (mPa*s)	EN 13302:2018	2767.7
Elastic Recovery (%)	EN 13398:2018	72.8

Template: Anonymized Document

Table 2. Properties of nano-ZnO and nano-SiO_2

Properties	Nanomaterials	
	nano-ZnO	nano–SiO_2
Particle size (nm)	30	15
Purity (%)	99.5	99.5
Density at 20 °C (g/cm^3)	5.606	-
Melting point (°C)	1975	>1700
Physical appearance	milky white powder	white powder

The effect of mixing time on the physical and mechanical properties of nano–ZnO and nano–SiO_2 modified bitumen were determined by penetration, softening point, dynamic viscosity, recovery and non-recoverable creep compliance. The penetration was determined according to European standard EN 1426:2015. The softening point was measured according to EN 1427:2015 by using a ring and ball apparatus. Brookfield viscometer was employed to measure the dynamic viscosity at 135 °C according to standard EN 13302:2018.

Mechanical properties of the modified bitumen were determined using Anton Paar M302 dynamic shear rheometer (DSR). Multiple Stress Creep and Recovery (MSCR) test was performed according to AASHTO TP 70:2013 and EN 16659:2016. Tests were conducted at 60 °C temperature and stress of 0.1 kPa and 3.2 kPa. Therefore, a 1 mm gap and 25 mm diameter plates were used.

2. Results and analysis

The effect of mixing time on the basic properties of nano–ZnO (6%) and nano–SiO_2 (1%) modified bitumen PMB 25/55-60 are shown in Figure 2–4. It was observed from Figure 2 that both mixing time and nanomaterial type only slightly affect the penetration. It varied from 30.5 dmm (1% nano-SiO_2 and mixing time – 90 minutes) to 32.3 dmm (1% nano-SiO_2 and mixing time – 60 minutes) while reference value was 31.2 dmm. The softening point of the nanomaterial modified bitumen increased compared to the reference bitumen from 65.0 °C to 67.6–69.0 °C. However, the effect of longer mixing time on softening point was different depending on the nanomaterial type: it decreased from 68.5 °C to 67.6 °C for nano-ZnO while for nano-SiO_2 – increased from 67.7 °C to 69.0 °C (Figure 3). As shown in Figure 4, the viscosity of nanomaterial modified bitumen increased by 5–8% irrespective of mixing time, except for nano-ZnO modified bitumen that was mixed for 90 minutes. For this one, it decreased by 2.2%. As nano–SiO_2 was added into bitumen PMB 25/55-60 and mixing time was prolonged from 60 minutes to 90 minutes, viscosity increased from 2977 mPa·s to 3008 mPa·s.

The MSCR test results are shown in Figure 5–6. It is obvious that the use of nanomaterials to modify bitumen PMB 25/55-60 leads to higher recovery and lower non-recoverable creep compliance irrespective of both stress level and nanomaterial type. At stress level of 0.1 kPa, recovery increased from 49% to 57–58% and 59–60% while non-recoverable creep compliance decreased from 0.19 1/kPa to 0.13–0.14 1/kPa and 0.12–0.13 1/kPa respectively for nano–ZnO and nano–SiO_2 modified bitumen. The increase in stress level from 0.1 kPa to 3.2 kPa resulted in the same trend: recovery increased from 37% to 44–46% and 46–48% while non-recoverable creep compliance decreased from 0.24 1/kPa to 0.18–0.20 1/kPa and 0.17–0.18 1/kPa respectively for nano–ZnO and nano–SiO_2 modified bitumen. As observed from Figures 5–6, prolonged mixing time slightly affected test results. In all cases, recovery slightly increased and non-recoverable creep compliance decreased.

In order to better evaluate the impact of nanomaterials (nano–ZnO and nano–SiO_2) and mixing time on the physical and mechanical properties of bitumen PMB 25/55-60, the change of each property in percentage was calculated (Table 3–4). The results show that nanomaterials have significantly higher effect on the mechanical properties of PMB 25/55-60 comparing to the physical properties. The change in penetration, softening point and viscosity was lower than 9% while recovery and non-recoverable creep compliance increased (decreased) by 15–35% depending on both nanomaterial type and mixing time. Focusing only on the effect of mixing time, it was observed that change in each property is lower than 6% and 7% for nano–ZnO and nano–SiO_2, respectively. This kind of change in bitumen properties is not significant and keeping in mind that longer mixing time may accelerate aging process in the bitumen, which is undesirable, mixing time of 60 minutes for bitumen modification with nanomaterials is assumed as optimal. As a result, slow addition of nanomaterials into bitumen at 180 °C and mixing with a high-speed mixer for 60 minutes at a speed of 4000 rpm is considered as the most rational technology to modify bitumen with nanomaterials. This technique ensures the homogenous dispersion of nanomaterials.

Template: Anonymized Document

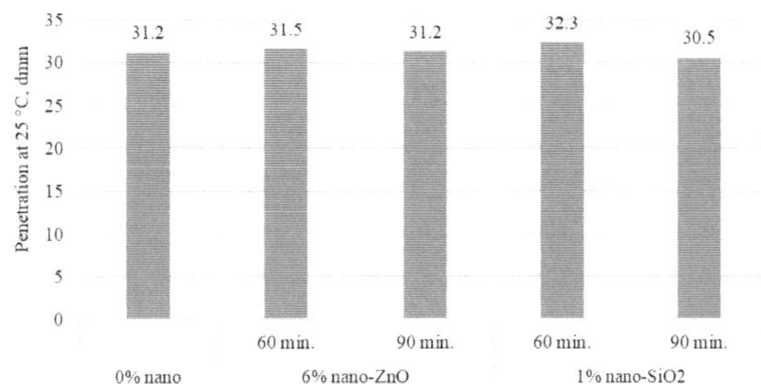

Figure 2. Penetration at 25 °C

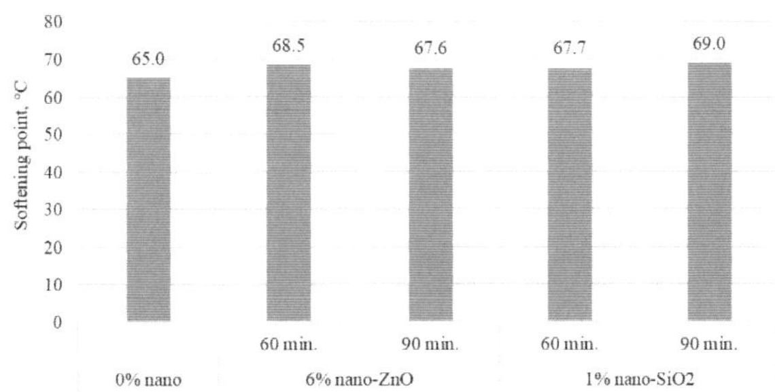

Figure 3. Softening point

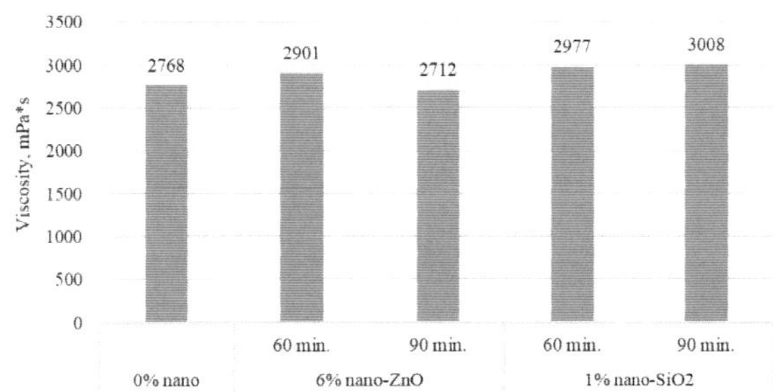

Figure 4. Dynamic viscosity

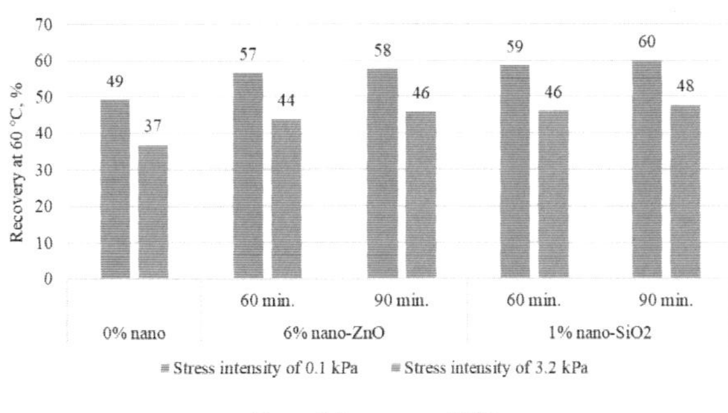

Figure 5. Recovery at 60 °C

Template: Anonymized Document

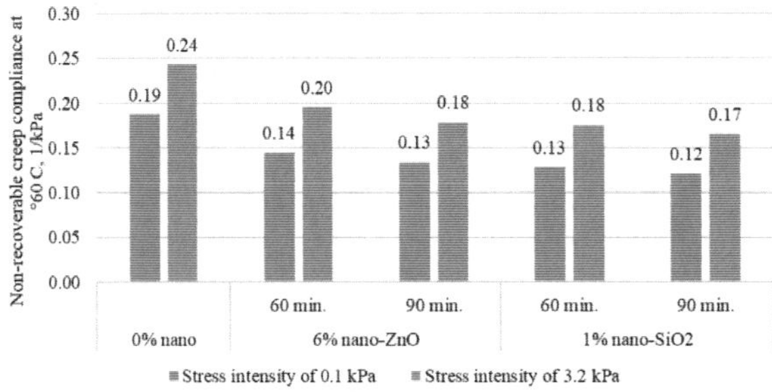

Figure 6. Non-recoverable creep compliance at 60 °C

Table 3. Effect of nano-ZnO and prolonged mixing time on bitumen performance

Change in property	6% nano-ZnO		Influence of longer mixing time
	60 min.	90 min.	
Penetration at 25 °C (%)	+1.1	+0.1	1.0
Softening point (%)	+5.4	+3.9	1.5
Viscosity at 135 °C (%)	+4.8	-2.0	6.8
Recovery at 0.1 kPa and 60 °C (%)	+15.2	+17.3	2.1
Recovery at 3.2 kPa and 60 °C (%)	+19.1	+24.6	5.5
Non-recoverable creep compliance at 0.1 kPa and 60 °C (%)	-22.7	-28.3	5.6
Non-recoverable creep compliance at 3.2 kPa and 60 °C (%)	-19.4	-26.4	7.0

Table 4. Effect of nano-SiO_2 and prolonged mixing time on bitumen performance

Change in property	1% nano-SiO_2		Influence of longer mixing time
	60 min.	90 min.	
Penetration at 25 °C (%)	+3.5	-2.2	5.8
Softening point (%)	+4.1	+6.1	2.0
Viscosity at 135 °C (%)	+7.6	+8.7	1.1
Recovery at 0.1 kPa and 60 °C (%)	+19.3	+21.8	2.5
Recovery at 3.2 kPa and 60 °C (%)	+25.2	+29.5	4.2
Non-recoverable creep compliance at 0.1 kPa and 60 °C (%)	-31.2	-35.2	4.0
Non-recoverable creep compliance at 3.2 kPa and 60 °C (%)	-27.6	-31.8	4.2

Conclusions

1. The conducted study showed that longer mixing time (90 minutes) only slightly (6–7%) changes bitumen characteristics (penetration, softening point, viscosity and resistance to permanent deformation) comparing to mixing time of 60 minutes.

2. The slow addition of nanomaterials to bitumen at 180 °C temperature and mixing with a high-speed mixer for 60 minutes at a speed of 4000 rpm is considered as the most rational technology to modify bitumen with nanomaterials. It leads to the homogenous dispersion of nanomaterials and prevents bitumen from premature aging that might happen because of too long mixing at such high temperature.

3. Laboratory tests with bitumen PMB 25/55-60 modified with either 6% nano-ZnO or 1% nano-SiO_2 showed that nanomaterials harden the bitumen and as a result, lead to higher softening point, viscosity and resistance to permanent deformation. In fact, the effect of nanomaterials on bitumen mechanical performance is significantly higher than on physical properties.

Template: Anonymized Document

4. Since this study was limited in terms of test methods and only one amount of each nanomaterial was tested, further comprehensive studies need to be carried out in order to get a full picture of effect of nanomaterial on bitumen performance especially on mechanical properties and identify the optimal content of nano-ZnO and nano-SiO_2 for bitumen modification. Moreover, a special attention has to be paid on resistance to low temperature cracking because as this study has shown nanomaterials harden the bitumen and it is the main cause for this type of cracking.

Disclosure Statement

No potential conflict of interest was reported by the authors.

References

Alhamali, D. I., Wu, J., Liu, Q., Hassan, N. A., Yusoff, N. I. M., Albrka, S. I., & Ali, A. S. I. 2016. Physical and Rheological Characteristics of Polymer Modified Bitumen with Nanosilica Particles, *Arabian Journal for Science and Engineering* 41: 1521–1530. https://DOI: 10.1007/s13369-015-1964-7

American Association of State Highway and Transportation Officials. (2013). *AASHTO TP:2013: Standard Method of Test for Multiple Stress Creep Recovery (MSCR) Test of Asphalt Binder Using a Dynamic Shear Rheometer (DSR).*

Azarhoosh, A. R., Nejad, F. M., & Khodaii, A. 2018. Nanomaterial and fatigue cracking of hot mix asphalt, *Road Materials and Pavement Design* 0629(19:2): 353–366. https://DOI: 10.1080/14680629.2016.1261724

Du, P. F., Ke, N. X., & Zhang, H. L. 2015. Effect of Nano-zinc Oxide on the Morphology and Ultraviolet Aging Properties of Various Bitumens, *Petroleum Science and Technology* 6466(33:10): 1110–1117. https://DOI: 10.1080/10916466.2015.1043739

European Standard. (2015). *EN 1426:2015: Bitumen and bituminous binders – Determination of needle penetration.*

European Standard. (2015). *EN 1427:2015: Bitumen and bituminous binders – Determination of the softening point - Ring and Ball method.*

European Standard. (2015). *EN 12594:2015: Bitumen and bituminous binders – Preparation of test samples.*

European Standard. (2016). *EN 16659:2016: Bitumen and Bituminous Binders – Multiple Stress Creep and Recovery Test (MSCRT)*

European Standard. (2018). *EN 13302:2018: Bitumen and bituminous binders – Determination of dynamic viscosity of bituminous binder using a rotating spindle apparatus.*

Ezzat, H., El-badawy, S., Gabr, A., Zaki, & E.I. 2016. Evaluation of Asphalt Binders Modified with Nanoclay and Nanosilica, *Procedia Engineering* 143(Ictg): 1260–1267. https://DOI: 10.1016/j.proeng.2016.06.119

Khairul, Y., Mohd, N., Ezree, A. M., Khairul, A. M., Nurul, K., Mohd, H., & Haniffah, T. 2018. Physical and rheological properties of nano zinc oxide modified asphalt binder, *MATEC Web of Conferences* (January). https://DOI:10.1051/matecconf/201825002004

Lazzara, G., & Milioto, S. 2010. Dispersions of nanosilica in biocompatible copolymers, *Polymer Degradation and Stability* 95(4): 610–617. https://DOI: 10.1016/j.polymdegradstab.2009.12.007

Liu, H. Y., Zhang, H. L., Hao, P. W., & Zhu, C. Z. 2015. The Effect of Surface Modifiers on Ultraviolet Aging Properties of Nano-zinc Oxide Modified Bitumen, *Petroleum Science and Technology* 6466(33:1): 72–78. https://DOI: 10.1080/10916466.2014.948119

Rezaei, S., Ziari, H., & Nowbakht. S. 2016. High-temperature functional analysis of bitumen modified with composite of nano-SiO_2 and styrene butadiene styrene polymer, *Petroleum Science and Technology* 34(13): 1195–1203. https://DOI:10.1080/10916466.2016.1188112

Rezaei, S., Ziari, H., & Nowbakht, S. 2016. Low temperature functional analysis of bitumen modified with composite of nano-SiO_2 and styrene butadiene styrene polymer, *Petroleum Science and Technology* 34(5): 415–421. https://DOI: 10.1080/10916466.2016.1143841

Saltan, M., Terzi, S., & Karahancer, S. 2017. Examination of hot mix asphalt and binder performance modified with nano silica, *Construction and Building Materials* 156: 976–984. https://DOI: 10.1016/j.conbuildmat.2017.09.069

Sezavar, R., Shafabakhsh, G., & Mirabdolazimi, S. M. 2019. New model of moisture susceptibility of nano silica-modified asphalt concrete using GMDH algorithm, *Construction and Building Materials* 211: 528–538. https://DOI:10.1016/j.conbuildmat.2019.03.114

Yang, J., & Tighe, S. 2013. A review of advances of Nanotechnology in asphalt mixtures, *Procedia - Social and Behavioral Sciences* 96(Cictp): 1269–1276. https://DOI: 10.1016/j.sbspro.2013.08.144

Yao, H., You, Z., Li, L., Lee, C. H., Wingard, D., Yap, Y. K., Shi, X., & Goh, S. W. 2013. Rheological Properties and Chemical Bonding of Asphalt Modified with Nanosilica, *Journal of Materials in Civil Engineering* (November): 1619–1630. https://DOI: 10.1061/(ASCE)MT.1943-5533.0000690

Yusoff, N. I. M., Alhamali, D. I., Ibrahim, A. N. H., Rosyidi, S. A. P., & Hassan, N. A. 2019. Engineering characteristics of nanosilica /polymer-modified bitumen and predicting their rheological properties using multilayer perceptron neural network model, *Construction and Building Materials* 204: 781–799. https://DOI: 10.1016/j.conbuildmat.2019.01.203

Zhang, H., Gao, Y., Guo, G., Zhao, B., & Yu, J. 2018. Effects of ZnO particle size on properties of asphalt and asphalt mixture, *Construction and Building Materials* 159: 578–586. https://DOI: 10.1016/j.conbuildmat.2017.11.016

Zhang, H., Zhu, C., Yu, J., Shi, C., & Zhang, D. 2015. Influence of surface modification on physical and ultraviolet aging resistance of bitumen containing inorganic nanoparticles, *Construction and Building Materials* 98: 735–740. https://DOI: 10.1016/j.conbuildmat.2015.08.138

Ziari, H., Farahani, H., Goli, A., & Galooyak, S. S. 2014. The Investigation of the Impact of Carbon Nano Tube on Bitumen and HMA Performance, *Petroleum Science and Technology* 6466(32:17): 2102–2108. https://DOI: 10.1080/10916466.2013.763827

VIASTRUCTURA – A NEW WAY OF PAVEMENT STRUCTURE DESIGN

Audrius VAITKUS (ORCID ID 0000-0001-5103-9747), Rita KLEIZIENĖ (ORCID ID 0000-0001-7478-3069), Martynas KARBOČIUS (ORCID ID 0000-0002-3904-3214)

Road Research Institute, Vilnius Gediminas Technical University, Vilnius, Lithuania
Corresponding author: Martynas Karbočius, martynas.karbocius@vilniustech.lt

Received day month year; accepted date

Abstract. The catalogues of standard pavement structures are common way to design road pavements. In 2019, new regulation for the design of standard pavement structures KPT SDK 19 was issued in Lithuania. One of the new requirements require verification of layer thickness of high-class pavement structures. Such verification should be done by internationally approved mechanistic-empirical methods. In addition, it is recommended to use the same methods to adjust the layer thickness of the selected standard pavement structure for lower classes. These calculations are particularly applicable when the design load (ESAL) is at the lower or upper limit of the class range. Vilnius Tech Road Research Institute experts and outsource IT specialists spent two years for the design model ViaStructura development. Web software based on mechanistic-empiric approach include the boundary conditions, based on Austria, the United States and Germany experience and laboratory test results of construction materials. Materials can be selected from created database, which can be simply expanded with the new materials by the user. as Additional function allow comparison of separate designed pavement structures. The article present the concept of the ViaStructura model for the design of flexible pavement structures, reveals its main principles and advantages comparing to the pavement structure selection by the standard catalogue..

Keywords: ViaStructura, pavement structure, mechanistic-empirical method, design software, asphalt fatigue, permanent deformation, pavement design.

Introduction

Typically, the pavement structures for road are designed according to catalogues (or guidelines) of standard pavement structures. The standard pavement structure performs well at standard (empirically approved) conditions, however its is not adapted to special purpose pavements, do not take into account the impact of special loads that occurs when:

- heavy traffic travels on the same tracks;
- heavy vehicles travel exclusively one after the other;
- heavy traffic occurs in short radius turns;
- there is slow heavy traffic;
- heavy transport brakes and accelerates often;
- heavy traffic is in the intersection area;
- at heavy vehicle parking lot.

Therefore, standard pavement structures limit the application of new and non-standard materials to the construction of the pavement structure. The concept of a standard pavement structure catalog simplifies the design process, however the decision is not always optimal due to many approximations associated with design load, hydrothermal condition, climate condition (pavement temperature and frost heave), material properties. Vaitkus, Žalimienė, Židanavičiūtė and Žilionienė (2019) found that one of the most significant factors for pavement structure strength is the bearing capacity of subgrade and hydrothermal impact which are highly interconnected. Mshali and Steyn (2020) investigated the effect of truck speed on the response of flexible pavement structures. It was found that increased vehicle speed results in the decrease in elastic surface deflection response which results in lower stresses and strains and affects the further performance of the pavement structure. Tran and Hall (2007) found that the spectrum of axle loads on the studied roads differed and significantly influenced the performance of the pavement structure. The study showed that the predicted pavement service life differed by up to 25 percent compared to the values determined using the default load values.. This shows that underestimation of such factors can lead to unsatisfactory performance of the pavement structure, faster degradation or inefficient use of funds.

Also, other currently valid normative technical documents in Lithuania do not encourage the design and substantiation of the suitability of design solutions, as they are focused on the application of standard materials, indirectly assessing the impact of their mechanical properties on the wear rate of the pavement structure. There has been a lot of research in recent years on the use of new materials in road construction. Researchers have explored the potential of using materials such as waste (Choudhary, Chattopadhyay, Kumar and Julaganti, 2017, Vaitkus, Gražulytė, Vorobjovas, Šernas and Kleizienė, 2018), geogrid (Zofka, Maliszewski and Maliszewska, 2017, Šiukščius, Vorobjovas, Vaitkus, Mikaliūnas and Zarinš, 2019), noise-reducing pavement (Vaitkus, Vorobjovas, Jagniatinskis, Adriejauskas and Fiks, 2014, Vaitkus, Šernas, Vorobjovas and Gražulytė, 2019), and a design model based on a mechanistic-empirical method could be a useful tool to assess the performance of pavement structures using such materials and possibly accelerate the practical application. Therefore, it is necessary to evaluate the possibilities of adapting fast non-destructive bearing capacity determination methods to Lithuanian conditions. For these purposes, by order of the Lithuanian Road Administration under the Ministry of Transport and Communications, a design model for asphalt pavement structures with application software ViaStructura was developed, based on a mechanistic-empirical approach. This article reviews the concept of the ViaStructura model for the design of flexible pavement structures, reveals its main principles, highlights the possibilities and advantages of application compared to the application of standard pavement structures, and takes into account the challenges of the model application.

1. ViaStructura pavement structure design model

ViaStructura is the design model of flexible pavement structures based on the mechanistic-empirical approach. The design result, in this case, is not the thickness of the layers of the pavement structure, but the effect on the trial pavement structure and its resistance to defined type of distress. Figure 1 presents the procedure of pavement structure design according to ViaStructura.

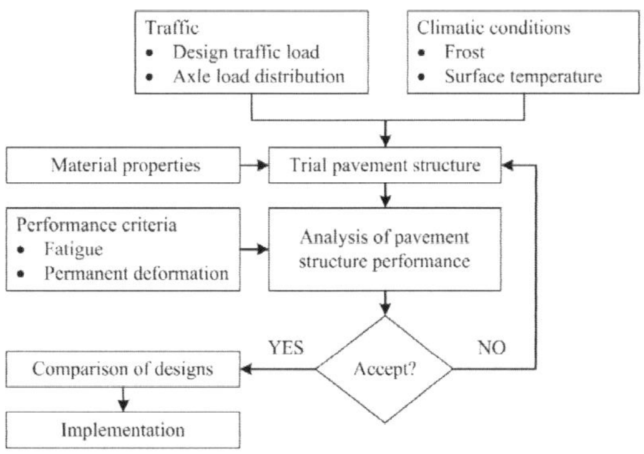

Figure 1. ViaStructura design procedure of pavement structure

The model first evaluates traffic loads, climatic conditions, subgrade soils, then proceeds to the selection of layer thicknesses and materials for the trial pavement structure. The trial pavement structure is then analyzed through performance criteria taking into account user input parameters. ViaStructura model is subject to the performance criteria for resistance to fatigue in bound layers and permanent deformations in unbound layers and subgrade. If the design does not meet the desired performance criteria, it has to be revised and the evaluation process repeated. The user can comprehensively analyze the interaction of input parameters and results, compare different designs and thus make optimal decisions. Each stage of the design process is discussed in more detail in the subsections below.

1.1. Design traffic load

The first requirement for a designed pavement structure is that it must withstand the cumulative traffic loading in the design lane over the selected design period. Estimation of this loading requires the calculation of the cumulative number of heavy vehicle axles over the design period. Design model ViaStructura gives two different methods to calculate the design load, which are based on Lithuanian Design rules for standard road pavement structures KPT SDK 19. These methods are:

- According to the average daily traffic of heavy vehicles;
- According to the axle loads of particular classes of heavy vehicles.

The calculation method, based on average daily traffic when data of vehicle axle loads is absent, evaluates the average quantity of axles and their effect on pavement structure expressed in equivalent standard axles through empirical coefficients related to the category of the road. In contrast, the calculation method, based on data of axle loads of heavy vehicles, evaluates the effect on pavement structure of each axle expressed in equivalent standard axles through fourth-order exponential function. To bring the design conditions as close as possible to the real operating conditions of the structure, the load distribution has been applied in the model. The distribution consists of 11 load ranges from 0 to 22 tons, for each of which the stresses and strains of the pavement structure are calculated later. The percentage distribution of axle loads intervals by road category in Lithuania is presented in Figure 2.

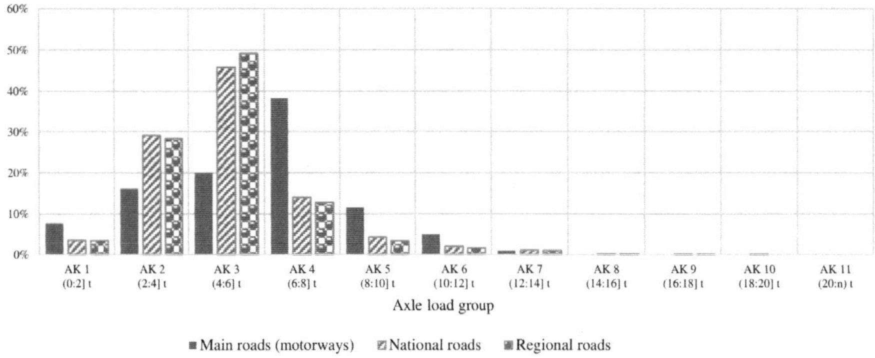

Figure 2. Percentage distribution of axle loads by road category in Lithuania (Kleizienė, Vaitkus and Čygas, 2015)

The flow of heavy vehicles, as well as the loads, are unique to each region and even the road, so it is recommended to perform periodic weighing of heavy vehicles on the roads. It is also recommended to perform heavy axle load analysis on priority, strategic roads or road structures before adopting pavement structure design solutions and to assess additional threats and impact on pavement structure deterioration.

1.2. Climatic conditions

The performance of the road structure depends not only on the traffic loads but also on the effects of climate. Climate (hydrothermal conditions and pavement temperature) significantly influences the effects of transport loads by reducing or increasing them. It is important to ensure that precipitation has no or little effect on the load-bearing capacity of the pavement structure, i. properly designed transverse and longitudinal profiles of the road, and ensure good drainage. It is also very important to reduce the negative impact of temperature on the pavement structure. Estimation of climatic conditions in the ViaStructura model includes assurance of sufficient resistance to frost damage through the minimum total thickness of pavement structure, and determination of hydrothermal condition seasonality effect as well as pavement surface temperature distribution.

Simonsen and Isacsson (1999), Thomson et al. (1987) and Huang (1993) found that the load-bearing capacity of the pavement structure increases to 15% in winter, returns to its original state in summer, and the load-bearing capacity of the subgrade decreases from 15% to 70% in autumn and spring. Sufficient frost resistance of the pavement structure and protection against possible damage due to repeated exposure to freezing and thawing cycles are ensured by the thickness of the pavement structure. If no special tests have been carried out or there is no experience in determining the thickness of the frost-resistant pavement structure, then this thickness shall be calculated taking into account:

- maximum frost depth;
- the frost sensitivity of subgrade soils;
- design traffic load;
- the provisions for adjusting the thickness of the pavement structure.

The thickness of the frost-resistant pavement structure shall include the stabilized top layer of frost-sensitive subgrade soil, but shall not include the improved top layer of frost-sensitive subgrade soil, even if tests have shown that it has become non-frost susceptible.

During the design of the new pavement structure, it is assumed that the hydrogeological conditions are neutral in all cases, due to the requirements for construction to ensure good drainage. However, during the design of the pavement rehabilitation, there may be a need to assess unfavorable hydrothermal conditions due to seasonal climate change. The seasonal decrease in soil bearing capacity is estimated when the pavement temperature changes from -10 to + 10 ° C. The level of decrease of bearing capacity is based on RDO Asphalt 09 (FGSV 498, 2009) assuming a 0% reduction in F1 class (non-frost susceptible) soil bearing capacity, a 45–55% reduction in F2 class soil (low to moderately frost susceptible) and an 65–80% reduction in F3 class soil (highly frost susceptible) bearing capacity. It is assumed that in the case of qualified improvement or stabilization of subgrade soils, seasonal fluctuations of hydrogeological conditions do not affect the bearing capacity of the pavement structure.

The load-bearing capacity of a flexible pavement structure depends on the temperature due to the composition and physical properties of the asphalt mix. Therefore, when designing a flexible pavement structure, possible temperature and load scenarios during operation must be considered. Studies by Nunn and Smith (1997) have shown that the stiffness modulus of an asphalt mix usually decreases from 0.12 GPa to 0.06 GPa with increasing temperature by 1 °C. It has also been found that the sensitivity of the stiffness modulus of an asphalt mix due to a change in temperature decreases as an asphalt mix has higher stiffness modulus (Nunn and Smith, 1997).

Typical pavement structures designed and constructed for the average daily temperature decay faster than the expected service life due to the prolongation and heating of the summer season, which results in a change in the top asphalt layer volume, binder rise to the pavement surface, and faster plastic deformation (rutting) (Qiang, Mills and Mcneil, 2011; Mundt, Marano, Nunes, and Adams, 2009). Due to extreme climatic factors (extremely negative or positive temperatures, heavy rainfall and excess humidity, deeper than average freezing depth and a higher number of freezing/thawing cycles), the service life of the pavement structure is significantly reduced in terms of degradation and wear factors (Meyer, Amekudzi and O'Har, 2010; Meyer and Wiegel, 2011). Therefore, the methodology of modern pavement construction and material selection must be flexible and allow to estimate temperature and its variation.

The influence of temperature on the performance of the pavement structure depends on the thickness of the pavement structure. When modeling the temperature change of a pavement structure, it is assumed that the temperature in changes (increases or decreases) only in the vertical direction (it is assumed that the temperature does not change in the horizontal direction) (Herb, Marasteanu, and Stefan, 2006).

The temperature change in the pavement structure is modeled by dividing each layer of the pavement structure into sublayers of smaller thickness. In the ViaStructura model, the upper and lower layers of asphalt are divided into 1 cm thick sublayers, and the asphalt base layer is divided into 4 parts as the temperature change gradient is higher in the upper part of the pavement.

The ViaStructura model is based on a simplified method for determining the temperature change in the pavement used in the calculations of the pavement structure in Germany (FGSV 498, 2009). In Germany, the surface temperature of the pavement surface varies from -15 to +50 ° C. This pavement temperature range is divided into 13 ranges every 5 ° C. Based on the statistical distribution of the surface temperature in the intervals, the territory of Germany is divided into 4 zones. When developing the ViaStructura model, it was taken into account that the surface temperature in Lithuania drops below -15 ° C, therefore the temperature distribution has been expanded. According to the statistical temperature distribution, Lithuania was divided into 3 zones.

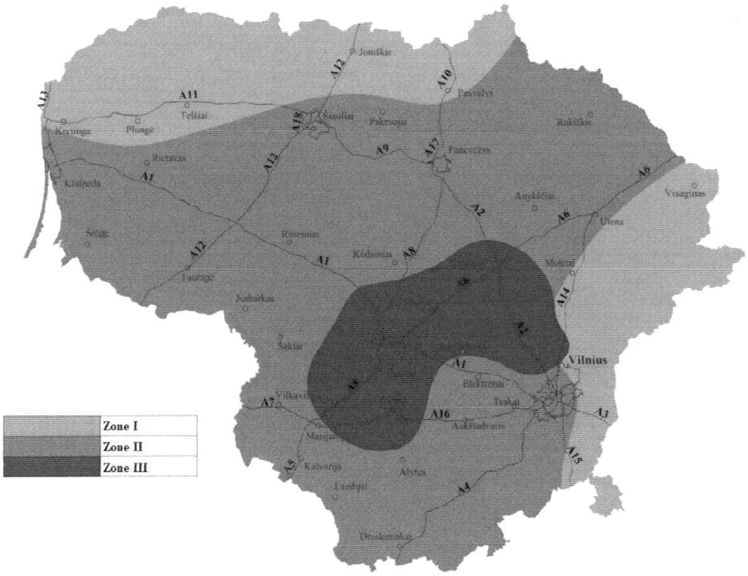

Figure 3. Zones according to the percentage distribution of pavement surface temperature intervals in Lithuania (Kleizienė, Vaitkus, Židanavičiūtė and Marcinkevičius, 2017)

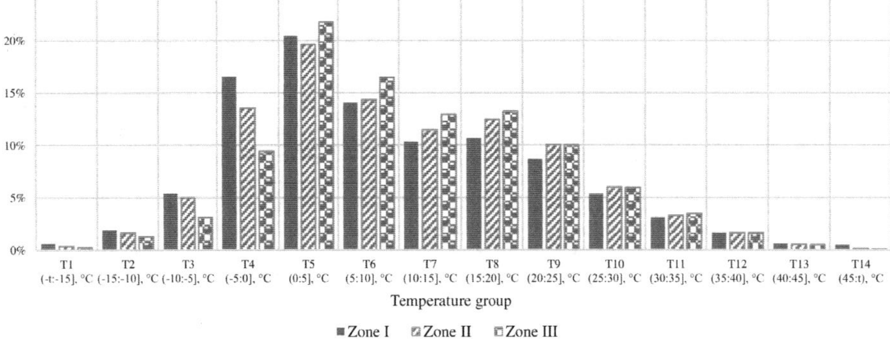

Figure 4. Percentage distribution of pavement surface temperature intervals by zones in Lithuania (Kleizienė, Vaitkus, Židanavičiūtė and Marcinkevičius, 2017)

Knowing the pavement surface temperature, the asphalt pavement temperature at any depth can be calculated. The temperature gradient (change) function in asphalt layers under different conditions can be estimated according to the equation given by Speth (1985) or Hess (1996):

$$y = b \cdot ln(0{,}01 \cdot x + 1{,}00) + T, \qquad (1)$$

where: y – temperature at the depth x of the asphalt layer ° C; x – depth at which the temperature is to be determined, mm; T – temperature of the pavement surface, °C; b – parameter depending on the temperature of the pavement surface.

The calculated temperatures are used to calculate the stiffness modulus of each asphalt sublayer, which is used as input to the MnLayer (Khazanovich & Wang, 2007) algorithm to determine the reactions of the pavement structure.

1.3. Trial pavement structure and materials

When designing the pavement structure by the ViaStructura, first of all, the computational model of the structure is created, the loads and configuration of the vehicle wheels are defined, and the stresses and strains at critical layers of structure are calculated. The input data used to calculate the response of the pavement structure to loads are:

- tire-pavement contact area (fixed) and magnitude (dependent on road significance) of the wheel load;
- thickness of each layer.
- stiffness modulus and Poisson's coefficients of each layer;
- bonding (friction) between the separate layers.

The ViaStructura model offers three methods for entering the properties of asphalt mixtures:

- stiffness modulus at different temperatures;
- Hirsch method;
- Francken/Verstraeten method.

In the first method, the stiffness modulus of the asphalt mixture is entered at 15 different temperatures. Based on the entered stiffness modules, the software recalculates the stiffness modulus of the asphalt layers according to the pavement temperature ranges and the temperature at a specific depth. The stiffness modules can be determined according to standard LST EN 12697-26, as well as the data from the back-calculation using the falling weight deflectometer measurement data if the rehabilitation of the pavement structure is planned. The Hirsch and Francken / Verstraeten methods use input data determined in the laboratory. The stiffness modulus is calculated from the shear stiffness of bitumen using a complex shear modulus and the volumetric properties of the aggregate mixture or an asphalt complex modulus.

Tests to determine the mechanical properties of asphalt mixes are time-consuming, so the models for calculating these properties are relevant to this day. The application of these models allows calculating with sufficient accuracy the dynamic modulus of the asphalt mixture according to the granulometric composition of the asphalt mixture, the volume of air voids, the amount of bitumen and the mechanical properties of bitumen. Pellinen, Zofka, Marasteanu and Funk (2007) conducted a comparison study of three prediction models for the modulus of elasticity of an asphalt mix by Olard-DiBenedet, Andrei-Witczak, and Hirsch. The researchers found that all three models correlated with each other and with laboratory-determined values, but the Hirsch model is closest to the behavior of the asphalt mix because it was developed using an elastic viscous liquid model. Pellinen et al. (2007) emphasize that these models must be applied with great caution when assessing the characteristics of asphalt mix at high temperatures, where the mean error of prediction can be up to 40%, while the error of experimental test results is 20%.

The stiffness modulus of unbound materials is considered to be temperature independent. The design of the pavement structure may take into consideration that the modulus is influenced strongly by the stress level. For unbound granular materials, modulus increases significantly with increasing mean normal stress and decreases with increasing shear stress (Jameson, 2012).

1.4. Analysis of pavement structure performance

Multilayer elasticity theory is used to model flexible pavement structure and to calculate the reaction parameters (stresses and strains). Multilayer elastic theory is applied both in the design and analysis of the performance of the asphalt pavement structure. The multi-layer elastic theory is based on the following assumptions:

- all layers are linearly elastic;
- all layers are unlimited in the horizontal direction;
- all layers except the last one are of constant thickness;
- the layers have the same adhesion/friction conditions over the entire surface area;
- the calculations do not take into account the own weight of the structure and the initial stresses/deformations;
- the load on the surface is distributed over the circled.

The high-performance analysis program MnLayer, developed by professor Lev Khazanovich and Qiang Wang at the University of Minnesota in 2008, is used to calculate the stresses, strains and displacements of the pavement structure. A computational algorithm developed exclusively for the ViaStructura software that generates input data files, submits it to the MnLayer and returns the output files to the software with the calculated load response parameters of the pavement structure. The MnLayer computational algorithm is designed specifically for ViaStructura software and is covered by the patent protection of the ViaStructura brand and the design model of the ViaStructura.

The degradation (fatigue) effect of the pavement structure is assessed according to the ratio of the total design load and limit load for the intended design period. This method is called the Miner hypothesis and its basic principle is that the total number of passes of the design load cannot exceed the limit total number of passes. The Miner's rule is checked for each layer of the pavement structure separately, using boundary state equations for critical points. Miner's rule is expressed as:

$$Impact = \sum \frac{N_{design,i-j}}{N_{lim.,i-j}}, \tag{2}$$

where: $N_{design,i-j}$ – design load for the load range i and temperature range j; $N_{lim.,i-j}$ – limit load for the load range i and temperature range j.

To estimate the fatigue resistance of asphalt layers, the boundary state model based on Austrian pavement structure design methodology RVS 03.08.68 (FSV, 2016) is used. The limit load is calculated as follows:

$$N_{lim.,i-j} = \frac{k_1(T)}{F_{(\varepsilon 6)}} \cdot \left(\frac{S_{mix}(T)}{\sigma_{v,i-j} \cdot \gamma_{AC}}\right)^{k_2(T)}, \tag{3}$$

where: $S_{mix}(T)$ – temperature-dependent stiffness modulus of asphalt mix, MN/m2; $\sigma_{v,i-j}$ – vertical stress under load level i and temperature range j, MN/m2; γ_{AC} – safety factor; $F_{(\varepsilon 6)}$ – fatigue factor depending on type of binder and strain of the asphalt layer at 106 load cycles determined by a four-point bending test according to EN 12697-24; $k_1(T)$, $k_2(T)$ – temperature factors.

To estimate the permanent deformation resistance of unbound base/subbase layers and subgrade, the boundary state model based on German pavement structure design methodology RDO Asphalt 09 is used. The limit load is calculated as follows:

$$N_{lim.,i-j} = 10^{\frac{1}{5}\left(84 - \frac{100 \cdot \gamma \cdot \sigma_{zz,i-j}}{\beta_{Bz}}\right)}, \tag{4}$$

where: $\sigma_{zz,i-j}$ – vertical stress under load level i and temperature range j, MN/m2; γ – safety factor; β_{Bz} – the flexural strength of the hydraulically bound layer (MPa).

$$N_{lim.,i-j} = 10^{\frac{1}{0.7}\left(\frac{0.00875 \cdot E_{v2}}{\sigma_{zz,i-j} \cdot \gamma}\right)}, \tag{5}$$

where: E_{v2} – the deformation modulus of unbound layer/subgrade, MPa; $\sigma_{zz,i-j}$ – vertical stress under load level i and temperature range j, MN/m2; γ – safety factor.

It should be noted that the boundary condition models are calibrated for Lithuanian conditions, but each user has access to the system parameters associated with a specific user account. This provides a wide range of model applications and versatile analysis capabilities.

2. Comparison of ViaStructura-designed and standard pavement structures

One of the advantages of applying the developed model is the testing and optimization of standard pavement structures by assessing the climatic and traffic load characteristics of the specific area. Figure 5 shows a graph of the performance of a standard and individually designed pavement structure using the ViaStructura model, showing the change in the effect on the pavement structure according to Miner's rule. A common case where the design load is in the beginning of the range for which standard pavement designs are selected from the catalog is represented in Figure 5 a. According to the catalog, the designer must choose a pavement design that is adapted to a load of 2 million to 3 million ESAs. It can be seen that the standard design is not optimal for the selected design period, as the service life of the structure is almost 38 years in terms of fatigue resistance. By choosing a special design and reducing the thickness of the asphalt base layer by 2 cm, it can be seen that the bearing capacity of the structure ideally corresponds to the selected design period of 20 years Figure 5 b shows the case where the calculated design load is at the end of the range. In this case, the standard and ViaStructura-designed pavement structures are identical. The difference in performance is determined by the load distribution used, as only a single axle load of 10 t is used to calculate the standard design. It can be seen that in some cases the real impact on the pavement structure may be greater and may even lead to pavement failure several years earlier than expected.

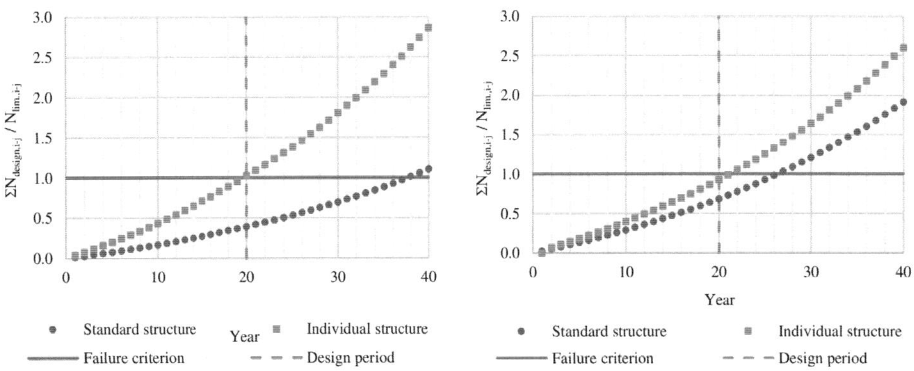

a) b)

Figure 5. Performance comparison of standard and ViaStructura-designed pavement structures at design load of a – 2.05 mill. ESA's, b – 2.95 mill. ESA's

As the load-bearing capacity of a flexible pavement structure depends on the temperature due to the composition and physical properties of the asphalt mix, it is crucial to assess the climatic condition under which the designed pavement structure will operate. Figure 6 represent how the performance of the same pavement structure can differ at different pavement surface temperature distribution.

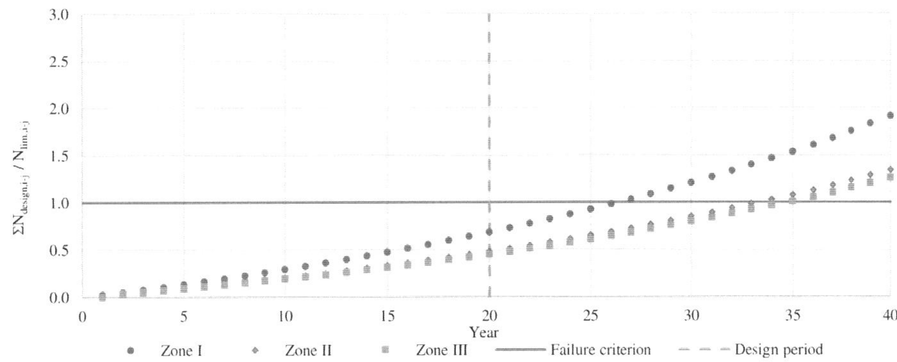

Figure 6. Performance comparison of pavement structure at different pavement surface temperature zones in Lithuania

Compared to other zones, zone 1 is more likely to have lower temperatures, but also extremely high temperatures are approximately 2 times more likely to occur. As asphalt mixtures are very sensitive to extremely high temperatures, in this case, it becomes an essential factor for differences in the fatigue resistance of the pavement structure. This shows how important is to have reliable data on the actual operating conditions of the pavement structure and it is necessary to properly assess the composition and materials for the designed pavement structure. The calculation results also show that the application of the same standard pavement structure regardless of location is not always effective and the possibility of individual design should be considered.

Conclusions

ViaStructura model represents a significantly different design concept for pavement structures than selection of standard pavement structures. ViaStructura provides a direct tie between structural design, degradation prediction, materials performance based characteristics, climate and traffic factors. This leads cost savings and pavement thickness design for particular design load instead of load range.

The major principles of the ViaStructura model include:

- Design traffic load evaluation through the axle load spectra in range between 2 t ant 22 t;
- Frost depth data and asphalt surface temperature intervals dependant on climate zone;
- Calculation of stress and strain in critical points of structure using multilayer linear elastic theory;
- Degradation analysis including asphalt and hydraulically bound base layer fatigue and permanent deformation (rutting) of unbound base layers and subgrade.

Compared to the pavement structure selection by standard catalogues, the following advantages of the model can be distinguished:

- Allows to design structure using asphalt layers stiffness characteristic based on particular pavement temperature range and load spectra instead of ESA ;
- Allows to evaluate pavement structure performance with newly composed construction materials by adding them to model database;.
- Allows a better understanding of the principles and factors affecting pavement structures performance and thus making more efficient decisions;
- Allows to combine the selection of pavement rehabilitation solutions with non-destructive bearing capacity measurements and thus obtain reasonable and expectation-meeting results.

On the other hand, the application of the model also poses some challenges related to the need for data. A larger amount of input data is required, covering not only usual traffic intensity data but also such as pavement surface temperatures, frost depth and axle load distribution, which require long-term monitoring data, which makes the design process more reliable. The applicability of the model is greatly facilitated by the development of a network of road traffic and climate monitoring systems.

Funding

This research was funded by Lithuanian Road Administration under the Ministry of Transport and Communications fund

Author Contributions

MK conceived the study and were responsible for the data collection, analysis and preparation of the first draft of the article. RK was responsible for data analysis and interpretation and review of first draft of article. AV was responsible for planning of the data collection, finalization of article and conclusions.

Disclosure Statement

The authors declare no conflict of interest.

References

Choudhary, R., Chattopadhyay, D., Kumar, A., & Julaganti, A. (2017). Use of industrial wastes as filler in open-graded friction courses. *The Baltic Journal of Road and Bridge Engineering*, 12(2), 106-116.

FGSV 498. (2009). *Richtlinien Für Die Rechnerische Dimensionierung Des Oberbaus von Verkehrsflächen Mit Asphaltdeckschicht RDO Asphalt 09* [Guidelines for Mathematical Dimensioning of Foundations of Traffic Surfaces with a Course Asphalt Surface]. Köln, Germany.

FSV. (2016). *Rechnerische dimensionierung von asphaltstrassen RVS 03.08.68* [Mathematical dimensioning of asphalt roads]. Vienna, Austria

Herb, William, Mihai Marasteanu, and Heinz G. Stefan. (2006). Simulation and Characterization of Asphalt Pavement Temperatures. Project Report No. 480. Minnesota.

Huang, Y.H. (1993). *Pavement Analysis and Design*. Prentice-Hall, Englewood Cliffs, NJ.

Jameson, G. (2012). Guide to Pavement Technology. Part 2 : Pavement Structural Design (AGPT02/10). Austroads Ltd.

Khazanovich, L., & Wang, Q. (2007). MnLayer: High-performance layered elastic analysis program. *Transportation Research Record*, 2037, 63–75. https://doi.org/10.3141/2037-06

Kleizienė, R., Vaitkus, A., & Čygas, D. (2015). Axial Load Distribution Corresponding to Vehicle Type and Gross Weight. *International Conference on Applied Mechanics and Mechatronics Engineering* (AMME 2015), 25-26 October, 2015, Bangkok, Thailand, Amme, 99–104.

Kleizienė, R., Vaitkus, A., Židanavičiūtė, J., & Marcinkevičius, E. (2017). Classification of surface temperature for the flexible pavement design. *10th International Conference on Environmental Engineering*, ICEE 2017, April, 27–28. https://doi.org/10.3846/enviro.2017.139

Li, Qiang, Leslie Mills, and Sue Mcneil. (2011). The Implications of Climate Change on Pavement Performance and Design. Delaware.

Lietuvos automobilių kelių direkcija. (2019). *Automobilių kelių standartizuotų dangų konstrukcijų projektavimo taisyklės KPT SDK 19* [Design rules for standard road pavement structures KPT SDK 19]. Vilnius, Lithuania.

LST EN 12697-26. (2020). *Bituminiai mišiniai. Bandymi metodai. 26 dalis. Standis* [Bituminous mixtures – Test methods – Part 26: Stiffness].

Meyer, M., A. Amekudzi, and J.P. O'Har. (2010). Transportation Asset Management Systems and Climate Change. *Transportation Research Record: Journal of the Transportation Research Board* 2160: 12–20.

Meyer, M., and B. Wiegel. (2011). Climate Change and Transportation Engineering. Preparing for a Sustainable Future. *Journal of Transportation Engineering* 137: 393–403.

Michael R. S. Mshali & Wynand JvdM. Steyn (2020): Effect of truck speed on the response of flexible pavement systems to traffic loading, *International Journal of Pavement Engineering*, DOI: 10.1080/10298436.2020.1797733

Mundt, Diane J., Kristin M. Marano, Anthony P. Nunes, and Robert C. Adams. 2009. A Review of Changes in Composition of Hot Mix Asphalt in the United States. *Journal of occupational and environmental hygiene* 6(11): 714–25.

Nunn, M.E., and T. Smith. (1997). Improvements to the Indirect Tensile Stiffness Modulus Test. In *Proceedings of the Second European Symposium on Performance and Durability of Bituminous Materials*, , 411–26.

Pellinen, Terhi, Adam Zofka, Mihai Marasteanu, and Noa Funk. (2007). The Use of Asphalt Mixture Stiffness Predictive Models. *Journal of the Association of Asphalt Paving Technologist* 76: 575–626.

Simonsen, E., and Isacsson, U. (1999). Thaw weakening of pavement structures in cold regions. *Cold Regions Science and Technology* 29:135-151.

Šiukščius, A., Vorobjovas, V., Vaitkus, A., Mikaliūnas, Š., & Zariņš, A. (2019). Long Term Behaviour of An Asphalt Pavement Structure Constructed on a Geogrid-Reinforced Subgrade Over Soft Soils. *The Baltic Journal of Road and Bridge Engineering*, 14(3), 384-404.

Tran, N. H., & Hall, K. D. (2007). Development and influence of statewide axle load spectra on flexible pavement performance. Transportation Research Record, 2037(1), 106-114.

Vaitkus, A., Grazulyte, J., Vorobjovas, V., Šernas, O., & Kleiziene, R. (2018). Potential of MSWI bottom ash to be used as aggregate in road building materials. *The Baltic Journal of Road and Bridge Engineering*, 13(1), 77-86.

Vaitkus, A., Šernas, O., Vorobjovas, V., & Gražulytė, J. (2019). Selection of constituent materials for asphalt mixtures of noise-reducing asphalt pavements. *The Baltic Journal of Road and Bridge Engineering*, 14(2), 178-207.

Vaitkus, A., Vorobjovas, V., Jagniatinskis, A., Andriejauskas, T., & Fiks, B. (2014). Peculiarity of low noise pavement design under Lithuanian conditions. *The Baltic Journal of Road and Bridge Engineering*, 9(3), 155-163.

Vaitkus, A., Žalimienė, L., Židanavičiūtė, J., & Žilionienė, D. (2019). Influence of temperature and moisture content on pavement bearing capacity with improved subgrade. Materials, 12(23), 3826.

Zofka, A., Maliszewski, M., & Maliszewska, D. (2017). Glass and carbon geogrid reinforcement of asphalt mixtures. *Road Materials and Pavement Design*, 18(sup1), 471-490.

ASSESSMENT OF SURFACE CHARACTERISTICS OF COARSE AGGREGATES BY FLOW COEFFICIENT METHOD

Arturs RIEKSTINS https://orcid.org/0000-0002-5786-1333[12], Janis BAUMANIS[1], Kristaps KRASTINS[1], Kristaps KALINKA[1]

[1]*SLLC "Latvian State Roads", 3 Gogola Street, Riga, LV-1050, Latvia*
[2] *Department of Roads and Bridges, Riga Technical University, Riga, Latvia*
Arturs.Riekstins@lvceli.lv, Janis.Baumanis@lvceli.lv, Kristaps.Krastins@lvceli.lv, Kristaps.Kalinka@lvceli.lv

March 14, 2021

Abstract. In this study, for assessment of surface characteristics, the flow coefficient of coarse aggregates was evaluated as an alternative to the widely used crushed and broken surfaces test. It has been proved that visual assessment is not only time consuming but also inaccurate. The European standard EN 933-5 allows to use of a flow coefficient method according to EN 933-6 as an alternative for the assessment of coarse aggregates, but it does not specify any requirements. Therefore, this study aimed to assess the flow coefficient test and if it could replace tradition visual assessment. In total, 28 gravel samples were tested. Following properties and their influence to flow coefficient were evaluated - flakiness, particle size distribution and amount of crushed and rounded surfaces. The results show that the flow coefficient test could be used as an alternative if the boundaries of the granulometric curves are set.

Keywords: coarse aggregate angularity, flow coefficient, crushed and broken surfaces, flakiness index, gravel, assessment of surface characteristics

1. Introduction

Angularity and surface texture of aggregates are important properties that influence the strength of bound and unbound mixtures (1,2). It is known that a structural layer has to consist of interlocking aggregates to ensure internal friction. Thus, a higher shear strength in the structural layer is ensured. Various direct and indirect test methods are used to assess these properties under laboratory conditions (see Table 1).

In Latvia crushed and broken surfaces test is used to evaluate the angularity and surface texture of mixtures in accordance with EN 933-5. The test is performed organoleptically (visual inspection), so it is a subjective assessment. The results of interlaboratory testing approve large distribution between the same sample assessed by different operators. Therefore, the potential of different alternatives has been evaluated.

As technologies develop, it is also possible to assess the surface properties of aggregates using digital image methods (3–6). Currently the most known tool in the market is Aggregate Image Measurement System (AIMS) that evaluates aggregate shape properties by using digital imaging (4). It uses a high-resolution camera to picture aggregates, then uses data processing software to calculate the shape and surface texture of aggregate (7). The test is fully automated, thus excluding the human factor. The main disadvantage of this method is the high price of the equipment.

In some countries, the uncompacted void content of coarse aggregate test is widely used to evaluate the angularity of materials. The test is done according to AASHTO TP 56. The idea of the test is simple: a certain amount of material is dropped into a cylinder and based on the weight of the aggregate in the cylinder the volume of voids as a percentage of total volume may be calculated (8). The higher the void content, the higher the assumed angularity and rougher the surface (8). The equipment is inexpensive in comparison to digital image processing methods (2). Swamy concluded that uncompacted void content may be effectively used to capture changes in aggregate morphology (9).

The flow coefficient test is widely used to evaluate angularity of fine aggregate fraction of 0.063-2.0 mm but little is known about its use in the evaluation of coarse aggregates. As far as we know, no previous research has investigated the correlation between the crushed and broken surfaces test and the flow coefficient test. Standard EN 933-6 describes the test procedure of this test. Special equipment is needed for the test (see Figure 1). The main advantage of this method is the testing range, which may be 4.0/6.3; 6.3/10; 10/14; 4/10; 4/20. The smallest fractions allow evaluating materials that are used to produce asphalt mixtures. On the other hand, a fraction range of 4/20 would allow the estimation of angularity of fractions of gravel samples such as 0/32, 0/45, 0/56, 0/63. Another advantage is the repeatability of the results (10).

Table 1. Breakdown of methods used for assessment of surface characteristics.

Direct method	Indirect method
Crushed and broken surfaces	Uncompacted voids in coarse aggregate
Digital image processing	Flow coefficient

2. Objective

The overall goal of this study was to analyse the correlation between two tests – crushed and broken surfaces (EN 933-5) and the flow coefficient of coarse aggregates (EN 933-6) tests, including, what aspects should be taken into account in order to replace crushed and broken surfaces test with a flow coefficient of coarse aggregates test.

3. Materials and methods

3.1. Materials

Given that there is a lack of information about the flow coefficient method and interpretation of results, 28 gravel samples from different quarries with fractions 0/32 mm, and 0/45 mm were obtained for this study. The following particle size fractions were obtained: 0/32 mm and 0/45 mm. The distribution of the obtained materials by fractions may be seen in Table 2.

Table 2. Distribution of obtained materials by fractions.

Particle size distribution, mm	Number of mixtures
0/32	12
0/45	16

3.2. Methods

3.2.1. Particle size distribution

Determination of particle size distribution was done according to EN 933-1. The weight of the samples depends on the maximum aggregate size in the mixture. The compliance of the particle size distribution of all materials was assessed according to the LSR Road Specifications (11). The ones that did not meet the particle size limits were not further tested and marked as invalid.

3.2.2. Flakiness index

Determination of flakiness index was done according to EN 933-3. The test was done for all samples that fulfil the requirement of particle size distribution. The following bar sieves were used: 25, 20, 16, 12.5, 10, 8, 6,3, 5, 4, 3,15, 2,5. Material test portions larger than 4 mm were tested. The result for this test is the sum from each bar sieve, expressed as the percentage of the total mass of the mixture.

3.2.3. Crushed and broken surfaces

Determination of percentage of crushed and broken surfaces in coarse aggregate particles was done according to EN 933-5. Usually, the test is done for fraction 4/63 mm but for this study fraction 4/20 mm was used, so that results would be directly compared to those obtained from the flow coefficient test. For each gravel mixture at least 100 particles were tested. Each test was done by two operators (the final result is the average of both) to improve the reliability of the obtained results. The basic principle of the test is to manually segregate each aggregate into one of four categories – (1) total crushed (TC) aggregate that has more than 90% crushed or broken surfaces, (2) crushed (C) aggregate that has 50-90% crushed and broken surfaces, (3) total rounded (TR) aggregate that have more than 90% rounded surfaces, (4) rounded (R) aggregate that has 50-90% rounded surfaces. An example of each group may be seen in Table 3. The mass of each group was recorded and calculated as a percentage of the total sample mass.

IOP Conf. Series: Materials Science and Engineering 1202 (2021) 012019 doi:10.1088/1757-899X/1202/1/012019

Table 3. Conditions for the distribution of aggregates by categories based on surface characteristics.

Classification	Total crushed (TC)	Crushed (C)	Total rounded (TR)	Rounded (R)
Criterion	>90%	50-90%	>90%	50-90%
Visualization				

3.2.4. Flow coefficient of course aggregates

Determination of flow coefficient of course aggregates was done according to EN 933-6. The flow coefficient of an aggregate is the time, expressed in seconds, for a specified volume of aggregate to flow through a given opening, under specified conditions using a standard apparatus. The test may be used for the following particle size fractions: 4/6.3 mm, 6.3/10 mm, 10/14 mm, 4/10 mm, or 4/20 mm. For this research particle size fraction 4/20 was chosen because it was wider and represented a higher percentage of material. Before testing device has to be calibrated to make sure that everything works properly. Calibration is done by a special reference mixture with particle size fraction 6.3/10 mm. The final result of the test is an average value of five consecutive measurements, expressed in seconds. The mass of the test portion is preprepared according to the first equation.

$$M = 10 \times \frac{\rho_{rd}}{2.70} \quad (1)$$

M – the mass of the test portion;

ρ_{rd} – particle density of the aggregate, in accordance with EN 1097-6, in megagrams per cubic metre;

2.70 – fixed value for the particle density of the reference material, in megagrams per cubic metre;

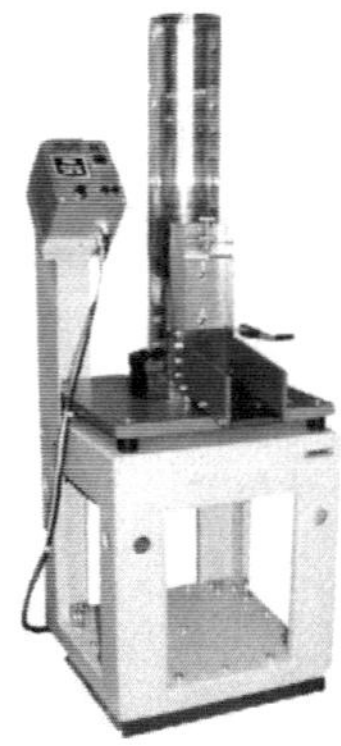

Figure 1. Device of flow coefficient of course aggregates.

4. Results and discussion

4.1. Particle size distribution of mixtures

Conformity assessment of particle size distribution of all 28 gravel mixtures may be seen in Figure 2. In the figure, the particle size distribution of fraction 4/22.4 may be seen. On the x-axis, material identification numbers are displayed. Each x denotes the percentage of material of the given mixture on the corresponding sieve. The coloured dashed lines (blue, red, green, purple) show boundaries for each sieve. The red squares on the x-axis highlight the non-compliant samples, which were further excluded from the calculation.

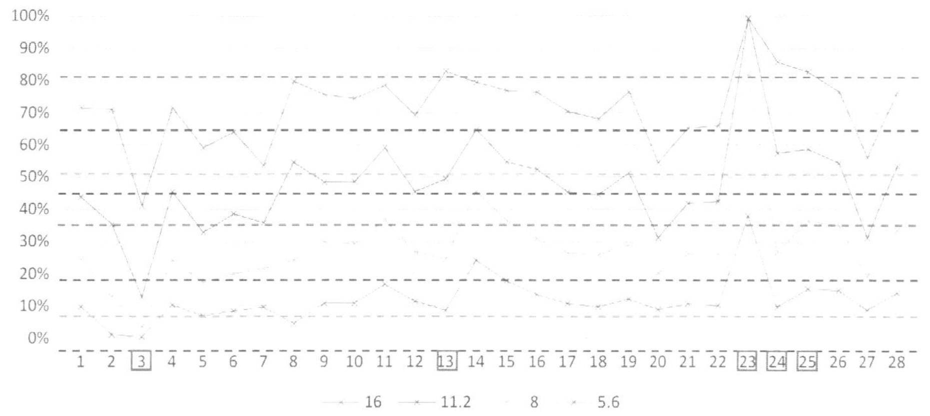

Figure 2. Compliance assessment of particle size distribution of gravel mixtures.

4.2. Distribution of crushed and rounded particles

Results of the crushed and broken surfaces test are shown in Figure 3. The results show that 4 samples (14, 17, 26, 27) do not meet the requirement of less than 30% of TR particles in the mixture. What is more, 13 mixtures do not meet the minimum requirement of 50% of TC and C particles. Thus, in total 13 of the 24 mixtures do not meet the requirements of the LSR road specifications (11).

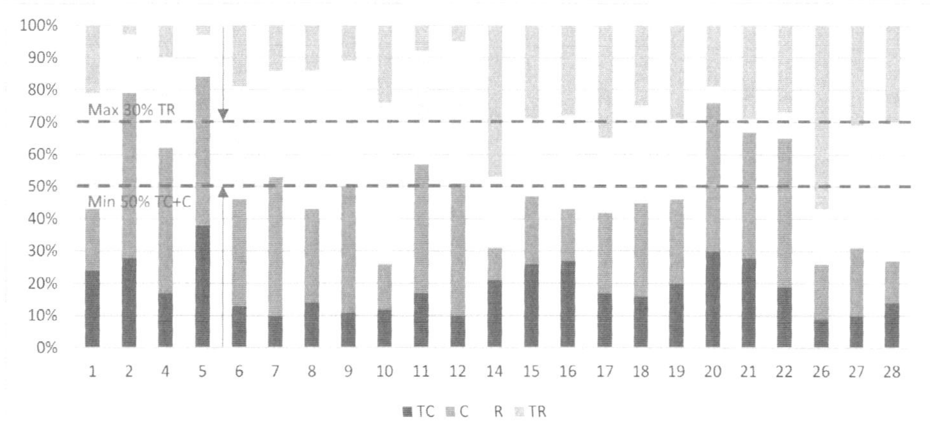

Figure 3. Summary of the results for the crushed and broken surfaces test. The red dashed lines show the maximum and minimum limits for the specific category.

4.3. Correlation between flakiness index and flow time

The correlation of the results between the flakiness index and flow coefficient may be seen in Figure 4. All mixtures fulfil the national requirement value of the flakiness index of 20. Almost no correlation (R^2=0.049) may be observed if the value of the flakiness index is between 4 and 13. However, a good correlation is achieved (R^2=0.5573) when mixtures with a flakiness index of 8 or higher are taken into an account (see dark blue dots). Obtained results lead to assume that a relatively small flakiness index less than 8 does not impact flow time and correlation may not be observed, but when the index is 8 or higher then a correlation between two values significantly increases. In line with results, a higher flakiness index reduces the flow time.

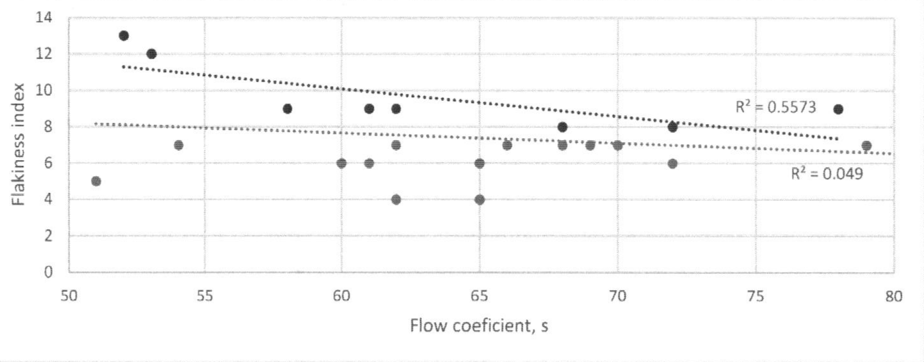

Figure 4. Comparison of flakiness index and flow time.

4.4. Correlation between crushed and broken surfaces and flow time

4.4.1. Total crushed and crushed surfaces versus flow time

A comparison of the results of the TC and C surfaces and flow coefficient may be seen in Figure 5. From the obtained results, only 10 out of 23 (43.5 %) fulfil the minimum requirement TC+C ≥50%. A wide range of flow coefficient results was obtained, from 51 to 83. A good correlation of R^2=0.5105 may be observed between the amount of TC and C particles and the flow coefficient. From the results, it is clear that the amount of TC+C affects the flow time. In order to better understand the existence of correlation, it would be necessary to have more samples with TC+C more than 70% and TR+R less than 30%.

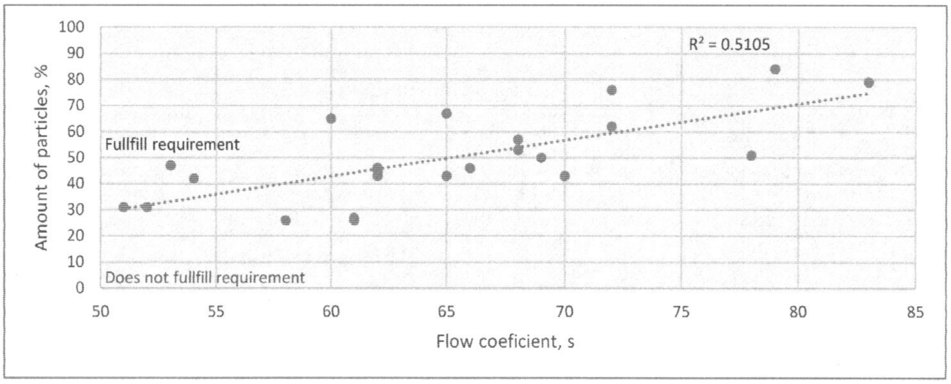

Figure 5. Comparison of total crushed and crushed surfaces and flow time.

4.4.2. Total rounded versus flow time

A comparison of the results of the TR and flow coefficient may be seen in Figure 6. According to LSR specifications quantity of TR has to be less than 30%. From obtained results, 4 out of 23 failed to fulfil this requirement. In line with results, a strong correlation R^2=0.6684 may be observed between the percentage of TR particles and flow coeficient. Thus, the amount of TR particles greatly affects the flow time.

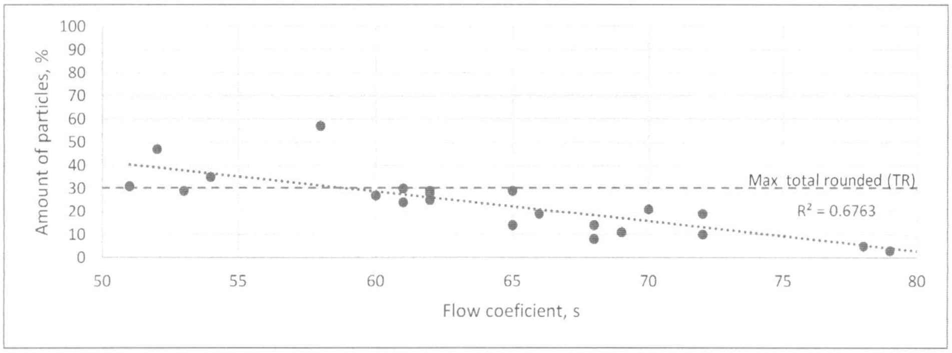

Figure 6. Comparison of TR and flow coefficient. The red line shows the maximum percentage of TR that is allowed in the mixture.

5. Conclusions

In total 28 gravel mixture were obtained for this study, and only 23 that fulfilled the requirements of the granulometric curves were tested further. In this study correlation between percentage of crushed and broken surfaces and flow coefficient test was evaluated. Based on the analysis of results, the following conclusions were drawn:

- It is recommended to determine the flow coefficient only for materials that fulfil the requirements of particle size distribution as it is greatly impacting the flow time. Coarser mixtures have a longer flow time than finer ones;
- A small flakiness index (<8) does not impact flow time, however, when it is higher (≥8) good correlation between the two parameters may be observed. A higher flakiness index reduces flow time.
- There is a good correlation between the percentage of total crushed and crushed particles and the flow coefficient. This also confirms the logic that as the percentage of total crushed and crushed particles increases, so does the flow time.
- A strong correlation may be observed between the percentage of total rounded particles and flow time. From the results, it is clear that as the percentage of total rounded particles increases, the flow time decreases.
- Obtained results lead to assume that for gravel mixtures the minimum requirement of flow coefficient value of 60 to 65 might be optimal.
- The flow coefficient method is a good alternative to the crushed and broken surfaces test, as it is possible to obtain good repeatability of the results. At the same time, the disadvantage of this method is longer testing time (from sampling to testing).
- It will be important that future research investigates the mixtures that have a wider range of percentage of crushed and rounded aggregates so that a better understanding of the correlation between the crushed and broken surfaces test and flow coefficient test is obtained. It would also be necessary to understand the impact of particle size distribution on the flow coefficient.

Acknowledgements

State Limited Liability Company "Latvian State Roads" gratefully acknowledge VAS "Latvijas Valsts meži" for their cooperation in providing the gravel samples that were used for this study.

Funding
This work was supported by State Limited Liability Company "Latvian State Roads" under the internal research program.

Disclosure Statement

No potential conflict of interest was reported by the authors.

References

1. Nikolaides A. Highway Engineering: Pavements, Materials and Control of Quality. 2015. 1–821 p.
2. Rahimi Z, Muniandy R. Quantification of Coarse Aggregate Angularity by a Newly Developed Auto Grader Machine. Curr World Environ. 2016;11(Special Issue 1(2016)):45–53.
3. Zhang S, Li R, Pei J. ScienceDirect Evaluation methods and indexes of morphological characteristics of coarse aggregates for road materials : A comprehensive review. J Traffic Transp Eng (English Ed [Internet]. 2019;6(3):256–72. Available from: https://doi.org/10.1016/j.jtte.2019.01.003
4. Leslie Gates, Eyad Masad, Roger Pyle and DB. Aggregate Image Measurement System (AIMS 2): Final Report. Vol. 2. 2011.
5. Little D, Button J, Jayawickrama P, Solaimanian M, Hudson B. Quantify Shape, Angularity and Surface Texture of Aggregates Using Image Analysis and Study Their Effect on Performance. Vol. 7, Texas Transportation Institute. 2003.
6. Rajan B, Singh D. Understanding influence of crushers on shape characteristics of fine aggregates based on digital image and conventional techniques. Constr Build Mater [Internet]. 2017;150:833–43. Available from: http://dx.doi.org/10.1016/j.conbuildmat.2017.06.058
7. Kong D, Chen M, Xie J, Zhao M, Yang C. Geometric Characteristics of BOF Slag Coarse Aggregate and its Influence on Asphalt Concrete. 2019;
8. Pavement Interactive. Coarse Aggregate Angularity [Internet]. Available from: https://pavementinteractive.org/reference-desk/testing/aggregate-tests/coarse-aggregate-angularity/
9. Journal AI, Swamy AK, Matolia V, Ramana G V, Matolia V, Ramana G V. Interrelationship between uncompacted void content of aggregates and asphalt concrete properties concrete properties. Part Sci Technol [Internet]. 2018;0(0):1–9. Available from: https://doi.org/10.1080/02726351.2017.1414906
10. Caro AR. Contribution of the coarse aggregates to rheology - effects of flow coefficient , particle size Distribution , and volume fraction. 2018;(March).
11. VAS "Latvijas Valsts ceļi". Ceļu specifikācijas 2019. 2019;

ROAD DESIGN AND CONSTRUCTION ON LOW BEARING CAPACITY SOILS USING PILING METHOD: EXPERIENCE OF SLLC "LATVIAN STATE ROADS"

Mārtiņš ĀBOLS, Ojārs DĀRZNIEKS, Māris SPILVA

Construction Division, SLLC "Latvian State Roads", 3 Gogola Street, Riga, LV-1050, Latvia
Corresponding author: Ojārs Dārznieks, ojars.darznieks@lvceli.lv

March 15, 2021

Abstract. The road system in Latvia has evolved over time to a complex network of physical structures that include roads, bridges and overpasses, however, in large it consists of narrow roads of local significance that follow the existing topography and consist of thin foundation and pavement layers. In-depth geotechnical research of road sections before construction has only been carried out in recent decades, thereby with a certain regularity SLLC "Latvian State Roads" has to deal with low bearing capacity soils under the road foundation. As the recent experience shows, there are sections of roads that in the past were constructed on peat. In 2018 during the reconstruction works of the regional road P32 Augšlīgatne – Skrīveri a low bearing capacity soil under the road foundation was determined. An additional in-depth geotechnical research showed a bog section with a peat layer at a depth of 10m in a 320 metres long section. Although at that point it was possible to continue the work using simple soil stabilization methods, there were concerns about the longevity of the investment. To stabilize the road foundation, a combination of gravel columns and unreinforced concrete pile columns was used. A total of 952 columns were constructed. The aim of this paper is to share technical information and our good practice of road design and construction on low bearing capacity soils using piling method, and it gives a summary of field observations, geotechnical research, design considerations and risk management assessments that were carried out in this specific case. The positive outcome of this case led to a number of future projects where similar methods for load bearing capacity improvement were planned and used.

Keywords: roads, peat, construction, piling.

Introduction

The territory of Latvia is characterized by diverse geological conditions. Technical properties of some of the existing soils are not sufficient to ensure the construction and safe operation of transport structures. Temperate climate typical for Latvia with high amounts of precipitation rainfall, topographical features, as well as widespread plastic soils are very good conditions for the formation and development of bogs. An area may be called a bog if the peat layer in drained condition is at least 30 cm thick, and such areas cover more than 10% of Latvia (Kalniņa, 2020). Because of this fact some roads are crossing swampy areas where technical properties of natural soils do not provide sufficient bearing capacity to ensure the necessary stability for road pavements throughout their forecast lifetime. Therefore, the construction design must provide solutions for special road subgrade strengthening which would ensure its compliance with the specified requirements. In cases where low bearing capacity soils are determined at great depths, a traditional method like weak soil replacement is not cost-effective and involves many risks – difficulties in excavation and placing fill below water table, effects on adjacent lands and structures and future settlements caused by unexcavated soft material below embankment. Design and construction of road embankment in such situations is a challenge for a geotechnical engineer. Methods that leave the low bearing capacity soil in place use the strength of the in-situ weak soil to support the intended loads avoiding the disadvantages of bulk earthworks. These methods are more attractive to the client`s as construction budgets reduce and more cost-effective solutions are sought. The goal is to utilise the underlying soil as a load bearing layer by improving its strength (preloading, surcharging), modifying the load (lightweight fill, offloading, profile lowering) or reinforcing with geosynthetics (Munro 2004).

But in some cases where the settlement control is very important and a rapid execution is needed a piling method can be used. Piling does not belong to that group that rely on strength of in-situ soil because the entire embankment load is on supportive piles. The method has high mobilizing costs, setting up and driving costs but has a quick time of installation, does not require lots of earthworks, has a limited site disturbance and most important – minimal settlement. In 2018 during the reconstruction works of the regional road P32 Augšlīgatne – Skrīveri a low bearing capacity soil under the road foundation was determined. An additional in-depth geotechnical research showed a bog section with a peat layer at a depth of 10m. The concluded construction contract indirectly set a time limit for the development of the solution so a quick action was needed. To stabilize the road foundation, a combination of gravel columns and unreinforced concrete pile columns was used.

1. Reconstruction of road P32

The development of road P32 Augšlīgatne - Skrīveri is typical for a regional road in Latvia: the original pavement structure was constructed from layers of unbound aggregates, the thickness of with, as a result of repeated replenishment, reached up to four meters. First layer of asphalt (6 cm) was laid in the 1970-ies. In the following years the asphalt pavement was renewed several times by adding new layers of asphalt to the existing pavement - in the early

IOP Conf. Series: Materials Science and Engineering 1202 (2021) 012020 doi:10.1088/1757-899X/1202/1/012020

1980-ies the second 6 cm thick asphalt layer and at the end of the 1980-ies the third 6 cm thick asphalt layer was laid. Asphalt defects were repaired from time to time. In 2009 the wearing course was renewed and until the preparation of construction design for reconstruction works no other repairs were done.

When the designing of road reconstruction project was commenced, road maintainers informed the engineers that in some sections of the road additional repairs were performed several times. However, the depth of the boreholes made during the initial geotechnical survey proved to be insufficient and, as it turned out in a later detailed study, good results were obtained as only the mineral material used for road embankment during the initial road construction was tested. In 2018 during the reconstruction of road P32 the old road pavement was demolished and the bearing capacity of soil base was measured. Measurements showed decreasing values under the impact of the dynamic load of working equipment. It was suspected that low bearing capacity soils could be under the pavement structure. The cross-section of the road structure also revealed an uneven deformation of the pavement structure. After this fact was revealed, additional geotechnical research works were carried out in order to obtain more precise information about the properties of the soil and the depth and area of soil layer with poor bearing capacity. Initially, a couple of additional boreholes were drilled to ascertain possible low bearing capacity soils in the deeper layers. Additional bored samples were taken, materials were tested and cone penetration tests were performed. The obtained results showed the presence of peat and mineral sludge in the deepest layers (from 5 m to 10.5 m depth).

The engineers proposed to use simplified methods of road subgrade strengthening:

- Complete replacement of poor soil and peat, as well as use of temporary corrugated walls to exclude the possibility of further settling;
- Use of geosynthetic materials in pavement structure, without excluding the possible deformation of road pavement because of further settling of road embankment but with more even potential impact of road subgrade deformations on pavement structure.
- Use of geosynthetic materials in road pavement structure with additional initial loading of road subgrade, draining water from soil layers with poor bearing capacity by means of vertical drains (see Figure 1). Such technology would significantly reduce the stabilization time of embankment soils, but without guaranteeing their complete consolidation during construction.

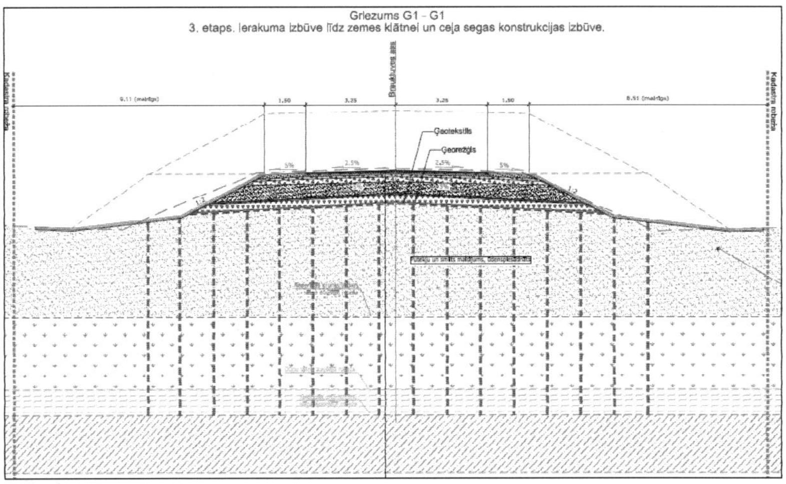

Figure 1. Vertical drainage system for draining water from the lower layers.

It was decided to perform more complex geotechnical research as well as calculations for road pavement solidifying in order to obtain data of better quality and quantity for geotechnical design. The designer was requested to determine the total bearing capacity of the whole soil mass in the respective road section and provide the forecast of values of potential deformations (vertical and horizontal, displacement, depression). In addition to that the analysis of hydrogeological conditions was requested in order to determine to what extent hydrological conditions influence the respective soil mass. Within the scope of complex geotechnical research in the respective road section the works were performed in 44 research spots in 12 road cross sections in order to cover the whole research area. The distance between cross sections was 30 m in average. The following tests were performed in order to determine the deposition of natural and artificial soils, as well as physical and mechanical properties of these soils: 27 borewells in the depth of 3.0 –

10.0 m below the surface, 22 cone penetration tests (CPTu) in the depth of 2.6 – 12.4 m, 8 flat dilatometer tests (DMT) in the depth of 6.0 – 11.0 m and 6 field vane tests (FVT) in the depth of 3.5 – 7.5 m. Further analysis (modelling) of the situation was carried out for specific geotechnical conditions which were determined during research. Potential unfavourable hydrological and hydrogeological conditions, as well as foreseen traffic load was taken into consideration. It was determined that the ground water level had to be restored in its original level by filling water drainage ditches along the embankment. The base of the embankment must be strengthened, as well. As a result of modelling the geotechnical engineers proposed the following three options for road subgrade strengthening in which the thick layer of mineral material of the existing road pavement structure is optimally preserved and used, and the construction time of the low bearing capacity soil section is minimized:

1. Construction of soil (sand, gravel or crushed stone) columns with geotextile shell

Placement of soil columns: diameter of columns is 0.60 m. Columns need to be placed every 2 m in the form of a square. Sand layer with geogrid need to be constructed on top of the columns. Placement of columns is commenced with inserting a steel jacket pipe with vibration equipment down to the depth specified in the design. Bottom end of the columns need be placed at least in the depth of 0.5 m in the bearing layer. As soon as steel jacket pipe is inserted, geotextile shell is inserted in the pipe and is filled with frost resistant sand or crushed stone.

2. Construction of soil (gravel or crushed stone) columns without geotextile shell

Placement of soil columns: diameter of columns is 0.60 m. Columns need to be placed every 2 m in the form of a square. Sand layer with geogrid need to be constructed on top of the columns. Placement of columns is commenced with inserting a steel jacket pipe with vibration equipment down to the depth specified in the design. Bottom end of the columns must be placed at least in the depth of 0.5 m in the bearing layer. As soon as steel jacket pipe is inserted, it is filled with frost resistant sand or crushed stone and then slowly removed.

3. Construction bi-modulus columns - gravel or crushed stone and concrete columns

Construction of gravel or crushed stone and concrete columns: diameter of soil columns is 0.60 m (for the depth of 2 – 3 m), diameter of concrete columns – 0.32m. Columns need to be placed every 2 m in the form of a square. Sand layer with geogrid need to be constructed on top of the columns.

Considering the time needed for the construction of all three options, a decision was made to continue with option 3 (see Figure 2).

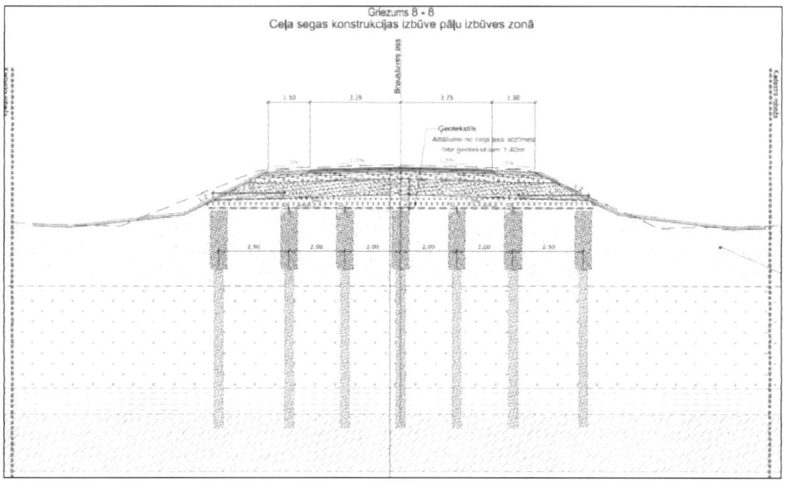

Figure 2. Road pavement structure.

Construction of bi-modulus columns was preferred as it was possible to construct asphalt pavement immediately after the concrete had hardened. In case of other two alternative solutions it was recommended to wait for 3 – 4 months for embankment to solidify. To speed up the construction process it was decided to divert the traffic on a detour from the construction area. Such a decision made it possible to plan the completion of the 300 m long section in 30 days. The contractor completed the construction of 952 bi-modulus columns in less than two weeks by building an average of

100 columns per day. During construction works all data on each column was monitored and recorded thus ensuring continuous control over the execution of works. The depth of piles varied from 5.5 to 10.5 metres. The total length of the constructed bi-modulus columns piles is 2732 metres; 537 m3 of concrete and 2005 m3 of crushed stone were consumed. During the following 14 days road subgrade and base layers were built and on the 25th day after the commencement of works traffic was opened along the reconstructed road section.

2. Conclusions and implemented measures after the case of P32

After the P32 project was completed it was concluded that the existing legal framework was not sufficient for the design and rational application of this type of solution. In-depth research has to be carried out and information on methods for the improvement of soils with poor bearing capacity and limited use around the world, as well as for their possible application in Latvia has to be compiled.

The requirements for the design of road subgrade are specified in the Latvian State Standard LVS 190-5 "Road Design Regulations. Part 5: Road Subgrade", which describes requirements and design principles in situations where it is practically impossible to ensure the defined road subgrade properties (bearing capacity, compaction, geometrical parameters, frost resistance, ensured water drainage, etc.) or their provision is not economically viable (LVS 190-5, 2011). One of the special cases analysed is the situation described in this paper - design of road subgrade in bogs or swampy areas. Two solutions for the construction of subgrade are offered - road subgrade structure on a compacted massif of peat or road subgrade structure on the mineral soil base of the bog (LVS 190-5, 2011). It should be noted that an earlier version of this standard also illustrated other solutions, such as the construction of piles under the embankment and widening of existing roads to cover the existing structure. Detailed solution characteristics and application criteria are not described. The standard specifies that the solution with the construction of subgrade structure on top of a solidified layer of peat is suitable for:

- Bogs of type 1 and 2 (bogs where no lateral displacement of soil occurs due to the load of the embankment);
- Up to 3 m high embankments;
- Road pavements with unbound pavement (in case of bituminous pavements additional compaction has to be carried out), traffic intensity of trucks <500;
- Construction of road pavement has to be carried out only after the consolidation of peat base and settlement of embankment (1-2 years after the construction of embankment) (LVS 190-5, 2011).

The second described solution for bogs is suited for roads with traffic intensity of trucks ≥500 and with the thickness of swampy layer up to 4m (LVS 190-5, 2011). In case of deeper bogs special trench wall reinforcements, e.g. corrugated walls must be applied. The road design norms and standards until 2019 described only two solutions for the construction of a road route in bog areas: consolidation of the existing peat layer and building of road subgrade on top of that and draining of the bog and filling it with some draining material. The possibilities to apply the first solution are significantly limited by the significant consumption of time, as well as due to the remaining layer of organic material that may lead to certain deformations in the future. The implementation of the second solution is safer and more predictable from the engineering point of view, but its costs are significantly higher. Despite the shortcomings of both solutions, their application has been the basic practice for the construction of road subgrades across bogs. As a result, in 2019 "Latvian State Roads" ordered the development of a manual called "Development of Solutions to Ensure the Bearing Capacity of Road Subgrade". During the development of this manual both foreign and Latvian experience of soil reinforcement and stabilization was researched - application criteria, technologies, binders, mix design, specifications, descriptions of constructed projects. The document describes the solutions potentially applicable in Latvia, recommendations for conducting geotechnical research, methodology for selection of the most appropriate solutions (incl. life cycle cost analysis) (Ceļa zemes klātnes, 2019).

3. Present practice

Initially the methods proposed in the geotechnical design for P32 project were not economically comparable and the choice was done basing on the foreseen schedule of road subgrade strengthening works for each of the proposed methods. The best design solution must be chosen, where the engineering and calculations of road subgrade and pavement take into account both strength and durability, as well as longevity and operation requirements of the structure in order to ensure rational capital investments, therefore, a life cycle cost analysis is done at the design stage.

The described methods for the time being are applied in three road construction projects:

- Road P32 (reconstruction of state road P32, section Augšlīgatne – Skrīveri from km 47.20 to km 60.29): construction of combined soil and concrete columns;
- Road P8 (reconstruction of state regional road P8, section Inciems – Sigulda – Ķegums from km 21.00 to km 21.40): construction of concrete columns;

- Road A13 (reconstruction of state main road A13 Russian border (Grebņeva) – Rēzekne – Daugavpils – Lithuanian border (Medumi), section from km 53.40 to km 57.80) (works are still underway): construction of concrete columns.

In 2021 the project V10 (reconstruction of state local road V10 Babītes stacija – Vārnukrogs, section from km 0.70 to km 3.82) will be commenced and it will include the construction of combined soil and concrete columns for subgrade strengthening. The project P8 (reconstruction of state regional road P8 Inciems - Sigulda – Ķegums, section from km 8.30 to km 12.50) will include local strengthening of soils with deep soil stabilisation. Similar solution is proposed for the project P30 (reconstruction of state regional road P30 Cēsis – Vecpiebalga - Madona, section from km 61.13 to km 83.81).

Soils with poor bearing capacity were found in the above mentioned projects during geotechnical research (peat and mud with ground water up to the depth of 12 m). The height of columns to be constructed varies from 5 to 12 m. Columns are based on sand or clay soil. Road subgrade in these projects is strengthened in 130 - 450 m long road sections. The construction technology mostly allows to open road sections for traffic with the organisation of traffic flow along one lane with traffic lights.

Conclusions

1. Where low bearing capacity soils are determined at great depths, road construction using piling methods can be a feasible alternative solution.

2. Extensive input data is essential to determine an effective solution for construction on low bearing capacity soils, which can only be provided by performing an in-depth geotechnical research.

3. Due to lack of experience from both the client and the contractor, it is necessary to properly and timely establish requirements for the materials and technologies in a way that construction participants have a clear understanding about the work to be done and the responsibilities that falls on them.

Disclosure Statement

The authors declare that they have no competing financial, professional, or personal interests from other parties.

References

Ceļa zemes klātnes grunts nestspējas nodrošināšanas risinājumu izstrāde (2019) https://lvceli.lv/celu-tikls/tehniskie-noteikumi-metodiskie-noradijumi/rokasgramatas/
Kalniņa, L., (2020). Purvi Latvijā, Nacionālā enciklopēdija https://enciklopedija.lv/skirklis/27677-purvi-Latvij%C4%81
Latvijas Kūdras asociācija (LKA), http://www.latvijaskudra.lv
Latvijas Valsts standarts LVS 190-5 Ceļu projektēšanas noteikumi. 5.daļa: Zemes klātne, 2011.g.
Munro, R. 2004. Dealing With Bearing Capacity Problems On Low Volume Roads Constructed On Peat. Including case histories from roads projects within the ROADEX Partner Districts. ROADEX II Northern periphery.

FALLING WEIGHT DEFLECTOMETRE AND PLATE LOAD TEST: DIRECT AND INDIRECT COMPARATIVE TESTING

Andrejs TARANOVS
"Latvian State Roads", 3 Gogola Street, Riga, LV-1050, Latvia
Andrejs.taranovs@lvceli.lv

February 22, 2021

Abstract. Plate load test is a widely used method in Latvia both in quality control and in road design process. This test is performed according to the standard DIN 18134. Such test usually takes at least 30 minutes and requires certain load weight. Considering the relatively long time needed for this test, alternatives were sought and a potential alternative was defined to perform testing with Falling Weight Deflectometre (FWD). In order to check this assumption both direct and indirect testing was performed and correlation between the results of both tests was defined.

In the first case the test was performed in the same location with both pieces of equipment on a surface of unbound pavement. In the second case the test with Falling Weight Deflectometre was performed on the surface of bituminous pavement but plate load test was performed in the same location on the surface of base course with prior demolition of bituminous layers. In order to compare the results of indirect comparative testing, the backcalculation for the data acquired with Falling Weight Deflectometre was performed according to German calculation method.

Results acquired with direct testing showed that the testing with Falling Weight Deflectometre and plate load test are interchangeable if no characterization of the layer compaction is required. The German method of backcalculation (FGSV, 2014) is very simple. Despite positive references from other specialists this method in comparative testing did not show sufficiently good correlation with the results acquired in plate load test.

Keywords: plate load test, falling weight deflectometre (FWD), comparative testing, bearing capacity, unbound base course,

Introduction

For efficient use of funding when designing the reconstruction of an existing road, it is important to assess the bearing capacity of unbound road pavement materials. Such an assessment may be necessary for both paved roads and gravel roads where it is planned to improve the driving comfort by laying asphalt pavement.

In Latvia the plate load test in accordance with DIN 18134 is used to assess the bearing capacity of unbound road pavement layers, and the test is used both during designing the roads and during quality control in construction process.

Plate load test has several disadvantages. To perform the test on an existing road, it is necessary to demolish and then restore a small area of asphalt pavement. Such a test may take a long time (more than 1 hour), and this also impairs road traffic safety. For the test it is necessary to ensure the presence of load weights (truck or heavy road construction equipment) at the test site. Because of several disadvantages of the plate load test, an alternative was sought to evaluate the bearing capacity of the existing asphalt pavement base courses or the bearing capacity of the gravel pavements. In the framework of this study the Falling Weight Deflectometre was evaluated as a possible alternative to the plate load test. Two hypotheses were tested in the study. The first hypothesis is that by performing measurements on the asphalt surface with a Falling Weight Deflectometre and by performing a certain calculation using the Grätz's evaluation method, it is possible with sufficient accuracy to determine the value of the deformation modulus E_{v2} that could be acquired by removing the asphalt and performing a plate load test at the same location. The second hypothesis is that when measurements are performed on unbound materials, e.g., gravel, the Falling Weight Deflectometre test and the plate load test are interchangeable. Indirect and direct comparative testing was performed to test the first and second hypotheses.

IOP Conf. Series: Materials Science and Engineering 1202 (2021) 012021 doi:10.1088/1757-899X/1202/1/012021

1. Methodology and measurements

1.1. Methodology and measurements of indirect comparative testing

In order to check whether there was any correlation between the results of the plate load test performed on an unbound pavement and the results of the Falling Weight Deflectometre test on the asphalt pavement, an indirect comparative testing of these devices was performed. Indirect comparative testing was performed on the road A10, section from km 13.3 to km 19.2 in July 2017, in a closed parking lot in Saurieši near the road A5 in October 2019 and in Rīga avenue in Ādaži in November 2019.

The order of the indirect comparative testing was the following:

1. Measurements with a Falling Weight Deflectometre on the asphalt surface;
2. Marking of the measured spot;
3. Demolition of asphalt layers in the marked spot;
4. Performing of plate load test.

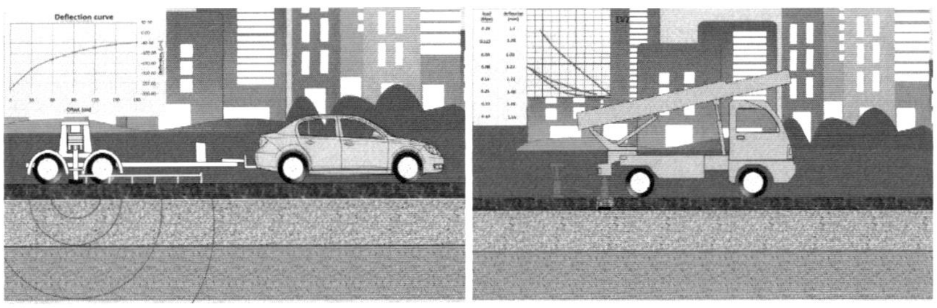

Figure 1. Visualization of Falling Weight Deflectometre and plate load test

Due to certain circumstances 9 out of 18 comparative testing measurements performed in 2017 were performed in a different order. Initially, a plate load test was performed and on the following day testing with a Falling Weight Deflectometre was performed in the distance of approximately 1 metre from the plate load test spot. In further data processing, it was assumed that the conditions in the distance of 1 m are the same as in the spot of plate load test.

Three loadings with Falling Weight Deflectometre were performed in each measurement spot in order to avoid seating errors. Only data from the last loading were used for further analysis of road pavement.

Settings for FWD test were the following:

1. Plate diametre – 300(mm) ;
2. Installed load – 50(kN) ;
3. Number of sensors – 7 ;
4. Placement of sensors (in 2017) – distance from plate(D1-0; D2-30; D3-60; D4-90; D5-120; D6-150; D7-180);
5. Placement of sensors (in 2019) – distance from plate(D1-0; D2-20; D3-30; D4-45; D5-60; D6-90; D7-150).

Plate load test was performed in accordance with DIN 18134.

The results of comparative testing were not compared directly, as the bearing capacity was measured with the equipment on different pavement layers. Testing with Falling Weight Deflectometre was performed on the pavement wearing course and the obtained bearing capacity value would characterize the entire road pavement. In order to obtain comparative results from the Falling Weight Deflectometre data, certain calculations have to be performed. Grätz's evaluation method described in (Čičković, 2017) and (FGSV, 2014) were used in this study.

The basic principle of Grätz's evaluation method and its advancement is the concept of a slab on an elastic-isotropic halfspace (see Figure 2). This concept is not suited for bodies with finite dimensions. Plate theory includes also the following simplifications (Čičković, 2017):

1. Plate is stiff in comparison to the halfspace;
2. Plate thickness h is insignificant in comparison with the remaining dimensions of the plate.

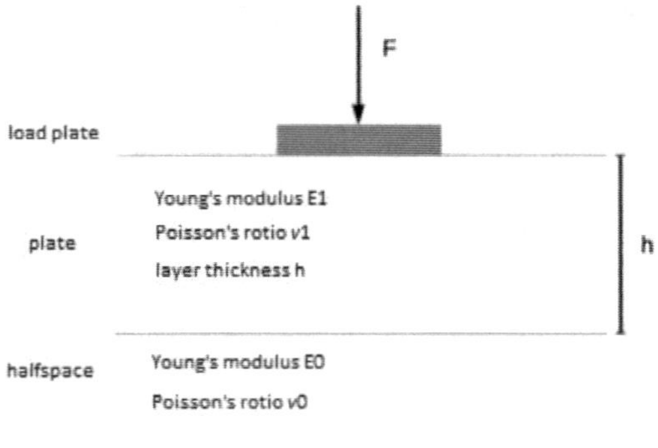

Figure 2. Model 'slab on elastic-isotropic halfspace' without layer bonding between slab and halfspace.

In order to acquire the strain modulus E_{v2c} from Falling Weight Deflectometre measurements according to this method, regression parameters A and B have to be obtained initially in equation (1) so that the theoretical deflection curve would be maximally close to the measured one.

$$w(r) = A \cdot (a_0 \cdot e^{B \cdot a_1 \cdot r} + a_2) \tag{1}$$

where w – deflection (mm); r – distance to load centre (mm);A – regression parameter (mm); B– regression parameter (mm^{-1}); a_0 – regression coefficient 0.392948; a_1 – regression coefficient -0.398483; a_2 – regression coefficient 0.0137024 (FGSV, 2014).

Methods (FGSV, 2014) and (Čičković, 2017) do not describe how to find the most suitable regression parameters A and B. Within the scope of this study it was done with the application of Sum of Squared Deviations Q and "Solver" function in "Microsoft Excel".

$$Q = \sum_{i=1}^{n} (x_{ci} - x_{mi})^2 \tag{2}$$

where Q – sum of Squared Deviations; n – number of sensors; x_{ci} – calculated deflection at i^{-th} sensor; x_{mi} – measured deflection at i^{-th} sensor.

In order to perform this on the basis of (FGSV, 2014), the following regression coefficients were assumed: A=0.6 mm and B= 0.002 mm^{-1}. With the application of equation (1) with such regression coefficients, the theoretical deflection was calculated in the same distances as the measurements were performed with deflection sensors of the Falling Weight Deflectometre.

Table 1. Calculation of Sum of Squared Deviations

Calculated deflection w (μm) at different distance r to centre (mm)							Load (N) Stress
0	300	600	900	1200	1500	1800	
Calculated deflection x_{ci} (μm)							
244	194	154	123	99	80	64	49737.6
Measured deflection x_{mi} (μm)							
322	241	162	103	67	46	34	
Deviation (μm)							Sum of Squared Deviations Q (μm)
-78	-47	-8	21	32	34	30	
Deviation squares(μm)							
6023	2223	61	424	1032	1126	899	11789

A calculation algorithm in "Microsoft Excel" has to be created in such manner that when the regression parameters A and B are changed, the calculated surface deflection values and the Sum of Squared Deviations Q change automatically. "Solver" function in "Microsoft Excel" with the objective of minimum Sum of Squared Deviations Q at variable regression parameters A and B was used to find the regression parameters A and B that would approximate the calculated deflection values to the measured deflection values.

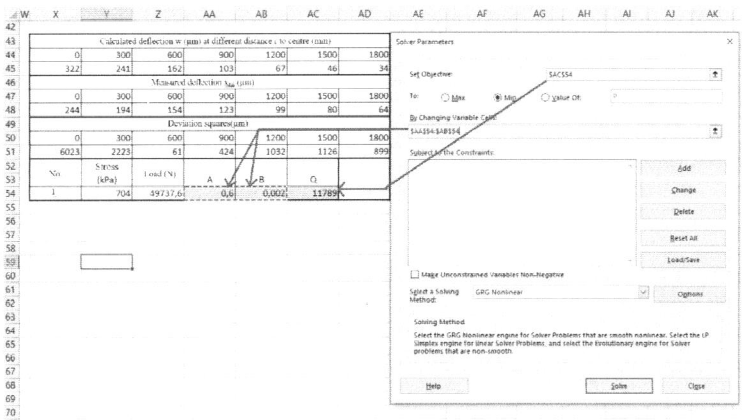

Figure 3. Use of Solver function

This regression approach is closely related to mechanical properties of road pavement. Both regression parameters (A un B) are related to layer modulus of the halfspace M_0 or elastic length l (Čičković, 2017):

$$l = \frac{1}{B} \tag{3}$$

where l – elastic length (mm);B– regression parameter (mm^{-1}).

$$M_0 = \frac{Q}{A \cdot l} \tag{4}$$

where M_0 – layer modulus of the halfspace (MPa); Q – impact force (N); A – regression parameter (mm); l – elastic length (mm) (Čičković, 2017).

With the application of (FGSV, 2014) the following approximation of the strain modulus Ev_{2c} and the layer modulus of the halfspace M_0 may be assumed:

$$E_{v2c} = \frac{M_0}{1{,}25} \tag{5}$$

where E_{v2} strain modulus (MPa); M_0 – layer modulus of the halfspace (MPa).

1.2 Methodology and measurements of direct comparative testing

To check whether there was any correlation between the results of plate load test and the Falling Weight Deflectometre test, a direct comparative testing of these devices was performed. Testing was performed with both pieces of equipment on a gravel surface. Direct comparative testing was performed on roads V59 and V60 2018 in April 2018 and on road V6 in October 2019. Due to certain circumstances the spots for comparative testing on roads V59 and V60 were already determined, but spots for comparative testing on road V6 were determined according to preliminary research.

To choose the spots for comparative testing, measurements with Falling Weight Deflectometre were performed on the previous day in road V6, section from km 3.6 to km 7.1.

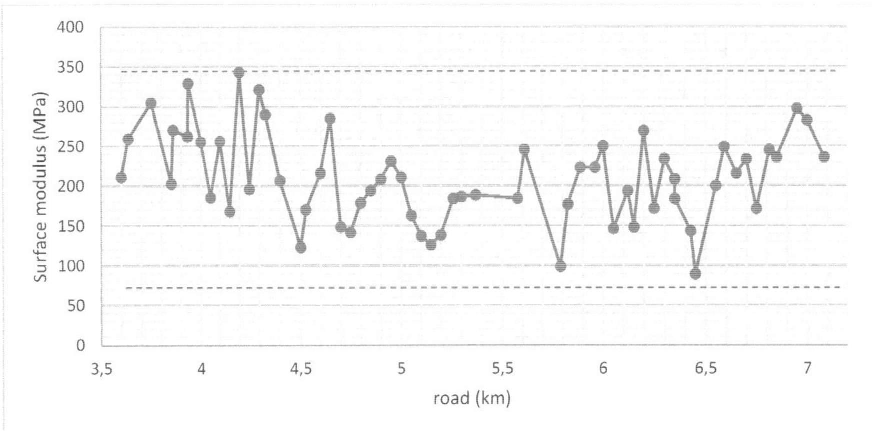

Figure 4. Surface modulus values from road V6

Surface modulus values from 89 MPa to 343 MPa were acquired in this road section. Measurement spots were chosen so that the surface modulus values acquired in comparative testing would be scattered between the values of 89 MPa and 343MPa.

Figure 5. Performance of measurements with Falling Weight Deflectometre and plate load test

The order of comparative testing in 2019 was the following (testing in 2018 was done in completely opposite order):

1. Measurements with Falling Weight Deflectometre in a predefined spot;
2. Marking of measurement spot;
3. Plate load test in the marked spot.

Three loadings with Falling Weight Deflectometre were performed in each measurement spot in order to avoid seating errors. Only data from the last loading were used for further analysis of road pavement.

Settings for FWD test were the following:

1. Plate diametre – 300(mm);
2. Installed load – 50(kN);
3. Number of sensors – 7;
4. Placement of sensors – distance from plate (D1-0; D2-30; D3-60; D4-90; D5-120; D6-150; D7-180);

Plate load test was performed in accordance with DIN 18134.

Surface modulus acquired in measurements with Falling Weight Deflectometre was compared with the acquired E_{v2} values from plate load test.

2. Results and analysis

2.1. Results and analysis of indirect comparative testing

In accordance with Grätz's evaluation method the results of Falling Weight Deflectometre measurements acquired in indirect comparative testing were used to determine the regression coefficients A and B for equation (1). Formula (4) was applied with these coefficients to find the values of M_0 and E_{v2c} was calculated according to formula (5).

Table 2. Calculation and measurement strain moduli

No.	M_0	E_{v2c}	E_{v2}	No.	M_0	E_{v2c}	E_{v2}	No.	M_0	E_{v2c}	E_{v2}
1	202	162	144	12	182	146	169	22	234	187	331
2	171	137	203	13	222	178	161	23	208	167	374
3	148	118	117	14	217	174	287	24	216	173	385
4	151	121	152	15	207	166	314	25	227	181	162
5	227	182	147	16	206	165	184	26	178	143	298
6	196	157	216	17	266	213	258	27	258	206	330
7	174	139	244	18	235	188	300	28	178	142	121
8	202	161	247	19	196	157	273	29	165	132	201
9	184	147	269	20	246	197	345	30	251	201	256
10	174	140	194	21	180	144	202	31	232	186	226
11	197	158	215								

Correlation between the calculation E_{v2c} acquired from the results of Falling Weight Deflectometre measurements and the E_{v2} acquired in plate load test is determined.

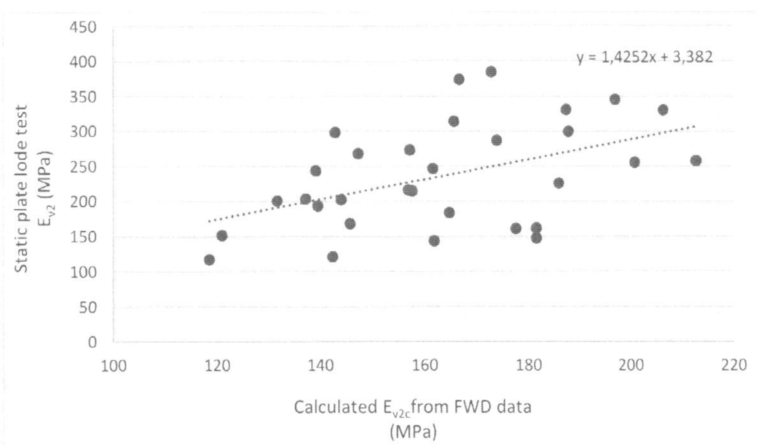

Figure 6. Falling Weight Deflectometre and plate load test

Pearson correlation coefficient between the calculation Ev2c and Ev2 is r = 0.47. Such correlation coefficient shows a moderately close correlation between these values and it is not sufficient enough in order to recommend the replacement of plate load test with backcalculation with the use of Falling Weight Deflectometre and application of the method described in the study.

2.2. Results and analysis of direct comparative testing

The following results of direct comparative testing were compared: surface modulus acquired from the results of Falling Weight Deflectometre measurements and E_{v2} acquired in plate load test.

Table 3. Results of Falling Weight Deflectometre measurements and plate load test

No.	D1 (μm)	E_{surf} (MPa)	E_{v2} (MPa)
1	628	297	215
2	738	250	220
3	980	186	170
4	577	318	327
5	952	190	176
6	536	348	263
7	1907	92	103
8	918	198	166
9	725	249	221
10	1187	149	146
11	1275	140	110
12	1983	87	83
13	828	227	167
14	979	191	115
15	992	187	124
16	485	386	273
17	527	354	205

Correlation between surface modulus and E_{v2} acquired in plate load test was determined.

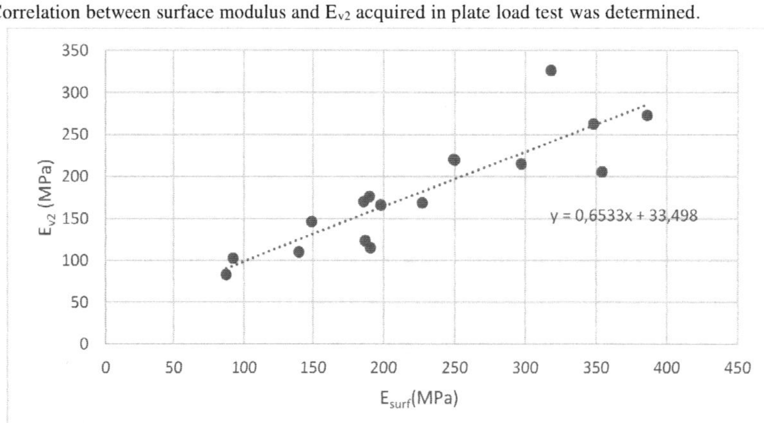

Figure 7. Strain modulus E_{v2} & surface modulus E_{surf}

Conclusions

Grätz's evaluation method is very simple in comparison with other backcalculation methods. In order to calculate the layer modulus of the halfspace M_0 that characterises the rigidity of pavement base courses, only the data from Falling Weight Deflectometre is required. Despite good references from other specialists, this method in comparative testing did not show sufficiently good correlation with the results of plate load test.

On the basis of research on backcalculation methods where one of the methods includes the creation of a number of regression equations from deflection curve databases and specific regression equation is determined for each type of road pavement, it is assumed that regression calculation equation in Grätz's evaluation method was not well suited for most of the pavements tested during comparative testing. Considering that asphalt layers on roads tested in indirect comparative testing were thin or old, it is assumed that such regression equation would show better results for road pavements with thicker and newer asphalt layers.

Results from direct comparative testing showed that plate load test and Falling Weight Deflectometre test may be interchangeable if no compaction properties have to be determined. Considering the speed of testing with Falling Weight Deflectometre, it could be used, e.g. during the construction works in combination with plate load test if the bearing capacity has to be determined in small intervals in a longer road section. Falling Weight Deflectometre might be also used as an instrument for determining the bearing capacity in road reconstruction projects in case asphalt pavement is paved on gravel roads to improve driving comfort.

References

Čičković, M. (2017). *Analysing FWD data — from the deflection bowl to material parameters.* BCRRA 2017 - 10th International Conference on the Bearing Capacity of Roads, Railways and Airfields. Athens, Greece.

Forschungsgesellschaft für Straßen- und Verkehrswesen. (2014). *Arbeitspapier Tragfähigkeit von Verkehrsflächenbefestigungen. Teil C 2.1: Falling Weight Deflectometer (FWD): Auswertung und Bewertung - Asphaltbauweise (AP Trag Teil C 2.1).*

INVESTIGATION OF COMPARABILITY OF TSRST AND SCB CRACKING TESTS FOR EVALUATION OF LOW-TEMPERATURE PROPERTIES IN ASPHALT MIXTURES AND USE IN QUALITY CONTROL

Janis BAUMANIS[1], Arturs RIEKSTINS[1,2], Andris BALODIS[1]

[1]*SLLC "Latvian State Roads", 3 Gogola Street, Riga, LV-1050, Latvia*
[2]*Riga Technical University, 6A/6B–345, Ķīpsalas Street, Riga, LV-1048, Latvia*
Janis.Baumanis@lvceli.lv, Arturs.Riekstins@lvceli.lv, Andris.Balodis@lvceli.lv

March 10, 2021

Abstract. Mix design procedure for asphalt mixtures in the Baltic region requires to ensure resistance to low temperatures due to climatic conditions. Thermal Stress Restrained Specimen Test (TSRST) has been considered as the most precise direct test method to determine the thermal behaviour of asphalt mixtures. As the TSRST test is time-consuming and the equipment is much more expensive, therefore the possibility to use Semi-Circular Bending (SCB) as a preliminary test was evaluated and the potential threshold was recommended. This study presents the evaluation of low-temperature properties with SCB and TSRST methods and the test suitability assessment for use in quality control. The supplementary rating was made by analysing Fraass breaking point test results of asphalt binders. In total 36 different asphalt samples were tested to investigate fracture test methods and to assess the influence of bitumen type and composition on resistance to low-temperature cracking. The results displayed an acceptable correlation between both test methods that allow using SCB for pre-screening purposes. At the same time, the results indicated that the type of used bitumen has a crucial influence on asphalt mixtures resistance to low-temperature cracking.

Keywords: Asphalt mixtures, Thermal Stress Restrained Specimen Test, Semi-Circular Bending, Low-temperature cracking, Fracture properties, Asphalt binders

Introduction

Despite the increase in the average value of global temperature, climatic conditions in the Baltic and other northern countries still present severe winters with extreme daily low temperatures. Low temperature cracking is one of the most dominant distress that occurs in asphalt pavements built in such climatic conditions. With the propagation of low temperature cracks through the pavement structure, a passage is created for intrusion of water in the pavement, thus accelerating the deterioration of road structure. At the same time during summertime pavement temperatures rise even up to 60°C, therefore when designing asphalt concrete mixtures in such climate, it is important to combine the strength of this mixture at low and high temperatures. In Latvia since 2010 the resistance of asphalt concrete to plastic deformations at high temperatures has been controlled using a wheel track test in accordance with the requirements of the standard EN 12697-22. Today rutting is much less frequent and occurs to a lesser extent, however, to improve this property, asphalt producers have made asphalt concrete masses tougher, which has led to a possible decrease in the resistance of these mixtures to thermo-cracking.

One of the most common laboratory methods to evaluate low-temperature properties of asphalt mixtures is the Thermal Stress Restrained Specimen Test (TSRST), presented for the first time by Monismith, G.A Secor & K.E. Secor (1965). This research showed that TSRST may be successfully used to evaluate low-temperature cracking resistance of asphalt mixtures. It is concluded that TSRST test is a test method which may simulate the low temperature cracking process of asphalt in pavement well according to the analysis. Therefore, this method has been chosen to be introduced in Latvian road specifications as a performance-based test in contrary to Fraass breaking point value which characterizes only bitumen properties, but not the properties of the whole asphalt mixture. Requirements proposed and adopted in the Latvian Road Specifications 2019 regarding TSRST value for asphalt surface layers -22.5°C and for binder and base layers -20°C, although initially they were intended to be even higher: -25°C and -22.5°C respectively. In Austria, the proposed achievable cracking temperature for surface layers ranges from -30°C to -25°C and for base courses from - 25°C to -20°C depending on the type of mixture proposed by Karcher & Wittenberg (2010). In Germany (FGSV, 1994), on the other hand, the recommended requirements for the achievable TSRST temperature are -20°C or -25°C, depending on the climatic zone. It is important to note that in both countries the test sample size and test conditions are the same as in Latvia.

The TSRST method has its drawbacks, but it is considered to be the closest to real conditions compared to other similar tests according to Gražulytė, Vaitkus, Andrejevas & Gribulis (2017). In various sources, the main disadvantage of

TSRST methods is the accuracy and rate of temperature reduction (Pszczola, Szydlowski & Jaczewski, 2019). The standard initial temperature is set to 20°C and the rate of cooling is constant and equal to 10 °C/h. However, the information available in the literature shows that the temperatures obtained in real conditions are in the range of 1 to 2°C/ h (Gražulytė et al., 2017). The main reason why the rate of temperature drop is so high is the test time. A reduced rate of temperature drop would significantly extend the test time by more than 24 hours. Therefore, it may be considered that the rate of temperature decrease used in laboratory conditions is several times higher than the values obtained in real conditions.

Another disadvantage is the high price of equipment and components. It is reasonable to assume that contractors and independent laboratories will not make such an investment in the purchase of equipment. Therefore, it is necessary to propose another test method as a preliminary test to TSRST. A cheaper and universally available test should be used for self-monitoring to assess thermal crack resistance. The semi-circular bend (SCB) is mentioned in many literature sources as a cheaper and faster alternative to the TSRST test and is considered by many scientists to be a potentially successful test for characterizing mixtures for practical purposes (Huang, Shu & Tang, 2005; Mohammad, 2006). This is due to simple and fast performance of the test, relatively inexpensive equipment required, high repeatability, as well as relatively high correlation with the actual cracking results (Wistuba, Mollenhauer & Metzker, 2009). As stated in research by Zaumanis & Valters (2018), TSRST and SCB should be considered as complementary and not competing test methods as each method applies different principles. The SCB specimen geometry is also suitable for laboratory compacted specimens and field cores, therefore increasing the possible field of application of the method.

Objective

The primary purpose of this study is to evaluate the most used asphalt mixture types for low temperature cracking using various test methods and to determine possible correlation between mixture fracture toughness (SCB), tensile strength (TSRST) and bitumen brittleness (Fraass breaking point) for the use in quality control. The secondary aim was to evaluate characteristics of typical asphalt mixes in Latvia. This evaluation was necessary to determine if the currently used mixes were adequate to resist low temperature cracking and if the set requirements in Latvian road specifications were sufficient.

1. Methods and materials

1.1. Materials

Laboratory tests were conducted on most typical types of asphalt mixtures used in road construction which included mixtures used for wearing course, binder course and base course. All mixes were designed in compliance with the Latvian technical specifications. As the aim of this research was not to evaluate different aspects of asphalt mixture properties that influenced mixture resistance to low temperature cracking, the characteristics such as volumetric properties, density, void content, reclaimed asphalt content and others were excluded from the assessment. All mixtures had to comply with Latvian road specifications for the use in construction sites on roads with minimum annual average daily traffic intensity above 500 cars, thus ensuring that evaluated mixtures represented the actual situation. In total 36 different samples were tested, including open graded SMA type mixtures and dense grades AC type mixtures. 26 mixtures were produced in plant and 10 were produced in plant laboratory according to approved mix designs and were prepared using laboratory mixer equipment in accordance with EN 12697-35 standard. Type of mixture production is indicated as ageing conditions vary significantly between plant-produced and laboratory produced samples. The laboratory asphalt mixtures were not subjected to additional ageing procedures. Presented asphalt binder content for each mixture complies with the requirements of Latvian road specifications that define the minimum binder content in correspondence with the pavement course type and traffic intensity.

1.2. Fraass breaking point test

Fraass breaking point temperature was determined according to EN 12593. This test is used to indicate the behaviour of the bitumen at low service temperatures according to Latvian road specifications. A sample of bituminous binder is applied to a metal plate at an even thickness. This plate is submitted to a constant cooling rate of 1°C/min and flexed repeatedly until the binder layer breaks; the temperature at which the first crack appears is reported as the Fraass breaking point. All Fraass breaking point tests were performed on extracted bitumen as it was required in Latvian road specifications.

1.3. Thermal stress restrained specimen test

The resistance of asphalt mixtures to low-temperature cracking was assessed by means of the TSRST method according to EN 12697-46 standard and was used to determine the low temperature cracking susceptibility of asphalt concrete. The testing procedure is based on the hypothesis that contraction of the surface layer of the pavement structure during cooling is longitudinally restrained as the length of the road may be considered to be infinite (Carter & Paradis, 2010). The test procedure regulates that the rectangular shaped sample is held at a constant length, while temperature is decreased at a constant rate of 10°C/h. The test starts at the temperature of +20°C. Due to the prohibited thermal shrinkage, cryogenic stress is built up in the specimen. At sample failure, critical stress and cracking temperature are recorded. Samples for a TSRST were cut from slabs compacted with roller compactor according to standard EN 12697-33. Figure 1 shows setup of the used TSRST equipment and fractured sample.

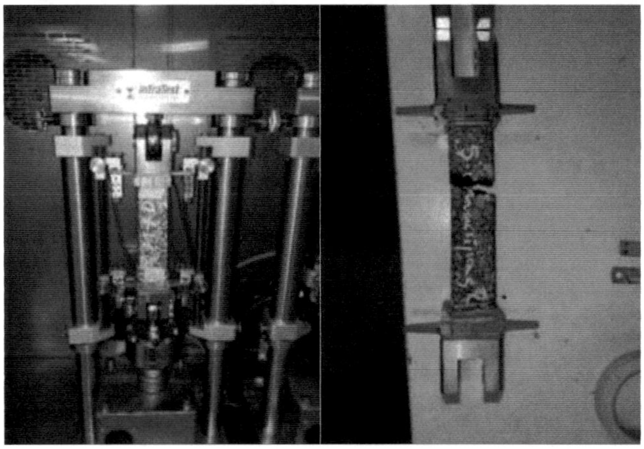

Figure 1. TSRST sample testing device and specimen after fracture

1.4. Semi-circular bending test

The SCB test is used to determine the fracture energy of asphalt mixtures from a load-displacement curve. Semi-circular samples were cut from compacted asphalt slab in 150 mm cores that were afterwards cut in halves. For every type of asphalt mixture four specimens were prepared. Samples were cut from the same slabs as TSRST samples, compacted with the use of roller compactor. Notch of 10 mm depth is cut in the middle of flat side. The specimens are positioned with the flat side on two rollers and a load is applied along the vertical diameter of the specimen. Load and displacement are measured to calculate test results. Other studies have described and proposed multiple possible loading rates and test temperatures for SCB test procedure, but in this research the test was run at a constant deformation rate of 5.0 mm/min at 0°C according to EN 12697-44 standard procedure. Figure 2 shows the setup of used SCB equipment and fractured sample.

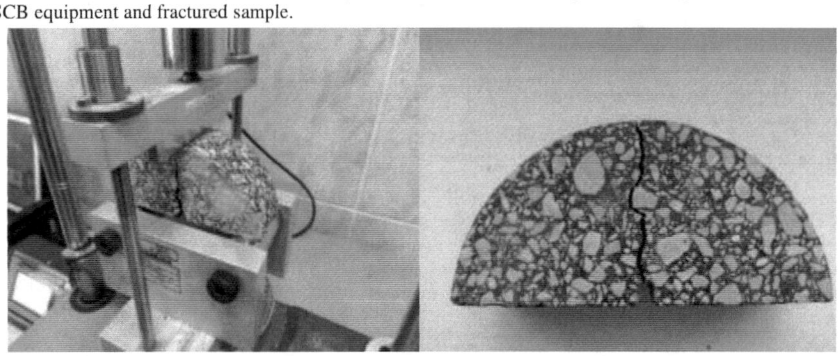

Figure 2 SCB sample testing device and specimen after fracture

2. Results and discussion

The summary of all tested asphalt mixtures and corresponding TSRST, SCB and Fraass breaking point test results is presented in Table 1. In the table the values of TSRST cracking temperature, fracture stress and SCB fracture toughness represent average obtained results for each mixture.

Table 1. Summary of tested mixtures and test results

Mixture type	Bitumen type	Mixture production	TSRST Cracking temperature °C	TSRST Fracture stress MPa	SCB $N/mm^{3/2}$	Bitumen content %	Fraass breaking point °C
AC-11	50/70	Plant	-21.8	1.922	25.8	-	-14
AC-11	50/70	Plant	-22.3	2.707	30.4	-	-15
AC-11	45/80-55	Plant	-30.6	4.258	38.4	-	-16
AC-11	45/80-55	Plant	-28.2	3.940	31.7	-	-17
AC-32	70/100	Plant	-24.0	2.704	17.5	3.69	-16
AC-16	70/100	Plant	-16.8	0.861	18.9	4.45	-16
AC-11	45/80-55	Plant	-24.9	2.526	24.7	5.17	-14
AC-11	70/100	Plant	-29.9	3.656	27.4	5.08	-20
SMA-11	70/100	Plant	-26.4	2.742	24.5	6.39	-16
AC-22	70/100	Plant	-25.4	1.949	24.1	3.88	-12
AC-32	70/100	Plant	-24.6	1.943	25.3	4.06	-13
AC-11	50/70	Plant	-27.3	3.663	34.6	5.45	-22
AC-16	70/100	Plant	-26.2	3.975	25.9	5.15	-14
AC-11	50/70	Plant	-23.6	4.985	22.9	4.92	-17
AC-16	50/70	Plant	-19.4	3.177	22.3	3.54	-22
AC-22	50/70	Plant	-23.1	3.818	17.6	3.61	-13
AC-32	50/70	Plant	-17.4	4.462	23.0	3.41	-12
AC-16	50/70	Plant	-24.1	3.756	17.7	4.08	-17
AC-32	50/70	Plant	-22.7	2.403	21.2	3.95	-13
AC-16	45/80-55	Plant	-33.1	1.901	40.6	4.78	-18
AC-16	50/70	Plant	-24.0	1.053	32.0	5.03	-14
AC-32	50/70	Plant	-24.5	2.790	22.3	3.72	-12
AC-32	70/100	Plant	-29.4	2.175	29.6	4.28	-23
SMA-11	45/80-55	Plant	-26.3	5.325	38.7	5.74	-21
AC-22	50/70	Plant	-21.8	4.451	15.2	3.60	-13
AC-22	50/70	Plant	-26.7	3.323	21.8	3.85	-15
SMA-11	45/80-55	Lab	-24.5	4.503	33.0	6.02	-17
AC-11	50/70	Lab	-24.5	3.008	19.6	4.86	-19
AC-11	50/70	Lab	-22.0	2.859	15.5	4.84	-28
AC-11	50/70	Lab	-23.1	3.826	25.4	5.53	-18
AC-11	50/70	Lab	-20.9	4.649	34.2	5.25	-14
AC-11	70/100	Lab	-23.7	4.867	26.4	5.20	-16
AC-11	50/70	Lab	-21.9	4.561	25.7	5.45	-19
AC-11	50/70	Lab	-21.1	4.706	31.4	5.80	-18
AC-11	45/80-55	Lab	-23.5	4.687	30.6	5.25	-18
AC-11	50/70	Lab	-23.4	4.854	33.2	5.37	-19

Figure 3 shows the results of low-temperature cracking resistance depending on the bitumen grade in the asphalt concrete mixture. Although bitumen usually represents only 4-6% of the total composition of the asphalt mixture, as it is in this study, the bitumen grade has a significant role in the resistance of the whole pavement to thermal cracking. As expected, the poorest results are for mixtures with bitumen 50/70, while all samples with PMB bitumen meet the defined requirements. Almost all samples of mixtures with bitumen 70/100 provide the required resistance to cracking, only in one case an inadequate result was obtained which was considered as a separate exception rather than tendency. Thus, it may be stated that under the existing requirements, non-compliance of the results is expected for asphalt concrete mixtures with grade 50/70 bitumen. It is necessary to bear in mind that the use of PMB or soft bitumen grade does not guarantee that the requirements are met as the thermal crack resistance may be affected by parameters summarized in Table 2 (Marasteanu et al. 2007; Zaumanis & Valters, 2018; Pszczola, Szydlowski & Jaczewski, 2019; Stienss & Szydlowski, 2020).

Table 2. Summary of parameters affecting resistance to thermal cracking.

Component of mixture	Characteristics or properties
Bitumen	Bitumen hardness Properties of added modifiers Aging of bitumen
Mineral material	Quantity of the filler Gradation Angularity of course aggregate Fineness of fine mineral material Mechanical strength of mineral material
Properties of the designed mixture	Volumetric properties RAP content of the mixture Presence of WMA additives Binder content Presence of crumb rubber or fibres

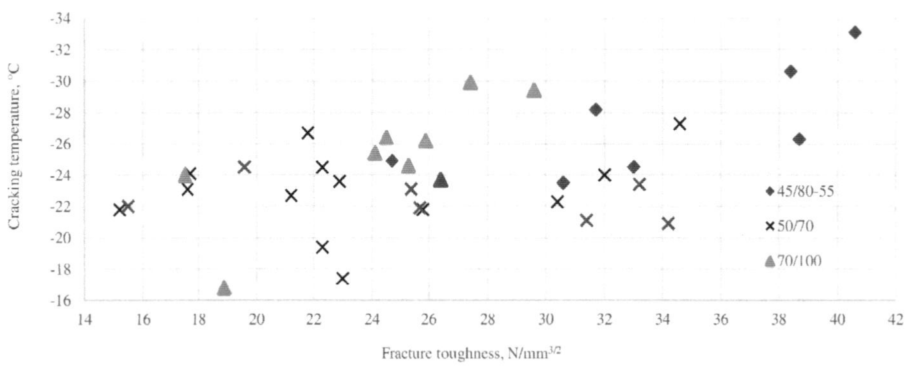

Figure 3. TSRST cracking temperature versus SCB fracture toughness with lab produced samples marked in red.

The correlation of the results between the TSRST and SCB test results is moderate, however, if the results are distributed separately by bitumen grades, then this correlation increases for mixtures with bitumen grade 70/100 (R^2=0.55) and 45/80-55 (R^2=0.50). In case with 50/70 grade bitumen, it is possible that the results of SCB test may be more affected by the type of aggregate used (sedimentary or magmatic) and the type of mixture (AC-11 or AC-32) than the bitumen grade because a harder binder is less elastic, consequently more vulnerable to fracture-mechanics based tests like SCB. Depending on the grade each bitumen may react differently to the rate of temperature drop in the TSRST test. It should also be noted that the information gathered by Marasteanu et al. (2012) confirms that as the rate of temperature drop decreases, so does the temperature at which the samples break. This shortcoming may be addressed by setting slightly lower requirements as inadvertently has been done by decreasing initially proposed specification of -25°C for surface layers and -22.5°C for base and binder courses.

Given that 80% of the tested samples meet the requirements for resistance to thermal cracking, and the results of SCB tests are widely distributed, it is not possible to objectively set a selection criterion, according to which TSRST test for asphalt mass may not be performed. At the same time, it is possible to suggest that mixtures with SCB result below 24 N/mm$^{3/2}$ should be always tested for TSRST criteria. This is supported by the given data as 38% of samples below this threshold do not meet TSRST requirement, in contrast mixtures with SCB result above 24 N/mm$^{3/2}$ only in 12% cases do not meet the same criteria. From specimens with 70/100 grade bitumen and polymer modified bitumen only two presented results below aforementioned criteria. The limited number of tests performed in this study is not conclusive and further research is necessary to investigate a wider pool of different mixtures before procedure of pre-selecting mixtures for TSRST based on SCB criteria is implemented in specifications and quality control. Nevertheless, this principle may be useful in mix design stage when it is necessary to test numerous modifications before optimal composition of mixture is set for type-testing.

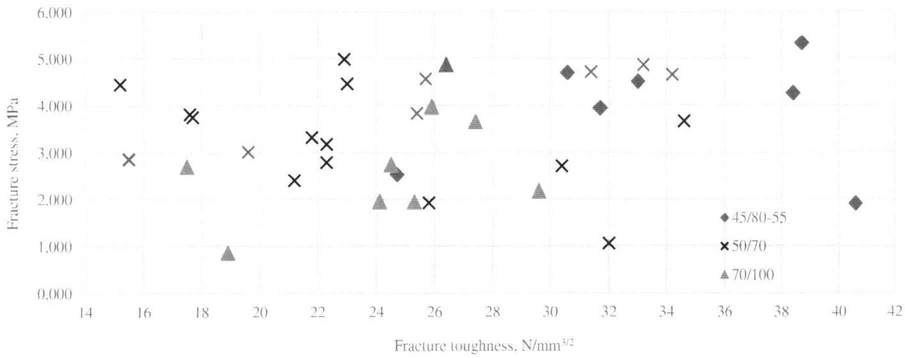

Figure 4. TSRST fracture stress versus SCB fracture toughness with lab produced samples marked in red.

For asphalt mixture evaluation purposes usually TSRST fracture temperature is used instead of fracture stress as its coefficient of variation is lower (Marasteanu et al. 2007). Results in this study confirm this statement since the fracture stress data is widely distributed as illustrated in Figure 4. For example, AC-32 sample at failure temperature of -17.4 °C indicated 4.462 MPa of internal stress, but different AC-32 sample failed at 2.175 MPa in recorded temperature of - 29.4 °C. Furthermore, the fracture stress values for specimens for the same mixtures exhibited significant disparity even if the failure temperatures were similar. This phenomenon may be explained with the fact that during test procedure specimen may bend thus creating non-uniform stress distribution. It may also be observed that no correlation exists between SCB fracture toughness and TSRST fracture stress, therefore this parameter is not suited for quality control purposes.

One of the most important factors that may have influenced the relationship was the accuracy of TSRST and SCB sample preparation. The process of preparing of specimens is delicate and requires accurate equipment, especially in case for SCB specimens, since the sample geometry according to research results by Wagoner, Buttlar, Paulino & Blankenship (2005) may play a role in data analysis because of relatively small fracture surface created when testing in standard configuration.

The lab produced samples in theory should have given "better" results considering that the aged bitumen is more brittle consequently shifting the TSRST failure temperature to a higher temperature as suggested in research conducted by Mollenhauer & Tušar (2016). The lab produced samples are presented in Figure 3 and Figure 4 marked in red colour. Collected data does not indicate difference between plant produced mixtures and laboratory produced mixtures, nor it demonstrates increased correlation to any of test methods, with exception of TSRST fracture stress correlation to SCB fracture toughness for samples produced in lab. At the same time, this observation does not indicate that such aspect did not have effect on gathered data, rather suggests that possible differences between lab and plant mixtures are not that significant as assumed previously. This conclusion is not definitive and requires further research as the previous studies also have shown that ageing of asphalt mixtures might affect the results of each mixture cracking test differently (Zaumanis & Valters, 2018). Nevertheless, the importance of aging should not be overlooked since the cracking mostly occurs in aged samples, which would require to implement ageing assessment to the type test of asphalt mixtures.

The relationship between TSRST results and Fraass brittleness temperature results (Figure 5) presents a low correlation, however, a general trend may be observed that lower TSRST cracking temperatures are reached when Fraass test results also show lower temperatures. Such observations have been made with other similar studies (Wistuba, Mollenhauer & Metzker, 2009; Zaumanis & Valters, 2018).

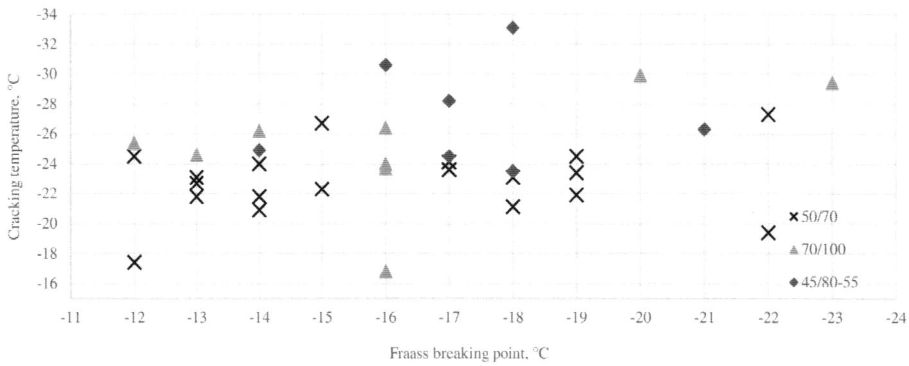

Figure 5. TSRST cracking temperature versus Fraass temperature.

Obtained correlation could be related to poor repeatability and reproducibility of the Fraass test method as the given uncertainty for recovered bitumen is ±4°C. It is also criticized for being unreliable in cases where polymer modified bitumen is used (Bueno, Hugener & Partl, 2015). This was indirectly confirmed in this study since the results vary significantly among the mixtures, for example -14°C for PMB 45/80-55 type bitumen to -28°C for 50/70 bitumen. The TSRST results for respective samples were -24.9°C and -22°C while both mixtures were designed for the use in wearing coarse, thus mixture with polymer-modified bitumen would not meet Fraass breaking point requirement but would meet TSRST criteria according to Latvian Road Specifications. The directly opposite condition would occur with mixture that used 50/70 bitumen. In total five samples which met the requirements of the Road Specifications for the Fraass breaking point temperature of ≤-15°C, did not meet the TSRST criteria. In total 7 samples with insufficient Fraass breaking temperature results did comply to TSRST requirements according to national specifications.

The low correlation confirms that the resistance of asphalt concrete mix to low temperatures is not only affected by the properties of the bitumen used, but rather a combination of multiple bitumen properties and characteristics and the use of brittleness tests is insufficient and, in many cases, could give misleading results as demonstrated in this study.

Conclusions

Based on the obtained results and analysis, the following conclusions can be formulated:

1. According to the investigations performed, there is an average relationship between TSRST and SCB results, as there is distinct tendency that lower TSRST cracking temperatures transfer to increased SCB fracture toughness. Concurrent, if the results are divided according to bitumen type used in mixtures, the relation between both methods increases significantly.
2. Most of asphalt mixtures in this study meet the proposed requirements defined in Latvian Road Specifications for the assessment of thermal cracking for asphalt surface layers (-22.5°C) and asphalt binder and base courses (- 20°C). At the same time, it is recommended to revise implemented requirements after approbation period since the actual temperatures on roads may exceed the defined values.
3. As suspected, the type of bitumen used in each mixture had the most crucial effect on TSRST and SCB results. Samples with polymer modified bitumen and 70/100 grade bitumen showed the highest results both for TSRST and SCB test. On the other hand, mixtures with bitumen 50/70 displayed the poorest results as harder bitumen is less elastic.
4. Based on tested samples, a weak relationship between the brittleness of conventional bitumen and resistance to low-temperature cracking of bituminous mixtures was found. This suggests that it is not possible to draw correct conclusions about the resistance of the whole asphalt mixture to low temperature cracking only from the results of the Fraass breaking point temperature, as the correlation to direct cracking tests as TSRST or SCB is poor. For this reason, it is suggested to abandon Fraass breaking point requirement in quality control as soon as TSRST method is fully approbated.
5. Categorization by mixture production type did not increase the correlation between TSRST and SCB results in contrast to indications of other studies.
6. Further investigation should be considered to evaluate the possibility to reduce the test temperature of SCB samples from 0° to lower, because the information found in the literature shows that it would more accurately reflect the laboratory assessment with road conditions and improve correlation to TSRST.

Funding

This work was supported by State Limited Liability Company "Latvian State Roads" under internal research program.

Disclosure Statement

No potential conflict of interest was reported by the authors.

References

Bueno, M., Hugener, M., & Partl, M.N., (2015). Fracture toughness evaluation of bituminous binders at low temperatures. *Materials and Structures*, 48 (9), 3049–3058

Carter A. & Paradis M., (2010), Laboratory Characterization of the Evolution of the Thermal Cracking Resistance with the Freeze-thaw Cycles, *11th International Conference on Asphalt Pavements 2010,* Nagoya, Aichi, Japan, August 1-6, 2010.

Forschungsgesellschaft für Straßen und Verkehrswesen (1994) *Technische Prüfvorschrift – Verhalten von Asphalten bei tiefen Temperaturen*; Ausgabe 1994; e.V.; Köln, Germany

Gražulytė J., Vaitkus A., Andrejevas V. & Gribulis G., (2017), Methods and criteria for evaluation of asphalt mixture resistance to low temperature cracking, *The Baltic Journal of Road and Bridge Engineering*, 2017, 12(2): 135–144

Huang B., Shu X. & Y. Tang (2005). Comparison of semi-circular bending and indirect tensile tests for HMA mixtures, *Advances in Pavement Engineering*, Austin, TX, USA

Karcher C. & Wittenberg A. (2010). Low-temperature Flexibility of Mastic Asphalt and Asphalt Concrete with Polymer-modified Bitumen, *11th International Conference on Asphalt Pavements 2010,* Nagoya, Aichi, Japan, August 1-6, 2010.

Marasteanu M., Zofka A., Turos M., Li X., Velasquez R., Li X., Buttlar W., Paulino G., Braham A., Dave E., Ojo J., Bahia H., Williams Ch., Bausano J., Gallistel A. & McGraw J. (2007), Investigation of Low Temperature Cracking in Asphalt Pavements, *National Pooled Fund Study 776. Report No. MN/RC 2007-43*, Minnesota Dept of Transportation.

Marasteanu M., Zofka A., Turos M., Li X., Velasquez R., Li X. & Buttlar W., (2012), Investigation of Low Temperature Cracking in Asphalt Pavements. *National Pooled Fund Study -Phase II. Report No. MN/RC 2012-23*, Minnesota Dept of Transportation.

Mohammad L. (2006), Fatigue Growth Analysis of Hot-Mix Asphalt Employing Semicircular Notched Bend Specimen, *Transportation Research Board 85th Annual Meeting*, Washington, DC, USA, January 22-26, 2006.

Mollenhauer K. & Tušar M., (2016), Correlations between bitumen and asphalt properties, *Functional Durability-related Bitumen Specification (FunDBitS)*, Deliverable D.2c, CEDR Call 2013: Energy Efficiency – Materials and Technology.

Monismith C. L., Secor G. A. & Secor, K. E. (1965). Temperature Induced Stresses and Deformations in Asphalt Concrete, *in Proc. of Association of Asphalt Paving Technologists* 248–285.

Pszczola M., Szydlowski C. & Jaczewski M., (2019), Influence of cooling rate and additives on low-temperature properties of asphalt mixtures in the TSRST, *Construction and Building Materials* 204 (2019) 399–409

Stienss M., & Szydlowski C, (2020), Influence of Selected Warm Mix Asphalt Additives on Cracking Susceptibility of Asphalt Mixtures, *Materials* 2020, 13(1), 202

VAS "Latvijas Valsts ceļi". Ceļu specifikācijas 2019. (2019)

Wagoner M., Buttlar W., Paulino G., & Blankenship P., (2005), Investigation of the Fracture Resistance of Hot-Mix Asphalt Concrete Using a Disk-Shaped Compact Tension Test, *Transportation Research Record Journal of the Transportation Research Board* 1929(1):183-192

Wistuba M., Mollenhauer K., & Metzker K., (2009). Assessing low temperature properties of asphalt materials by means of static testing techniques. *8th international conference bearing capacity of roads, railways and airfields*. Champaign, IL, USA, June 29 - July 2, 2009.

Zaumanis M. & Valters A. (2018), Comparison of two low-temperature cracking tests for use in performance-based asphalt mixture design, *International Journal of Pavement Engineering*, 21(80):1-9

RESEARCH OF ASPHALT CONCRETE FROST RESISTANCE ACCORDING TO FATIGUE RESISTANCE AND STIFFNESS CRITERIA

Nauris PAULOVSKIS

[1]*SLLC "Latvian State Roads", 3 Gogola Street, Riga, LV-1050, Latvia*
Nauris.Paulovskis@lvceli.lv

March 13, 2021

Abstract. The formation of snow cover has been unsustainable in recent decades, which means that freezing and thawing are becoming more frequent in Latvia and that is one of the reasons why asphalt concrete frost resistance matters. Asphalt concrete beams were exposed to multiple freeze-thaw cycles and afterwards tested on a 4-point bending test for stiffness and fatigue resistance. Frost resistance testing for aggregates and concrete is a popular and widely described and used test method worldwide, but there is no such testing standard for asphalt concrete. This study is aimed to develop a methodology for determining the frost resistance of asphalt concrete by adapting the requirements of cement concrete standart AASHTO C666, as well as to experimentally evaluate the effect of frost resistance on the fatigue strenght and stiffness of asphalt concrete. The results show significant correlation between asphalt concrete void content and fatigue life after multiple freeze-thaw cycles and correlation between the amount of freeze-thaw cycles and asphalt concrete fatigue life.

Keywords: freeze-thaw cycle, asphalt concrete, air void content, 4-point bending test, asphalt fatigue, frost resistance

Introduction

Latvia's climate is largely determined by the location of its territory in the temperate climate zone on the shores of the Baltic Sea and the Gulf of Riga. Asphalt concrete layer is the most sensitive of bituminous road structural layers, as it is in direct contact with the environment (weather conditions) and traffic loads. Premature wear of bituminous pavements may cause irreversible damage to the road surface, as well as high additional costs due to water entering the road structure and freezing-thawing processes. Today's drastic temperature changes in combination with heavy precipitation in Latvia and the filling of the road pores with water are major threats to bituminous road surfaces.

That is the reason why freeze-thaw cycles are inevitable and therefore, frost resistance testing is important not only for aggregates, but also for the entire asphalt concrete layer, as it will more clearly show the overall performance of the material behaviour. However, at the moment there is no standard for testing the asphalt concrete frost resistance anywhere in the world, therefore various studies around the world are related to modifying the concrete frost resistance standard, which with small corrections has proved to be suitable for asphalt concrete testing. Also fatigue resistance testing is becoming more and more popular and is included into multiple pavement design calculation procedures, showing that it is important and reliable indicator for asphalt concrete layer longevity.

The aim of this research is to determine the performance properties of asphalt concrete mixtures often used in Latvia using such test methods as fatigue resistance and frost resistance, as well as to evaluate their interaction and find relationships between various parametres that affect these performance properties. Finally, it will determine the effect of porosity on the fatigue strength and stiffness of asphalt concrete, with and without freeze-thaw cycles.

1. Objective

The main goal of this study was to analyse the impact and correlation between two tests – freeze-thaw resistance (modified AASHTO C 666) and fatigue resistance (EN 12697-24). Concrete freeze-thaw standard adjustments for asphalt concrete testing and air-void content impact on asphalt concrete fatigue resistance and stiffness were included, as well.

As a small side experiment ultrasonic testing was made for asphalt concrete beams before and after freeze-thaw testing and fatigue resistance testing. This is also a concrete testing method adjusted for asphalt concrete testing.

2. Materials and methods

2.1. Materials

The maximum density and bulk density for the produced mixture of AC11 and AC11 PMB were determined in the laboratory, thus the samples may be divided by calculating the air void content.

When preparing samples for frost resistance testing, the number of samples with low air void content (2-4%) and the number of samples with high air void content (6-8%) were distributed as proportionally as possible for each type of testing.

AC11 samples were tested without freeze-thaw cycles, after 50 frost resistance cycles, after 100 freeze-thaw cycles and after 250 freeze-thaw cycles.

AC11 PMB samples were tested without frost resistance cycles and after 100 frost resistance cycles. They were not tested after 50 frost resistance cycles because the polymer modified bitumen is known to have a much higher stiffness and in this condition the fatigue test would take a disproportionate amount of time and a small number of such low frost cycles would have some effect on AC11 PMB 45 / 80-55 samples.

Table 1. Distribution of the obtained materials by asphalt concrete type and number of samples.

Asphalt concrete type	Freeze-thaw cycle count	Number of samples
AC11 surf B50/70	0	18
AC11 surf B50/70	50	9
AC11 surf B50/70	100	9
AC11 surf B50/70	250	9
AC11 PMB 45/80-55	0	5
AC11 PMB 45/80-55	100	5

2.2. Methods

2.2.1. Resistance to freezing and thawing

Determination of asphalt concrete beam resistance to freezing and thawing was done according to AASHTO C 666, but as it is concrete testing method, the author adjusted it for asphalt concrete testing.

Adjustments:

Frost resistance of asphalt concrete samples was tested in the temperature range from -20°C to + 20°C (in AASHTO C 666 standard from -20°C to + 4°C)

A change has been made in the lengths (durations) of freeze-thaw cycles. For testing asphalt concrete beams, one cycle was 8 hours long and the samples withstood 3 freeze-thaw cycles every day. These are more aggressive conditions than those specified in AASHTO C 666 so that the sample may freeze completely and thaw completely during cycling.

The AASHTO C 666 standard stipulates that defrosting the sample takes up at least ¼ part (25%) of the cycle time, while for asphalt concrete samples the time of the refrigeration and defrost cycles is the same - the refrigeration time is 4 hours and the defrosting time is 4 hours to fully influence asphalt concrete beam.

Figure 1. Asphalt concrete beams moisturized in specially made tin baths before freezing in a cooler.

2.2.2. Fatigue resistance and stiffness

Determination of fatigue resistance was done according to EN 12697-24. The samples were divided into equal groups for every type of fatigue resistance testing. Testing was made using 150;200;250;300(microstrains) relative deformation for the AC11 surf 50/70 and for the AC11 PMB beams 200;250;300 microstrain were used.

By using the same equipment stiffness testing procedure was done according to EN 12697-26.

These tests were made in a 4-point bending testing chamber.

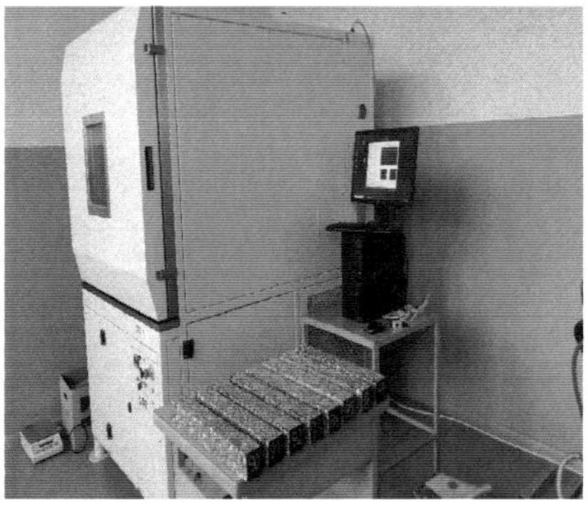

Figure 2. Four-point bending testing machine and asphalt concrete beams after freeze-thaw testing.

2.2.3. Ultrasonic testing

During the ultrasonic testing, the applied method was for testing the hardness of the concrete monolithic structure with the ultrasonic method based on GOST 17624-87 standard "Concrete. Ultrasonic method for hardness assessment". According to the standard, testing of monolithic structures may only be performed by non-destructive (e.g. ultrasonic) testing. GOST 17624-87 standard contains the main rules for determining the compressive strength of concrete on the surface of the sample using the ultrasonic method.

The concrete test specimens are cubic in shape, but the longitudinal dimensions of the asphalt concrete beams prepared for the four-point bending test are much larger than their cross-section, so there is not much variation in how ultrasonic testing is performed. GOST 17624-87 instruction states that within one test area several measurements may not differ by more than 2%, which was also not observed when testing asphalt concrete beams and the results may be considered objective. Testing was made using Ultrasonic tester UK1401.

3. Results and discussion

3.1. Fatigue resistance average results

3.1.1. Fatigue resistance averages

Figure 3 shows the fatigue strength of each series of AC11surf samples - the number of cycles up to a 50% reduction in the stiffness value, depending on the number of freeze-thaw cycles. AC11 reference samples that were not subjected to frost cycles show a higher average number of fatigue cycles, but samples after 50 and 100 frost cycles show similar results. The samples after 100 cycles of frost resistance show almost the same averages compared to the samples after 50 cycles, and the author's explanation is the small number of samples. Most of the damage was done to the samples after 250 freeze-thaw cycles, they have 4 times weaker fatigue resistance than reference samples.

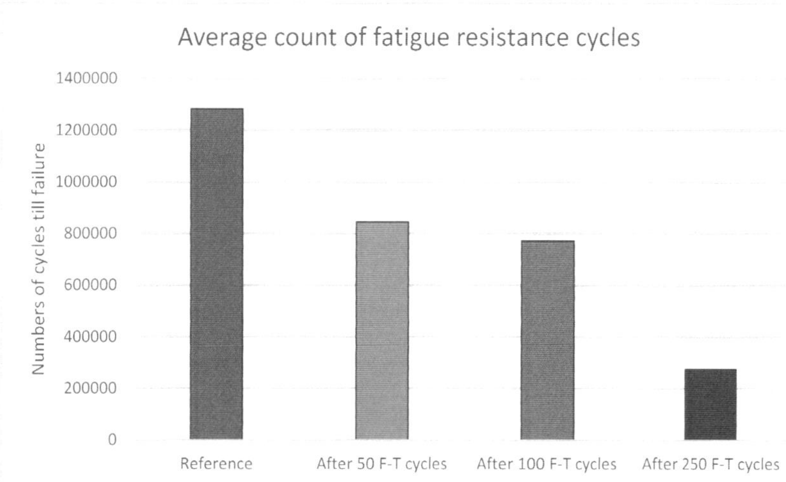

Figure 3. AC11 surf sample fatigue resistance averages after various freeze-thaw cycles.

3.1.2. Fatigue resistance tendency curves

Figure 4 shows the tendency curves of AC11 surf samples after 0;50;250 microstrain testing. The chart shows the effect of freeze-thaw cycles.

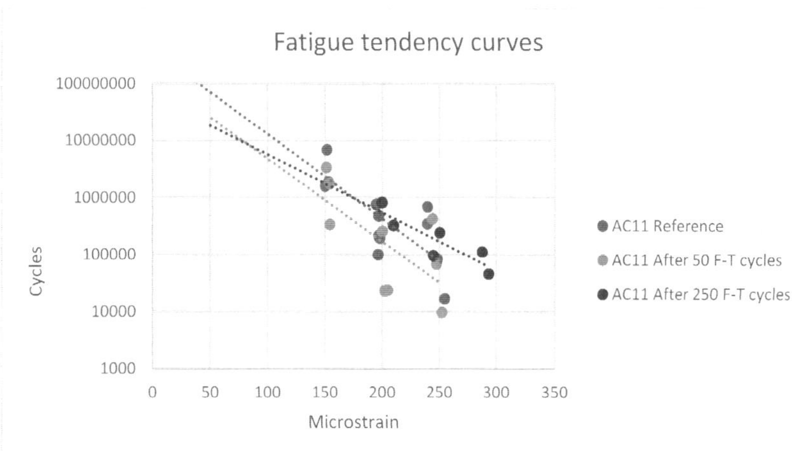

Figure 4. Fatigue tendency curves

3.2. Air void content impact on fatigue resistance

3.2.1. Air void content above 7%

Figure 5 shows a significant difference in the number of fatigue test cycles for samples after 100 frost resistance cycles, by separating samples with the air void content above 7%. Samples with an air void content above 7% reach 50% decrease in stiffness significantly faster.

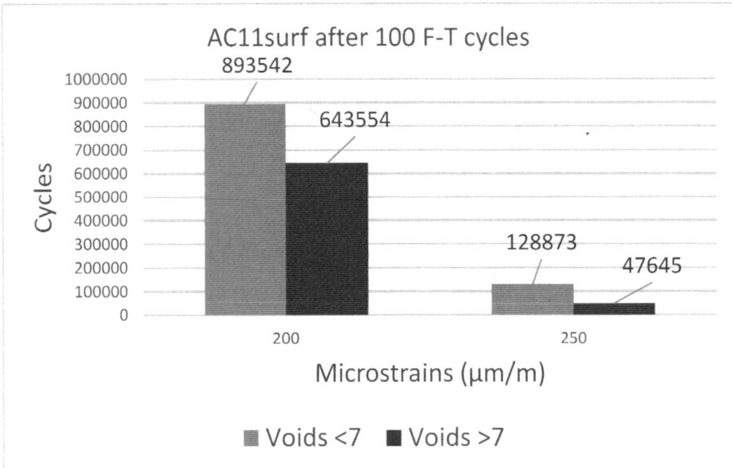

Figure 5. Average fatigue cycle count according to air void content over and under 7%.

3.2.2. Air void content 2-4% and 6-8% results.

After dividing samples by their air void content the results show that samples with small air void content may reach almost 8 million cycles while the same asphalt concrete type samples with big air void content (6-8%) may reach only 2 million.

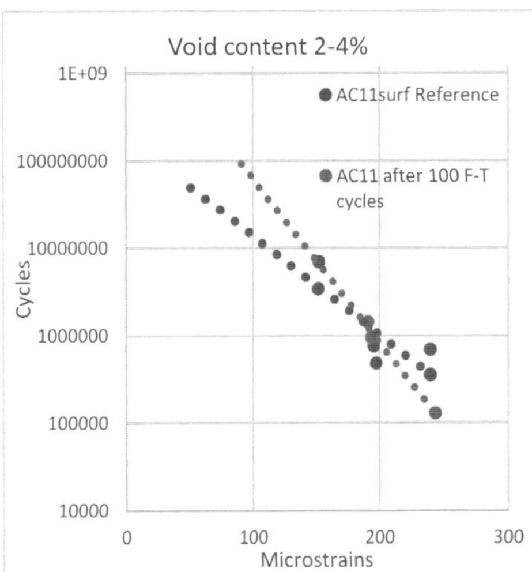

Figure 6. Fatigue tendency curves from samples with 2-4% air void content

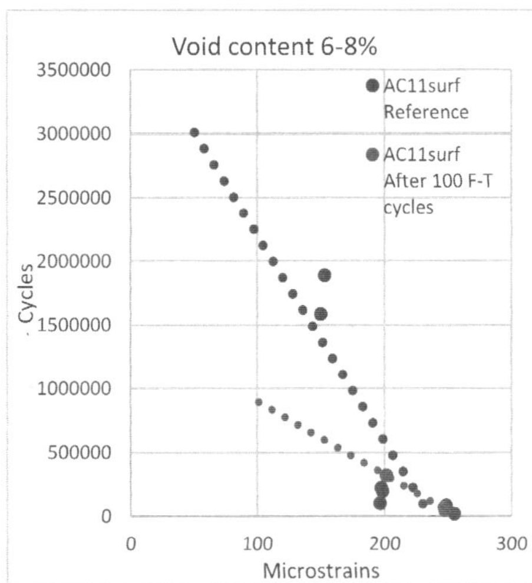

Figure 7. Fatigue tendency curves from samples with 6-8% air void content

3.3. Correlation between stiffness and air void content

As the criterion for the assessment of fatigue resistance is the change in stiffness under the influence of cyclic loading, this study found a correlation between the porosity of AC 11 asphalt concrete compositions and stiffness indices. Figure 8 shows the correlation between the air void content (%) and the stiffness of the samples (MPa), namely, the lower the air-void content, the higher the stiffness.

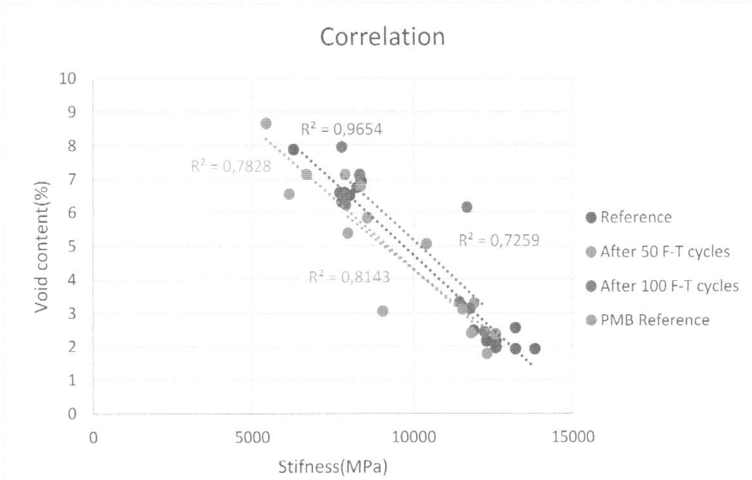

Figure 8. Comparison of stiffness index and air void content.

3.4. Correlation between ultrasonic and void content.

Figures 9 and 10 show the correlation between the air void content of AC11 PMB and AC 11 B50 / 70 (reference) and the ultrasonic velocity. The lower the void content, the higher the ultrasonic speed. Thus, ultrasonic measurements may be used not only to determine compaction, but also to characterize mechanical properties.

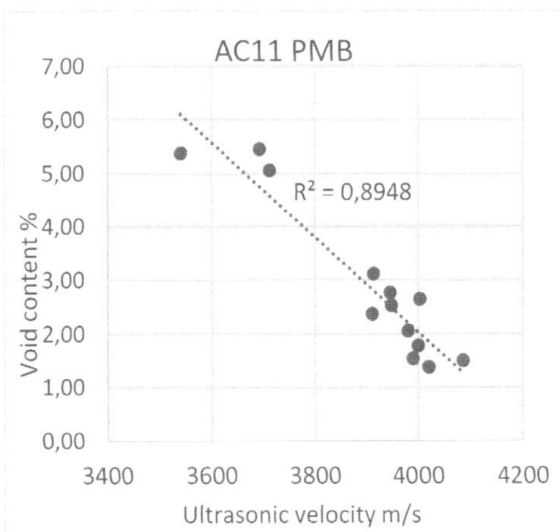

Figure 9. Correlation between ultrasonic speed and air void content

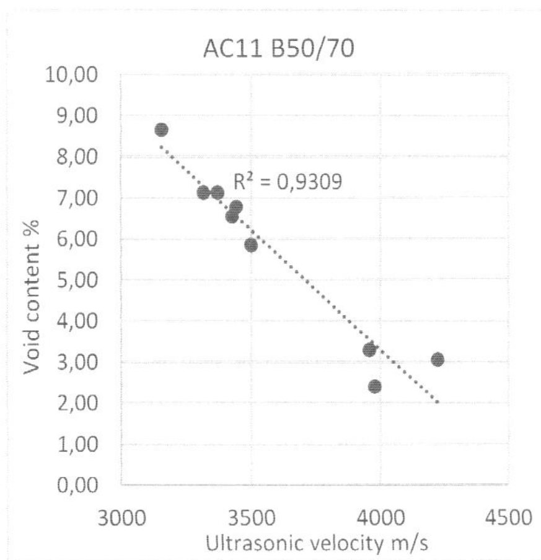

Figure 10. Correlation between ultrasonic speed and void content

3.5. Freeze-thaw cyclicality

Maintaining a uniform cyclicality during the test is of great importance, because the latest information from the sensors embedded in the asphalt concrete shows that in spring, when the road surface begins to heat up, the temperature range and cyclicality increase significantly.

Figures 11 and 12 perfectly reflect the uniformity of the operation and amplitude of frost resistance cycles, which at the same time confirms the compliance of the freezing-thawing equipment with the specific test.

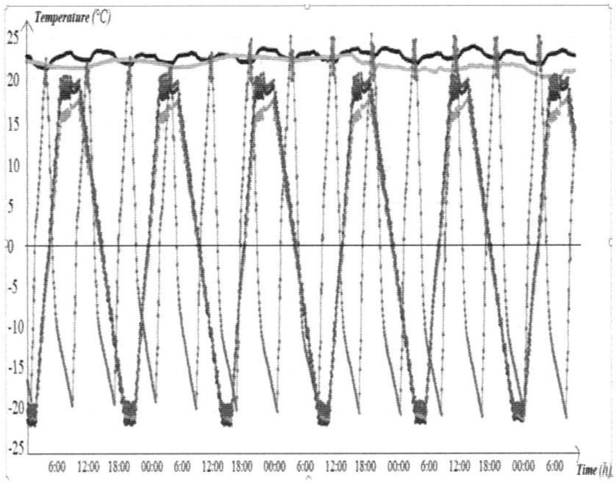

Figure 11. Temperature change chart from freezing chamber

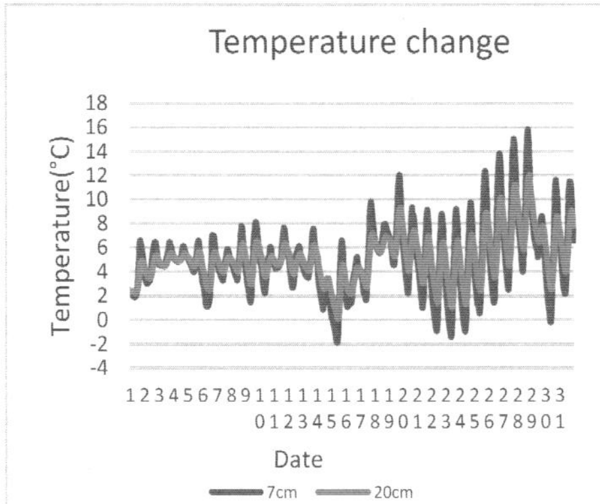

Figure 12. Temperature change data from embedded sensors in the Latvian state road A7 from March 2020

4. Conclusions

Based on the analysis of results, the following conclusions were drawn:

- As the number of freeze-thaw cycles increases, the fatigue resistance and stiffness of the samples decrease, and by doing more freeze-thaw cycles the impact grows more quickly.
- Fatigue resistance of AC11 B50 / 70 samples after 100 frost resistance cycles with relative deformation of 200 and 250 µm / m for the samples with air void content above 7% achieved a 50% decrease in stiffness significantly faster than the samples with air void content below 7%.
- The obtained results show that the average number of fatigue resistance cycles for the samples subjected to frost resistance (50; 100 cycles) compared to the reference samples decreased by about 20%.
- It has been experimentally determined that the air void content has significant impact on stiffness. AC 11 with B50 / 70 stiffness at 6-8% porosity is 6000-8000MPa, but at porosity 2-4% stiffness value is from 10000 to12000MPa.
- In the fatigue resistance test of asphalt concrete type AC11 PMB45 / 80-55 without frost resistance cycles and with 100 frost resistance cycles, no significant changes in the number of fatigue resistance cycles were observed, which means that this type of asphalt concrete is not significantly affected by this number of frost resistance cycles.
- A close correlation between ultrasonic velocity and sample void content has been experimentally determined. The concrete hardness test method may also provide useful data on the structure and stiffness properties of asphalt concrete samples.

Disclosure Statement

No potential conflict of interest was reported by the authors.

References

1. Bituminous mixtures – Test methods for hot mix asphalt – Part 24: Resistance to fatigue
2. Daniel Balageas, Structural Health Monitoring, 2006.
3. Effect of Freeze-Thaw Cycles on Mechanical Characteristics of Bitumens and Stone Mastic Asphalts. Available from: https://www.mdpi.com/2076-3417/9/3/458
4. Ultrasonic pulse velocity analysis in concrete specimens. Available from:https://www.researchgate.net/publication/237322793_Ultrasonic_pulse_velocity_analysis_in_concrete_specimens
5. Freez-thaw Durability of Concrete / J.Marchand, M.Pigeon and M.J.Setzer
6. Review of Techniques, Approaches and Criteria of Hot-Mix Asphalt Fatigue.
7. Laboratory evaluation of stiffness and fatigue susceptibility of asphalt paving materials incorporating environmental factors.Available from: https://www.researchgate.net/publication/318853978_Laboratory_evaluation_of_stiffness_and_fatigue_susceptibility_of_asphalt_paving_materials_incorporating_environmental_factors
8. Resistance of Concrete to Freezing and Thawing. Available from: http://onlinepubs.trb.org/onlinepubs/shrp/SHRP-C-391.pdf
9. THE PREDICTION OF FATIGUE LIFE OF ASPHALT MIXTURES USING FOUR-POINT BENDING TESTS. Jorge Pais, M. Minhoto
10. AASHTO, "AASHTO Standard Specifications For Transportation Materials And Methods Of Sampling And Testing." 23rd Edition, Part 2B. T321-03: Determining the Fatigue Life of Compacted Hot-Mix Asphalt (HMA) Subjected to Repeated Flexural Bending, AASHTO, Washington,D.C, 2003.
11. VAS "Latvijas Valsts ceļi". Ceļu specifikācijas 2019. 2019;
12. Ali Mohamed Zaltuom, CEST-2018, AIJR Proceedings 4, pp.618-625, 2018. Available from: https://doi.org/10.21467/proceedings.4.29

PAVEMENT OPTIMISATION WITH AGGREGATE BASE OR ASPHALT LAYERS STABILISED WITH HEXAGONAL GEOGRIDS

Michał GOŁOS[1], Piotr MAZUROWSKI[1]

[1]*Tensar Polska Sp. z o.o., Gdańsk, Poland*

Received day month year; accepted date (use style Received dates, or Alt + Ctrl + R)

Abstract. The use of hexagonal geogrids in pavement structures results in the increase of the life of designed structure. This offers the possibility to reduce the thickness of layers without reduction of pavement life, or to increase the traffic capacity of a pavement without the need to increase its thickness. This way of using geogrids in pavements design was introduced to the pavement industry by one producer of hexagonal geogrids as a Pavement Optimisation (PO) concept. It can be transformed into both economic and environmental benefits, and obviously results in savings of natural resources and reduction of carbon footprint of a project. PO with geogrids can be used both in the newly designed pavement structures, and in the asphalt overlays of the existing old pavements. Asphalt overlays enhancement with a geogrid either increases the fatigue life of overlays or allows the reduction of overlays thickness to achieve the same pavement life. In new pavements, stabilisation of aggregate base with geogrids increases the stiffness of aggregate, which increases the performance of a whole pavement. This paper presents several tests results, which confirm beneficial effects of using hexagonal geogrids in asphalt overlays and aggregate base layers, from laboratory to full scale accelerated pavement tests. Also, modifications of Mechanistic-Empirical pavement design method, which allow to implement the geogrid benefits into the design process, are discussed. Finally, case studies of pavements – newly designed and reconstructed – optimised with hexagonal geogrids are presented.

Keywords: pavement design, hexagonal geogrid, geocomposite incorporating hexagonal grid, pavement optimisation, pavement tests, fatigue life increase, asphalt overlay

Introduction

Different types of geogrids or geocomposites are used in pavement structures for a relatively long time. Typically, they are used in two main applications: improvement of soft soil bearing capacity under new pavements, or protection against reflective cracking in asphalt overlays on existing pavements. However, in both these cases the design with geogrids do not address the fatigue life of the pavement itself.

One producer of stiff, punch and drawn hexagonal proposed new approach to the use of geogrids in pavement structures. It is to take the influence of geogrids into account in calculation of pavement life, thus enabling to deliver more economic and environmentally friendly designs. The use of such geogrids in pavement layers improves parameters of these layers, and it affects the life of a whole pavement. This offers the possibility to reduce the thickness of layers without reduction of pavement life, or to increase the traffic capacity of a pavement without the need to increase its thickness. While improvement of soft soil and protection against reflective cracking remain main and important part of geogrids applications in pavement, this new concept, named "Pavement Optimisation" (PO), offers new opportunities to pavement designers.

In US the concept of using geosynthetics in new pavements to reduce its thickness or increase its life is covered within AASHTO R50-09 Standard. According to it: "Geosynthetics are used in the pavement structure for structural support of traffic loads over the design life of the pavement. The geosynthetic is expected to provide one or both of these benefits: (1) improved or extended service life of the pavement, or (2) reduced thickness of the structural section".

Two different approaches to PO with geogrids concept are presented in this paper. First is to use hexagonal geogrid to stabilise aggregate base or sub-base layer of new pavement. Second is to use geogrid in asphalt layers, mostly in asphalt overlays of existing pavements. Both enable designers to design pavements which are either thinner (thus cheaper) or offer a substantial increase of pavement life compared to traditional pavements without geogrids.

1. Pavement Optimisation of new pavements

Aggregate layer stabilised with hexagonal geogrid has increased parameters compared to non-stabilised layer. This increase has been confirmed by laboratory and field tests. Kwon et al. (2012) observed a modulus increase of 5% to 20% when testing stabilized and non-stabilized silty gravel samples in a test conducted with combination of AASHTO

T307 and NCHRP 598. In-situ resilient modulus tests performed with Automated Plate Load Tests (APLT) exhibited 5% to 30% modulus increase of geogrid stabilized sections compared to non-stabilized (White 2014a, White 2014b). Increased stiffness on aggregate layer(s) enhances the performance of a whole pavement. This can be utilized in two ways: to increase the life of a pavement compared to typical, non-stabilised section, or to reduce thickness of layer(s) – the stabilised layer, or other, including asphalt.

To be able to include the geogrid in the pavement design process, it is necessary to modify pavement design methods. The development of these modifications should not be based on theoretical considerations only. According to AASHTO R50-09: "Because the benefits of geosynthetic reinforced pavement structures may not be derived theoretically, test sections are necessary to obtain benefit quantification" [AASHTO 2009]. ARA suggests a number of laboratory and field tests, including full scale Accelerated Pavement Tests: "The purpose of the Accelerated Pavement Testing (APT) is to gather the data needed for developing the design inputs described in the previous sections (i.e., adjusted modulus and transfer function coefficients)" [Lee 2017].

A series of APTs with Heavy Vehicle Simulator (HVS – Figure 1) device were performed at U.S. Army Engineer Research and Development Center (ERDC) facility to quantify the benefits of one type of hexagonal stabilisation geogrid [Jersey et al. 2012, Norwood et al. 2014, Robinson et al. 2017]. In total, 8 sections were tested. Test sections dimensions were 2.44 m (8 ft) by 15.2 m (50 ft). The subgrade made of locally available clay was prepared to achieve a bearing capacity of either 3% or 6% CBR. The granular base layer consisted of either 15 cm or 20 cm of crushed aggregate, on four of the sections base was stabilized with hexagonal geogrids. Six sections had surfacing of 5, 7.5 or 10 cm of dense-graded hot mix asphalt (HMA), and two sections had a surfacing of double bituminous surface treatment (DBST).

Figure 1. HVS device in ERDC hangar during the test

All sections with base stabilised with hexagonal geogrids heavily outperformed their non-stabilised counterparts. Five to ten higher number of axle loads was applied to stabilised sections to reach the same critical rut depth (12.5 mm) compared to non-stabilised sections.

Results of these tests were a basis of modifications of both empirical and mechanistic-empirical pavement design methods. In case of mechanistic-empirical design, the influence of geogrid on a pavement is taken into account by two mechanisms, used simultaneously. First is the increase of elastic modulus of aggregate layer(s) stabilised with geogrid. Second is to apply a Life Shift Factor to calculated life of a pavement. The development of these modification and discussion of tests results are discussed by Mazurowski at al. [2019].

Another confirmation of hexagonal geogrid influence on pavement life was obtained in research conducted by Nevada Department of Transportation [2016]. A series of cycling loading tests of pavement sections constructed in 1.8 m diameter tank was performed (Figure 2). Six sections were tested: two control, two with biaxial geogrids and two with triaxial geogrids. Sections consisted of 7.6 cm thick asphalt layer and 30 or 40 cm thick aggregate base layer. 3 mln 40 kN load cycles were applied to each section by an actuator and 305 mm diameter steel plate.

Figure 2. Tank and steel loading plate used in Nevada DOT test

Test results are presented on Figure 3. After 3 mln load cycles rut depth on sections with hexagonal geogrids (marked T16 and T12 of Figure 3) was ~30% lower compared to control section (C16 and C12) and ~17% lower compared to section with biaxial geogrid (B16 and B12).

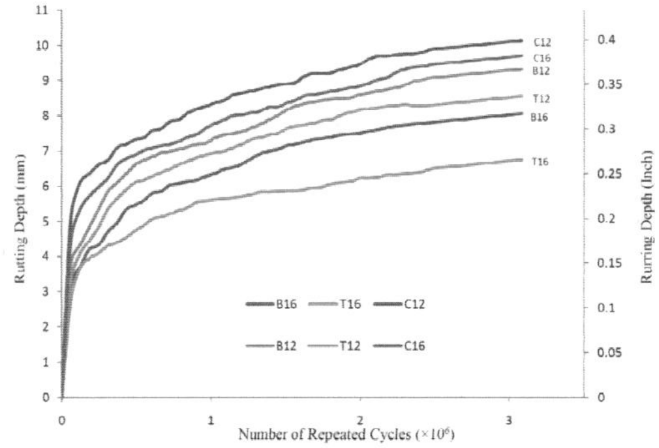

Figure 3. Results of Nevada DOT test

2. A Pavement Optimisation of new pavement Case Study: the Voivodeship Road 507 Braniewo – Pieniężno, Poland, 2018-2019

The Voivodeship Road 507 is located in the north of Poland, along Polish – Russian border (with Kaliningrad), a few kilometres from Vistula Lagoon and Baltic Sea. Reconstruction of 29 km long section of this road, between towns Braniewo and Pieniężno, started in 2018. The project was divided into two parts: reconstruction of the first part (15 km) was finished in 2019, reconstruction of the second part (14 km) is planned to be finished in 2021.

Existing pavement was in very bad condition, with a lot of different damages, and the Designer, with the approval of the Client, decided to design its complete demolition and construction of a new one for the whole section. A new pavement with base made of Recycled Asphalt with bitumen emulsion and cement mix (RAP) was chosen to reduce costs and utilize recycled materials.

On the first, 15 km long section, about 8 km of the road had poor ground conditions, with high plasticity clay of estimated bearing capacity of E2 = 15 MPa directly under the pavement. Pavement design on this section has been optimised with the use of hexagonal geogrids.

Typical pavement for prognosed medium traffic category KR3 and stiff clay subgrade designed according to Polish Catalogue of Typical Flexible and Semi-Rigid Pavement Structures is as follows:

- 12 cm of asphalt;
- 20 cm of RAP base;
- 24 cm of CBR 60% aggregate subbase;
- 40 cm of CBR 20% aggregate capping layer.

The catalogue does not cover the cases of subgrade consisting of high plasticity cohesive soils and requires individual design in such situation. For this project an additional capping layer has been added: 15 cm of in-situ chemically stabilised layer. The full structure of the conventional pavement on high plasticity clay for this project is presented in Figure 4, the left side.

Then Pavement Optimisation with hexagonal geogrid was done for this project. It allowed for substantial reduction of thickness of pavement. Asphalt layers thickness was reduced by 1 cm, RAP base by 2 cm, and CBR 20% aggregate capping layer by 20 cm. It was also possible to remove whole 15 cm in-situ chemically stabilised layer. This optimisation resulted in savings of about 2.5 mln PLN (~0,55 mln euro) on the first section.

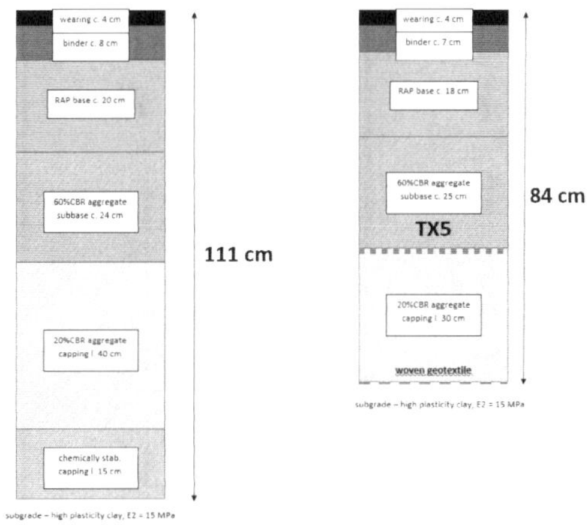

Figure 4. Comparison of catalogue and optimised with geogrid pavement sections on road no. 507

Figure 5. Installation of hexagonal geogrid on road no. 507

3. Asphalt interlayers development. Geocomposites

3.1 Geocomposite incorporating stiff monolithic grids

Pavement Optimisation is usually used in design of new pavement structures. There is, however, a second way of optimizing pavements by the use of geocomposites in the new asphalt overlays on existing pavements. In this approach the asphalt overlays are enhanced by a geocomposite, which either (1) increases the fatigue life of asphalt overlays of or (2) allows for reduction of asphalt overlays thickness to achieve the same pavement life as a thicker overlay not incorporating geocomposite. Geocomposites, also known as asphalt interlayers, are a proven solution for extending the life of the repaired and reconstructed pavements.

The process of manufacturing asphalt interlayers based on a stiff monolithic grids and geocomposites incorporating such grids has been developed within the last decades. The initial laboratory tests and several site applications incorporated a biaxial grid alone. A stiff, monolithic polypropylene grid punched and stretched in two directions was used. After some years of its application the process has been modified and the biaxial grid has been mounted on a paving fabric to create a geocomposite. There have been a massive and successful applications of this product, which is available in two versions: as a both large (65x65 mm) and small (39x39 mm) square aperture size grids (see Figure 6).

Recently the interlayer system has been further developed, which resulted in creation of new type a geocomposite, based on a triaxial grid, with triangular apertures, whilst the stiff and monolithic structure of a grid itself remained unchanged. The idea of using multiaxial isotropic grid to strengthen the asphalt bound layers has been transmitted from the concept of using the similar type of a grid for unbound granular layers, which is present on the market for more than a decade.

The paving hexagonal grid is orientated in three directions such that the resulting ribs have a high degree of molecular orientation which continues through the area of the integral node. Ribs in all directions - longitudinal and transverse - have a rectangular cross section. This stiff hexagonal grid is thermally bonded to a paving fabric backing, and the resultant product is then called a geocomposite (see Figure 7).

Figure 6. Geocomposites incorporating biaxial grids. Large version of 65x65 mm (left) and small version of 39x39 mm aperture size (right)

Figure 7. Geocomposite incorporating triaxial grid with the hexagonal pitch of nominal value of 80 mm

3.2 Functions of geocomposite and its components

The fabric is a polypropylene non-woven needle punched geotextile, which has an adequate residual bitumen retention to absorb the bitumen once the underlayment is sprayed with either straight run bitumen or bitumen emulsion. According to EN 15381:2008 the minimum bitumen retention to assure installation integrity should be not less than 0.9 kg/sqm. The primary role of the fabric when used in conjunction with bitumen tack coat is to stick it enough to the underlayment and hold the grid in place during paving operations. Once installed and fully saturated with bitumen, the fabric functions as a stress relief system (STR) and interlayer barrier (B). Stress Relief is a function of an asphalt interlayer provided by a bitumen saturated fabric which allows for slight differential horizontal movements between two layers of asphalt (under and overlays) and thus provides the stress relief, which delays or stops (arrests) crack propagation in the asphalt overlay. The Interlayer Barrier is also a function provided by a fabric, which acts as a barrier to the ingress of water (EN 15381:2008). Thus, both functions provided by a bitumen saturated fabric prevent or delay the deterioration of the entire pavement. In turn, the performance of a stiff Polypropylene (PP) geogrid in asphalt overlay is based on two main principles, i.e.:

1. The mechanism of interlocking particles of an asphalt mixture within the hexagonal apertures of a monolithic grid, taking over the horizontal tensile stresses after initial decay of the bitumen (Andrews, 2013). This creates the lateral confinement of bitumen bound particles, by which structure of the hexagonal grid restrains the asphalt mixture;
2. The high fatigue resistance, increasing designed pavement life (Andrews, 2013). Distribution of wheel load in all directions in the plane of geocomposite of the bitumen bound particles of an asphalt overlayer improves the fatigue performance by increasing the isotropy of the resulting composite structure (asphalt overlay + geocomposite).

Due to its stiff and monolithic structure as well as because it is typically used at lower structural level, under base course and binder course, this material is also called a *structural grid* or *structural geocomposite*. Therefore, the main range of structural geocomposite's applications in asphalt is an increasing of the fatigue life of pavement and delaying the cracking occurrence in a new asphalt overlay.

3.3 Inter-layer bonding concept with geocomposite

As both STR and B functions are important from the geocomposite's performance perspective, it means that - if the stress relief is to occur between the under bound layers and asphalt overlays - the control de-bonding of these layers is needed. It has been observed that the use of a geocomposite in the asphalt overlay causes reduction in shearing strength and inter-layer bonding. However, it does not mean that the asphalt overlay reinforced with geocomposite will be susceptible to its pre-mature distress. The maximum horizontal shear force in pavement, which may occur during e.g. emergency braking of heavy vehicles, is about 0.25 ÷ 0.3 MPa (Jaskuła & Rys, 2017; Jaskuła, 2018). If the shearing strength of the inter-layer bonding within the asphalt layers is at least this high, then the pre-mature distress will not happen, even if the shearing strength at the plane of geocomposite would be lower compared to shearing strength between asphalt layers only.

4. Delay of fatigue cracking propagation and fatigue life improvement by using asphalt geocomposites

4.1. Pavement surface deterioration

According to (Code of Practice for Geosynthetics and Steel Meshes, 2012) deterioration of the pavement surface is mainly caused by weathering, movement, and fatigue (caused by a failure under trafficking), accelerated by the asphalt's susceptibility to bottom-up cracking leading to ingress of water, then to potholes and a finally, total breakdown of the surface (Figure 8).

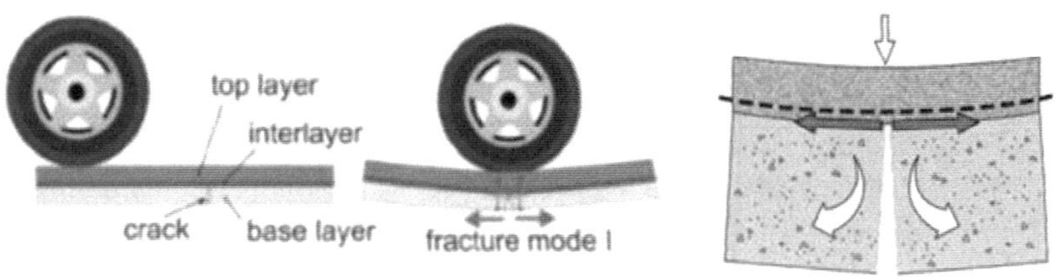

Figure 8. The process of crack development under wheel loading. Model of cracking caused by bending mechanism

The fatigue life of asphalt layers is defined as a number of standard axle loads which can be applied before a given percentage of a pavement surface area will be covered with cracks. In mechanistic-empirical design approach it is calculated with the help of an appropriate Transfer Function. Fatigue cracking development depends on the magnitude of horizontal tensile strain at the bottom of lowest asphalt layer. This approach is used for a bottom-up mode of fatigue cracking.

4.2 Mechanism of asphalt fatigue cracking

Fatigue cracking (called also as *alligator cracking*) is a series of cracks caused by fatigue failure of the asphalt surface under repeated traffic loading. Cracking begins at the bottom of the lowest asphalt layer, where both tensile stress and strain are the highest under wheel loading. They propagate to the surface - initially as a series of parallel longitudinal cracks - and later they connect, forming sharp-angled pieces (less than 0.5m on the longest side) that develop a pattern resembling the skin of an alligator. At the highest severity of its development fatigue cracking creates well defined pieces which are spalled at the edges. Some of them may rock under trafficking (ASTM D 6433-07). Another negative effect of deterioration process is water penetration throughout the cracking network and its freezing within the voids created by the cracks. This process is presented on the scheme illustrated in Figure 9. (Błażejowski & Wójcik-Wiśniewska, 2016).

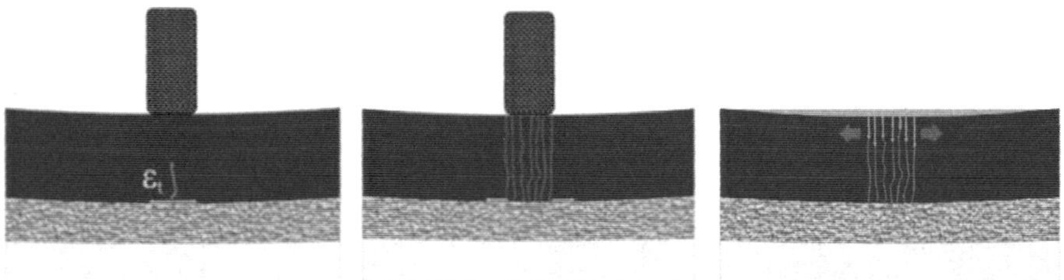

Figure 9. Fatigue cracking process of creating and developing

4.3 Two approaches of using geocomposites for fatigue life increase

It is important to provide efficient performance of an interlayer between existing pavement and a new asphalt overlay package to arrests and delay fatigue crack propagation and to provide a barrier to the ingress of water. Hexagonal geocomposite was developed for those purposes and may ideally solve the issue. In this approach the asphalt overlay is reinforced by a geocomposite, and the asphalt interlayer either increases the fatigue life of this overlay or allows to reduce the overlays thickness, while achieving the same pavement life as thicker overlay not incorporating the geocomposite. This approach is illustrated on the Figure 10:

- Section no. 1: Typical section with new asphalt overlay (H) which is not strengthened by any geocomposite designed for a respective fatigue life of a pavement (N0);
- Section no. 2: Section with new asphalt overlay (H same as in Section no.1) strengthened by hexagonal geocomposite gives a massive benefit of an increased fatigue life of a pavement (N >> N0);
- Section no. 3: Section with new asphalt overlay (h) strengthened by hexagonal geocomposite designed for at least the same fatigue life as a Section 1 ($N \geq N0$) gives a benefit of a thinner overlay thickness ($h < H$).

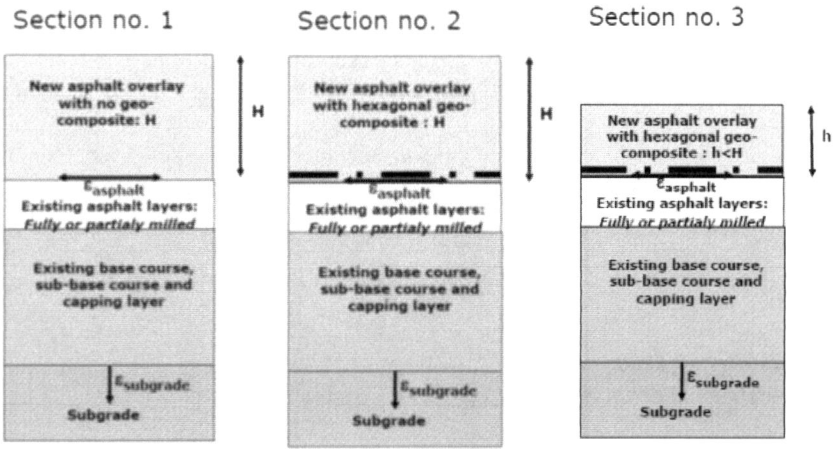

Figure 10. Fatigue life improvement approach. Comparison of three sections of reconstructed pavement

Additionally, to use a geocomposite as much efficiently as possible, it is important to locate it at the bottom of the lowest asphalt course, which usually is a bottom of the new asphalt overlay.

This approach has been already utilized in several applications, where the existing pavements required reconstruction and strengthening by new asphalt overlay. One of the case studies is presented below.

5. A Pavement Optimisation of asphalt overlay Case Study: the Aleksandra Caka Street, Riga, Latvia, 2020

After decades of service the surface of the Caka Street in Riga, Latvia, was in very poor condition. City Council's Transport Department needed a long-term solution to carry the heavy traffic by the reconstructed pavement with the new relatively thin asphalt overlay. It was finally decided to reconstruct approx. 2.5 km length of the street.

Aleksandra Caka Street is one of the main streets in the Riga centre. Investigations revealed structural and surface issues on the full length of the road, requiring reconstruction of the entire street section. There was an extensive network of cracking which were formed on the entire pavement surface with an intensive 'alligator cracking' mode close to the edges of the roadway. Ruts have been formed in some places of approx. 4 cm deep down with the part of the surface being pushed to the sides. The subgrade was in a good ground-water conditions, as it consisted of non-cohesive soils. The existing structure had a following cross section:

- existing asphalt layers, remaining after assumed milling depth of ~10÷12.5 cm, 5 cm;
- existing base layer of an unbound crushed aggregate, 25 cm;
- existing capping layer of a sand blanket, 70 cm;
- subgrade, non-cohesive soils of $E2 \geq 50$ MPa.

The design assumed restoration of the pavement structure to the existing width and to mill the existing surface in variable thickness (12.5 cm on average). In order to create a suitable crossfall of the pavement surface, a hot asphalt AC11 levelling course of min 2.5 cm thick was paved, on which the geocomposite with hexagonal grid was installed. It has been then overlaid with 2 layers of hot asphalt, i. e. binder course AC 22 of 6 cm and wearing course AC11 of 4 cm, respectively. Figure 11 illustrates the designed cross section.

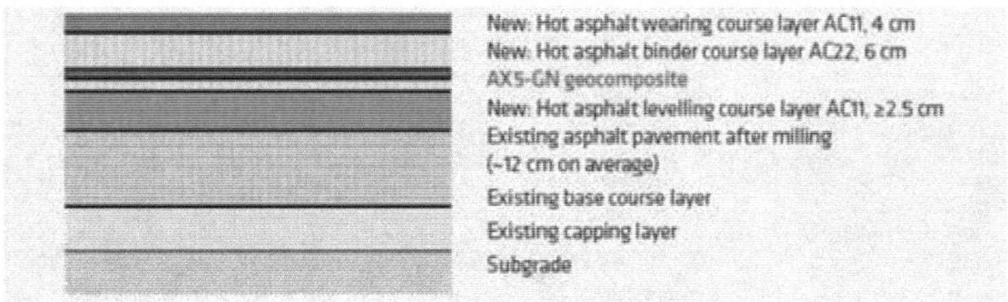

Figure 11. Cross section of the existing pavement overlaid by new asphalt enhanced with geocomposite

The designed traffic based on the investigation carried out and provided by the Client, was calculated as >34 mln ESALs of 100 kN axles. The main goals of the pavement reconstruction formulated by the Client were:

- to carry prognosed heavy traffic;
- to extend the fatigue life of a reconstructed pavement, and finally;
- to retard the reflective fatigue cracking to the asphalt surface.

The calculations based on the Mechanistic-Empirical design method have been carried out. The first approach assumed no grid nor geocomposite application within the asphalt overlay. Based on that the calculated pavement life was assessed as approx. half of a total assumed pavement life (>16 mln ESALs). The increase of the pavement life due to implementation of a geocomposite with a hexagonal grid has been incorporated to the method by a relevant fatigue improvement factor. It was based upon the results of Prof. S. F. Brown at the University of Nottingham in the UK in the mid of 80's last century which have been conducted for the bi-axial grids. In the laboratory conditions, (Brown et al., 1985) an increase of asphalt fatigue life by a shift factor of 10 in a strain-controlled beam testing at the grid positioned at the base of the asphalt layer was achieved. Based on this research and a long-term performance of projects incorporating geocomposites with stiff PP grid it was assumed that a fatigue improvement factor of between 2 and 3 of the relevant fatigue transfer function is acceptable and conservative (IB/Grid_Fatigiue, 2012).

Finally shift factor of 2.0 for hexagonal geogrid has been assumed for the Caka Street reconstruction project. As a result, the calculation for the 2-layered asphalt overlay reinforced by geocomposite with a structural hexagonal grid, delivered the anticipated fatigue life of a reconstructed pavement.

The works of milling the existing asphalt layer, paving the hot asphalt levelling course and installing the geocomposite started in August 2020. Rolls of 3.8 m width were mostly installed by the purpose-built interlayer installation machine, while some of them were installed also manually. Immediately after installation pressure with brooms or brushes was applied to geocomposite to get rid of the folds and wrinkles and – what is also important - to assist the fabric to be initially soaked with the bitumen. The bitumen spray rate of 1.2÷1.5 kg/m2 of residual bitumen was recommended by producer's installation guideline. As the Contractor used a bitumen emulsion with a bitumen solid content of ≥65%, the amount of freshly sprayed bitumen emulsion had to be within the range of 1.85÷2.30 kg/m2. Figures 12 and 13 illustrate the mechanical installation process which efficiently and successfully has been carried out between August and October 2020. The paving works were finished in November 2020. Figure 14 reflects the existing condition of A. Caka Street in Riga (January 2021).

Conclusions

The use of geogrids, in aggregate or asphalt layer(s), can help in optimisation of pavement structures, meaning reduction of pavement thickness or increase of pavement life. Such Pavement Optimisation offers a wide range of benefits for clients, designers and contractors, like: reduction of construction costs and time, construction traffic, maintenance costs, damage to access roads and construction risks, and an increased life of the pavement. Another important aspect, not discussed in this paper, are the environmental benefits associated with the use of geogrids. Use of geogrids in pavements can result in substantial savings in carbon emission compared to traditional technologies, especially cement and other hydraulically bound mixtures.

Modified pavement design methods, based on extensive research, are available to assist designers with designs of pavements incorporating hexagonal geogrid, both in aggregate base on new pavements and asphalt overlays of existing ones.

Figure 12. Process of geocomposite installing by the purpose-built asphalt interlayer installation machine

Figure 13. Geocomposite with hexagonal grid installed in place

Figure 14. Fully trafficked A. Caka Street. Section strengthened by asphalt overlay with a geocomposite, January 2021

Disclosure Statement

Authors declare that they have no competing financial, professional, or personal interests from other parties.

References

American Association of State Highway and Transportation Officials. (2009). Standard Practice for Geosynthetic Reinforcement of the Aggregate Base Course of Flexible Pavement Structures AASHTO R 50-09, Washington, USA.

Andrews, C. M. (2013). The Contribution of Asphalt Reinforcement Geocomposites in a Pavement Structure; A Long-term Performance Review, UK. The XXVIII International Baltic Road Conference, Vilnius, Lithuania.

ASTM D 6433-07. Standard Practice for Roads and Parking Lots Pavement Condition Index Surveys.

Błażejowski, K., Wiśniewska-Wójcik, M. (2016). Odporność na zmęczenie i pękanie mieszanek mineralno asfaltowych z różnymi asfaltami (Fatigue and cracking resistance of asphalt mixtures with adding diverse types of bitumen). The IV Silesian Road Forum, Katowice, Poland, 13 April 2016.

Brown, S. F., Brunton J. B., Hughes, D. A. B., Broderick, B. V. (1985). Polymer Grid Reinforcement of Asphalt, AAPT Annual Meeting, San Antonio, Texas, U.S.A.

Code of Practice for Geosynthetics and Steel Meshes. (2012). RSTA Code of Practice for Geosynthetics and Steel Meshes (for Inhibiting Cracking in Bituminous Bound Layers), UK.

European Committee for Standardization. (2008). EN 15381:2008: Geotextile and geotextile-related products – Characteristics required for use in pavements and asphalt overlays.

Fatigue behaviour of AR1 and AR-G reinforced asphalt, Tensar Information Bulletin, IB/Grid_Fatigue. (2012).

Ghafoori, N., & Sharbaf, M. (2016). Use of Geogrid for Strengthening and Reducing the Roadway Structural Sections, Nevada Departament of Transport, Report No. 327-12-803, 2016, Carson City, Nevada, USA.

Jaskuła, P., Ryś, D. (2017). Effect of interlayer bonding quality of asphalt layers on pavement performance, IOP Conference Series: Materials Science and Engineering 236 012005.

Jaskuła, P. (2018). Interlayer bonding of asphalt layers in multilayer structures of road pavements. Gdansk University of Technology, Gdansk, Poland.

Kwon, J., Wayne, M., Norwood, G. J., & Tingle, J. S. (2012). The Implementation of Findings From Accelerated Pavement Testing in Pavement Design And Construction Practice, in Advances in Pavement Design through Full-scale Accelerated Pavement Testing – Jones, Harvey, Mateos & Al-Quadi (Eds.), Taylor&Francis Group, London, UK.

Lee, H. S., (2017). Recommended Practice for Incorporating Geogrids in ME Pavement Design, Applied Research Associates, Champaign, USA.

Mazurowski, P., Buckley, J., Kawalec, J. (2019). Modification of mechanistic-empirical pavement design method to incorporate the influence of hexagonal stabilisation geogrids. AAPA International Flexible Pavements Conference, 18th, 2019, Sydney, Australia.

Norwood, G. J., & Tingle, J. S. (2014). Performance of Geogrid-Stabilized Flexible Pavements. Final Report. EDRC/GSL TR-14-28. U. S. Army ERDC, USA.

Robinson, W. J., Tingle, J. S., & Norwood, G. J. (2017). Full-Scale Accelerated Testing of Multi-Axial Geogrid Stabilized Flexible Pavements. Draft Final Report. EDRC/GSL TR-17-X. U.S. Army ERDC, USA.

Scale Evaluation of Geogrid-Reinforced Thin Flexible Pavements. Transportation Research Record, Journal of the Transportation Research Board, No. 2310, TRB of the National Academies, Washington, USA.

White, J. D. (2014a). In Situ Performance Verifiaction of Geogrid-Stabilized Aggregate Layers: Dewitt County, Cuero, Texas – Schlinke Rd., Geomatters Technologies Inc., Ames, USA.

White, J. D. (2014b). Preliminary APLT Comparision of Ev Strain Modulus and In-situ Resilient Modulus/Permanent Deformation, Geomatters Technologies Inc., Ames, USA.

A CRITICAL REVIEW ON MIXING PARAMETERS FOR HIGH CONTENT RECLAIMED ASPHALT MIXTURES

Received 01 March 2021

Mukul Rathore[a,*], Viktors Haritonovs[b], Martins Zaumanis[c]

[a]Celu Eksperts, Ikskile-5052, Latvia. Email: mukul.rathore@edu.rtu.lv

[b]Faculty of Civil Engineering, Riga Technical University, Riga, Latvia. Email: viktors.haritonovs@rtu.lv

[c]EMPA Swiss Federal Laboratories for Materials Science and Technology, Dübendorf-8600, Switzerland. Email: martins.zaumanis@empa.ch

*Corresponding author

Abstract. High content reclaimed asphalt (RA) mixtures have been identified as one of the options to reduce the environmental and economic impacts of pavements construction. However, the process of designing and producing high content RA mixtures is challenging and the asphalt industry have serious concerns towards quality and long-term performance these mixtures. In laboratory, several parameters affect mixture characteristic, and if not controlled, may results into inaccurate estimation of performance. This state-of-the-art study aims to identify critical parameters for high content RA mixture production and highlight the effects of these parameters on mixture performance. The mixing parameters adopted in several laboratory studies have been highlighted and compared. The best practices to mix recycled asphalt in laboratory are reviewed in order to optimize the laboratory mixing. Based on review, important considerations for evaluating laboratory performance have been discussed.

Keywords: reclaimed asphalt pavement; high RAP; laboratory mixing, plant production

1. Introduction

Recycling of asphalt pavements is not a new concept. Although the use of reclaimed asphalt (RA) from old pavement for building and maintaining new roads is now considered as an environment friendly alternative to conventional hot mix asphalt (HMA). It originated in 1970s primarily to improve the economics of pavement construction after the Arab oil embargo when the cost of asphalt was increasing significantly (Kandhal and Mallick, 1997). Until this time, the cost of virgin asphalt was lower than the cost for processing RA, which provided no incentive for using RA materials (J. Don Brock, and Jeff L. Richmond, 2010). During the past few years, there has been a major breakthrough in recycling of pavements and increasingly large amount of reclaimed asphalt (RA) material is being re-used in production of new asphalt. The use of RA material in new pavements not only reduces the consumption of natural resources but also addresses problems like disposal of milled RA. It truly creates a cycle that sustains the asphalt pavement industry (Copeland, 2011).

A logical approach to enhance the cost-effectiveness and mitigate the environmental impact of pavement construction is to use larger amounts of reclaimed asphalt (Mogawer Walaa S. et al., 2015). This can only be possible when asphalt mixtures are designed with high content of RA material. Although possible, but the use of reclaimed asphalt (RA) in base layers or shoulder material may not be the most economical available option (Willis et al., 2012). A recent life cycle cost assessment study shows that agency cost was reduced by about 18% when 40% RA material was used in HMA (Qiao et al. 2019). Several other studies have also shown that the amount of savings can increase exponentially when a greater percentage of RA material is incorporated in the asphalt mixtures (Franke et al. 2014; Hong et al. 2016). However, the process of designing and manufacturing high RA mixtures is challenging compared to conventional asphalt and requires much more experience (Zaumanis and Mallick, 2015). Moreover, the asphalt industry has some serious concerns about the quality and long-term performance of pavements with high content RA materials.

The binder available in reclaimed asphalt is excessively stiff as a result of oxidation of bitumen during its lifetime. Hence, the resultant mixture with high content RA may be "overly stiff" and experience issues such as low-temperature cracking failure (Copeland, 2011). Therefore, for high RA content, it becomes necessary to change the mix design by using rejuvenators, additives, or softer binder, etc (Rathore et. al., 2019; Izaks et. al. 2020). AASHTO M323, 2012 guidelines suggest no change in binder when RA content is less than 15%, one grade softer binder to be used for 15-25% RA and using blending charts for more than 25% RA mixtures. However, these guidelines assume complete blending of oxidized RA binder with the virgin binder. As observed by from different studies, 100% blending does not happen (Bowers et al., 2014; Gottumukkala et al., 2018; W. S. Mogawer et al., 2013). Assuming complete blending can cause problems in predicting the pavement performance in laboratory, especially for long-term properties, such as fatigue (Carpenter and Wolosick, 1980). On the other hand, assuming lower blending than actual may overrate the binder quantity requirement and can ultimately lead to plastic deformation of the mix (Al-Qadi et al., 2007). Therefore, it is important to conduct extensive blending characterisation study of RA binder and develop mix design for high content RA mixtures.

As RA percentage is increased (>50%), incorporating recycling agents, commonly known as "rejuvenators" into the mix offer many unique benefits as compared to the use of softer binders (Zaumanis and Mallick, 2013). These recycling agents replace the oils lost during aging process and balance the bitumen proportions such that it is no longer brittle. Some studies also show that oxidation effect can be less severe on the rejuvenated bitumen than on the virgin one as rejuvenators have the ability to mechanically restore the aged binder (Bocci et al., 2019; Cavalli et al., 2018). But it is important to determine the quantity of rejuvenator required for a particular RA mixture as incorrect design may lead to premature distresses in pavement. A rational procedure should be developed to determine dosage of rejuvenator in order to balance the softening effect of rejuvenator (W. Mogawer et al., 2013). Some studies have developed the methods for determination of rejuvenator dosage for RA mixtures (Im et al., 2016; Zaumanis et al., 2014).

There has been wide research showing that these mixes can perform equal to or sometimes even better than conventional hot mix asphalt (Mhlongo et al., 2014; Zaumanis et al., 2018a). However, the performance of mixtures in the laboratory depends upon various parameter that must be carefully controlled for accurate estimation (Rathore et al. 2020). For example, the parameters: plant type, production temperature, mixing time, discharge temperature, storage time, RA source, RA properties, and virgin binder grade etc. can impact the properties of high content RA mixtures (Mogawer et al., 2012). It is clear that the mix results obtained in laboratory cannot be directly applied to predict the performance of mixture in field. The studies should focus on simulating the plant mixing conditions in the laboratory for an accurate evaluation of performance.

2. Objective and Scope

The main objective of this paper is to present a critical review highlighting the importance of laboratory mixing parameters in performance evaluation for high content RA mixture. This study will identify the parameters involved in plant production of high content RA mixtures and form a comparison with mixing process adopted in the laboratory. In the first section, several parameters linked with plant production of high content RA mixture were highlighted and their impact on mixture properties were summarized from available literature. In the next section, the effect of laboratory mixing parameters on performance of high content RA mixtures is presented. Finally, a conclusion is made recommending the important consideration to be taken for mixing high content RA mixtures in the laboratory.

3. Plant production parameters

The plant production parameters of conventional hot mixing are generally not affected when low amount of RA is added to asphalt mixtures. Since, at low RA content (10-20%), there is not enough aged binder present in mixture to change the total binder properties (NCHRP Report 452). However, at high RA content (>25%), various considerations and/or modifications are required to the asphalt plant (Stroup-Gardiner, 2016). Some of these production parameters for high RA content mixtures are described in this section.

3.1. RA heating process during production

The aggregates are required to be heated at high temperature to allow complete drying and to reach uniform temperature throughout the aggregate structure for mixing. For hot mix asphalt production, all the constituent materials are heated at their desired production temperature typically around 150°-190°C (Rubio et al. 2012). The recycling of asphalt can be categorized as warm/hot recycling and cold recycling depending upon the heating temperature of RA material. The warm/hot recycling uses a dual-drum system to dry and preheat RA materials at temperatures range of 110–160°C and virgin aggregates at the temperature range of 190–250°C simultaneously before mixing them together (Zhang et. al., 2019). On the other hand, cold feed recycling uses RA material to be added at ambient temperature using a double drier drum to avoid direct contact of RA material to the flame. In asphalt plants when low RA contents are used, RA is either used as unheated aggregates or heated at very low temperature to remove the moisture. Though, using unheated RA has shown lower stiffness and poor blending of oxidized and virgin binder (Pérez Madrigal et al., 2016). On the contrary, it is also recommended not to heat RA to very high temperatures in order to avoid extra oxidation of binder as well as emissions from RA (Rathore et al. 2019). The batch plant and drum plant can incorporate up to 35% RA and 50% RA into asphalt mixtures, respectively depending on the superheating capacity and emission regulations (Brock 2007; Liu et. al. 2017). Figure 1 shows the batch plant with double drum dryer and Double Barrel® combination dryer/mixer.

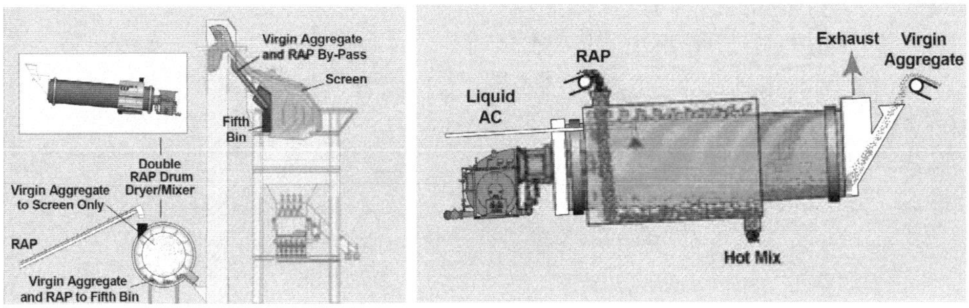

(a) (b)

Figure 1. (a) Batch plant with double RA dryer; (b) Double Barrel® combination dryer/mixer (Brock and Richmond 2007)

3.2. RA mixing process during production

In a conventional drum plant, a center entry is used to introduce RA material to the superheated virgin aggregates. It is expected that the superheated aggregates will heat the RA material and melt its binder. However, due to short mixing time available, complete melting of binder may not be possible (Dedene et al., 2014). Moreover, overheating of RA material may result into excessive gases and sticking of RA on preheating facilities. The above technical and environmental concerns led to the development of new drying and processing technologies, such as the counterflow drum mixer, (NAPA, 1996). Most modern plants now are designed to handle higher percentages of RA by adding the RA downstream of the burner in a counter-flow dryer arrangement thus providing longer mixing time (West, 2015). Further, double and triple drum mix plants were developed to prevent the virgin and RA materials from being exposed to hot gases or steam of the drying process (FHWA, 2019).

In asphalt plants, the RA material mixing with the virgin aggregates and binder takes places in two stages that are known as "dry mixing" and "wet mixing" process. In the first stage, RA material at ambient temperature is introduced with superheated virgin aggregates, and later the combination of RA and virgin aggregate is mixed with virgin binder (Brock and Richmond 2005). The dry mixing stage enhances the RA binder activation, and the wet mixing process distributes the binder uniformly and increases the diffusion between virgin binder and RA binder. The above two mixing process can take place either in a single drum of parallel/counterflow mixer or in a mixer accompanied with an outer coating chamber. The total mixing time in a pugmill is typically around 30–60 s and longer in the outer mixing chamber of the drum mixers approximately 40–90 s (Zhang et al. 2019). Depending on the plant type the total production cycle varies between 60 s, for both batch and drum plants, and 90 s for double barrel drum plants (Antunes et. al. 2019).

3.3. RA mixing temperature during production

Mixing of aggregates and asphalt has to be done at an appropriately selected temperature. The mixing temperature controls the dryness of the aggregates, the quality of the mixture, the time it takes for the mix to cool down during laying and the ease of compaction during paving (Shenoy, 2001). It will not be possible to have a uniform coating of bitumen over aggregates at lower mixing temperature. On the contrary, mixing temperature that is too high will make the bitumen to rapidly age and increase the cracking susceptibility of mix. There are several methods for determination of mixing temperatures including equiviscous method, phase angle procedure, and steady shear flow procedure (Asphalt Institute, 2014a, West C. Randy, Donald E. Watson, , Pamela A. Turner, 2010). It was observed that the linear temperature-viscosity relationship assumed for unmodified binders may not be valid for modified asphalt binders (Gayle E. Albritton, PE William F. Barstis, PE Alfred B. Crawley, 1999). The researchers hypothesized that this was due to the fact that most of the modified binders were sensitive to shear rate and therefore did not meet the current assumption of all binders as Newtonian fluids (Asphalt Institute, 2014b)

When high RA content is used, generally a higher virgin aggregates temperature is required to attain sufficient mix temperature in order to compensate the lower heating temperatures of RA. A laboratory study showed that for 50% RA mixtures, it was required to heat the virgin aggregates at 220°C to reach a mix temperature of 150°C where RA was heated at 120°C (Yu et al., 2017). The required superheating temperatures of virgin aggregates for 50% RA mixture production for different moisture contents of RA are given in Table 1. The shaded region shows excessively high temperatures that are not realistically achievable.

Table 1. Virgin Aggregates temperature requirement at asphalt plant corresponding to discharge mix temperature for 50% RA: 50% aggregate mix* (NAPA, 1996).

RA Moisture Content	Recycled mix discharges temperature, °C			
	104°C	116°C	127°C	137°C
	Virgin aggregates temperature requirement, °C			
0	210	235	257	282
1	241	268	288	310
2	271	293	318	343
3	302	327	349	374
4	338	360	377	404
5	366	391	413	438

*20°F loss between dryer and pugmill assumed in these calculations

3.4. Rejuvenator application during production

®A method is widely used in Japan, where rejuvenators are mixed with heated RA in a small pugmill and then transferred to a surge bin to give additional conditioning time (2–3 hours) (West and Copeland, 2015). The finished mix with typical temperature of 160°C is very well coated and uniform even before it is transferred to the silo (West and Copeland, 2015). In a study by Zaumanis et al., 2018b, a conventional approach of adding rejuvenator on hot RAP into mixer was compared to a rather innovative approach- spraying rejuvenator on cold RA at conveyer belt. No significant difference was observed from extracted binder properties in both the cases (Zaumanis et al., 2018b). However, the mixture test results demonstrated potentially improved fatigue life in sprayed rejuvenator mix due to improved blending of the materials (Zaumanis et al., 2019).

4. Laboratory mixing parameters

The key elements in mixing process of high content RA mixtures are similar to conventional hot mix asphalt which includes the mixture design (selection of aggregate gradation, asphalt content, additive dosage etc.), heating and mixing of raw materials, conditioning, and compaction of mixtures. However, when high content of RA material is used in mixture, the mixing parameters can have an important effect on mixture properties (Rathore & Zaumanis 2020). The important parameters of mixing in the laboratory have been described in the sections below.

Table 2 summarizes the information mixing parameters adopted in the laboratory studies for high content RA mixtures. The RA content considered in these studies was from 40% and upto 100%. Based on these studies, the most commonly used RA material heating temperature was 110°C and heating time was 2 h. However, when high content upto 100% RA was considered, the heating temperature of RA material was increased even above 155°C (Moniri et. al. 2019; Daryaee et. al., 2020). The mixing temperature for HMA mixtures was around 155-165°C and for WMA mixtures was around 120-130°C. The information about the mixing time was not reported in most of the studies. Another important information that was not reported by most of the authors was the rejuvenator/ additive application method. One study considered two mixing methods i.e. adding rejuvenator directly to RA material vs adding rejuvenator. It was found for 50% RA mixtures; moisture resistance was enhanced when rejuvenator was added directly to the RA material.

Table 2. Laboratory mixing parameters adopted in recent high RA content studies

Reference	RA % upto	RA material heating temperature	RA material heating time	Mixing temperature	Mixing time	Rejuvenator/ additive
Colbert et. al., 2012	50%	135°C	2 h	155°C	No information	No information
Lopes etl al. 2014	50%	110°C	No information	HMA: 160°C WMA: 130°C	2.5 min	No information
Dinis-Almeida et. al. 2016	100%	130°C	No information	100-120°C	No information	Added to RA material
Lu et. al. 2016	70%	No information	No information	110°C	No information	Added to virgin binder
Fakhri et. al., 2017a	50%	125°C	2 h	125°C	No information	Added to mixture
Fakhri et. al., 2017b	40%	110°C	2 h	135°C	No information	Added to mixture
Kumari et. al. 2018	100%	110°C	1.5 h	156-159°C	No information	No information
Song et. al. 2018	50%	No information	No information	HMA: 150°C WMA: 125°C	No information	No information
Lizzarga et. al. 2018	100%	95-100°C	No information	No information	No information	No information
Kumari et. al. 2019	70%	110°C	No information	HMA: 163°C WMA: 130-140°C	No information	No information
Lu et. al. 2019	50%	No information	No information	115°C	No information	Two cases: Added to virgin binder; added to RA material
Moniri et. al. 2019	100%	163°C	2 h	163°C	5 min	Added to RA material
Daryaee et. al., 2020	100%	155°C	2 h	155°C	No information	Added to virgin binder

4.1. RA heating in laboratory

In laboratory, heating temperature of 110°C (230°F) for a time of no more than 2 h is recommended for sample sizes of 1 to 2 kg as higher temperatures and longer heating times have been shown to change the properties of RA (Rebecca Mcdaniel; R. Michael Anderson, 2001). Mogawer et al. [2013] observed when RA material was heated for 2 hours prior to mixing, it resulted into lower air voids and heating time had to be increased to 4 hours to reach the desired air voids. The reason stated was that the heating time was not enough for commingling of binders. Results of another heating experiments showed that an appropriate method is to heat RA in oven at the mixing temperature for 1½ to 3 hours. Heating RA samples for more than 3 hours may cause excessive aging of the RA binder (Willis et. al. 2013).

In a study, four different RA heating process (cold, heated in a microwave, heated in an oven in covered pan, and heated in an oven in a non-covered pan) were adopted to prepare 25% RA mix in a laboratory and results shown no difference in stiffness and thermal cracking resistance (Basueny et al., 2014). A heating temperature of 110°C for 3 hours in an oven was recommended as the best option. It may be possible that the effect of conditioning process was not captured due to low content of RA. Microwave heating could also be used while heating RA at higher temperatures as microwave heat is more easily absorbed by the aggregates as compared to the binder, thus reducing its susceptibility to aging during production (Al-Qadi et al., 2007). In a study, to simulate the plant heating conditions, RA was not heated in an oven, but rather heated only by contact (conduction) from the superheated virgin aggregate. It was observed that the high conductive heat was sufficient to significantly age the binder (Willis et. al. 2013).

4.2. RA mixing time in laboratory

A sufficient mixing time is required for the aggregate-bitumen mix to form a uniform coating over the aggregates. AASHTO, (2008) has set down procedures to establish the accurate mixing time in laboratory which involves separating coarse aggregate particles from the mix on a selected sieve size and examining 200 to 300 particles are under a strong light. Usually, the minimum coating percentage required is around 90 and 95% for base and surface course respectively (Shenoy, 2001) The least time needed for the pugmill to achieve these minimums is taken as the most desirable mixing time.

When softer binder is used in high RA mix, diffusion between the two binders is better at higher mixing time, and that changes the properties of the mix (Abed et al., 2018). Results from the optical microscopy showed that by increasing the mixing time to double the normal time for 15% RA mixes, improves blending of RA and virgin binder (Hassan et al., 2015). Increased mixing time also has an impact on the physical and mechanical properties of laboratory mixtures. Longer mixing times tend to reduce the air voids in mixtures containing high RA material (Hassan et al., 2015; Pérez Madrigal et al., 2016). This is attributed to increase in blending between virgin and oxidized binder and thus having a better compactibility.

4.3. RA mixing temperature in laboratory

With the increase in mixing temperatures, the air voids in RA mixes were found to be reduced (Abed et al., 2018). This is due to the fact that at higher temperature, the particle clustering is lowered which in turn leads to increased compaction. Increasing mixing temperature improves the blending which leads to greater modulus. This change is more significant when RA percentage is increased (Pérez Madrigal et al., 2016). (Navaro et al., 2012) observed if the mixing temperature is reduced by 30°C then, generally to obtain a recycled asphalt concrete with a binder of the same homogeneity, the mixing time has to be multiplied by factor of 2.5–3. Therefore, in some cases where high mixing temperatures are not possible, a longer mixing time will be required. Warm mix additives are also used to reduce the mixing temperature while maintaining the same workability of mix. The use of WMA technologies- foaming, organic or wax usually allows the incorporation of higher RA amounts than for HMA and appears to provide a synergetic effect on improving both the WMA and high RA mix performance (Zaumanis and Mallick, 2015). Xie et al. (2016) observed that WMA with 20% and 30% RA mixes showed similar compactibility as corresponding HMA mixes with the production temperature being approximately 30°F higher. Warm mixes are generally more susceptible to moisture damage due to lower mixing temperature.

Overheating the aggregates without preheating RA and extending mixing time has also proven to be good expedients to improve the cracking resistance of the mixtures (Pérez Madrigal et al., 2016). But this is not always possible with higher RA contents and to heat the aggregates at very high superheated temperatures becomes practically difficult in plant. Where high mixing temperatures cannot be attained, mixing time can be increased upto a certain level to achieve desired air voids. For example, in a study on mix containing 50% RA, mixing at 135°C for 3 min was sufficient to achieve the target air voids of 5%, whereas at 115°C, a longer mixing time was required and the target air voids were barely obtained after 5 min of mixing. At 95°C, the target air voids were never achieved, even after 5 min of mixing (Abed et al., 2018).

4.4. Rejuvenator incorporation method in laboratory

Conventionally, for production of HMA in laboratory, all mineral aggregates are mixed together (dry mixing), followed by pouring the bitumen on this aggregate mixture (wet mixing) which is continued till the desired mixing time (Roberts et al., 1996). When high content of RA is being used in the mix, the incorporation of rejuvenators is also required in the process. Recycling agents offer many unique benefits as compared to the use of softer binders (Zaumanis and Mallick, 2013). In laboratory, the rejuvenators are generally incorporated by blending rejuvenator with virgin bitumen. This is the most convenient and widely used rejuvenator application method in laboratory. The rejuvenator is pre-blended into bitumen with the required dosage and then added directly to the mix. Though, this method simulates the plant operation of adding rejuvenator, it could be less efficient than other methods discussed below as some part of added rejuvenator may not be available to RA due to lower blending between virgin bitumen and RA bitumen.

Another method is to add rejuvenator directly to RA material. In this method, the rejuvenator is added directly to RA material (with or without preheating the RA material) in order to activate the stiff binder. This method facilitates the diffusion of the rejuvenator as rejuvenator is in direct contact with the RA binder maximizing the binder activation. However, in the previous method of blending with binder, the rejuvenator is diluted by virgin bitumen and therefore weakens the modifying effect (Yu et al., 2017). A sufficient rest period is likely to cause some diffusion of rejuvenator into RA binder and ultimately restore the original properties of binder.

4.5. Mixture conditioning in laboratory

To have an accurate assessment of performance of asphalt mixes, conditioning in laboratory should be able to simulate the field conditions. Therefore, it is important to quantify the effect of short term and long-term ageing on mixes and compare it with plant produced mixes. Ageing during mixing and construction is referred as short-term ageing and ageing during the service life of the pavement is referred to as long-term ageing. The short-term conditioning of asphalt mixtures also affects the long-term ageing, and the effect is more significant for some mixtures than the others (Azari 2013). For mixes containing RA, it becomes much more important as RA binders often age differently than the virgin binder. Two ageing methods have been described in a technical specification released by the CEN recently: (CEN/TS 12697-52, 2017) one with the loose sample and other with the compacted specimen. The short-term ageing is done at 135°C for 4h and long-term aging is done at 85°C for 216 h. AASHTO R30 specifies mixture to be conditioned for 2 hours at compaction temperature for volumetric mix design. Short term ageing is carried out at a temperature of 135°C for 4 h and long-term ageing is done at 85°C for 5 days.

5. Summary and discussion

This paper highlights the current scenario for production of high content reclaimed asphalt mixtures and identifies the important associated parameters. The current scenarios of high RA content mixture production at plant shows that most of modern plants are now designed to handle higher percentages of RA material by having RA material drying mechanism. The main reason of limiting the heating temperatures of RA material was to avoid the oxidation of binder and sticking of binder to heating facility. However, to produce mixes with high RA content, it will become necessary to heat the RA material to a higher temperature so as to allow higher activation of RA binder and avoid heating virgin aggregates at unrealistically high temperature. Recently, double and triple drum mix plants were developed which prevent the RA binder exposure to hot gases and thus prevents "blue smoke". This indirect heating principle allows heating RA without direct contact with the flame, to a conventional hot-mix asphalt production temperature of 160 °C.

It was observed that laboratory mixing parameters are often not reported which makes it difficult to compare various studies. As laboratory mixing parameters were found to significantly affect the performance of mixtures. For example, very high mixing temperature will cause oxidation of bitumen hard and will increase the cracking susceptibility of mix. However, it might not be possible to have a uniform coating of bitumen over aggregates at lower mixing temperature. In some cases where high mixing temperatures are not possible to achieve, a longer mixing time will be required or warm mix asphalt (WMA) technology can be incorporated. Providing longer mixing times tend to reduce the air voids in mixtures containing high RA material until a minimum mixing temperature below which the target air voids cannot be achieved. Microwave heating was also given as one of the options to heat RA material as at higher temperatures as microwave heat is more easily absorbed by the aggregates comparing to the binder, thus reducing its susceptibility to aging during production. The sequence of mixing material can be another important mixing parameter in the laboratory as it can affect the workability and compactibility of mix. This has not been much explored and thus future studies must investigate the effect of sequencing the materials while mixing. The method of addition of rejuvenator is another important parameter that affects the degree of binder activation and blending with virgin bitumen. Ideally, rejuvenator should be added directly to RA material to increase the binder activation. However, more studies are required in this direction to determine the optimized procedure for rejuvenator application so that this can be extended to asphalt plants. Finally, the laboratory conditioning is very important to predict the actual performance of high RA mixes in field. Therefore, there is a need to develop laboratory aging procedure for high RA mixtures (upto 100%) to simulate the plant aging of materials.

Acknowledgements

This study is part of SAFERUP!, an Innovative Training Network devoted to "Safe, Accessible and Urban pavements". SAFERUP! project has received funding from the European Union's Horizon 2020 research and innovation programme under the Marie Skłodowska-Curie grant agreement No 765057.

References

AASHTO, 2008. Standard specification for Transport materials and methods of sampling and testing 28th Editi.

AASHTO M323-12, 2012. Standard Specification for Superpave Volumetric Mix Design. Stand. Specif. Superpave Vol. Mix Des. 1–15.

Abed, A., Thom, N., Lo Presti, D., 2018. Design considerations of high RAP-content asphalt produced at reduced temperatures. Mater. Struct. Constr. 51, 1–16. https://doi.org/10.1617/s11527-018-1220-1

Al-Qadi, I.L., Elseifi, M., Carpenter, S.H., 2007. Reclaimed asphalt pavement - A literature review. Fed. Highw. Adm. 07, 23. https://doi.org/http://hdl.handle.net/2142/46007

Antunes, V., Freire, A. C., Neves, J., (2019). A review on the effect of RAP recycling on bituminous mixtures properties and the viability of multi-recycling, Construction and Building Materials, Volume 211, 2019, Pages 453-469, ISSN 0950-0618, https://doi.org/10.1016/j.conbuildmat.2019.03.258

Asphalt Institute, 2014a. MS-2: 7th Edition Asphalt Mix Design Methods 7th Editio, 1–197.

Asphalt Institute, 2014b. MS-2: 7th Edition Asphalt Mix Design Methods 7th Editio, 1–197.

Basueny, A., Perraton, D., Carter, A., 2014. Laboratory study of the effect of RAP conditioning on the mechanical properties of hot mix asphalt containing RAP. Mater. Struct. Constr. 47, 1426–1450. https://doi.org/10.1617/s11527-013-0127-0

Bocci, E., Mazzoni, G., Canestrari, F., 2019. Ageing of rejuvenated bitumen in hot recycled bituminous mixtures: influence of bitumen origin and additive type. Road Mater. Pavement Des. 0, 1–22. https://doi.org/10.1080/14680629.2019.1587492

Bowers, B.F., Huang, B., Shu, X., Miller, B.C., 2014. Investigation of Reclaimed Asphalt Pavement blending efficiency through GPC and FTIR. Constr. Build. Mater. 50, 517–523. https://doi.org/10.1016/j.conbuildmat.2013.10.003

Brock, J. D., Richmond, S., Jeff, L., (2007). Milling and Recycling, Astec Industries, 2007.

Brett A. Williams, Audrey Copeland, T.C.R., 2018. Asphalt Pavement Industry Survey on Recycled Materials and Warm-Mix Asphalt Usage: 2017. Natl. Asph. Pavement Assoc. 8th Annual.

Carpenter, S.H., Wolosick, J.R., 1980. Modifier Influence in the Characterization of Hot-Mix Recycled Material. J. Transp. Res. Board.

Cavalli, M.C., Zaumanis, M., Mazza, E., Partl, M.N., Poulikakos, L.D., 2018. Aging effect on rheology and cracking behaviour of reclaimed binder with bio-based rejuvenators. J. Clean. Prod. 189, 88–97. https://doi.org/10.1016/j.jclepro.2018.03.305

CEN/TS 12697-52, 2017. Bituminous mixtures - Test methods - Part 52 : Conditioning to address oxidative ageing.

Copeland, A., 2011. Reclaimed Asphalt Pavement in Asphalt Mixtures : State of the Practice. Fed. Highw. Adm.

Colbert, B., You, Z., (2012). The determination of mechanical performance of laboratory produced hot mix asphalt mixtures using controlled RAP and virgin aggregate size fractions, Construction and Building Materials, Volume 26, Issue 1, 2012, Pages 655-662, ISSN 0950-0618, https://doi.org/10.1016/j.conbuildmat.2011.06.068.

Daryaee, D., Ameri, M., Mansourkhaki, A., (2020). Utilizing of waste polymer modified bitumen in combination with rejuvenator in high reclaimed asphalt pavement mixtures, Construction and Building Materials, Volume 235, 2020, 117516, ISSN 0950-0618, https://doi.org/10.1016/j.conbuildmat.2019.117516.

Dedene, C.D., Voller, V.R., Marasteanu, M.O., Dave, E. V, 2014. Calculation of particle heating times of reclaimed asphalt pavement material Calculation of particle heating times of reclaimed asphalt pavement material. Road Mater. Pavement Des. 15, 721–732. https://doi.org/10.1080/14680629.2014.909324

Dinis-Almeida, M., Castro-Gomes, J., Sangiorgi, C., Zoorob, S. E., Lopes Afonso, M., Performance of Warm Mix Recycled Asphalt containing up to 100% RAP, Construction and Building Materials, Volume 112, 2016, Pages 1-6, ISSN 0950-0618, https://doi.org/10.1016/j.conbuildmat.2016.02.108.

EAPA, 2018. Asphalt in Figures 2017. Eur. Asph. Pavement Assoc.

Fakhri, M., Ahmadi, A., (2017). Recycling of RAP and steel slag aggregates into the warm mix asphalt: A performance evaluation, Construction and Building Materials, Volume 147, 2017, Pages 630-638,

ISSN 0950-0618, https://doi.org/10.1016/j.conbuildmat.2017.04.117.

Fakhri, M., Hosseini, S. A., (2017). Laboratory evaluation of rutting and moisture damage resistance of glass fiber modified warm mix asphalt incorporating high RAP proportion, Construction and Building Materials, Volume 134, 2017, Pages 626-640, ISSN 0950-0618, https://doi.org/10.1016/j.conbuildmat.2016.12.168.

FHWA, 2019. Hot-Mix Asphalt Recycling - Drum Plant (Construction Methods And Equipment) [WWW Document]. URL https://www.fhwa.dot.gov/pavement/recycling/98042/06.cfm

Franke. R., & Ksaibati. K., (2015). A methodology for cost-benefit analysis of recycled asphalt pavement (RAP) in various highway applications. International Journal of Pavement Engineering, 2015 Vol. 16, No. 7, 660–666, http://dx.doi.org/10.1080/10298436.2014.943217

Gayle E. Albritton, PE William F. Barstis, PE Alfred B. Crawley, P., 1999. Polymer Modified Hot Mix Asphalt Field Trial.

Gottumukkala, B., Kusam, S.R., Tandon, V., Muppireddy, A.R., 2018. Estimation of Blending of Rap Binder in a Recycled Asphalt Pavement Mix. J. Mater. Civ. Eng. 30, 04018181. https://doi.org/10.1061/(asce)mt.1943-5533.0002403

Hassan, N.A., Khan, R., Raaberg, J., Lo Presti, D., 2015. Effect of mixing time on reclaimed asphalt mixtures: An investigation by means of imaging techniques. Constr. Build. Mater. 99, 54–61. https://doi.org/10.1016/j.conbuildmat.2015.09.009

Hellriegel, E.J., 1980. Recycling Asphalt Pavements. Fed. Highw. Adm. 1, 325–342.

Hong, F., & Prozzi, J. A., (2017): Evaluation of recycled asphalt pavement using economic, environmental, and energy metrics based on long-term pavement performance sections, Road Materials and Pavement Design, DOI: 10.1080/14680629.2017.1348306

Im, S., Karki, P., Zhou, F., 2016. Development of new mix design method for asphalt mixtures containing RAP and rejuvenators. Constr. Build. Mater. 115, 727–734. https://doi.org/10.1016/J.CONBUILDMAT.2016.04.081

Izaks, R., Rathore, M., Haritonovs, V., Zaumanis, M., (2020). Performance properties of high modulus asphalt concrete containing high reclaimed asphalt content and polymer modified binder. International Journal of Pavement Engineering, https://doi.org/10.1080/10298436.2020.1850721

Kandhal, P.S., Mallick, R.B., 1997. Pavement recycling guidelines for state and local governments. Report No. FHWA-SA-98-042 301.

Kumari, M., Ransinchung, G.D. R.N., Singh, S., (2019). Effect of long-term ageing on properties of RAP inclusive WMA mixes, Construction and Building Materials, Volume 206, 2019, Pages 483-493, ISSN 0950-0618, https://doi.org/10.1016/j.conbuildmat.2019.02.087.

Kumari, M., Ransinchung, G.D. R.N., Singh, S., (2018). A laboratory investigation on Dense Bituminous Macadam containing different fractions of coarse and fine RAP, Construction and Building Materials, Volume 191, 2018, Pages 655-666, ISSN 0950-0618, https://doi.org/10.1016/j.conbuildmat.2018.10.017.

Lopes, M., Gabet, T., Bernucci, L. *et al. (2015).* Durability of hot and warm asphalt mixtures containing high rates of reclaimed asphalt at laboratory scale. *Mater Struct* **48,** 3937–3948 https://doi.org/10.1617/s11527-014-0454-9

Lu, D. X., Saleh, M., (2016). Laboratory evaluation of warm mix asphalt incorporating high RAP proportion by using evotherm and sylvaroad additives, Construction and Building Materials, Volume 114, 2016, Pages 580-587, ISSN 0950-0618, https://doi.org/10.1016/j.conbuildmat.2016.03.200.

Lu, D. X., Saleh, M. Nguyen, N. H. T., (2019). Effect of rejuvenator and mixing methods on behaviour of warm mix asphalt containing high RAP content, Construction and Building Materials, Volume 197, 2019, Pages 792-802, ISSN 0950-0618, https://doi.org/10.1016/j.conbuildmat.2018.11.205.

Mhlongo, S.M., Abiola, O.S., Ndambuki, J.M., Kupolati, W.K., 2014. Use of Recycled Asphalt Materials for Sustainable Construction and Rehabilitation of Roads. https://doi.org/10.15242/iicbe.c0314157

Mogawer, W., Austerman, A., Mohammad, L., Kutay, M.E., 2013. Evaluation of high RAP-WMA asphalt rubber mixtures. Road Mater. Pavement Des. 14, 129–147. https://doi.org/10.1080/14680629.2013.812846

Mogawer, W., Bennert, T., Daniel, J.S., Bonaquist, R., Austerman, A., Booshehrian, A., 2012. Performance characteristics of plant produced high RAP mixtures. Road Mater. Pavement Des. 13, 183–208. https://doi.org/10.1080/14680629.2012.657070

Mogawer, W.S., Booshehrian, A., Vahidi, S., Austerman, A.J., 2013. Evaluating the effect of rejuvenators on the degree of blending and performance of high RAP, RAS, and RAP/RAS mixtures. Road Mater. Pavement Des. 14, 193–213. https://doi.org/10.1080/14680629.2013.812836

Mogawer Walaa S., Austerman Alexander, Roque Reynaldo, Underwood Shane, Mohammad Louay, Zou Jian, 2015. Aging and rejuvenators: Evaluating their impact on high RAP mixtures fatigue cracking characteristics using advanced mechanistic models and testing methods. Road Mater. Pavement Des. 84, 1–42. https://doi.org/10.1080/14680629.2015.1076996

Moniri, A., Ziari, H., Aliha, M. R. M., & Saghafi, Y., (2019). Laboratory study of the effect of oil-based recycling agents on high RAP asphalt mixtures, International Journal of Pavement Engineering, DOI: 10.1080/10298436.2019.1696461

NAPA, 1996. Recycling Hot Mix Asphalt Pavemets. Natl. Asph. Pavement Assoc. - Qual. Improv. Ser. 123 1–28.

Navaro, J., Bruneau, D., Drouadaine, I., Colin, J., Dony, A., Cournet, J., 2012. Observation and evaluation of the degree of blending of reclaimed asphalt concretes using microscopy image analysis. Constr. Build. Mater. 37, 135–143. https://doi.org/10.1016/j.conbuildmat.2012.07.048

Pérez Madrigal, D., Iannone, A., Martìnez, A.H., Giustozzi, F., 2016. Effect of Mixing Time and Temperature on Cracking Resistance of Bituminous Mixtures Containing Reclaimed Asphalt Pavement Material. J. Mater. Civ. Eng. 29, 04017058. https://doi.org/10.1061/(asce)mt.1943-5533.0001831

Qiao, Y., Dave, E., Parry, T., Valle, O., Mi, L., Ni, G., Yuan, Z., et al. (2019). Life Cycle Costs Analysis of Reclaimed Asphalt Pavement (RAP) Under Future Climate. Sustainability, 11(19), 5414. MDPI AG. Retrieved from http://dx.doi.org/10.3390/su11195414

Rathore, M. , Haritonovs, V. , and Zaumanis, M., (2019). Asphalt recycling technologies: a review on limitations and benefits. IOP Conference Series: Materials Science and Engineering, 660, 012046. https://doi.org/10.1088/1757-899X/660/1/012046

Rathore, M., Zaumanis, M., (2020). Impact of laboratory mixing procedure on the properties of reclaimed asphalt pavement mixtures. Construction and Building Materials, Volume 264, 2020, 120709, ISSN 0950-0618, https://doi.org/10.1016/j.conbuildmat.2020.120709.

Rebecca Mcdaniel; R. Michael Anderson, 2001. Recommended Use of Reclaimed Asphalt Pavement in the Superpave Mix Design Method: Technician's Manual. Natl. Coop. Highw. Res. Progr. REPORT 452.

Roberts, F.L., Prithvi, S., Kandhal, E., Brown, R., Dah-Yinn, L., Kennedy, T.W., 1996. Hot Mix Asphalt Materials, Mixture Design, and Constrution. Hot Mix Asph. Mater. Mix. Des. Constrution.

Shenoy, A., 2001. Determination of the Temperature for Mixing Aggregates with Polymer-Modified Asphalts. Int. J. Pavement Eng. 2, 3347. https://doi.org/10.1080/10298430108901715

Song W, Huang B, Shu X, (2018), Influence of warm-mix asphalt technology and rejuvenator on performance of asphalt mixtures containing 50% reclaimed asphalt pavement, *Journal of Cleaner Production* doi: 10.1016/j.jclepro.2018.04.269.

Stroup-Gardiner, M., 2016. Use of Reclaimed Asphalt Pavement and Recycled Asphalt Shingles in Asphalt Mixtures, Use of Reclaimed Asphalt Pavement and Recycled Asphalt Shingles in Asphalt Mixtures. https://doi.org/10.17226/23641

West C. Randy, Donald E. Watson, , Pamela A. Turner, J.R.C., 2010. Mixing and Compaction Temperatures of Asphalt Binders in Hot-Mix Asphalt, Mixing and Compaction Temperatures of Asphalt Binders in Hot-Mix Asphalt. https://doi.org/10.17226/14367

West, R.C., 2015. Best Practices for RAP and RAS Management. Natl. Asph. Pavement Assoc. 44.

West, R.C., Copeland, A., 2015. High RAP Asphalt Pavements: Japan Practice — Lessons Learned. Natl. Asph. Pavement Assoc. 139, 62.

Willis, J.R., Turner, P., Julian, G., Taylor, A.J., Tran, N., Padula, F. de G., 2012. Effect of Changing Virgin Binder Grade and Content on RAP Mixture Properties.

Xie, Z., Trail, N., Taylor, A., Julian, G., West, R., Welch, J., 2016. Evaluation of foamed warm mix asphalt with reclaimed asphalt pavement: Field and laboratory experiments. Asph. Paving Technol. Assoc. Asph. Paving Technol. Tech. Sess. 86, 449–479. https://doi.org/10.1080/14680629.2017.1389077

Yetkin Yildirim, Mansour Solaimanian, T.W.K., 2000. Mixing and Compaction Temperatures for Hot Mix 1250–5, 100.

Yu, B., Gu, X., Wu, M., Ni, F., 2017. Application of a high percentage of reclaimed asphalt pavement in an asphalt mixture: blending process and performance investigation Application of a high percentage of reclaimed asphalt pavement in an asphalt mixture: blending process and performance inv. Road Mater. Pavement Des. 18, 753–765. https://doi.org/10.1080/14680629.2016.1182941

Zaumanis, M., Boesiger, L., Kunz, B., Cavalli, M.C., Poulikakos, L., 2019. Determining optimum rejuvenator addition location in asphalt production plant. Constr. Build. Mater. 198, 368–378. https://doi.org/10.1016/j.conbuildmat.2018.11.239

Zaumanis, M., Cavalli, M.C., Poulikakos, L.D., 2018a. How not to design 100% recycled asphalt mixture using performance-based tests. Road Mater. Pavement Des. 0, 1–13. https://doi.org/10.1080/14680629.2018.1561381

Zaumanis, M., Cavalli, M.C., Poulikakos, L.D., 2018b. Effect of rejuvenator addition location in plant on mechanical and chemical properties of RAP binder. Int. J. Pavement Eng. 0, 1–9. https://doi.org/10.1080/10298436.2018.1492133

Zaumanis, M., Mallick, R.B., 2015. Review of very high-content reclaimed asphalt use in plant-produced pavements: State of the art. Int. J. Pavement Eng. 16, 39–55. https://doi.org/10.1080/10298436.2014.893331

Zaumanis, M., Mallick, R.B., 2013. Finite Element Modeling of Rejuvenator Diffusion in RAP Binder Film – Simulation of Plant Mixing Process. Multi-Scale Model. Charact. Infrastruct. Mater. 407–419. https://doi.org/10.1007/978-94-007-6878-9

Zaumanis, M., Mallick, R.B., Frank, R., 2014. Determining optimum rejuvenator dose for asphalt recycling based on Superpave performance grade specifications. Constr. Build. Mater. 69, 159–166. https://doi.org/10.1016/j.conbuildmat.2014.07.035

Template: Anonymized Document

SURFACE TEXTURE AND LAYER PERMEABILITY OF AQUAPLANING RESISTANT ASPHALT PAVEMENTS

Audrius VAITKUS (ORCID ID 0000-0001-5103-9747)[1], Dovydas SKRODENIS (0000-0002-7120-3724)[2], Ovidijus ŠERNAS (ORCID iD 0000-0002-0019-4242)[3], Viktoras VOROBJOVAS (0000-0001-9420-0668)[4]

[1,2,3]*Road Research Institute, Vilnius Gediminas Technical University, Vilnius, Lithuania*
[4]*Road Department, Vilnius Gediminas Technical University, Vilnius, Lithuania*
Corresponding author: Dovydas Skrodenis, dovydas.skrodenis@vilniustech.lt

Received day month year; accepted date (use style Received dates, or Alt + Ctrl + R)

Abstract. Tire/pavement noise is one of the biggest environmental problems caused by the contact between the car tire and the pavement surface. It is known that porous asphalt (PA) pavements has good properties in noise absorption, however these mixtures could also solve another important problem which appear on roads – aquaplaning. This phenomenon reduces traffic safety and driving comfort. Aquaplaning appears when tires become separated from the pavement surface by thin water film and the ability to increase braking force or control the vehicle motion is almost entirely lost. Although, PA pavements have relatively low durability properties. This research aims analyse surface texture and permeability characteristics of aquaplaning resistant asphalt pavements. Four different mixtures with different largest particle size (AT 5, AT 8, AT 11 and AT 16) were tested. Large-scale laboratory testing was performed to evaluate their surface texture and permeability properties The research revealed, that mixtures with 8 % activated mineral limestone powder (AMLP) showed better mechanical and physical properties than comparing to other mixtures with 4 % AMLP and 4 % granite screenings or just 4 % AMLP.

Keywords: aquaplaning, porous asphalt, skid resistance, mean profile depth, splash & spray

Introduction

Road traffic noise is a widespread problem, especially in the densely populated cities of Europe. Different type of asphalt mixtures are used for noise-reducing asphalt pavements. Optimisation of asphalt mixtures for tyre/road noise reduction mainly depends on the potential application area of these mixtures (Vaitkus *et al.* 2019). If asphalt pavements are not adequately designed, they will require frequent repair because of premature failures of the asphalt wearing layer (Vaitkus *et al.* 2017a). Moreover, to ensure a lower tire/road noise level and proper friction, exposed aggregate concrete widely used for the construction of major highways in Europe (Šernas *et al.* 2020).

In case of rain or wet road surface, the main factor of traffic safety is the sufficient grip between the road surface and the car wheel, which can be lost due to the resulting aquaplaning phenomenon, when the water accumulating on the top of the road surface does not run to the curb. Various traffic safety studies from indicate that approximately 20% of all road traffic accidents occur in wet weather conditions (Ivan *et al.*, 2012; Mayora and Pina, 2009; McGovern *et al.*, 2011). The formation of aquaplaning phenomenon is influenced by three factors: road surface characteristics (water film thickness, wearing course material); tire condition (load, tread depth, pressure); driving speed. When driving on the road, the grooves (tread) in the tire collect water from the surface of the pavement and push it out, thus ensuring good grip between the tire and the pavement. The aquaplaning phenomenon occurs when the tread of a tire no longer manages to displace water that has entered the grooves and the tire rises from the surface of the tread on a thin layer of water. This creates dangerous driving conditions (especially when driving over 80 km/h (Herrmann, 2008)) and exponentially increases the risk of an accident. Two of the three factors that determine the formation of the aquaplaning phenomenon can be controlled by the road users themselves – the condition of the tires and the driving speed. However, the condition of the road surface and the type of surface are controlled by the relevant authorities. Meaning that when the road surface is wet, road users must take safety measures and avoid aquaplaning.

Concept of aquaplaning asphalt pavements

Elimination of water from the road surface is possible in two ways: by the geometrical parameters of the pavement or by a certain pavement mixture (wearing course material). Usually in Lithuania, water is eliminated from the surface of the pavement by designing roads with a longitudinal and / or transverse slope, by which the water flows to the predetermined water collection points. This method of drainage usually uses conventional asphalt pavements, the surface of which flows with water, and when the unevenness of the pavement occurs, the water remains standing on the surface of the pavement, which can lead to the phenomenon of aquaplaning. In order to solve the formation of the aquaplaning phenomenon, various pavement structures have been used in other countries for a long time (about 50 years ago (Jacobson *et al.* 2016)), which drain the water "through themselves". Such structures are called permeable pavements and are designed that the water on the surface of the pavement runs directly through part of the pavement structure, meaning that no water accumulates on the surface of the wearing layer. Due to their ability to allow water to quickly infiltrate through the surface, permeable pavements allow to reduce the runoff quantity and peak runoff rates

Template: Anonymized Document

(Bratteo and Booth, 2003; Bean *et al.* 2005). This type of construction solves the problem of aquaplaning and the formation of water splashes. Different countries use different designs and technologies, however most common permeable pavement applications are: porous concrete, porous asphalt or permeable interlocking concrete pavers (PICP). The construction design and permeability properties also varies depending on the type of surface. The typical structures are shown in Figures 1 - 2.

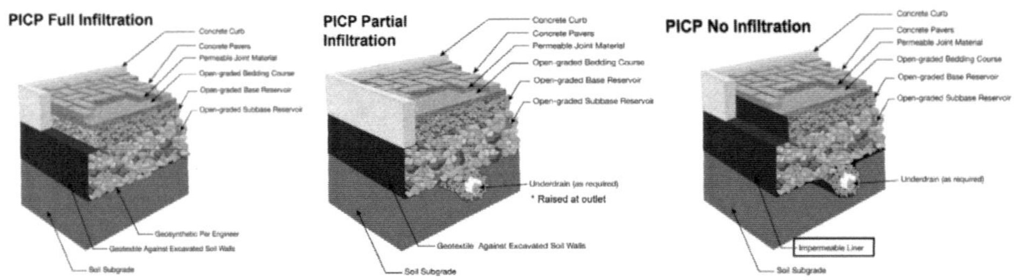

Figure 1. Principal permeable concrete pavers pavement structure (Biknel, 2016)

Figure 2. Principal permeable asphalt concrete pavement structure

Permeable structures vary in use. Often, in porous asphalt or concrete pavement structures, only the top (wearing) layer is porous. Additional waterproof layers are installed to prevent water from penetrating into the other layers of the structure below. In this way, the water entering through the surface layer of the pavement structure due to the transverse and longitudinal slopes in the pavement flows in the layer pores to the water collection point. This type of structure is the mostly used in roads. Fully permeable structures are the mostly used in parking lots or driveways to houses. Such structures are used in cases when it is difficult to transport water to the collection points with a transverse or longitudinal profile or a long water flow light is formed. These structures are unique in that the water on the surface penetrates through all layers of the structure all the way to the ground.

Permeable interlocking concrete pavers with a specific installation design can provide good water drainage, but such structures are most commonly used in parking lots. The top layer of such a structure is made of prefabricated concrete tiles, but the bottom layer is permeable to water. The gaps between the tiles make up 5–15% of the total surface area, so the water is drained throughout the pavement structure all the way to the subsoil. This type of structure can drain 10.2–15.2 mm/h of rainwater, depending on the environmental conditions and the composition of the structure (Haselbach *et al.* 2017).

Porous concrete is an uncommon alternative in Europe, but is widely used in the United States of America (USA). Porous concrete mixtures differs from conventional concrete mixtures in a similar way as conventional asphalt mixtures differ from porous asphalt mixtures. Porous concrete contains a higher fraction of aggregates, thus forming a concrete mixture with a voids content of 15 – 25%. Porous concrete can drain 7.6–50.8 mm/h of rainwater, depending on environmental conditions and the composition of the mixture (Haselbach *et al.* 2017). Porous concrete should theoretically absorb less heat from the environment due to the surface color, but no research has been done on this topic. In the USA, porous concrete pavements are most widely used on low volume roads, parking lots, and residential neighborhoods. Because this type concrete pavement has poor resistance to cracking, and it is not recommended to apply this type of pavement on primers with a low water conductivity. When choosing porous pavement, it is important to pay attention to two factors: bearing capacity and resistance to cold, depending on the climatic conditions of the area (Amirjani, 2010).

Porous asphalt (PA) is the most common and popular low noise as well as to reduce aquaplaning phenomenon pavement solution used across the world (Vaitkus *et. al.* 2017b) Porous asphalt is an asphalt mixture with a void content of more than 15%. This type of pavement is single-layer or double-layer. Porous asphalt has become most widespread as an alternative to reduce tire and asphalt pavement contact noise and to reduce the likelihood of aquaplaning. The large content of voids absorbs surface water, which also solves the problem of dripping and splashing, which results in better visibility for drivers and higher vehicle speeds in the rain. Porous asphalt, depending on environmental conditions and the composition of the mixture, can drain 4.3–12.7 mm/h of rainwater (Haselbach *et al.* 2017). Permeability, or otherwise, hydraulic conductivity, is the property of a material to pass fluids through its structure. Porous asphalt has this property and water permeability depends on the voids content and their relationship. The hydraulic properties of porous asphalt have been observed for a long time and the use of this type of pavement has been introduced precisely because of the water permeability to reduce the amount of water on the road surface. After some time, the acoustic properties of porous asphalt were also discovered. Also, the porous asphalt mixture acts as a filter to collect foreign particles that are in the water. This filtration function improves the quality of groundwater, but trapped foreign matter in the mixture impairs the properties of the

Template: Anonymized Document

binder, clogging the pores in the mixture (Ahmed, 2015). Clogged pores mean lower hydraulic conductivity of the mixture and poorer acoustic properties (Vaitkus *et al.* 2019). This type of pavement is not only clogged by surface water filtration, but is also affected by other external factors: dirt near the road or dirt spilled from car tires is compacted by vehicles and clogs the mixture pores, studded tires damage the microtexture of the pavement surface and the resulting dust is absorbed into open pores of the pavement. Due to the increased transport volume or increased traffic load, the aggregates in the mixture decompose inside the mixture, distorting the hydrological properties of the mixture.

Testing of prototypes in the laboratory

New asphalt mixtures (AT 5, AT 8, AT 11, and AT 16) have been designed under laboratory conditions in order to develop new pavement structures to drain water and eliminate the possibility of aquaplaning phenomenon. Asphalt mixtures that reduce aquaplaning differ from each other in the size of the largest particle, from the fine mixture of 5 mm maximum particle size to the course mixture of 16 mm maximum particle size. In each individually designed aggregate size distribution composition variant, three essential variables were used: the standard amount of activated mineral limestone powder (AMLP), the significantly increased amount of AMLP, and the granite fine aggregate fr. 0/2. The detailed compositions are presented in Table 1.

Table 1. Design compositions of porous asphalt mixtures

Asphalt mixture	Mix code	Binder content, %	AMLP content, %	Granite fine aggregate 0/2, %	Granite 2/5, %	Granite 5/8, %	Granite 8/11, %	Granite 11/16, %
AT 5	5-1	6.5	4.0	-	96.0	-	-	-
	5-2	6.5	8.0	-	92.0	-	-	-
	5-3	6.5%	4.0	4.0	92.0	-	-	-
AT 8	8-1	6.5%	4.0	-	-	96.0	-	-
	8-2	6.5%	8.0	-	-	92.0	-	-
	8-3	6.5%	4.0	4.0	-	92.0	-	-
AT 11	11-1	6.0%	4.0	-	-	-	96.0	-
	11-2	6.0%	8.0	-	-	-	92.0	-
	11-3	6.0%	4.0	4.0	-	-	92.0	-
AT 16	16-1	5.5%	4.0	-	-	-	-	96.0
	16-2	5.5%	8.0	-	-	-	-	92.0
	16-3	5.5%	4.0	4.0	-	-	-	92.0

Mixture codes in Table 1 denote different asphalt mixtures and their compositions. Codes 5-1, 8-1, 11-1 and 16-1 denote mixtures of normal composition. Codes 5-2, 8-2, 11-2 and 16-2 denote mixtures containing an increased amount of AMLP (0.063 mm) (8%). Codes 5-3, 8-3, 11-3 and 16-3 denote mixtures containing 4% of AMLP and additionally 4% of granite fine aggregates 0/2. The amount of mineral powder and the amount of fines were increased in order to increase the amount of mastic in the asphalt mixture and to evaluate the properties of the modified asphalt mixture composition.

The general trend in European countries shows that polymer modified binders (PMBs) are mostly used for porous asphalt mixtures in recent years. The additives contained in the binder prolong the durability of the coating by increasing the cohesive and adhesive properties of the mixture. It is also ensured that the rough aggregate particles are covered with a thicker layer of binder. Due to the properties of viscous additives, asphalt mixtures are more resistant to heat, but also provide sufficient flexibility in the cold season. For this reason polymer modified bitumen PMB 45/80-65 was selected for designing aquaplaning resistant asphalt mixtures (AT), since this binder have the most suitable physical and mechanical properties to design durable asphalt mixture for noise and aquaplaning reducing asphalt pavements. The properties of the selected aggregates and binder are presented in Tables 2–3.

The shape of granite course aggregates were determined in terms of the flakiness index and shape index. The flakiness index was determined by standard EN 933-3 and shape index – by standard EN 933-4.

Template: Anonymized Document

Table 2. Shape properties of the aggregates

Material	Flakiness index	Shape index
Granite 2/5	9	2
Granite 5/8	4	3
Granite 8/11	3	2

Table 3. Physical and mechanical properties of the binder

Material	Penetration at 25 °C, dmm	Softening point, °C	Elastic recovery, %	Critical temperature, °C	
				high	low
PMB 45/80-65	51.8	65.9	96.8	79.7	–20.5

The properties of PMB 45/80-65 evaluated according to the requirements of standard EN 14023. The penetration at 25 °C determined by standard EN 1426. The softening point determined by standard EN 1427:2015 and elastic recovery by standard EN 13398:2018.

The physical and mechanical characteristics that are given below were determined in the laboratory by testing 3 specimens for each asphalt mixture and for each test condition.

<u>*Air voids content*</u>

The air voids content determined by standard EN 12697-8. The air voids of the asphalt specimen calculated using the maximum density (standard EN 12697-5, Procedure A) of the mixture and the bulk density of the specimen (standard EN 12697-6, Procedure C: Bulk density - Sealed specimen)). The specimens for determining the bulk density were prepared by a Marshall compactor according to standard EN 12697-30 (2×50 blows).

<u>*Horizontal and vertical water permeability*</u>

Horizontal and vertical water permeability tests carried out according to standard EN 12697-19. The specimens for this test were prepared by a Marshall compactor according to standard EN 12697-30 (2×50 blows). For the determination of water permeability in the vertical direction, the sample placed in the device, and water allowed to flow through the sample at a distance of 300±1 mm (Figure 3). To prevent leakage of water along the wall of the tube, the tube covered in a rubber membrane, and a pressure of 50 kPa applied. At least 10 minutes of water flowing through the sample was required to saturate the specimen and to remove enclosed air. After that, an empty vessel placed, and water allowed to flow for 1 minute. After 1 minute, the water entering the vessel was determined, and the water permeability calculated in the vertical direction.

To determine the horizontal water permeability, the sample placed in the device, and water allowed to flow through the sample at a distance of 300±1 mm. To prevent leakage of water, the bottom of the sample insulated with paraffin, and an aluminium ring placed on the top of the sample and glued with silicone glue.

1. 2. 3. 4. 5.

Figure 3. Vertical water permeability method sequence of work (1 – beginning, 5 – end)

Template: Anonymized Document

Drainability

The drainability of asphalt mixtures is determined by a standardized method according to EN 12697-40 . For the method, 4 cm thick slabs were prepared according to the standard EN 12697-33 + A1. The device used to determine the drainability of the asphalt mix is shown in Figure 3. A device with a load and a leveling surface is placed on the plate. The tube was filled with 5 litres of water and waiedt until no air bubbles remain in the water. When the plunger is pulled out in the vertical direction, the time taken for 4 liters of water to flow is calculated. The test is performed 3 times and the time is averaged.

1. 2. 3. 4. 5. 6.

Figure 4. Drainability determination method sequence of work (1 – beginning, 6 – end)

Skid resistance

Determination of skid resistance of asphalt mixtures was performed according to standard EN 13036-4. Skid resistance was determined by pendulum on asphalt slabs which were made in the laboratory. Skid resistance was performed for each asphalt mixture - a total of 12 different specimens (Table 1). The measurement were performed with a pendulum and a type 57 slider. This type of slider was used for asphalt pavements intended for motor vehicle traffic. The asphalt samples were moistened before each test. The value of the slip resistance (PTV) was determined for each sample 5 times, from which the average value of the PTV coefficient was calculated.

Macrotexture

The macrotexture of each asphalt sample surfaces was evaluated based on the MPD (mean profile depth) parameter obtained by CTM (circular tester) measurements in accordance with standard EN ISO 13473-1. Measurement of MPD under laboratory conditions was performed on prepared asphalt slabs. Measurements were performed for each asphalt mixture and 5 times.

Texture and permeability characteristics of aquaplaning asphalt mixture prototypes

The test results for the void content, permeability and drainage properties of the four asphalt mixtures are presented in Table 4.

The analysis of tests results shows that void content varies from 16.3% to 29.5%, and the courser the mixture, the higher void content as expected. The lowest void content (16.3%) was determined for asphalt AT 5 mixture with increased amount of AMLP, while normal composition asphalt AT 16 mixture had the highest void content (29.5%). Comparing each type of asphalt mixture and their composition, it was found that the use of increased amount of AMLP reduces void content by about 12.3% (ratio) comparing to normal composition asphalt mixture AT, and the use of AMLP plus granite fine aggregates reduces void content by about 12.4% (ratio) comparing to normal composition asphalt mixture AT.

Horizontal and vertical water permeability test results shows that permeability values of normal composition asphalt mixtures AT are about 2 times higher than asphalt mixture AT with increased amount of filler aggregate (see Table 2). It was found that vertical water permeability of asphalt mixtures AT 5 and AT 8 varies from 0.101×10^{-3} m/s to 0.692×10^{-3} m/s, while values of asphalt mixtures AT 11 and AT 16 are from 1.130×10^{-3} m/s to 2.762×10^{-3} m/s. Or vertical water permeability of asphalt mixture AT 11 and AT 16 are from 4 to 11 times higher than those of AT 5 and AT 8. Such results are due to the increased amount of filler aggregate, which reduce voids content in the asphalt mixture. It was found that the higher content of voids, the higher the vertical and horizontal water permeability. Comparing the dependence of the vertical water permeability results on the maximum aggregate particle size of the asphalt mixture it was found that vertical

Template: Anonymized Document

water permeability values between asphalt mixtures AT 5 and AT 8, or AT 8 and AT 11 differ almost 3 times, and between AT 11 and AT 16 only 22%. The analysis of horizontal water permeability test results shows similar tendencies. The horizontal water permeability of asphalt mixtures AT 5 and AT 8 varies from 0.268×10^{-3} m/s to 1.173×10^{-3} m/s, while values of asphalt mixtures AT 11 and AT 16 are from 1.142×10^{-3} m/s to 2.730×10^{-3} m/s. The use of increased amount of AMLP reduces horizontal water permeability by twice comparing to normal composition asphalt mixture AT. The use of AMLP plus granite fine aggregates reduces horizontal water permeability by 2.4 times for asphalt mixtures AT 5 and AT 11, while for asphalt mixtures AT 8 and AT 16 by 1.2 times comparing to normal composition asphalt mixtures AT. It was found that horizontal water permeability values of asphalt mixtures AT 5 are 64% lower than that of AT 8, but values of AT 8 are 150% lower than that of AT 11, and the difference between values of AT 11 and AT 16 is insignificant. Thus, some values got in these water permeability tests appear to be sufficient ($1.130–2.762\times10^{-3}$ m/s), but lower than those found by Haselbach *et al.* (2017) ($4.3-12.7\times10^{-3}$ m/s) for porous asphalt.

Drainability test showed that water drainage values correlate similarly with water permeability in the horizontal and vertical directions (see Table 2). Drainage values varies from 0.083 s^{-1} to 0.500 s^{-1}, and the courser the mixture, the higher drainability. The lowest drainability (0.083 s^{-1}) was determined for asphalt AT 5 mixture with increased amount of AMLP, while asphalt AT 16 mixture with increased amount of AMLP showed the highest drainability (0.500 s^{-1}). The use of increased amount of AMLP reduces horizontal water permeability by 50% comparing to normal composition asphalt mixture AT, except asphalt mixture AT 16, where the use of increased amount of AMLP almost does not effected to reduction of drainability. The use of AMLP plus granite fine aggregates reduces drainability by 47% comparing to normal composition asphalt mixture AT, except asphalt mixture AT 5, where the use of AMLP plus granite fine aggregates increased drainability by 11%. There was not found the dependence of the drainability on the maximum aggregate particle size of the asphalt mixture.

The analysis of the average profile depth showed that the depth of the profile depends on the maximum aggregate particle size of the asphalt mixture. The larger aggregates are used for asphalt mixture, the greater the profile depth value. The analysis of profile depth results shows that mean profile depth increases by 33% using larger aggregates. The lowest mean profile depth (0.78–0.95 mm) was determined for asphalt AT 5 mixtures, and the highest (about 2.00 mm) for asphalt AT 16 mixtures.

Table 4. Results for the the air voids content and water conductivity

Asphalt mixture	Mix code	Maximum density (average). kg/m^3	Bulk density (average). kg/m^3	Air voids (average). %	Vertical water permeability. K_v(m/s) x10^{-3}	Horizontal water permeability. K_h (m/s) x10^{-3}	Drainage (HC). s^{-1}
AT 5	5-1	2.036	2.525	19.4	0.239	0.663	0.110
	5-2	2.100	2.509	16.3	0.103	0.344	0.083
	5-3	2.073	2.522	17.8	0.101	0.268	0.122
AT 8	8-1	1.974	2.575	23.3	0.692	1.173	0.326
	8-2	2.036	2.561	20.5	0.358	0.521	0.110
	8-3	2.054	2.567	20.0	0.407	1.006	0.149
AT 11	11-1	1.910	2.626	27.3	2.608	2.657	0.492
	11-2	1.967	2.618	24.9	1.130	1.434	0.197
	11-3	1.995	2.601	23.3	1.257	1.142	0.211
AT 16	16-1	1.844	2.616	29.5	2.762	2.730	0.484
	16-2	1.931	2.601	25.8	1.493	1.413	0.500
	16-3	1.940	2.616	25.8	1.620	2.231	0.337

The mean values from the measurements of the Pendulum test value (PTV) are presented in Figure 5. The PTV is used as a surrogate for microtexture. The slip speed of the British Pendulum Tester is very low and equals approximately 10 km/h. Therefore, the PTV is mainly dependent on microtexture (J.J. Henry, 2000). Microtexture is having direct influence to the friction between the tyre and pavement surface and with that to the traffic safety. Macrotexture is assuring waterless contact between tyre and pavement surface and decreasing aquaplaning danger but at the same time increasing tyre noise. The analysis of skid resistance measurement results shows that PTV varies from 68 to 83. These values are high as PTV for newly constructed asphalt pavement should be not less than 55.

Template: Anonymized Document

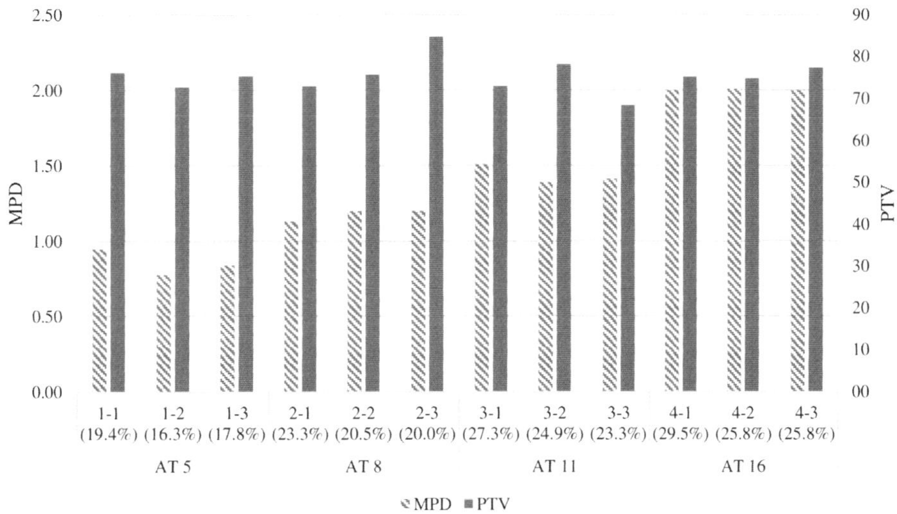

Figure 5.Dependency of MPD and skid resistance of AT asphalt mixtures

Conclusions

1. Vertical water permeability of asphalt mixtures AT 5 and AT 8 varies from 0.101×10^{-3} m/s to 0.692×10^{-3} m/s, while values of asphalt mixtures AT 11 and AT 16 varies from 1.130×10^{-3} m/s to 2.762×10^{-3} m/s. It means that vertical water permeability of asphalt mixture AT 11 and AT 16 are from 4 to 11 times higher than those of AT 5 and AT 8. However all tested asphalt mixtures meet the hydrological conditions of Lithuania, where it is recommended that the value of the vertical water permeability should be at least 0.08×10^{-3} m/s.
2. The analysis of vertical and horizontal water permeability results showed that permeability depends on aggregate particle size of asphalt mixture. Vertical water permeability values between asphalt mixtures AT 5 and AT 8, or AT 8 and AT 11 differ almost 3 times, and between AT 11 and AT 16 only 22%. It was found that horizontal water permeability values of asphalt mixtures AT 5 are 64% lower than that of AT 8, but values of AT 8 are 150% lower than that of AT 11, and the
3. The lowest drainability (0.083 s^{-1}) was determined for asphalt AT 5 mixture with increased amount of AMLP, while asphalt AT 16 mixture with increased amount of AMLP showed the highest drainability (0.500 s-1). The use of increased amount of filler aggregate reduces horizontal water permeability by about 48% comparing to normal composition asphalt mixture AT, except asphalt mixtures AT 5 and AT 16, where the use of increased amount of filler aggregate almost does not effected to reduction of drainability.
4. The analysis of profile depth results shows that mean profile depth increases by 33% using larger aggregates for asphalt mixture. The lowest profile depth was obtained for asphalt AT 5 mixtures (average MPD of three different composition asphalt mixtures is 0.86) and the highest results of profile depth were obtained for asphalt AT 16 mixtures (average MPD of three different composition asphalt mixtures is 2.00).
5. The analysis of skid resistance measurement results shows that surface of specimens prepared from tested asphalt mixtures is significant rough. Pendulum test values varies from 68 to 83. For newly constructed asphalt pavement PTV value should be not less than 55.
6. The purpose of porous asphalt mixtures is to drain the water on the surface of the pavement and to reduce the noise generated by the contact of vehicle tires with asphalt pavement surface. In addition, in order to investigate the applicability of these asphalt mixtures in cold climates, additional tests for the determination of frost resistance, stiffness, resistance to permanent deformations (ruts) and abrasion resistance should be performed.

References

Ahmed. F. 2015. Porous Asphalt Clogging Performance Under Swedish Conditions Porous Asphalt Clogging Performance Under Swedish Conditions.
Amirjani. M.. 2010. Clogging of permeable pavements in semi-arid areas. Delft University of Technology.

Template: Anonymized Document

Bean. E. Z.; Hunt. W. F. 2005 NWQEP Notes: NC State University permeable pavement research: water quality. water quantity. and clogging. No. 119.

Bicknel. J. 2016. Introduction to Pervious Pavement Systems. Santa Clara Valley Urban Runoff Program. SCVURPPP C.3 Workshop .

Brattebo. B. O.; Booth. D. B. 2003. Long-term stormwater quantity and quality performance of permeable pavement systems. Water research 37(18): 4369.

Haselbach. L.. Lurie. C.. Ung. C.. Ballou. L.. Costa. M.. Peshkin. D.. Eisenberg. B. 2017. Guidance for Usage of Permeable Pavement at Airports. Guidance for Usage of Permeable Pavement at Airports. https://doi.org/10.17226/24852

Henry. J.J.. 2000. Evaluation of pavement friction characteristics (Vol. 291). Transportation Research Board.

Herrmann, S.R. 2008. Simulationsmodell zum wasserabfluss-und aquaplaning-verhalten auf fahrbahnoberflächen.

Ivan. J. N.. Ravishanker. N.. Jackson. E.. Aronov. B.. Guo. S. 2012 A statistical analysis of the effect of wet-pavement friction on highway traffic safety. Journal of Transportation Safety & Security. 4(2). 116-136.

Jacobson. T.. Sandberg. U. and Viman. L.. 2016. June. How do we improve the durability of porous asphalt. In Proceedings of the 6th Eurasphalt Eurobitume Congress. Prague. Czech Republic (pp. 1-3).

Mayora. P. J. M.. Pina. J. R. 2009. An assessment of the skid resistance effect on traffic safety under wet-pavement conditions. Accident Analysis and Prevention. 41(4). 881-886.

McGovern. C.. Rusch. P.. Noyce. D. A. 2011 State Practices to Reduce Wet Weather Skidding Crashes. Publication FHWA-SA-11-21. federal Highway Administration. U.S. Department of Transportation. Washington D. C.

Šernas, O., Zofka, A., Vaitkus, A., Gražulytė, J.2020. The Effect of Exposed Aggregate Concrete Gradation On The Texture Characteristics And Durability, Construction and Building Materials, Volume 261, 119921, ISSN 0950-0618, https://doi.org/10.1016/j.conbuildmat.2020.119921 .

Vaitkus, A., Andriejauskas, T., Vorobjovas, V., Jagniatinskis, A., Fiks, B., Zofka, E. 2017b. Asphalt wearing course optimization for road traffic noise reduction, Construction and Building Materials, Volume 152, Pages 345-356, ISSN 0950-0618, https://doi.org/10.1016/j.conbuildmat.2017.06.130

Vaitkus, A., Andriejauskas, T., Vorobjovas, V., Šernas, O.,Skrodenis, D. 2019. February. Acoustical ageing of low noise pavements expressed by different measuring techniques. In 5th International Conference on Road and Rail Infrastructure.

Vaitkus, A., Šernas, O., Vorobjovas, V., Gražulytė, J.2019. Selection of Constituent Materials For Asphalt Mixtures of Noise-Reducing Asphalt Pavements. The Baltic Journal of Road And Bridge Engineering 2019/14(2) ISSN 1822-427X/eISSN 1822-42882019 Volume 14 Issue 2: 178–207 https://doi.org/10.7250/bjrbe.2019-14.439

Vaitkus, A., Vorobjovas, V., Kleizienė, R., Šernas, O., Gražulytė, J. 2017a. Modified asphalt mixtures for heavy duty pavement wearing layers, Construction and Building Materials, Volume 131, Pages 503-511, ISSN 0950-0618, https://doi.org/10.1016/j.conbuildmat.2016.11.055.

29: Collazos-Arias

ADAPTATION AND RESILIENCE FROM A MAINTENANCE PERSPECTIVE FOR SWING BRIDGE. LESSONS LEARNED IN RECENT RETROFITTING PROJECT EXPERIENCES

Felipe COLLAZOS-ARIAS ORCID iD (https://orcid.org/0000-0002-9536-5752)[1], David GARCÍA-SÁNCHEZ ORCID iD (https://orcid.org/0000-0002-4763-206X)[2]

[1]*Department of Transports and Technology of Projects and Processes, University of Cantabria, Santander, Spain*
[2] *TECNALIA, Basque Research and Technology Alliance (BRTA), Derio, Spain*
Corresponding author: Felipe, Collazos-Arias, felipecollazos.fc@gmail.com

Received 30 September 2020; accepted 11 January 2021

Abstract. This paper collects all the existing swing bridges around the world to emphasize the high-level technology performed in the earliest 20th century and analyzes the recent experience in the rehabilitation and retrofitting of a singular swing steel bridge. The bridge over the Asón estuary at Treto, in the North of Spain, showed significant problems and the structural condition level was critical previously to its intervention in 2015. The rehabilitation project including the restoration of the structural and functional safety level of the structure was deled after material studies and tests were performed. The new updated of the structure from the old nineteenth centuries to the new standard codes was also performed. A part from the historical value of the bridges and the "green" and safe-security aspects, the intervention took into account all the social requirements of the population in the area who recovered the pride in this emblematic and centenarian infrastructure and it is, itself, a successful intervention from the resilient point of view.

Keywords: Rehabilitation, Swing Bridge, Heritage, Repair, Reinforcement

Introduction

A swing bridge is a movable bridge that has as its primary structural support a vertical locating pin and support ring, usually at or near to its center of gravity, about which the turning span can then pivot horizontally.

In its closed position, a swing bridge carrying a road or railway over a river or canal, for example, allows traffic to cross. When a water vessel needs to pass the bridge, road traffic is stopped (usually by traffic signals and barriers), and then motors rotate the bridge horizontally about its pivot point. The typical swing bridge will rotate approximately 90 degrees, or one-quarter turn; however, a bridge which intersects the navigation channel at an oblique angle may be built to rotate only 45 degrees, or one-eighth turn, in order to clear the channel.

This bridge typology is not very common nowadays, but it has an enormous historical and social value because it really solved a complex traffic problem in a innovative way in the earliest XXth Century.

Finally, some thoughts about adaptation and resilience from a maintenance perspective have been included.

1. Swing bridges around Europe

1.1. Netherlands and Belgium

Almost European swing bridges identified were built at the end of XIX Century. They have short spans (around 20 meters) and narrow width (less than 5 meters). Most of them are located in the Netherlands, Figure 1 until Figure 10, and Belgium, Figure 11 and figure 12.

29: Collazos-Arias

Figure 1 - Swing bridge in Afsluitdijk (Netherlands).

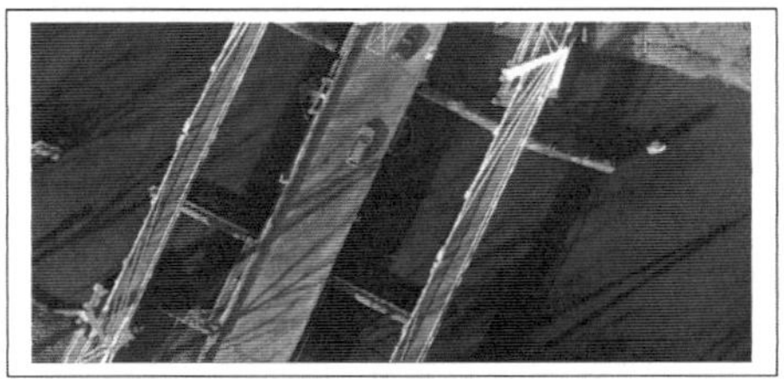

Figure 2 - Gerrit Krol bridge (Netherlands)

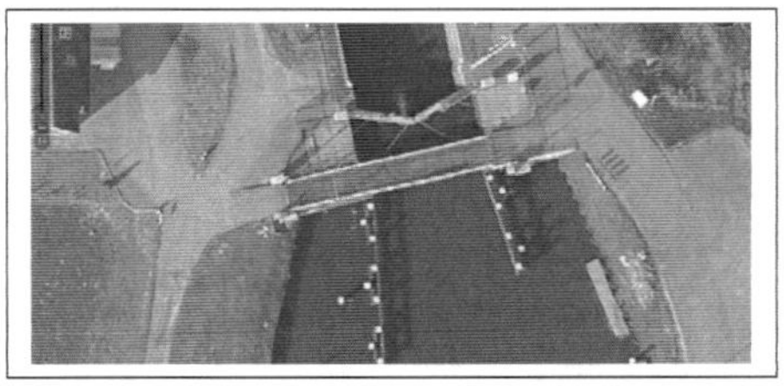

Figure 3 - Groevel bridge (Netherlands)

Figure 4 - Small lock bridge (Netherlands)

Figure 5 - Oisterwijksebaan bridge (Netherlands)

29: Collazos-Arias

Figure 6 - Gent bridge (Netherlands)

Figure 7 - Skulenboarch (Netherlands)

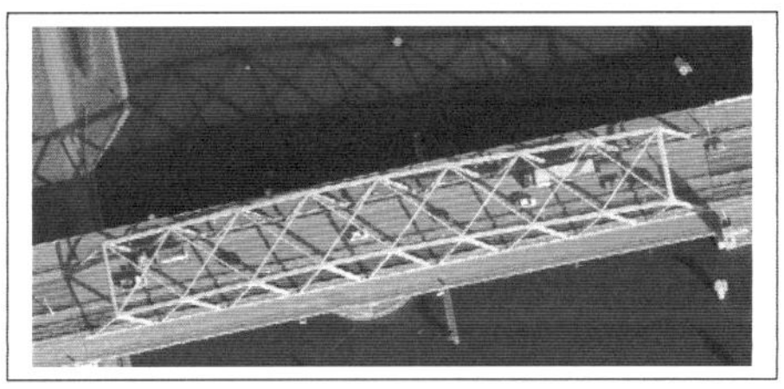

Figure 8 - Sluiskil (Netherlands)

Figure 9 - Stroobos (Netherlands)

Figure 10 - Zuidersluis (Netherlands)

29: Collazos-Arias

Figure 11 - Houdeng Aimeries (Belgium)

Figure 12 - Strepy Bracquegnies (Belgium)

2. Swing bridges around the world

2.1. Asia

Three bridges were identified in Japan: Shouten Bridge, Nishiki Bridge and Wada-Senkai Bridge in Japan. These three bridges are in service now.

Shouten Bridge, Figure13, is used for mainly pedestrians. This bridge was built in 1923 and was operated manually. In 1957, the bridge was replaced by a new bridge which was operated electrically.

Since the area near this bridge is a tourist destination, this bridge rotates although no ships go through this bridge.

Figure 13 - Shouten Bridge (Japan)

Nishiki Bridge, Figure 14, is completed in 1968. This bridge is currently in service for traffic including cars, bikes, and pedestrians.

29: Collazos-Arias

Figure 14 - Nishiki Bridge (Japan)

Wada-Senkai Bridge, Figure 15, was built in 1900. This bridge was used for a railway. It was rotated but not rotated right now. The bridge is currently used but fixed at the rotating pier.

Figure 15. Wada-Senkai Bridge (Japan)

2.2. USA

The National Bridge Inventory (NBI) contains summary data for 194 swing bridges in the U.S (see table below Table 1). This count is only for highway bridges so it does not include rail bridges of this type. Nine of these bridges were originally built before 1900 with the earliest being built in 1850. Of the 194, nearly half are located within Louisiana which is a State that contains a significant number of waterways due to the Mississippi River and delta, Figure 16.

Figure 16 - Frederick Douglass Memorial Bridge (USA)

Table 1. Movable-swing road bridges in USA (source: Author, year)

State	Movable-Swing
CALIFORNIA	16
CONNECTICUT	4
DELAWARE	2
DIST. OF COL.	1
FLORIDA	9
GEORGIA	1

29: Collazos-Arias

ILLINOIS	2
IOWA	2
LOUISIANA	83
MAINE	4
MARYLAND	3
MASSACHUSETTS	6
MICHIGAN	3
MINNESOTA	2
MISSISSIPPI	3
NEW JERSEY	8
NEW YORK	12
NORTH CAROLINA	7
OHIO	1
OREGON	3
RHODE ISLAND	1
SOUTH CAROLINA	6
TEXAS	5
VIRGINIA	5
WASHINGTON	5
TOTALS	194

2.3. South America

A total of seven companies were presented in 1905 for its construction, Carmelo Bridge, being awarded to the Society "Fábricas Unidas de Augsburg y Nürnberg", Figure 17.

Figure 17 – Carmelo Bridge (Colombia)

On the afternoon of December 2, 1911, in the midst of a great popular expectation, the test was made to make the bridge work. The official inauguration was held on May 1, 1912, with various celebrations that lasted for two days.

The years passed, the Rotating Bridge receives the daily passage of more than 10,000 people, and 26,000 on Sundays.

29: Collazos-Arias

3. Intervention experiences: Treto bridge (Spain)

3.1. Historical, Administrative and Technical background

Treto Bridge on the N-634 on its way through Colindres (Cantabria) was put into service in 1905 and, since then, it is an emblematic infrastructure of high traffic and pedestrian intensity. In 2016 his rehabilitation was faced. This rehabilitation required the development of very specific works in an environment of great natural and landscape singularity.

The structural non-uniformity of the bridge and the heterogeneity of the materials conditioned a rehabilitation clearly differentiated by the typology of the units and materials. This differentiated rehabilitation required the development of very specific works in an extremely worthy environment. Some innovative actions to mitigate the different sources of risks during the works on site are also explained. One of the most singular mitigation actions was the placement of a movable scaffold wearing an ad hoc encapsulation system in order to comply with the condition of no discharge to the Ason estuary of any sand blasting treatment for the cleaning of the rust on the entire surface of the elements.

Although the construction of Treto Bridge (Cantabria, Spain) is framed between 1893 and 1905, it was designed to satisfy a previous long term demand. Indeed, it was framed in a territorial conflict by the preponderance of an old trade route that communicated the fishermen towns in the North of Spain and that is why it was call to be a nexus of union between the most important provinces (Cantabria, Basque Country and Castilla y León).

Thus, the project of the puddled iron bridge with two isostatic spans (arches unit) and two continuous spans (swing unit) is approved in April 1890.

Most of the work for the fixed spans was carried out in Spanish workshops and those corresponding to the swing section by the Angler-Tilleur workshops from Liège (Belgium) Figure 18.

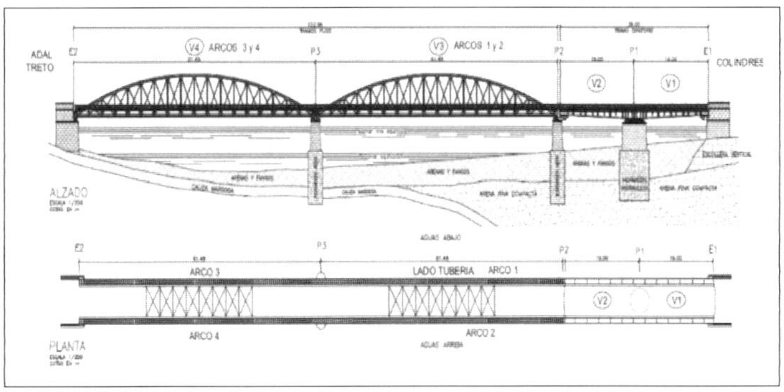

Figure 18. Elevation and plan view. Treto Bridge.

3.2. Bridge Description

Treto Bridge on the N-634 on its way through Colindres (Cantabria) was put into service in 1905 and, since then, it is an emblematic infrastructure of high traffic and pedestrian intensity. In 2016 his rehabilitation was faced. This rehabilitation required the development of very specific works in an environment of great natural and landscape singularity.

It is a four-span puddled iron bridge composed of first unit, with lower bow-string decks, and a continuous swing unit, very different structurally from the previous one. This superstructure is supported on piers and abutments made of masonry and hydraulic concrete cemented founded on the estuary.

The main structural elements of first unit are the top chords that basically are arches with a variable height along the spans according to a parabolic law of second degree that leads to the maximum use of the material. The arches are connected with transversal trusses limiting the vertical clearance. Thus, each fixed span defines a closed but permeable space through which the traffic circulates. The substructure is formed by abutments and masonry piers, two with an oval cross-section and a third one with a circular one on which the swing unit connected with the deck thanks to a wheel mechanism made of an improved iron comparing with the arches unit, close to the un-existing steel at this time. Originally and until the year 1940, the swing unit could rotate on the wheel gear housed in the circular pile and be arranged perpendicular to the road, allowing the passage of boats, with a height of 3.37 m above the level of PMVE, measured from the underside of the board in fixed unit.

The swing was possible because the unit was based on a circular pile of 6 m in diameter with a vertical axis of wrought puddled iron embedded in the center. To verify the movement, twelve conical casting wheels were arranged, on which the metal section was fixed by means of a circular wrought puddled iron disk of 2.40 m outside radius Figure 19.

The pavement of the swing unit was designed and made of timber. In the other unit it was a bituminous pavement.

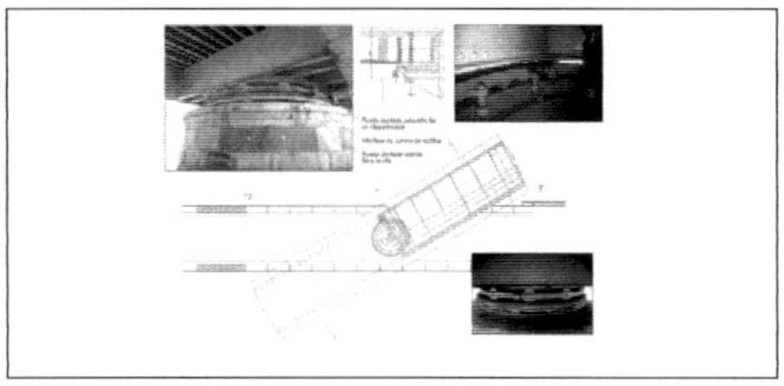

Figure 19. Mechanism for the swing span.

All structural elements are connected by rivets. According to the documentation related to the successive projects that gave rise to the Treto Bridge, the material used in it is wrought iron or puddled, 92.16%, which is one of the iron-derived materials that have been used successively in the construction of bridges, after the cast iron and before the steel.

The construction works of the bridge began in 1894, and it was inaugurated in 1905. In 1968, an intervention project was carried out as an emergency and in 1971. Since then, ordinary maintenance tasks have been carried out exclusively.

3.3. Inspections from the construction to 1960

There exist three reports that identified signs of corrosion, consequence of the insufficient protection of the metallic surfaces. They were solved with chopping and scraping of the rusted parts, and repainted with a metallic paint based on silver glitter. In these reports is also described the blocking of the swing span, which had not been revolved since the first years of the post Spanish Civil war period.

3.4. Inspection in 1995

The inspection works carried out in 1965 analyzed for the first time the conditions of resistance and stability and looked for a solution to reactivate the turn of the swing section, although finally it was blocked.

In order to know the mechanical behavior of the bridge, a load test was carried out, and it was observed that in all the sections the vertical displacements (3mm in the mobile and 12.5mm in the fixed ones) were recovered completely, from which it was deduced that the tensions were in the elastic range.

A chemical analysis of the steel of the bridge determined that the material was easily weld able by any procedure, and thus all the unions intervened, more than half, were repaired with this technique. Finally, in 1966 a project was drafted in which a large number of actions were collected, among which were , Table 2.

Table 2. List of relevant actions recommended

Relevant actions recommended in 1966
1.Union of the floor beams and bottom chords in the fixed units, which were in very bad conditions due to corrosion.
2.Stuffing of the bottom chords of a waterproofing material and lightweight to make them watertight, as they had insufficient drains and clogged. The proposed material was "termite " protected in its upper 5 cm with a lightweight concrete with aggregates.
3.The diagonals of the arches unit were not joined at the crossing points and they vibrated a lot when passing vehicles. To increase its stiffness, it was proposed to weld them to a sheet of 50x6 mm so that the cross section of each one was a T, and to join each pair of parallel diagonals together, in addition to welding them at the crossing point.

4.Repair of the metal formwork that constituted the resistant element of the forged of the parabolic sections. They were rusted due to seepage through the pavement and fill. Its repair was projected by sheet and welding and replacement of the filling by a rich concrete reinforced with a round mesh.
5.Replacement of timber sidewalks.

The consequences of this intervention has been observed and valued in the current intervention of 2016.

In 1986 the last rehabilitation project on the bridge was approved, which included only painting works.

In summary, the bridge built in 1905 did not receive any type of intervention until 1965. Twenty years after these works, 1986, it was repainted and no works were carried out on the bridge until the approval of the rehabilitation project in 2016.

3.5. Aim of the rehabilitation project in 2016.

The Division of State Highways and roads in Cantabria (Spain) promoted an emergency project in 2015 to retrofit, protect and structurally rehabilitate a bridge that, from its origins, has been fundamental for the socioeconomic development of the area. The condition in which the bridge was 30 years after its last intervention was severe enough to drive a comprehensive rehabilitation project that would return the bridge to levels of structural security and service suited to the needs of the 21st century and, at the same time, conserving its essence.

3.6. Inventory of most important damages

Corrosion, until the break, in many elements of the structure, highlighting the units between vertical and diagonal members, top and bottom chords, floor beams and bottom chords. It is a generalized damage caused by the deterioration and / or lack of protection, and it could be considered as a very risky damage because it endangers the safety of the pieces that are not easily accessible. Water accumulates easily, which intensifies corrosion, deteriorating the rivets and gussets and decreasing the strength of the union between structural elements Figure 20.

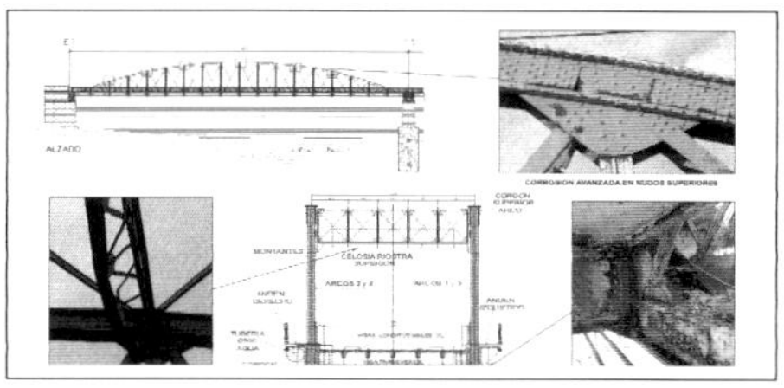

Figure 20. Corrosion damage location. General condition.

The most relevant damages are included in the Figure 20 and in Table 3.

Table 3. General list of relevant damages.

Relevant damages
1. Instability of the diagonals to the compression efforts.
2. The bearing capacity of the lost puddled iron formwork elements has been exceeded.
3. Loss of section in bottom chords.
4. Unexpected displacement of the retaining walls in their encounter with the abutments.
5. Scour in abutments.
6. Damages caused in diagonals, vertical members and by the impact of vehicles
7. Deformations in diagonals and vertical members and struts caused by vehicle impacts.
8. Damages, unexpected deformations and displacements in masonry parapets of abutments.
9. Leakage.

29: Collazos-Arias

3.7. Milestones during rehabilitation works

It can be found below the solutions undertaken to restore the level of structural and functional safety of the structure, defined after the diagnosis study and after analyzing the pathologies identified. The rehabilitation procedure is shown in Table 4.

Table 4. General list of rehabilitation procedures

General list of rehabilitation procedures
1. Blasting and cleaning of all truss members with sandblasting, grade Sa 2 ½. Initial primer of steel. Protection and greasing of bearing devices.
2. Calculation and evaluation of the remaining life of the bridge and definition of the retrofitting details to restore the resistant sections of steel, by means of the provision of new steel elements. Replacement of all lost, broken or loose rivets.
3. Replacement of the existing riveted unions by high resistance screws.
4. Reinforcement of all floor beam-bottom chord-vertical connections in fixed spans. The main connections of the bridge.
5. Reinforcement of all top chord-vertical connections in fixed units.
6. Rectification of diagonals in arches.
7. Dismantling of internal truss according to original version. Replacement of 18 portal struts and sway bracings in fixed units. Elimination of the existing subgrades (44 cm thick) on the original lost formwork supported by floor beams and stringers.
8. Complete removal of the pavement. Execution of reinforced concrete slabs connected to floor beams by bolts in fixed units. Execution of a waterproofing system on the new slabs. Cleaning and unblocking of the existing scuppers.
9. Construction of new scuppers to provide the deck greater drainage capacity.
10. Replacement of damaged verticals in fixed spans.
11. Replacement of expansion joints.
12. Stuffing of the bottom chords of a waterproofing material and lightweight to make them watertight, as they had insufficient drains and clogged
13. Retrofitting of stringers and floor beams in swing unit.
14. Anchoring of a new a traffic barrier system in the swing unit.
15. Widening of sidewalks, from 0.80 m to 1.90 m by means of metal brackets housing the water pipeline.
16. The original bridge railing was reproduced.
17. Replacing of the existing water pipeline with a diameter of 500 mm, by another with a diameter of 600 mm below the new sidewalks.
18. Protection of all the truss members of the bridge with a protective coating resistant to offshore conditions, (280 and 400 microns thick).

3.8. Rehabilitation of main connections

It has been considered as main connections those ones between bottom chord, vertical and floor-beam.

The rehabilitation of main connections consists in the replacement of old damaged gusset plates by new ones following an eight steps procedure Figure 21 and Figure 22.

29: Collazos-Arias

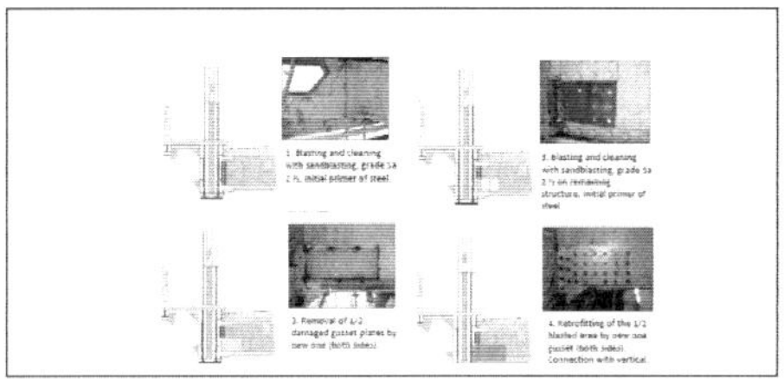

Figure 21. Rehabilitation of main connections.4 first steps

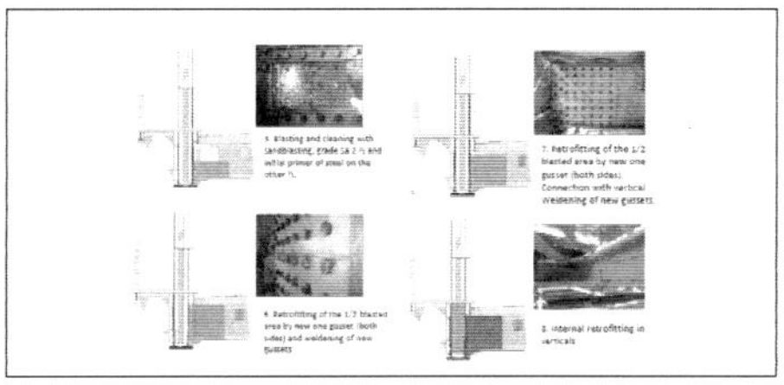

Figure 22. Rehabilitation of main connections.4 last steps.

4. Adaptation and resilience from a maintenance perspective

As presented before maintenance has not always taken the deserved attention even if it is a key factor on bridge performance. Currently, there is no a standardize protocol or procedure to solve maintenance problems and to increase infrastructure resilience. It is still a problem with solutions in development. Thus, the authors want to provide a standardized design tool to improve the design of new systems and remediation works on existing structures and elements. This tool is presented in Figure 23 in form of a flow chart. It summarizes the most important aspects of the maintenance procedure and the potential options with the aim of increase the resilience of the bridge as an infrastructure. It presents the different steps and aspects that should be followed to develop and design effective systems, adapted to new climate and operational conditions and therefore increase the infrastructure resilience. Two of the new aspects presented in this new design tool are: the resilience Key Performance Indicators (KPI) and the feedback from the continuous analysis. The main Key Performance Indicators (KPI) related with resilience that administrations should considered are the following five:

- Disruption
- Efficiency of use of the network
- Tracks with permanent speed restrictions
- Tracks with temporary speeds restrictions
- Maintenance costs

It is worth nothing that the design tool should be specific for each site as the soil properties, structural condition and administrations requirements are site-specific so there is a limitation on implementing successful solutions to different areas without the required analysis.

29: Collazos-Arias

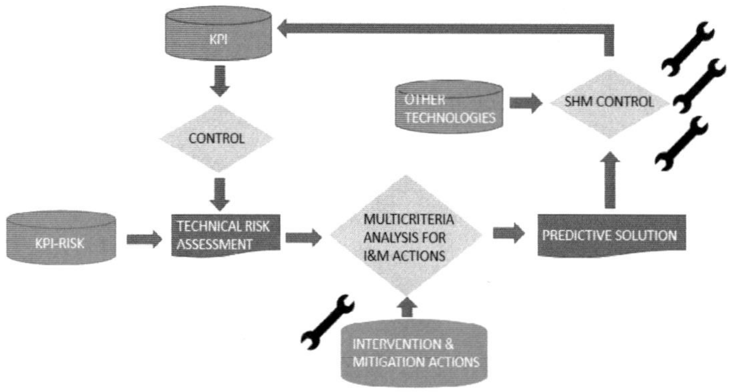

Figure 23. Maintenance strategy based on risk assessment and Key Performance Indicators (KPI).

Conclusions

On the one hand, all these bridges have in common that their construction were stream finally achieved what even the most remote dreams of its inhabitants had not devised: a bridge that, besides crossing it on its side, allows large-sized ships to openwork pass. In all the examples showed in this paper the dream became real and become part of the life of the citizens, who were so proud of "the work".

On the other one, the complete analysis of Treto Bridge shows that it is a very advanced technical and technological structure for its time. It is not coincidental that, despite the aggressive environment and the poor maintenance received since its construction, it has reached our days offering a fundamental service for the users of the road and its neighbors, and an iconic infrastructure for the region.

All works were performed in 10 months and traffic was disrupted just for 1 month.

Acknowledgements

Imai Kiyohiro, Pierre Guilles, Joey Hartmann, Gonzalo Arias Hoffman, Alvaro Navareño Rojo, Dick Schaafsma (PIARC Members), TC on Bridges D.3 members.

José Francisco Sánchez Cimiano (Director of the Project), Marcos Pantaleón Prieto (Head of the Project, Louis Berger International Design Center).

References

Ruiz Bedia, M. L. (2010). Hitos y Mitos de la Ingeniería Civil en Cantabria. El puente de Treto. Ponencia Universidad de Cantabria.
Sánchez Cimiano, J. F. Pantaleón Prieto, M. J. (2016). Proyecto Modificado de rehabilitación del puente metálico de 1901.
Sánchez Cimiano, J. F. Delgado, M. A. (2016). Rehabilitación integral del puente metálico entre Treto y Colindres, Ctra N-634.
García-Sánchez, D. Aurtenetxe, J. Zalbide, M. Socorro, R. Pérez-Hernando, A. Inaudi, D. "Technical management risks for transport infrastructures along whole life cycle: identification and intervention", CRC Press/Balkema–Taylor & Francis Group, 2018
Sañudo, R. Miranda, M. García, C. García-Sanchez, D. Drainage in Railways, Construction and Building Materials 210, 391-412, 2019

ANALYSIS OF QUALITY CONTROL PLANS FOR BRIDGE OVER THE RIVER MAZA JUGLA ACCORING TO COST TU 1406 METHODOLOGY

Roberts Auzins[1], Ainars Paeglitis[1]

[1]*Department of Roads and Bridges, Civil Engineering Faculty, Riga Technical University, Riga, Latvia*
Corresponding author: Roberts Auzins, email: roberts.auzins@rtu.lv

Abstract. Bridges are one of the most expensive elements of the road network, and therefore in the bridge management process, it is very important to make the most technically efficient and cost-effective decisions about planned actions such as maintenance, rehabilitation and reconstruction works. Decisions have to be based both on the current situation and possible future options and alternatives.
The European Cooperation in Science and Technology (COST) during the action TU 1406 "Quality specifications for roadway bridges, standardization at a European level (BridgeSpecs)" in the period from 2014 to 2019 has developed the framework for the development of bridge Quality Control Plans (QCP) including the system of data collection, data processing and outcomes.
This article analyses and compares different Quality Control Plans developed according to COST TU 1406 methodology for the existing bridge over the river Maza Jugla, located on regional road P10 at km 34.80 in Latvia.

Keywords: bridge, bridge stock, inventory, inspection, maintenance, condition, load model

1. Introduction

The European Cooperation in Science and Technology (COST) during the action TU 1406 "Quality specifications for roadway bridges, standardization at a European level (BridgeSpecs)" in the period from 2014 to 2019 has developed the framework for the development of bridge Quality Control Plans (QCP) including the system of data collection, data processing and outcomes.
Based on action workgroup technical reports and offered methodology two possible QCP for roadway bridge over the river Maza Jugla in Latvia are made. As the overall technical condition of the bridge is poor and it is planned to rehabilitate bridge in coming years, one of main outcomes of this article is to compare the strategy selected by State Limited Liability Company "Latvian State Roads" to rehabilitate bridge now versus the option "to do nothing" and reconstruct bridge when it becomes unusable. The preparation process of case study is done according to the scheme (Figure 3.1) described in WG4 Technical report "Preparation of case study" with the following main steps:

- Data collection of existing structure,
- Technical condition assessment and evaluation of performance indicators (PI) and key performance indicators (KPI),
- Material testing – carbonisation depth,
- Development of different QCP and their comparison.

Data used for case study is obtained from bridge inspection in 2020 and bridge management system LatBrutus of SLLC "Latvian State Roads".

2. General data of bridge

2.1 Basic information

The bridge over the river Maza Jugla is located in the Latvian regional road network on regional road P10 Incukalns-Ropazi-Ikskile at kilometre 34.80 (see Figure 1 and Figure 2). As the bridge is located in the urban area Tinuzi, the allowed driving speed is 50 km/h. According to last traffic data from 2019 the traffic intensity is 1677 cars per day including 11% of heavy traffic. According to data of Latvian bridge management system LatBrutus the bridge condition is evaluated as "Very poor condition".

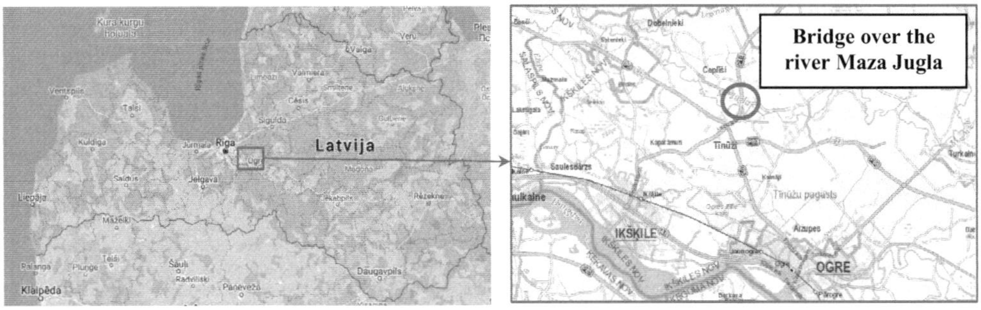

Figure 1. Location of the bridge

Figure 2. Bridge overview

2.2 Bridge structural elements and equipment

New bridge superstructure over the river Maza Jugla was built in 1962 on the existing cast-in-situ reinforced concrete abutments built in 1948. Bridge superstructure is made of four continuous reinforced concrete cast-in-situ cantilever beams with scheme 7.65+20.30+7.65m. Main beams are connected with cast-in-situ cross beams. The bridge is designed according to Soviet traffic loads N-13 and special vehicle loads NG-60. The original bridge design documentation is shown in Figure 3. In cross section the superstructure has two pedestrian sidewalks with corresponding width of 0.85m and carriageway with the total width of two lanes of 7.0 m.

Bridge substructure was constructed in 1948. The piers are massive cast-in-situ reinforced concrete supports based on solid rock. Main dimensions of piers are 9.61 x 1.22 metres. One pier has rigid steel bearings, but the another - reinforced concrete roller bearings. In the upstream side there are ice cutters in the form of a triangle.

The bridge is equipped with two reinforced concrete sidewalks, stone masonry slope covering, steel railings and steel drainage tubes. There are no special deformation joints as the bridge is designed with semi-integral superstructure. In both ends of superstructure there are saw cuts in asphalt pavement. Road accesses and superstructure itself is paved with asphalt. The bridge is equipped with galvanized steel road signs.

2.3 Load capacity

Bridge load bearing capacity is recalculated due to bridge special inspection caried out in 2020. Bridge bearing capacity is appropriate to load model of everyday traffic flow LM3 (52 tons) used in Latvia. The summary of bearing capacity recalculation results is shown in Table 1.

Table 1. Load bearing capacity calculation summary

Most laoded beam	Section	Force	Characteristic					Design	Section resistance	R_d/E_d
			Dead weight				Live Load			
			G1	G2	G3	$G_{k.sum}$	LM3 (52t)	Max. Design value $G_d.+Q_d.$	R_d	
	1-1	M, km*m	345,9	268,6	193,1	807,6	973,7	2318	2330	1,01
	3-3	M, km*m	392,9	450,3	359	1202,2	937,2	2790	3025	1,08
	1-1	V, KN	110	141	112,5	363,5	163	683	800	1,17
	3-3	V, KN	150,5	160,5	126,3	437,3	161	770	947	1,23

IOP Conf. Series: Materials Science and Engineering 1202 (2021) 012028 doi:10.1088/1757-899X/1202/1/012028

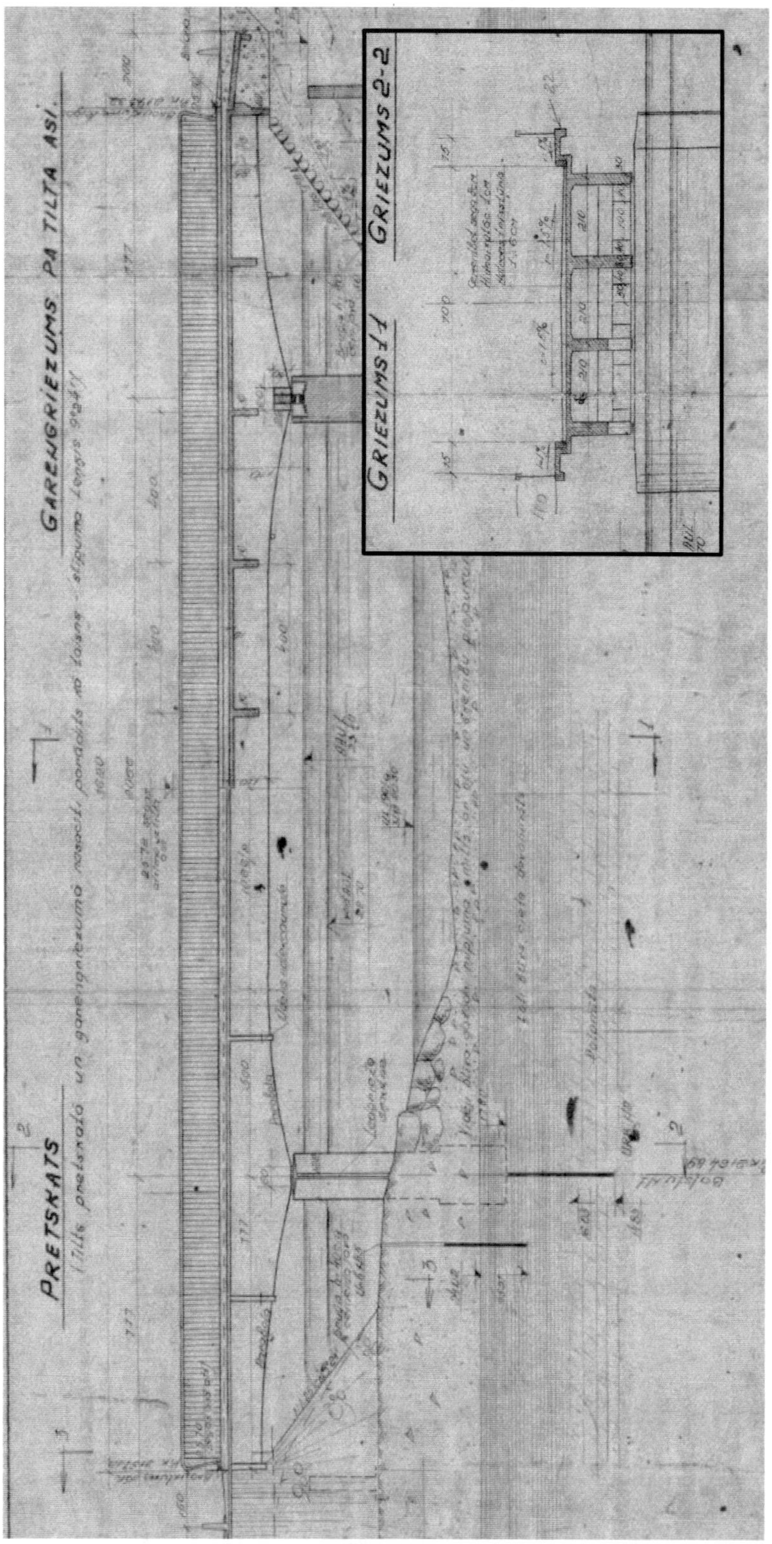

Figure 3. Original drawing of bridge over the river Maza Jugla

3. Technical condition

Technical condition of the bridge is evaluated based on existing bridge inspection data and visual inspection done by trained engineers. The results of technical condition data are processed according to WG3 framework ontology as shown in Figure 4 and summarized in Table 2 according to WG3 offered methodology. Damage process and performance indicators are selected from WG3 report, but derivation of KPI's is done based on engineering judgement. Observed defects of structural and non-structural elements are presented and described in Figure 5.

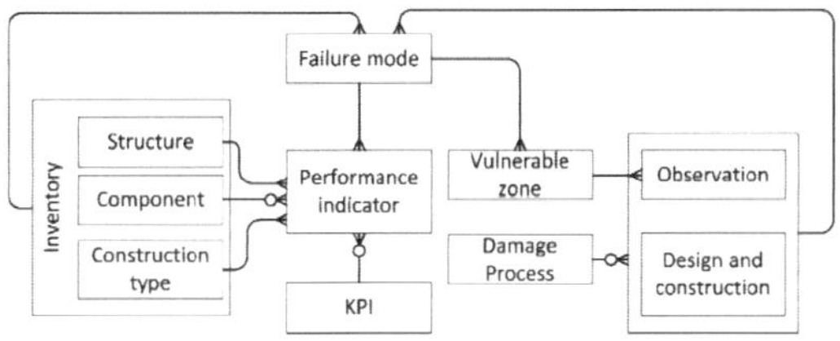

Figure 4. Framework ontology (WG3 report) [3]

In Table 2 damage processes are related to vulnerable bridge zones, as well. Vulnerable bridge zones are defined according to WG3 offered methodology. Vulnerable zones are presented in Figure 5.

Labels:

Orange circle – High moment zones (sagging and hogging),

Red rectangle – High shear region,

Blues diamonds – High compression zone, supporting zone

Figure 5. Vulnerable zones

Fig.6-1. Pavement, sidewalks and railings

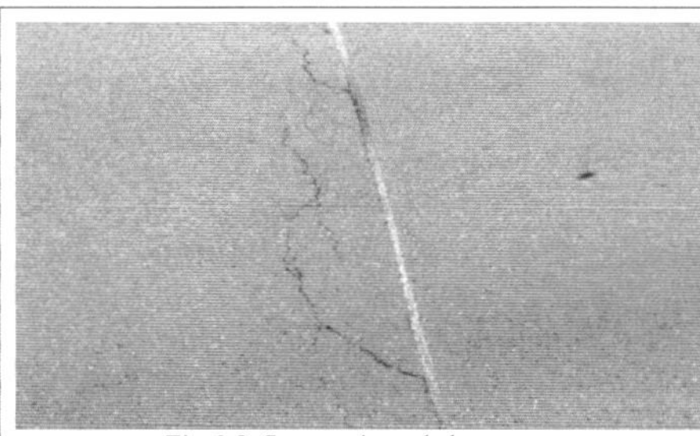

Fig.6-2. Saw cut in asphalt pavement

Fig.6-3. Middle span

Fig.6-4. Pier overview

Fig.6-5. RC plate between main beams

Fig.6-6. Integral end abutment

Fig.6-7. Main beam side surface

Fig.6-8. Main beam bearing area

Fig.6-9. Drainage tubes

Fig.6-10. Abutment top part

Figure 6. Overview of bridge main defects

Table 2. Overview of bridge main defects and vulnerable zones

Structure	Group	Element / Material	Observations			Damage process	Vulnerable zone	Failure mode	Primary KPI	Evaluation	KPI value	
			Damage	Closer specifications								
				QYT	Location						R	S
Coninious multiple span semi-integral beams	Structural elements	Main Beams / RC	concrete spalling, reinforcement corrosion	15%	side beams, outer side surface	corrosion, deterioration	HS, HMS, HMH	Bending and shear failure mode	Reliability (R)	2	4	4
			concrete spalling, reinforcement corrosion	100%	side beams, next to drainage pipes	corrosion, leakage	HS, HMS, HMH	Bending and shear failure mode	Reliability (R)	4		
			longitudinal cracks, concrete delamination	50%	side beams at supports	corrosion, deterioration, freeze-thaw	Bearing area	Shear failure mode	Reliability (R)	3		
			concrete spalling, reinforcement corrosion	15%	plate between main beams	corrosion, deterioration	collaboration	Local bending failure mode	Reliability (R)	2		
			micro cracks, concrete spalling	10%	all beam surface	corrosion, deterioration	-	Bending and shear failure mode	Reliability (R)	2		
		Cross beams / RC	small cracks, concrete spalling	10%	all beam surface	corrosion, deterioration	collaboration	-	Reliability (R)	1		
		Water proofing / bitumen layer	Efflarosence	90%	all below sidewalks	leakage	-	-	symptom	4		
		Bearings / RC roller	to much deformations	100%	all bearings	incorrect instalation	Bearing area	Compresion failure mode	Reliability (R)	2		
			small cracks, concrete spalling	10%	all bearings	corrosion, deterioration	Bearing area	Compresion failure mode	Reliability (R)	2		
		Integral end abutment / RC	small cracks, concrete spalling, reinf. corrosion	50%	bottom part	corrosion, deterioration, humidity	-		Reliability (R)	2		
		Abutment / RC	micro cracks, concrete spalling	50%	all abutment surface	corrosion, deterioration	High compression area	Compresion failure mode	Reliability (R)	2		
			grafiti	50%	long side surfaces	-	High compression area	-	Reliability (R)	1		
			Efflarosence	50%	short side top part	leakage	High compression area	Compresion failure mode	Reliability (R)	2		
	Equipment	Expansion joint (saw cut in asphalt) / bitumen	cracks next to cut, delamination	40%	all lenght	weak foundation, aging of material	-	driving comfort	Safety (S)	2		
		Drainage system / steel tubes	corrosion, loss of section, too short pipes	100%	all pipes	corrosion, inappropriate design solution	HMS	Bending failure mode	Safety (S) / Reliability (R)	3 /3		
		Overlay / asphalt pavement	ceacks	10%	near saw cuts, near center line	aging of material	-	driving comfort	Safety (S)	1		
		Railings / painted steel	flaking, deformations	100%	all surface	corrosion	-	falling from the bridge	Safety (S)	3		
		Sidewalks / RC	spalling, reinforcement corrosion, deformations	100%	surface	abrasion, freeze-thaw, corrosion, aging	-	walking comfort	Safety (S)	4		
			leakage	100%	under sidewalks	leakage	HMS, HS	Bending and share failure modes	Reliability (R)	2		
		Slope covering / stones	washout, vegetation	100%	minor washouts, overall vegetation	subsidence	-	Global stability failure mode	Safety (S)	1		
		Road signs / galvanized steel	no damage	-	-	-	-	driving safety	Safety (S)	1		

IOP Conf. Series: Materials Science and Engineering 1202 (2021) 012028 doi:10.1088/1757-899X/1202/1/012028

4. Material testing – carbonation depth

Carbonation is the reaction of carbon dioxide in the environment with the calcium hydroxide in the cement paste. Carbonation depth is the layer of concrete that is carbonated at its surface. The carbonation involves a decrease of pH in the pore solution which leads into steel depassivation, i.e. carbonation-induced corrosion. Therefore it corresponds to the diagnostic performance indicator whose corresponding performance goal is not reaching the rebar front (concrete cover). As the indicator is already systematically used in quality control checks and related decision-making, it is found already at the operational level (Indicator Readiness level, IRL=9) [8].

Within the scope of carbonation depth testing 35 carbonation tests at different places on the bridge were made. Test points were selected to cover bridge structural and non-structural elements as well as to obtain as extensive overview on the overall bridge carbonisation process as possible. The overview of carbonation test process, measures performed for Beam S1, B1 and carbonisation test results is shown in Figures 7 and 8 and in Table 3.

Carbonation test results show that there is significant scattering of results. It is observed that concrete material structure has major influence on carbonation depth. As bridge beams were cast-in-situ more than 60 years ago it was observed that beam high webs in bottom are poorly compacted, therefore carbonation depth is much higher than in shallow elements.

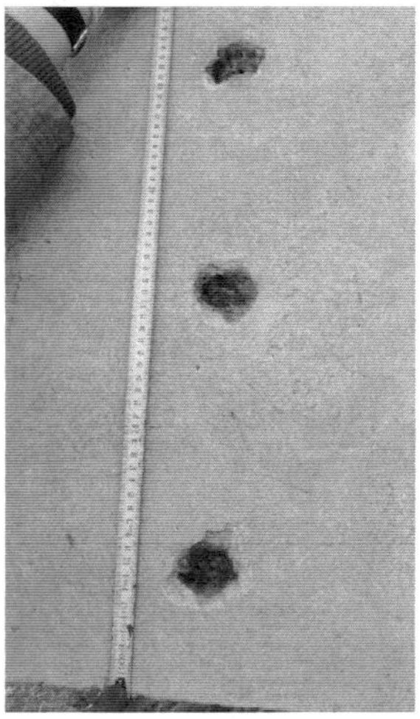
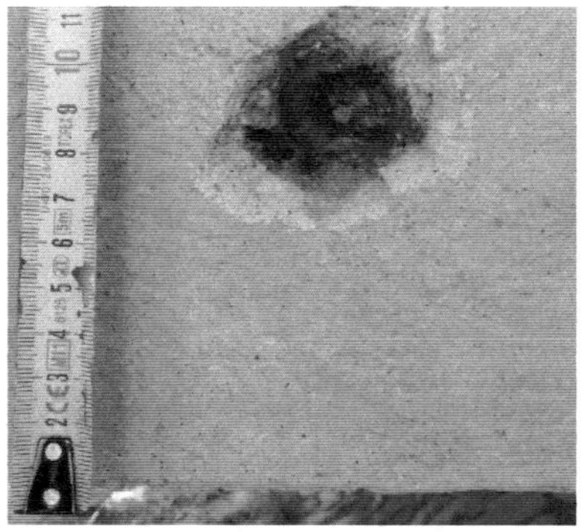

Figure 7. Carbonation tests

Right – set of tests for beam (S1, B4), west plane, location 2

Left – location 2-1, carbonation depth 5-10mm

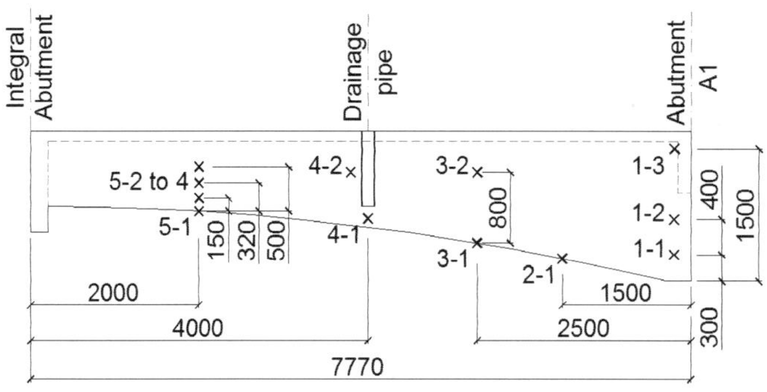

Figure 8. Performed carbonisation depth measures for Beam (S1, B1), east plane

Table 3. Overview of carbonation test results

Element	Location	Carbonisation depth (mm)	Material structure	Notes
Abutment, A1	upstrems, middle	10-15	solid structure	
Abutment, A1	upstrems, top	10-15	weak structure	very detoriated concrete, cracks, efflarocence
Beam (S1, B1), East plane	1-1	5-10	solid structure	
	1-2	>50	porous structure	
	1-3	5-10	solid structure	
	2-1	>50	porous structure	reached reinforcement
	3-1	>50	porous structure	reached reinforcement
	3-2	5	solid structure	
	4-1	>50	porous structure	Under drainage pipe, often wet conditions
	4-2	5	solid structure	next to drainage pipe
	5-1	>50	porous structure	reached reinforcement
	5-2	10-15	solid structure	
	5-3	5	solid structure	
	5-4	5	solid structure	
Beam (S1, B1), West plane	1-1	>50	porous structure	detached concrete cover, reached reinforcement
	1-2	5	solid structure	
	1-3	5	solid structure	
	1-4	>50	porous structure	
	1-5	>50	porous structure	
Beam (S1, B2), East plane	1-1	15-20	solid structure	
	1-2	5-10	solid structure	
	1-3	>50	porous structure	
	1-4	>50	porous structure	
Beam (S1, B4), West plane	1-1	5-10	solid structure	
	1-2	20	solid structure	
	1-3	<5	solid structure	
	1-4	5-10	solid structure	
	2-1	5-10	solid structure	
	2-2	15	solid structure	
	2-3	>50	porous structure	
Cross beam	between S1-S2, above A1	>50	porous structure	
Plate between main beams	between S1-S2	0-5	solid structure	
Integral abutment	middle, between S1-S2	0-5	solid structure	
	West side S1, B4	0-5	solid structure	under sidewalk
	West side S1, B4	0-5	solid structure	under sky

5. Quality control plans

To compare possible bridge management actions the following two possible scenarios were developed and compared: Reference scenario and Preventative scenario. Scenarios were compared on time basis where the bridge is reconstructed after its life-time ends according to Reference scenario.

Reference scenario or case "do noting" considers the lack of any repairs. It is assumed that bridge structural elements could deteriorate at a level when the bridge may not further perform as it is expected – mainly assumed as the lack of bearing capacity. Based on carbonation depth measurements and expert judgement this level is assumed as 30 years. In Reference scenario it should be accepted that the overall bridge condition until bridge replacement will be poor, including low values of Availability and Safety. According to actual new bridge construction prices in Latvia, it is assumed that a new bridge could be built approximately for one million Euros. Figure 9 shows the results in terms of Availability, Costs, Reliability and Safety for the Reference scenario.

Preventative scenario considers major rehabilitation immediately and periodical set of interventions later during life cycle to prevent further development of defects and overall damage to the structure [5]. According to actual bridge rehabilitation prices in Latvia it is assumed that the whole bridge rehabilitation costs approximately 320 thousand Euros. It should be noted that in Preventative scenario close to the year 35 the replacement of bridge water insulation layer and some equipment (parapets, safety barriers, expansion joints) should be planned very likely. Figure 10 shows the results in terms of Availability, Costs, Reliability and Safety for the Preventative scenario.

In both scenarios its assumed that daily maintenance of road including bridge deck is performed, but it is not taken into account in quality control plans.

Figure 11 shows a comparison of two considered scenarios in terms of "spider" diagram.

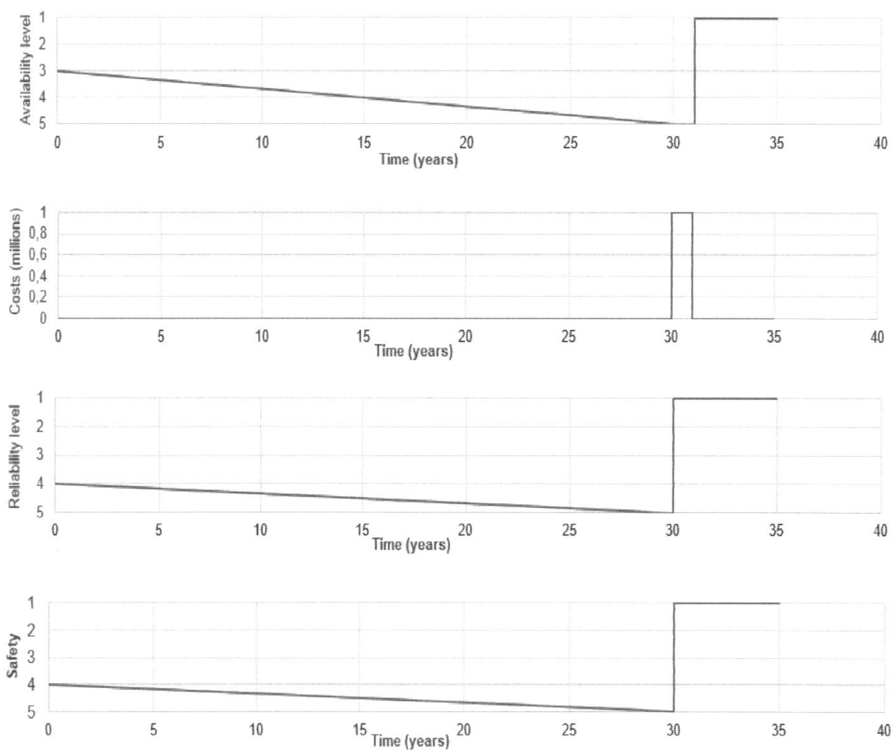

Figure 9. Preventative scenario

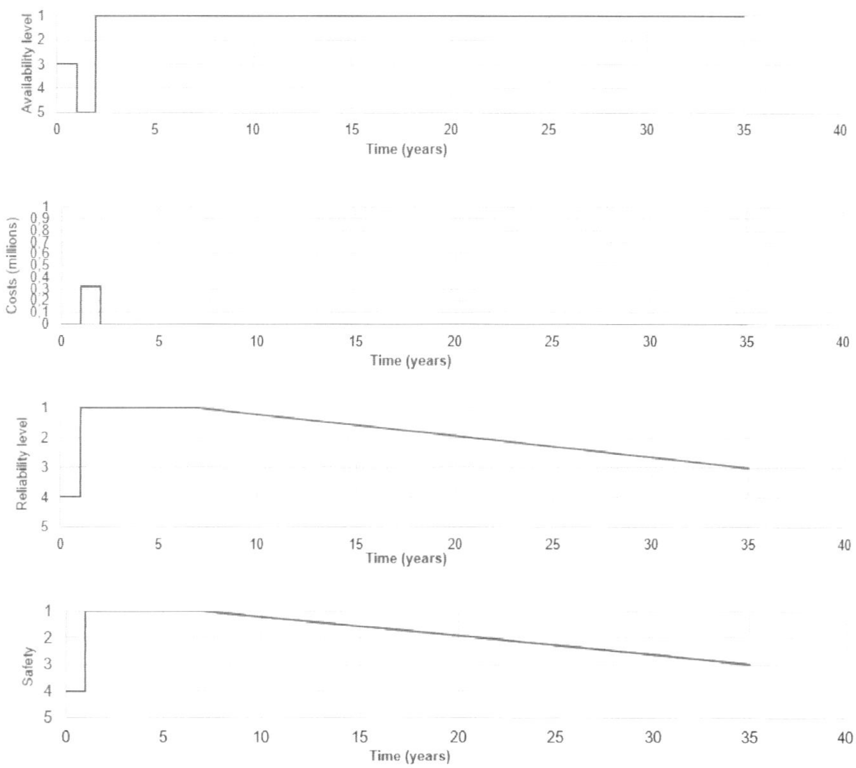

Figure 10. Reference scenario (''do nothing'' case)

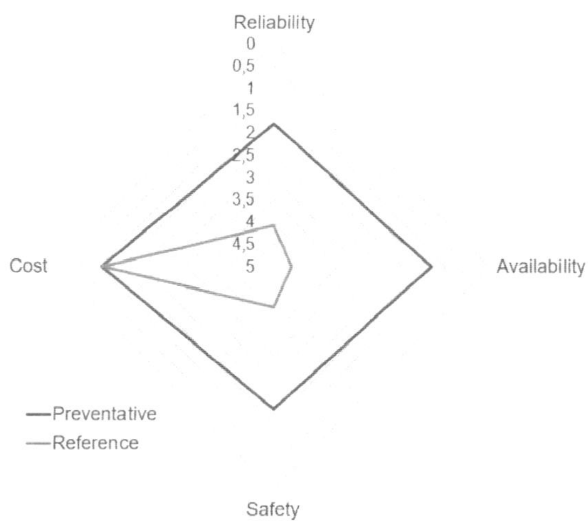

Figure11. Overview of bridge main defects

6. Conclusions and Suggestions

1. According to the analysis it may be concluded that Preventative scenario is more advantageous than Reference scenario. Though the overall costs are quite similar, other performance indicators are at much higher level.
2. The obtained results from the comparison of quality control plans coincide with the plans of SLLC "Latvian State Roads" to perform bridge rehabilitation.
3. For more accurate bridge life-time modelling the deterioration models based on carbonation measurements such as model described COST Action TU 1406 "Report of the Innovation Subgroup" should be used. To obtain more accurate model more data such as chloride content should be added.
4. It may be concluded that carbonation depth measurements made for this specific bridge may not be used to develop any overall bridge deterioration model because of significant data scattering.

7. Bibliography

1. WG1 Technical Report "Performance Indicators for Roadway Bridges of COST Action TU1406";
2. WG2 Technical Report "Performance Goals for Roadway Bridges of COST Action TU1406";
3. WG3 Technical Report "Establishment of a Quality Control Plan";
4. WG4 Technical Report "Preparation of a Case Study";
5. WG4 Technical Report "Guidelines for Preparation of a Case Study";
6. WG5 Technical Report "Drafting of Guideline / Recommendations of COST Action TU 1406"
7. COST Action TU 1406 Case studies from A1 to A17;
8. COST Action TU 1406 "Report of the Innovation Subgroup".

VIADUCT OVER THE LORUPE RAVINE

Indra DZIEDĀTĀJA
Latvian Road Museum, State Limited Liability Company "Latvian State Roads", 3, Gogoļa Street, Riga, LV-1050, Latvia
indra.dziedataja@lvceli.lv

February 24, 2021

Abstract. On the 20th July 2018, it was the 50th anniversary of the day when the Lorupe viaduct was opened. Its idea, design and construction were a significant turning point in bridge construction in the whole Soviet Union. In the comparatively small Republic of Latvia it was a unique project that required a lot of boldness and skills of road workers. Therefore, construction of this bridge is memorable, especially since several bridges were built using this technology later in the former Soviet Union. Latvian road workers with their courage, creative and meticulous work implemented a European scale project in difficult, constrained circumstances.

Keywords: Viaduct, ravine, longitudinal sliding, flexible piers, courage.

Introduction

More than fifty years ago, on the 20th July 1968, viaduct over the Lorupe ravine was opened. It is one of the greatest structures of Latvian bridge builders. In many aspects structure of the viaduct and its construction technology has been mentioned internationally. It has also started a new stage in the bridge construction in the former superpower Soviet Union.

The Lorupe ravine is 30 m deep and 200 m wide with very steep banks. A road was crossing the ravine already in the time of Russian Tsar in the middle of the 19th century. When automobilization developed, driving on this road was connected with numerous road traffic accidents, especially in winter.

Design work on the new road began in the early 1960s.

The overpass, 200 m long, was designed as two continuous girders of reinforced concrete with box cross-section. It was a novelty to slide this structure from one side onto bridge piers. It required additional bearings on which to slide the enormous concrete mass. A solution was found – plates of a super slippery material, fluoroplastic. However, this required a structure never seen before – structure of flexible piers that would react flexibly to vibrations during assembly and use. In that time, the structure assembled with this method was the first engineering work of this kind in the former Soviet Union and the second one in the world (Venezuela had the first).

Flexible piers, stressing, assembly, scenic research, methods of ecological approach and the construction process itself involving many road construction companies was a great teacher to the bridge builders of that time.

The viaduct was reconstructed in 2000.

1. Road history

Sigulda is a town in Latvia that attracts a lot of tourists. Latvians and travellers from other countries go there to enjoy the beautiful surroundings and views of the Gauja river valley. Not far from the Gauja is the Lorupe, a small tributary on the left bank of the Gauja river. It is 11 km long and begins in the Ummuri lake. The Lorupe has a considerable drop – 89 metres, i.e., 8.1m per km. Due to this fact, the river has formed a deep ravine with steep banks (Jemeļjanovs, 2018). Today, the Lorupe viaduct crosses this ravine, but once only a winding road in the bottom of the ravine was there for the travelers. The road that led from the Lorupe to Sigulda has a long history. It was used since the 12th century (Jemeļjanovs, 2018). However, the first images of bridges were discovered only from the end of the 18th century, when Johann Christoph Brotze, a remarkable researcher of local history, painted them in his pictures.

IOP Conf. Series: Materials Science and Engineering 1202 (2021) 012029 doi:10.1088/1757-899X/1202/1/012029

Figure 1. View of Starpas tavern and road from the Lorupe to Sigulda. Picture by J. Ch. Brotze. 1792 (Broce,2002).

In 1856, the Russian Empire constructed a highway Riga – Pskov in place of the old road. The highway was an important mail path. For crossing the Lorupe, a culvert from chiselled granite blocks was built [2]. Nowadays, this culvert and the old road section have been restored and may be seen.

The old road that crossed the Lorupe was winding. It was built by bypassing the old watermill of Kronnenberga Manor that was located 50 m above it.

There is an old legend that says that the miller had a beautiful daughter. A man fell in love with her, and he was designing the road. He wanted to spare the mill, so he deliberately made the road with so many bends. (Jemeljanovs, 2018), (Vecvagars,1998)

In the 20th century road transport was developing. Driving on the winding road, especially in winter, was difficult and there were many road traffic accidents.

Figure 2. Road in the Lorupe ravine. 1960. (Latvian Road Museum, n.d.)

This picture shows how the road went through the Lorupe ravine in the beginning of 1960-ties.

Figure 3. A traffic accident in the area of the Lorupe culvert in the beginning of 1960-ties. (Vecvagars, 1999)

2. The project of crossing the Lorupe ravine

In the beginning of the 20th century, the crossing of the Lorupe ravine has been preoccupying minds of several generations of Latvian engineers, according to information from Kārlis Gailis, professor of the University of Latvia, and documents of the former Department of Soil Roads and Highways (Vecvagars & Binde, 1998). To improve traffic in this road section, alignment research works had already begun in 1942, during the time of German occupation when one German construction company was financing it (Vecvagars, 1999). Due to political circumstances, the work did not continue.

The research works restarted for the second time in 1959, when specialists from Bridge Department of the institute "Ceļuprojekts" (Road Design) undertook the preparation for reconstruction of the Lorupe ravine crossing. By examining the new alignment, the specialists concluded that the best solution for straightening is the axis for ravine crossing outlined already in 1942 (Vecvagars & Binde, 1998).

In 1960, design works started and they had several stages. The first and most important question – how should a road profile in the Lorupe ravine be created? There were two technical options: to build a road over the ravine as a high embankment or to cross the ravine with an artificial structure – viaduct. In this matter, aesthetical architectural factors and technically economic factors were equally important. Therefore the architect Velta Reinfelde worked alongside road engineers in all the stages of research and design. The architect achieved synthesis of structural solutions and aesthetical architectural ideas. After a dendrological research of the area, the project provided for preservation of valuable and significant trees in the landscape, as well as, reconstruction of the current typical terrain (Vecvagars, 1999).

Figure 4. Viaduct sketch of the architect Velta Reinfelde, 1959 (Latvian Road Museum, n.d.)

For a 30 m high road embankment, almost 200,000 $m^3$of gravel would have to be brought in the ravine. The width of the embankment in its foot would reach 150 m. In addition, 140 m long culvert would have to be built for the river to pass. Extraction of soil in such proportions from the closest quarries would damage the beautiful surroundings that were included in the natural preserve foundation of the republic (currently the territory of the Gauja National Park). The huge mass of embankment would divide the ravine in two parts (Vecvagars & Binde, 1998), (Vecvagars, 1999).

However, a scaffold bridge of light construction with narrow, light piers would allow the ravine to preserve its unity and beauty.

The technical solution of the new engineering structure had to maximally preserve spatial unity of the beautiful ravine, and technology of construction works had to meet the requirements of environmental protection.

Considering these issues, the authors selected the artificial structure – a reinforced concrete viaduct– as the key solution. Construction costs were approximately the same for both options (Vecvagars & Binde, 1998), (Vecvagars, 1999).

Inspiring from crossings of ravines of mountain rivers in other countries, as seen in West Germany magazines, Ziedonis Vecvagars, Head of Bridge Department of the institute "Ceļuprojekts" at that time, considered that filling up the Lorupe ravine was unacceptable from the aesthetical point of view. However, to refer to foreign experience in the advocacy process for the viaduct solution was not appropriate and even dangerous on a personal level. It could mean losing a job or even personal freedom (Vecvagars, 1999).

After long discussions with functionaries of the Soviet Union, the scales tipped in the favour of the viaduct. The next issue was the selection of viaduct structure. Solution was made more complicated by the campaign "Combatting of excesses" initiated by Nikita Khrushchev. A. Afanasjevs, Deputy Minister of Road Transport and Highways of the Latvian SSR at that time, supported this campaign enthusiastically. He considered the Lorupe viaduct to be an example of excess. According to his views, ravine should be crossed with a bridge that complies with the standards of bridge construction of the USSR – several interrupted beams on 2 m thick, firm piers (Vecvagars & Binde, 1998).

To avoid this, Latvian engineers had to select an unconventional structure – a continuous girder – and to build it themselves basing on individual design (Vecvagars, 1966).

In technical magazines of that time, there were descriptions of bridge construction in Venezuela. In 1962, stressed continuous girder was built over the Caroni river. This girder, approximately 10,000 t heavy, was put on piers with a unique method in construction of reinforced concrete bridges – longitudinal sliding (VSL International LTD, 1977). If not for this prototype, bridge designers in Latvia or USSR would not have come to an idea to use sliding in the installation of reinforced concrete girder. This example determined the selection of technology for span construction and installation of the viaduct (Vecvagars, 1999).

Long, creative search for the best versions of viaduct had begun. After careful calculations and comparisons, the designers concluded that the most appropriate structure is a continuous, prefabricated, stressed girder that stretches as a concrete band from one ravine bank to another. However, stressed continuous girder was a very rare phenomenon in the practise of bridge construction, even rarer in prefabricated reinforced concrete structures.

Until that time, bridge piers were associated with something massive, mighty and indestructible. The piers of the Lorupe viaduct are not washed by a rapid stream, they do not have to take on the impacts of broken ice; therefore, they are very light and flexible. Already narrow bodies of the piers get even more narrower at the bottom to be able to bend freely in the connection point with support. In all the length of the viaduct, there is only one rigid point – anchor support in the Sigulda bank. Its big mass is hidden in an alcove carved out in a sandstone rock (Vecvagars, 1999).

It was decided to create continuous girder with five spans according to the following scheme: 33 + 3 x 43.25 + 33 m on 24 m high piers (Vecvagars & Binde, 1998).

3. Construction stages of the viaduct

Construction of the Lorupe viaduct took place from April 1965 until July 1968. Considering originality of the bridge structure and construction technology, speed of construction was indeed very fast. From today's perspective, construction costs seem small – 400,000 roubles (Kalnins, 1965), (Vecvagars, 1999)

The construction works may be divided into the following sets of works::

1. Construction of accesses was done in two stages: in 1965, alignment preparation, main earth works and partial construction of road foundation were done; from April until July 1968, construction of road foundation was completed, asphalt concrete laid and finishing works carried out.

2.Construction of viaduct piers – foundation was built in 1965, bodies of piers and headers – in 1966.

3.Span blocks were manufactured from January until June, 1967.

4.Assembling of span structure and sliding it on the piers – from June until November, 1967.

5.Rearrangement of stressed cable bundles for operation and canal injection was performed in two stages: September and October, 1967, for the first girder, and March and April, 1968, for the second one.

6.Pouring of concrete over stressed cable bundles and joining of both girders was carried out in April and May, 1968.

7.Construction of carriageway and sidewalks was performed in June and July, 1968.

8.Scaffolding and subsidiary buildings were demolished and the whole territory tidied out from May until July, 1968. (Vecvagars & Binde, 1998).

4. Challenges during the bridge construction

Due to the poverty and underdeveloped technical capacity of the USSR, there were many challenges to address.The 200 m long girder of the prefabricated, continuous span structure weights more than 1000 t. Installation of such structure could not be solved with the usual methods in these specific circumstances. Therefore, attention of designers was attracted to the sliding method. However, this issue was very complicated, almost unsolvable, when this method for assembling classical steel bridges is applied to reinforced concrete. During the installation, the girder during sliding reaches a condition that according to calculations is opposite to the one it will be later in during the time of operation. For a steel girder, such condition change is not dangerous, since they both resist tension and compression equally well. Steel girders may also be reinforced in a constructively simple way during installation. Reinforced concrete resists well only the compression, it takes on tensile stresses poorly. It is also very hard to reinforce the reinforced concrete girder during installation, since the cables that take on tension are hidden deep in the concrete mass (VSL Internationas Ltd, 1977).

Modification of sliding equipment was also a complicated matter. Steel spans are being slid on rollers. Spans are relatively light and rolling friction between rollers and sliding tracks formed in the steel is insignificant. Analogue method is not suitable for reinforced concrete spans. They are very heavy and concrete would crumble coming into contact with the circular surface of rollers (Vecvagars & Binde, 1998), (Vecvagars, 1999).

Notwithstanding, it was decided to install the spans of the Lorupe viaduct with sliding, even without temporary piers. The play of forces in the span, depending on the condition during installation, was regulated with special, movable stressed cable bundles. Therefore, it was possible to create artificially as big compressive stress as necessary in the respective installation moment on any side of the span (Vecvagars, 1999)

Issue of sliding equipment was also solved. Help came from the latest achievements in chemistry. Girders were not pushed but slid on plates of special polymer material – polytetrafluoroethylene. Friction between fluoroplastic and concrete surfaces was only few per cent. Therefore, the whole girder that weighted thousand tons could be easily slid with the help of several winches with the pulling capacity of few tons. In addition, there was an option to use a special material – concrete cylinders coated with neoprene cord – instead of fluoroplastic plates (Vecvagars & Binde, 1998).

5. Construction of the viaduct

Since a completely new technology for structures and construction works was used for the construction of the Lorupe viaduct, many and different auxiliary devices and special materials were necessary. Most of the auxiliary devices were non-standard. Therefore, companies in the industry manufactured them. Only the special equipment – hydraulic jacks and pumping stations – were brought in from other republics (Vecvagars & Binde, 1998).

Height of every bridge pier had to match the ravine terrain. The concrete was poured in a way that there would be no interruptions, so that all the mass from the foundation to the top would be poured without any possible gaps in between. It was not easy. Cranes for delivering the concrete upwards could not be used. The workers themselves have constructed an elevator-like concrete vat that could be lifted between the scaffolding. With its help, the concrete, that was prepared on site, was lifted in uninterrupted flow up to very top (30 m high) (Vecvagars & Binde, 1998).

A newspaper of that time "Latvijas Auto un Ceļu Darbinieks" (Latvian Road Worker) published the following: "Several scaffoldings of bridge piers rise as mighty skyscrapers over serrated spruce treetops. When one is looking from their tops and imagining how is it going to look like when the road is ready, the impression is magnificent" (Latvijas Auto un Ceļu Darbinieks, 1968).

Figure 5. A view of the Lorupe viaduct piers,1967 (Latvian Road Museum, n.d.)

Construction of the Lorupe viaduct is characterized by very high work culture and quality. Organization of works ensured that environmental damages to the Lorupe ravine were minimal. Only few trees and bushes that were included in the design and encumbered pier construction works were cut down. Preparation of blocks and assembly of girders were performed on access roads. The construction site did not take more space than a right-of-way. Making span blocks on construction site justified itself completely. Their quality was considerably better than that of mass production of reinforced concrete shop. This success was achieved by the conscientious and creative attitude of workers (Vecvagars & Binde, 1998).

The newspaper "Latvijas Auto un Ceļu Darbinieks" described the process of sliding the girder sections on piers:

"So the sliding of span girders on the bridge has already begun. Currently the speed is 6 m per hour or 1 m in 10 minutes.

Only when looking closely, it may be seen that the first twenty sections of span girders that weigh more than 3 t are indeed slowly moving in the direction of ravine. The first ones, right after the pushing mechanism, are lined right next to each other on support structure on a special rail track without any support.

Farther ones, already stressed and concreted in one section, are moving over abutment. In the front, the long iron launching girder has already passed the "fifth" pier. For a moment, the engine stops. Men transfer the next section on the tracks with a crane, and the peculiar "train" continues its slow movement.

At the top of the closest pier, workers are regulating sliding of the launching girder and span section over the pole with different mechanisms. Chromium-plated metal plate and a plastic plate is being placed between them so that the brake friction would not come into effect. The pier itself is carefully linked with iron cables. The workers are laughing: in the past, bridges were built; now they are crossing over themselves (Kaugurs, 1967).

Figure 6. Sliding of span girder section (Latvian Road museum, n.d.)

The workers implemented many small improvements during the sliding of span girders, as well. The biggest and the most modern project of the Latvian road workers at that time was created with intense, interesting and creative work.

Figures 7. and 8. Cross section of a span girder and pouring of concrete on stressed cable bundles (Latvian Road Museum, n.d.)

Before opening of the viaduct, it was tested with loading. The test was performed by the workers of the Riga Polytechnic Institute. The total weight in the test reached 109 – 120 % of the normative load – 2 columns of loaded heavy vehicles. The test results showed that the actual span deflection was ten times smaller than the permissible standard. Concrete of the viaduct has good elastic properties (Vecvagars, 1999).

Figure 9. Static loading of the Lorupe viaduct, 1968 (Latvian road Museum, n.d.)

On the 20th July, 1968, the new viaduct over the Lorupe ravine was opened.

6. Importance of construction of the Lorupe viaduct in the Soviet Union

The most qualified and objectively minded engineers and representatives of scientific organizations in the USSR were following the process of design and construction of the Lorupe viaduct with great interest. At the same time, there were specialist circles that did not like that an experimental structure of this magnitude is in the hands of representatives of a small national republic and not in theirs. The construction site was often visited by representatives of large bridge construction companies and design organizations subjected to the Ministry of Construction of Transport Structures of the USSR. However, when the German technical magazine "Die Strasse" started to publish materials about the viaduct construction, the Ministry of Construction of Transport Structures of the USSR called a special meeting where colleagues in Riga were criticized about "implementing the foreign experience". There were also other attempts to stop the expansion of the method of longitudinal sliding of bridge spans. However, its victory march through the whole territory of the USSR was unstoppable. Construction without any handles or cranes is organically connected with constructive nature of a continuous girder. Until the collapse of the USSR, 14 reinforced concrete bridges were built according to this technology, one of them in Latvia (Vecvagars, 1999).

Figure 10. The Lorupe viaduct, 1970-ties (Latvian Road Museum, n.d.)

Conclusions

It has been 50 years since the construction of the Lorupe viaduct. Generations have changed, and young bridge builders have little knowledge of challenges and hardships that workers of 1960-ties had to overcome during the viaduct construction. Nowadays bridge assembling with longitudinal sliding is a common practise. However, the principle of the project implemented for the first time in the Lorupe ravine has stayed the same. The purpose of this paper is not only to describe the idea and construction stages of this bridge, but to tell a wider society about the courage, initiative, creative and meticulous work of the Latvian road workers that allowed to implement a European scale project in difficult, constrained circumstances.

Author contributions

Laura Reble and Mārtiņš Dambergs translated the work in English.

Disclosure Statement

I do not have any competing financial, professional, or personal interests from other parties.

References

Broce, J.K. (2002). *Zīmējumi un apraksti 3. Sējums. Latvijas mazās pilsētas un lauki, Zinātne (pp. 138–139)* [Drawings and descriptions. Vol.3. Small towns and countryside of Latvia].
Jemeļjanovs, E. (2018). Senie ceļi Vidzemē un viadukts pār Lorupi Siguldā. [Old roads in Vidzeme and the Lorupe viaduct in Sigulda]. Retrieved from https://www.turaida-muzejs.lv/2018/10/senie-celi-vidzeme-un-viadukts-par-lorupi-sigulda/
Kalniņš, K. (1965). Lorupes estakādei likti pamati. *Latvijas Auto un Ceļu Darbinieks, October 28 (1965),1.* [Foundation of the Lorupe viaduct is completed].
Kaugurs, K. (1967). *Latvijas Auto un Ceļu darbinieks July 20 (1967), 1* .
Latvian Road Museum, State Limited Liability Company "Latvian State Roads", *The road history collection.*
Lorupes rītdiena top (1968), *Latvijas Auto un Ceļu Darbinieks, June 20 (1968), 2.*[The future of the Lorupe forms].
Vecvagars, Z. & Binde, J. (1998) *Viadukts pār Lorupes gravu.* Celi (pp 5 – 11). [Viaduct over the Lorupe ravine]
Vecvagars, Z. Lorupes gravas problēma atisināta (1966), *Latvijas Auto un Ceļu Darbinieks January 6 (1966),3.*
Vecvagars, Z. (1999) Viadukts pār Lorupes gravu Vidzemes šosejā, *Ceļi* (1999), pp 8 – 67 (in Latvian).
VSL INTERNATIONAL LTD. Bern/Switzerland. (1977) The Incremental Launching Method in Prestresses Concrete Bridge Construction. Retrieved from: https://www.structuraltechnologies.com/wp-content/uploads/2018/02/PT_Incremental_Launching_Method.pdf

IOP Conf. Series: Materials Science and Engineering 1202 (2021) 012030 doi:10.1088/1757-899X/1202/1/012030

METHODOLOGY FOR CALCULATION OF BRIDGE SAFETY FACTOR IN LITHUANIA

Saulius ZADLAUSKAS[1] Mindaugas AUGONIS[2]
[1]*State Enterprise Lithuanian Road Administration, Vilnius, Lithuania*
[2]*Kaunas University of Technology, Kaunas, Lithuania*
Corresponding author: Saulius, Zadlauskas, email (saulius.zadlauskas@lakd.lt)

Received day month year; accepted date

Abstract. The paper reviews the United States and Slovenian safety factor calculation methodologies and proposes a method for more accurate estimation of residual strength of bridges designed and built in Lithuania. For more detailed analysis, the main parameters and defects directly affecting the strength of the bridges were analysed in detail, and the flows of heavy vehicles, which have significantly increased for previously designed bridges, were assessed.

This article proposes to calculate the dynamic factor of bridges, not according to the empirical formulas used in the United States and Slovenian safety factor calculation methodologies, but after performing the bridge dynamic test, because the results of Lithuanian bridge dynamic tests show that the parameter strongly depends on smoothness and damage of carriageway wearing surface. In order to evaluate the suitability of the Lithuanian bridge safety factor calculation methodology proposed in this article, a real bridge was selected, and its safety factors calculated according to the above mentioned and proposed methodology and the results obtained were compared.

Keywords: bridges, safety factor, resistance, defects, dynamic factor.

Introduction

Bridges are sophisticated and expensive transport structures that are of the great importance for economical, political, and cultural relations. For a long time, reinforced concrete bridges have been considered as durable structures that require only regular maintenance, but in the recent decades in all countries it was observed intensive physical and moral aging of these structures. Heavy duty vehicle flows, higher speeds and axle loads are the factors that significantly affect the faster deterioration of the bridge deck.

The deterioration of the bearing structures of bridges is directly influenced by defects occurring during the operation stage of the bridge: carbonization of concrete, corrosion of reinforcement, shear and normal cracks in the structures (Augonis M., Zadlauskas S., Rudžionis Ž., Pakalnis A., 2012). In order to follow the deterioration process of bridges, it is necessary to constantly monitor changes in their condition, trends in the development of defects, evaluate and predict their durability and, if necessary, take appropriate measures such as limiting the gross weight of heavy vehicles. One of the ways to bridge condition assessment is the bridge "safety factor" application (AASHTO 2010, 2011; Žnidarič A., 2015; Peris, A. and Harik, I., 2016). When calculating the bridge safety factor, the condition of the bearing structures of the bridge, the effects of permanent and variable loads acting on it, and the effects of dynamic loads and overloads are assessed in detail.

Until now, no bridge safety assessment methodology has been applied in Lithuania, therefore, there is a need to do research, to analyze the calculation methods used in the United States and Europe, and to propose a methodology suitable for Lithuanian bridge safety assessment. Since about 95 % of bridges in Lithuania are reinforced concrete, therefore, the methods of calculating the safety factor of bridges analyzed in this work have been developed for safety assessment of reinforced concrete bridges.

1. Bridge safety factor calculation methodology

For the evaluation of safety of bridges used in Europe and the United States ("ultimate limit state"), a bridge safety factor is calculated considering the factors mentioned above. The bridge safety factor indicates whether the bridge is safe to operate under certain damages, defects, and traffic flows, or whether it is unsafe, and it is necessary to limit the total weight of heavy vehicles. Process to compose assumptions for calculating the safety factor of bridges have been started since 1970. American (Nowak, A. S. and Gruni, H. N., 1994; Frangopol, D. E. and Estes, A. C., 1997) and Canadian (Allen, D. E., 1992; Bartlett, F. M., Buckland, P. G., Kennedy, D. J., 1992) scientists have paid much attention to the preparation of assumptions in their research works.

The American Bridge Design Standard (AASHTO, 2010) provides the following formula for calculating the bridge safety factor:

$$RF = \frac{\phi_c \phi_s \phi R_n - (\gamma_{DC} DC + \gamma_{DW} DW)}{\gamma_{LL} (LL + IM)} \tag{1}$$

where: ϕ_c – condition factor (usually is assumed as 1.0); ϕ_s – system factor (usually is assumed from 0.85 till 1.0); ϕ – resistance factor (if no analysis carried out, assumed as 1.0); R_n – nominal resistance of bridge deck; γ_{DC} – load factor for structural components and attachments (usually is assumed as 1.25); DC – dead load of structural components and non-structural attachments; γ_{DW} – load factor for wearing surfaces and utilities (usually is assumed as 1.50); DW – dead load of wearing surfaces and utilities; γ_{LL} – load factor for live loads (1 table); LL – vehicular live load; IM – vehicular dynamic load allowance (2 table).

Table 1. Load factors for live loads (γ_{LL})

Heavy duty vehicles	Load factor for live loads	Number of heavy vehicles per day ≤1000	Number of heavy vehicles per day ≥1000	Unknown number of heavy vehicles
Routine vehicles	γ_{LL}	1.65	1.80	1.80
Special permits vehicles	γ_{LL}	1.40	1.50	1.50

Table 2. Dynamic load allowance (IM)

Bridge span	IM
Bridge span 40 m or less	33%
Bridge span more than 40 m	
– smooth carriageway wearing surface without defects in deck joints	10%
– small irregularities in carriageway wearing surface	20%
– rough carriageway wearing surface, high impulses when moving heavy vehicles	30%

Scientists at the National Building and Civil Engineering Institute ZAG, Slovenia (Žnidarič, A., 2010, Žnidarič, A., Lavrič, I., Kalin, J. 2010) have developed a methodology for calculating the safety factor of bridges. For bridges built in this country, their safety factor is calculated using the following expression:

$$RF = \frac{\Phi \cdot R_d - \gamma_G \cdot G_n}{\gamma_Q \cdot G_Q \cdot DAF} \tag{2}$$

where: Φ – bridge deck capacity reduction factor; R_d – design resistance of bridge deck; G_n – characteristic value of permanent action; G_Q – characteristic value of variable action; γ_Q – load factor for variable loads (assumed according to the flow of heavy vehicles over the bridge under investigation, but not less than 1.40); γ_G – load factor for permanent loads (assumed equal to 1.20); DAF – dynamic amplification factor.

In the developed methodology, the bridge deck dynamic factor is calculated according to the formula given in German bridge design standard (DIN 1072):

$$DAF = 1.4 - 0.008 \cdot L \tag{3}$$

where: L – bridge span length.

The capacity reduction factor of the bridge deck is calculated using the following expression:

$$\Phi = B_R \cdot e^{-\alpha_R \cdot \beta_C \cdot V} \tag{4}$$

where: B_R – the bias of carrying capacity, i.e., the ratio between the true and the design mean resistances of the critical section. In most cases assumed as 1.0, if no more detailed material analysis has been performed; α_R – the deterioration factor, accounting for the condition of the bridge (see Table 3); V – the coefficient of variation of the member resistance and is taken 10% when calculated from experiments, 15% when based on design information and 20% if less reliable information is used; β_C – the target value of the safety index, taken 3.5 for the normal service life (up to 20 years) and 2.5 for limited service life (up to 6 years or until the next main inspection).

Table 3. Values of deterioration factor (α_R)

Class	Inspected condition	Necessary intervention	Condition rating R_c	Deteriotiation factor, α_R
1	Very good	No maintenance/repair work required	<5	0.3
2	Good	Regular maintenance work needed	3 – 10	0.4
3	Satisfactory	Intensified maintenance/repair work within 6 years	7 – 15	0.5
4	Tolerable	Substantial repair work needed within 3 years	12 – 25	0.6
5	Inadequate	Immediate posting and repair required	22 – 35	0.7
6	Critical	Immediate closing and repair/strengthening required	>30	0.8

An overview of the methodologies used to calculate the safety factor for bridges in different countries of the world has shown that they have the same basic parameters, but differently measured. Each of the methodologies describes four main components: the bearing capacity of bridge deck considering various factors, the effects of permanent and variable loads, and the dynamic loads caused by heavy vehicles. Although these components are similar, in America and Slovenia different partial factors are used for permanent and variable loads, the impact of dynamic loads on the bridge is assessed differently, different system of assessment of the condition of the bridges, different operating conditions of the bridges. As the factors mentioned in Lithuania are also different, this paper presents a methodology for calculating the safety factor of bridges operating in Lithuania.

2. Condition of bridges assessment in Lithuania

In Lithuania, the condition of bridges administered by the Lithuanian Road Administration under the Ministry of Transport and Communications is assessed through the annual and substantive inspections, static and dynamic tests. During the annual inspection, bridges are evaluated using a five-point grading system. Each bridge element is visually inspected, and its condition assessed. The most important factor in the overall estimate of the bridge is the condition of the bearing structures. Because the static and the dynamic tests of the bridges are expensive, therefore they are carried out only for the problematic ones, and for the most of them only one or another type of inspection is carried out. Visual assessment of the condition of the bridge is not objective and sufficient to comprehensively evaluate the condition of the bridges and to plan investments for their repair, because: visual evaluation of the bridge does not always allow to avoid "human factor" errors and it is not clear what impact the defect has on the bearing capacity of the bridge, and how safe it is to operate such a structure in the future. Therefore, from 2015 to 2016, specialists from the Bridge Research Department of the Public Road and Transport Research Institute have developed a new bridge management system, in which appeared the parameter characterizing the bridge safety - the bridge safety factor.

2.1. Calculation of bridge safety factor in Lithuania

In the new bridge management system developed in Lithuania, it is proposed to calculate the bridge safety factor taking into account the safety factor expressions proposed by American and Slovenian scientists, however, by introducing the notations of permanent and variable loads, partial factor and dynamic factor used in Lithuania. It is proposed to calculate the safety factor of Lithuanian bridges by the following expression:

$$RF = \frac{\varphi \cdot R_d - (\gamma_{G1} \cdot G_{k1} + \gamma_{G2} \cdot G_{k2} + \gamma_{G3} \cdot G_{k3})}{\gamma_Q \cdot Q_k \cdot \mu_{din.}} \quad (5)$$

where: φ – bridge deck capacity reduction factor; R_d – design resistance of bridge deck; G_k – characteristic value of permanent action; Q_k – characteristic value of variable action; γ_Q – load factor for variable loads; γ_G – load factor for permanent loads; μ_{din} – bridge deck dynamic factor.

The deterioration of the bridge condition is evaluated by calculating the cross-section strength reduction index of the bridge load-bearing structures according to the following expression:

$$\varphi = \frac{1}{e^{\alpha_R}} \quad (6)$$

The deterioration factor of the bridge elements (α_R) directly depends on the condition rating of the bridge load-bearing structures. Since this parameter cannot be expressed in points, Table 4 presents the values of the deterioration factor of the bridge load-bearing structures as a percentage of their condition rating.

Table 4. The deterioration factor values of the bridge load-bearing structures

Bridge rating, in points	Deterioration factor of bridge load-bearing structures, α_R
5	0.05
4	0.10
3	0.20
2	0.25
1	0.35

In Lithuania, the Public Road and Transport Research Institute has carried out more than 150 dynamic tests of various types and conditions bridges, which accurately measured the static and dynamic deflections of the bridges decks and calculated the dynamic factors of the decks from the moving loads of heavy vehicles. According to the results of performed research, it is proposed to relate the dynamic factor of bridge decks to the rating of the condition of the bridge deck wearing surface. The approximate equivalent of the bridge deck dynamic factor is given in Table 5.

Table 5. Approximate evaluation of bridge deck dynamic factor based on carriageway wearing surface rating in points

Rating of bridge carriageway wearing surface	Bridge deck dynamic factor, μ_{din}
5	1.02
4	1.10
3	1.15
2	1.20
1	1.30

The rating of the bridge carriageway wearing surface is described in points:

- 5 points – smooth carriageway wearing surface, no defects;
- 4 points – smooth carriageway wearing surface, only with localised defects; no defects in deck joints;
- 3 points – rough carriageway wearing surface, ruts formed; small depressions formed in deck joints;
- 2 points – rough carriageway wearing surface, localised defects at the ends of the bridge or along the sidewalk, damaged deck joints;
- 1 point – very rough carriageway wearing surface, deep ruts formed; deep depressions formed in the wearing surface at the midspan of bridge deck; very damaged deck joints.

2.2. Evaluation of partial factor of permanent loads acting on the bridge

Bridges in Lithuania are designed according to various bridge design standards. Bridges designed in Lithuania can be divided into five different bridge design periods. Period I – bridges designed from 1945 to 1948. During this period, permanent loads of designed bridges were calculated without partial factors. Period II – from 1953 to 1962. The permanent loads of the bridges designed during this period were also calculated without partial factors. Period III – bridges designed from 1962 to 1984. During this period, the partial factor of designed bridge deck bearing structures, sidewalks, barriers, and handrails of the bridge was 1.10. The partial factor of bridge deck waterproofing, levelling layer, sidewalk protective coating and asphalt pavement was 1.50. Period IV – bridges designed from 1984 to 1997. During this period, the partial factor of designed bridge deck bearing structures, sidewalks, barriers, and handrails of the bridge was 1.10. The partial factor of bridge deck waterproofing and levelling layer was 1.30, and for sidewalk protective coating and asphalt pavement it was 1.50. Period V – bridges designed since 1997 until these days. The partial factor of bridge deck bearing structures and overlays designed during this period and currently being designed in Lithuania is assumed to be 1.35.

Because of the different bridge design periods, the methodology proposes to group the bridge deck permanent loads into three groups for each bridge design year, and apply different partial factors:

- partial factor of bridge deck bearing structures, sidewalks, barriers and handrails of the bridge (γ_{G1});
- partial factor of bridge deck waterproofing and levelling layer (γ_{G2});
- partial factor of bridge asphalt pavement and sidewalk protective coating (γ_{G3}).

2.3. Evaluation of partial factor of variable loads acting on the bridge

The European bridge design standards (Eurocode 1, 2000, 2004) assume partial factor for variable loads equal to 1.35, however American bridge design standards (AASHTO, 2010) and research of foreign scientists (Koteš, P., Vichan, J., 2012) indicate that the values of the partial factor of variable loads are directly dependent on the number of heavy vehicles per day. In this methodology it is proposed to calculate the partial factor of variable loads by considering the influence of heavy vehicle flows (see Table 6).

The analysis of traffic flow intensity in Lithuania is carried out by specialists of the Road Research Department of the Institute of Road and Transport Research and is an annually updated database of the number of heavy vehicles passing through one or another monitored bridge per day.

Table 6. Calculation of partial factor for variable loads

Total daily flow of heavy vehicles	Partial factor of variable load
<250	1.40
>250<1000	1.45
>1000<5000	1.55
>5000	1.60

3. Calculation of bridge safety factor according to Lithuanian proposed methodology by different bridge parameters

Non-continuous one spans reinforced concrete girder bridge (14.10 m) built in Lithuania was chosen to calculate the bridge safety factor. Ribbed bridge deck is made from 6 pcs prefabricated T-girders. The bridge was built in 1950, bridge length 16.10 m and width 8.68 m. The bridge is designed according to Russian bridge design standards, design loads – H-13 and HG-60. 58 heavy vehicles pass through the bridge per day. The overall condition of the bridge is bad (rating of load bearing structures – 2 points). Rating of carriageway coating – 3 points. Bearing capacity of bridge deck (resistance for one T-girder normal cross-section) R_d = 9546 kNm. Characteristic load of bridge deck bearing structures, sidewalks, barriers, and handrails of the bridge – G_{k1} = 1383 kNm. Characteristic load of bridge deck waterproofing and levelling layer – G_{k2} = 239 kNm. Characteristic load of bridge asphalt pavement and sidewalk protective coating – G_{k3} = 545 kNm. Effect of variable loads – Q = 167 kNm. Bridge deck dynamic factor – μ_{din} = 1.39.

The calculations of the bridge safety factor according to AASHTO 2010 methodology is given in Table 7, according to the methodology presented by the Slovenian researchers in Table 8, according to the methodology proposed in Lithuania in Table 9.

Table 7. Calculation of bridge safety factor according to AASHTO 2010 methodology

Titles of assumed/calculated parameters	Values of assumed/calculated parameters
1) Bridge condition factor, ϕ_c	1.0
2) Bridge system factor, ϕ_s	1.0
3) Resistance factor of bridge elements, ϕ	1.0
4) Nominal resistance of bridge deck, R_n	9546
5) Load factor for structural components and attachments, γ_{DC}	1.25
6) Dead load of structural components and non-structural attachments, *DC*	1928
7) Load factor for wearing surfaces and utilities, γ_{DW}	1.50
8) Dead load of wearing surfaces and utilities, *DW*	239
9) Load factor for live loads, γ_{LL}	1.65
10) Vehicular live load, *LL*	1715
11) Vehicular dynamic load allowance, *IM*	(0.33*1715) = 566
The safety factor of the bridge over the Gėluotas and Vašuokas lake	**1.80**

Table 8. Calculation of the bridge safety factor according to the methodology presented by Slovenian scientists

Titles of assumed/calculated parameters	Values of assumed/calculated parameters
1) Bridge deck capacity reduction factor, Φ	0.78
2) Design resistance of bridge deck, R_d	9546
3) Load factor for permanent loads, γ_G	1.20
4) Characteristic value of permanent action, G_n	2167
5) Load factor for variable loads, γ_Q	1.40
6) Characteristic value of variable action, G_Q	1715
7) Dynamic amplification factor, *DAF*	1.29
The safety factor of the bridge over the Gėluotas and Vašuokas lake	**1.56**

Table 9. Calculation of the bridge safety factor according to the methodology proposed in Lithuania

Titles of assumed/calculated parameters	Values of assumed/calculated parameters
1) Bridge deck load-bearing structures rating, in points	3
2) Bridge deck capacity reduction factor, φ	0.82
3) Design resistance of bridge deck, R_d	9546
4) Characteristic load of bridge deck bearing structures, sidewalks, barriers, and handrails, G_{k1}	1383
5) Characteristic load of bridge deck waterproofing and levelling layer, G_{k2}	239
6) Characteristic load of bridge asphalt pavement and sidewalk protective coating, G_{k3}	545
7) Load factor for variable loads, γ_Q	1.40
8) Partial factor of bridge deck bearing structures, sidewalks, barriers and handrails of the bridge, γ_{G1}	1.10
9) Partial factor of bridge deck waterproofing and levelling layer, γ_{G2}	1.50
10) Partial factor of bridge asphalt pavement and sidewalk protective coating, γ_{G3}	1.50
11) Characteristic value of variable action, G_Q	1715
12) Bridge deck dynamic factor, μ_{din}	1.39
The safety factor of the bridge over the Gėluotas and Vašuokas lake	**1.48**

Conclusions

1. After comparing three differ methodology for bridge safety factor calculations it was found that the main differences are with nodal loads coefficients, bridge capacity coefficients and dynamic coefficient.

2. The main differences comparing proposed methodology with Slovenian and AASHTO methodology's is that if there are problems with bridge deck capacity we do bridge dynamic testing and for safety factor coefficient calculation we use real bridge deck dynamic factor but not calculated by empirical formulas. Bridge dynamic coefficient calculated by empirical formulas depends by the length of the bridge, but in reality it depends by the asphalt surface and roughness.
3. After comparing bridge safety factors calculated in table 7, table 8 and table 9 it was found that the highest one is by AASHTO 2010 methodology and the lowest one is by methodology proposed in Lithuania.

References

Augonis, M.; Zadlauskas, S.; Rudžionis, Ž.; Pakalnis, A. (2012). The analysis of reinforced concrete box girder viaduct defects and their estimation // The Baltic Journal of Road and Bridge Engineering. Vilnius: Technika. ISSN 1822-427X. Vol. 7, iss. 1, p. 13-21.

AASHTO 2010. LRFD bridge design specifications, American Association of State Highway and Transportation Officials, 5h Ed., Washington, D. C., USA.

AASHTO 2011.The manual for bridge evaluation, American Association of State Highway and Transportation Officials, 2nd Ed., Washington, D. C., USA.

Allen, D. E. (1992). Canadian Highway Bridge Evaluation: Reliability Index. Canadian Journal of Civil Engineering. Vol. 19, p. 987-991.

Bartlet, F. M., Buckland, P. G., Kennedy, D. J. (1992). Canadian Highway Bridge Evaluation: Derivation of Clause 12 of CAN/CSA – 56 – 88. Canadian Journal of Civil Engineering, 1007 – 1016.

Bruni, S., Beretta, M., Simulation, S. 2003. Simulation of bridge heavy road vehicle interaction and assessment of structure durability, International Journal of Vehicle Design, 20 (1), p. 70-85.

Eurocode 1. (2000). (Eurokodas1. Projektavimo pagrindai ir poveikiai konstrukcijoms. 3 dalis. Tiltų eismo apkrovos LST L ENV 1991–3:2000).

Eurocode 1. (2004). (Eurokodas1. Poveikiai konstrukcijoms. 2 dalis. Tiltų eismo apkrovos LST EN 1991–2:2004).

Frangopol, D. M., Estes, A. C. (1997). System Reliability for Condition Evaluation of Bridges. In: IABSE Workshop "Evaluation of Existing Steel and Composite Bridges". Lausanne, p. 47-56.

Koteš, J. Vičan, P. (2012). Partial safety factor for evaluation of existing bridges according Eurocodes. 18th International Conference Engineering Mechanic., p. 681-690.

Lee, S. Y.; Yhim, S. S. 2005. Dynamic behaviour of long-span box girder bridges subjected to moving loads: Numerical analysis and experimental verification, International Juornal of Solids and Structures, 42, p. 5021-5035.

Lee, Y. S.; Kim, S. H.; Skibniewski, M. J. (2009). Analytical and experimental approach for assessing vibration serviceability of highway bridges due to heavy vehicle traffic. The Baltic journal of road and bridge engineering, Vol. 4, No. 3, p. 123-133.

Li, C. Y. 1996. Bridge vibration and impact under moving vehicles, In proceedings of the 3rd joint conference on engineering systems design and analysis, 81 (9), p. 17-23.

Nowak, A. S., Gruni, H. N. Calibration of the Ontario Highway Bridge Code 1991 Edition. Canadian Journal of Civil Engineering. 1994, Vol. 21, p. 25-35.

Paeglite, I., Smirnovas, J., Paeglitis, A. (2016). Dynamic behaviour of pre-stressed slab bridges. Modern Building Materials, Structures and Techniques, MBMST 2016, p. 831-838.

Paeglite, I., Paeglites, A. (2013). The dynamic amplification factor of the bridges in Latvia. 11th International conference on modern building materials, Structures and techniques, MBMST-2013, p. 851-858.

Peris, A., Harik, I. 2016. Load testing of bridges for load rating. The 7th International Conference on Sustainable Built Environment. ICSBE2016-116.

Shi, X.; Cai, C., S.; Chen, S., (2008). Vehicle induced dynamic behavior of short-span slab bridges considering effect of approach slab condition. Journal of Bridge Engineering, Vol. 13, No. 1, p. 83-92.

Zadlauskas, S.; Augonis, M. (2012). The influence of dynamic loads on prestressed concrete box girder viaduct deflection // Mechanika 2012: proceedings of the 17th international conference, 12, 13 April 2012, Kaunas University of Technology, Lithuania / Kaunas University of Technology, Lithuanian Academy of Science, IFTOMM National Committee of Lithuania, Baltic Association of Mechanical Engineering. Kaunas: Technologija. ISSN 1822-2951, p. 345-350.

Zadlauskas, Saulius; Augonis M. Dynamic testing of reinforced concrete bridges and the analysis of testing results // proceedings of the XXVIII International Baltic Road Conference, 26-28 August, 2013, Vilnius, Lithuania / Lithuanian Road Administration. Vilnius. ISBN 978-609-95526-0-6, p. 1-9

Žnidarič, A. (2010). Report P 901/09-670-1-Methodology to assess structural safety of bridges on state roads, in Slovene. Ljubljana: ZAG.

Žnidarč, A., Kreslin, M., Lavrič, I., Kalin, J. (2012). Simplified approach to modelling traffic loads on bridges. Proceedings of Transport Research Arena 2012. TRA2012, Athens, Greece: Elsevier publishing.

Žnidarič A. (2015). Slovenia – national report. Strategic direction session ST4 infrastructure. Management of bridges in Slovenia. XXVth world road congress in Seoul. Received from http://www.nc-piarc.si/dokumenti/gradivo_seoul/Slovene_national_report.pdf

The 30th International Baltic Road Conference

IMPROVING TRAFFIC SAFETY BY USING WAZE USER REPORTS

Raitis STEINBERGS[1], Maris KLIGIS[2]

[1]*SJSC Latvijas autocelu uzturetajs, Head of technological development department, raitis.steinbergs@lau.lv*
[2] *SJSC Latvijas autocelu uzturetajs, Head of IT Department, maris.kligis@lau.lv*

Abstract.

Road inspection regularity and existing types made by road maintenance crew have not been good enough to be aware what is really happening on the roads. Road users' contribution in road traffic safety is very important to ensure fast reaction on different road hazards.

It is important to ensure not only the most common ways to report road hazards on state roads by phone, by email and on social media, but also expand data sources options in modern and user-friendly way.

Waze navigation application already had functionality to report road hazards – to warn other application users, but no one acted to solve these road hazards until someone reported them through existing communication channels supported by Latvian State roads or Latvian road maintainer.

To ensure better road traffic safety and faster reaction time on road hazards solving, Latvian road maintainer gained access to Waze report feed, and, in corporation with Riga Technical university, made a system for analysing and processing Waze data. As the result - Latvian roads maintainer can improve road safety by faster reaction to road hazards reported by Waze users.

Today, up to 70 % from total reports processed by Latvian road maintainer are generated by Waze.

Keywords: traffic safety, road maintenance, road hazards, road user contribution, Waze

Introduction

To ensure better road traffic safety and faster reaction time, it's important to be aware what is really happening on the roads. There were two main ways how information about hazards on state roads was obtained – by phone when road users called and when performing inspections by road maintenance or road administration crews.

Regulations determines how often road maintenance requirement compliance monitoring must be done. The higher the maintenance class, the more frequently roads must be inspected. Inspection frequency for road supervisor – Latvian State roads are regulated by Regulations on the requirements for the daily maintenance (2010), and for Latvian road maintainer - Road specifications (2019). For both - road supervisor and road maintenance crew – inspection frequencies are the same. Detailed inspection regularity and total length for each maintenance class are displayed in Table 1.

Table 1. Advantages and disadvantages of communication channels [1], [2], [9]

Maintenance class	Total length, km	Inspection frequency
A	4 761.602	once a week
B	1 948.501	once every two weeks
C	11 439.867	once a month
D	2 004.396	once a quarter
Total	20 154.366	

As visible in Table 1, more than half of roads are with C maintenance class and inspection regularity is only once a month.

As there are many hazard types that cannot be predicted, like, obstacle (for example tree) or dead animal, even inspections once a week for A maintenance class roads is very low to be aware of what's happening on roads. Of course, road inspection also has different aims, not only discovering hazards, that needs to be solved within few or 24 hours.

On high maintenance class roads, where hazard reaction time is the shortest, also the traffic flow is the highest, it's important that drivers contribute in ensuring traffic safety, but to do that, appropriate communication channel is needed. For state road administration most commonly used and most advertised is Traffic information center's support line +371 8000 5555. Typical Road user can forget the number or they haven't heard it, as the result there is still great possibility that hazard won't be reported or will be reported later. The lower the maintenance class, the lower possibility that hazard will be reported.

The Main idea of developing and implementing system for analyzing and processing Waze data was: creating a new, modern and user friendly way for road users to report hazards on roads would be ineffective and the best strategy to improve awareness of what's happening on roads is to use already existing data source, like Waze navigation application. Special Waze data analyzing and processing system was developed and implemented in winter season 2019/2020. Hazards reported by Waze users have been registered and solved as hazard from any other communication channel.

1. Waze application

Waze is one of the most popular Navigation application worldwide. It's second most popular Navigation app for iOS [3]. Application have been downloaded from Android OS official application store "Google Play" more than 100 million times [4]. Waze is not just a regular navigation app with route planning and giving turn-by-turn guidance, but also Waze can tell a lot about traffic, construction on roads, police, crashes and more in real time, thanks to very high road user's contribution [4]. Waze is used even when travelling very well-known routes just to be aware what's happening on road ahead.

For a long time, Waze has been very popular in Latvia, there is not an official Waze announced statistics, but few years ago 100 000 active users was the most often used number [5],[6]. If the total number of registered and insured cars (up to 3.5 tons) on 1st of February, 2021 was almost 669 thousand, then almost every 7th car driver is using Waze.

Latvian State roads and Waze have been cooperating since 2014, information about road construction, road condition in winter and road maintenance unit location was shared with Waze and only reports about potholes was used to improve pothole repair planning in spring [7].

In Waze application it's possible to warn other drivers of different types of hazards, but only few of them can actually be solved by performing road maintenance work. Waze hazard types that meets the specifics of road maintenance work are:

- Pothole;
- Road kill (dead animal);
- Obstacle;
- Missing road sign;
- Floods;
- Ice on road;
- Snow on road.

Waze reports are also available on the Waze public map at https://www.waze.com/livemap, but the display of reports are not suitable for effective prevention of dangerous traffic situations because:

- All notifications are displayed, regardless of their type and location;
- Only partial hazard information is displayed;
- Reports are not "stored" - the report may disappear from the map, although not solved;
- It is not possible to filter reports according to the required specifics;
- Unable to merge messages for a single event;
- Unable to mark messages as processed.

Based on the above-mentioned shortcomings, it was not possible to work effectively with Waze reports as a stand-alone and equivalent to other communication channels and it was necessary to develop a specific Waze report processing tool.

Waze has been open for collaboration and offered access to the Waze report stream online through the Connected Citizens program. In 2019, within the framework of the program mentioned Latvian road maintainer cooperation was approved and an online flow of data on all events in the territory of Latvia was ensured.

2. Waze report analyse tool

To develop a special Waze data flow processing and analysis tool, cooperation was started with Riga Technical University. The data flow provided by Waze included not only reports about hazards, but also information on congestion, road construction and other public information. Each hazard report contains information about the time of publication, location (coordinates), hazard type, subtype, location (city) and street / road and several hazard reliability parameters.

The main task of the specialized processing and analysis tool was to filter only necessary information from the Waze data flow and display the reports in the desired way. The basic requirements of the solution were defined:

- Display of reports on the map which must update information every 60 seconds and displays current situation;
- Report filtering by certain types of hazards, location, time and other parameters;
- Ability to define which hazards types does not need to display or save for later analysis at all;
- Determining the belonging to the Latvian road maintainer district;
- Analysis of historical data by visually mapping their concentration;
- Ability to combine reports and mark them as processed.

Desired analysis functions could be implemented as Waze already provided various additional information to the report. The only information that was additionally added was belonging to the Latvian road maintainer district, by drawing their polygons on map.

Waze data flow processing and analysis tool was developed in 2019 and the analysis of Waze data flow as a full-fledged communication channel was started at the beginning of the winter season of 2019/2020 – on November 1, 2019.

3. Impact of Waze reports

Since 2018, incoming information about the situation on the roads in Latvian road maintainer has been processed and managed centrally by the Customer Support Service. In public, Latvian State roads Traffic information center still was positioned as the main communication center. But inevitably road users also called Latvian road maintainer directly and this process had to be managed, so the Customer Support Service was established. For more convenient communication between Latvian road maintainer and Latvian State roads and to determine the action status regarding the received information, a digital journal was created for registration and processing of incoming information.

Waze reports were recorded in the digital journal in the same way as information received through any other communication channel (calls, sms, e-mail, Facebook or Twitter entry) and passed on to regional districts for execution.

The last two types of hazards relate to road maintenance in winter and, given that the condition of roads is monitored 24/7 during the winter season, these reports provide an additional source of information for responsible staff but are not recorded in the incoming report log as a specific event.

In the first 12 months since the start of Waze reports processing, the number of reports received by the Latvian road maintainer Customer Support Service (not including Latvian State roads Traffic information center) more than doubled. Total share of Waze reports reached 64%. See the Figure 1 for a detailed info report distribution among different communication channels and Figure 2 for the most common types of hazards by Waze:

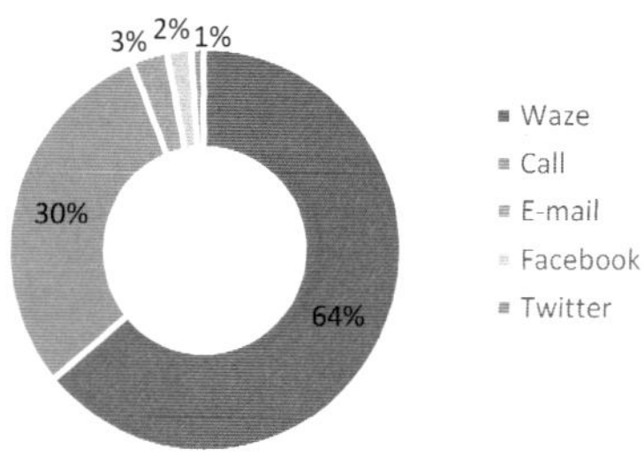

Figure 1. Communication channels shares

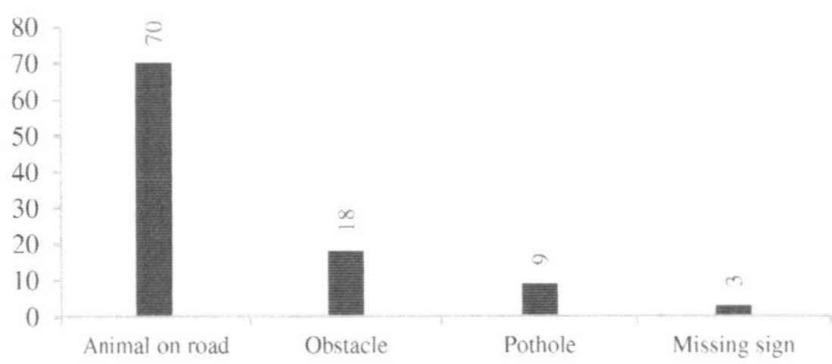

Figure 2. Most common types of hazards

Most often reported hazard by Waze users were animals on road - 70% and obstacles on the road - 18%.

The main communication channel for Latvian State roads Traffic information center is the support line (calls). Including the reports received by Latvian State roads, in 2020 they were total 9000 reports. Of these, 70% were received by phone, 26% by Waze, 4% by social networks or e-mail.

In essence, all reports (regardless of the communication channel) are the same, but there are some differences. Advantages and disadvantages of the information received via the support line (calls) and Waze are gathered in Table 1.

Table 1. Advantages and disadvantages of communication channels

Communication channel	Advantages	Disadvantages
Support line (call)	- The most accessible communication channel - Ability to describe the event in more detail - Ability to report unclassified message types	- Often problems describe the exact location of the event - More time-consuming report processing
Waze app	- More accurate location - Ability to automate the assignment and classification of additional information - The report also serves as a warning to other drivers	- It is not possible to provide additional information to describe the event - Reporting is only possible at the scene

4. Future development plans

After the implementation of the solution, the first improvements in the report processing automation were made. The need for improvement was identified in two directions:

- Hazard reports also often appeared on sections of roads that have been handed over to the Contractor during road construction;
- Waze data flow also contained reports generated through the LVC winter service system, which was originally generated by Latvian road maintainer winter service crew.

To reduce the time required for report processing, based on the Latvian State roads geospatial data about construction sites and the data flow submitted by the Latvian State roads winter service system, improvements were made:

- Additional information was assigned to the report - belonging to the construction site, as a result it is possible to filter the reports;
- Reports whose coordinates coincide with the coordinates of the reports in the Latvian State roads data stream were classified and can also be filtered.

In the near future we develop and implement the automation of Waze report flow processing, supplementing the solution with the addition of more accurate geospatial information (road addresses) and implementing a machine learning solution that combines reports that are actually about one event based on historical report processing data.

Conclusion

The opening of the Waze communication channel has been a successful way to complement existing / classic communication channels with a new way of digital reporting, without even developing the reporting platform itself, but using one that already exists.

By using Waze reports as an equal report to classic communication channel, it has been possible to increase the number of actual hazards treated and eliminated by more than ¼, which is reason to believe that it is possible to increase road safety due to more information on the actual road situation and hazards which needs attention.

The future lies in the ability to connect communication channels where information about what is happening on the road already exists, rather than creating new ones.

Acknowledgements

Waze data processing and analysis tool was developed based on agreement between Latvian road maintainer and Riga Technical university. This manuscript was created to inform about Waze project as a case study.

References

1. Regulations of the Cabinet of Ministers of March 9, 2010 No. 224 "Regulations on the Daily Maintenance Requirements of State and Municipal Roads and Control of Their Execution", Paragraph 17. https://likumi.lv/ta/id/206467#p17
2. Road specifications 2019. (2020). 369-374. http://lvceli.lv/wp-content/uploads/2015/06/CS2019_ar_Grozijumiem_Nr1un2-1.pdf
3. Waze Navigation & Live Traffic on App Store. https://apps.apple.com/us/app/waze-navigation-live-traffic/id323229106
4. Waze - GPS, Maps, Traffic Alerts & Live Navigation. https://play.google.com/store/apps/details?id=com.waze&hl=lv&gl=US
5. The Waze application has received more than 2,500 reports of pothole. Leta. 21.04.2015. https://www.la.lv/aplikacija-waze-sanemti-vairak-neka-2500-zinojumi-par-bedrem
6. Latvia is the seventh best country in the world to drive a car, according to a study by the Waze application. Portal skaties.lv. 18.02.2017. https://skaties.lv/auto/latvija-ir-septita-labaka-valsts-pasaule-brauksanai-ar-auto-liecina-aplikacijas-waze-petijums/
7. The winter maintenance season ends on state roads; please mark the holes in the Waze application. 03.04.2017. https://lvceli.lv/aktualitates/uz-valsts-autoceliem-beidzas-ziemas-uzturesanas-sezona-aicinam-atzimet-bedrites-aplikacija-waze/
8. Latvian State Roads. https://lvceli.lv/
9. Road maintenance. Latvian State Roads. https://lvceli.lv/celu-tikls/buvnieciba-un-uzturesana/celu-uzturesana/

Template: Anonymized Document

IDENTIFICATION AND ANALYSIS OF POTENTIAL RISK FACTORS INFLUENCING THE ROAD SAFETY LEVEL AT DESIGNATED PEDESTRIAN CROSSINGS

Gábor PAUER (https://orcid.org/0000-0002-3768-5328)[1], Nóra KRIZSIK [2], Szilárd SZIGETI [3], Zsolt HAMZA[4],

[1,2,3,4]*Road Safety Research Centre, KTI Institute for Transport Sciences Non-profit Ltd., Budapest, Hungary*
Corresponding author: Gábor, Pauer, pauer.gabor@kti.hu

Received day month year; accepted date

Abstract. According to the data of the European Road Safety Observatory, around 21% of all road fatalities are suffered by pedestrians in the EU. In 2019 in Hungary, road accidents of pedestrians have had a share of 14,6% in all road accidents with personal injuries, which meant 2535 accidents in which one or more pedestrians were injured. A significant proportion of the accidents occurred at designated pedestrian crossings (43,1% of pedestrian accidents in 2019), and this trend increased over the last 5 years. To account the problem, Institute for Transport Sciences Non-profit Ltd. conducted a research focusing on the identification of potential risk factors which may have a negative impact on the level of traffic safety of designated pedestrian crossings in urban areas. Analysis and ranking of the risk factors have been carried out based on experts' evaluation and scoring, with the use of statistical methods. The results of the work explore the risks that need to be addressed with special attention during the review of existing, and the establishment of new pedestrian crossings.

Keywords: road safety, pedestrian crossing, risk assessment, KIPA analysis, cluster analysis

Introduction

The road safety of vulnerable road users is a key area of recent and future strategic documents. In the framework of the European Union on road safety, actions have been determined in seven focus areas, including vulnerable road users such as pedestrians, cyclists and motorbike riders (EC, 2010). Based on the Working Paper on the EU Road Safety Policy Framework for 2021-2030, the safety of vulnerable road users seems to be highlighted also in the next decade, especially as a wider range of different automated/connected vehicles will appear in a mixed traffic with this group (EC, 2019).

Pedestrians, as part of the vulnerable road users are characterized by high probability of injuries in a road accident (Stutts & Hunter, 1999; Eilert-Petersson & Schelp, 1999). To highlight the importance of the investigations related to the safety level of pedestrians and pedestrian crossings, a short road safety analysis has been provided mainly focusing on to the country of the authors (Hungary).

In the European Union, pedestrians make up 21% of total fatalities in road accidents. If we look only at urban are as, where pedestrians mainly travel, this proportion is as high as almost 40%, accounting for the largest share of victims. The pedestrian fatalities in urban areas fell at a slower pace than the decline in the overall number of road traffic victims (EC, 2020).

In Hungary, between 2014 and 2019 pedestrian accidents accounted for about 15% of the number of total road accidents. In 2019, 2419 accidents happened in which pedestrians were injured. The proportion of pedestrian accidents has not changed significantly in this years (see Figure 1).

Based on the data of 2014-2019, the most frequent type of pedestrian accidents was "traffic accidents with pedestrian at designated pedestrian crossings at intersections". The rate of this type of accident has increased almost continuously over the last 6 years. In 2019, 22,7% of pedestrian accidents were of this type. In addition, the proportion of 3 other types of accidents increased significantly between 2014 and 2019:

- pedestrian crashes outside of intersections (16,9 % of pedestrian crashes in 2019);
- traffic accidents with pedestrians at designated pedestrian crossings outside of intersections (15,5% in 2019);
- other traffic accidents with pedestrians, (16,5% in 2019).

Examining the most common types of pedestrian accidents, we can conclude that a high proportion of pedestrian crashes happened at designated pedestrian crossings.

Template: Anonymized Document

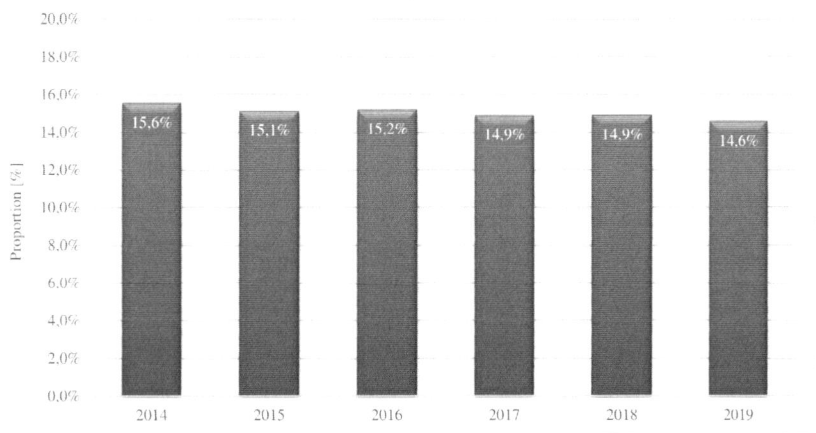

Figure 1. Proportion of pedestrian crashes within all road accidents in Hungary (2014-2019)

The rate of accidents at designated pedestrian crossings are especially high in urban areas. In Hungary in 2019, 43,1% of accidents involving pedestrians (971 accidents) has happened at these kind of locations. As can be seen in Figure 2, the rate of accidents at designated pedestrian crossings has shown an increasing trend urban areas (increasing from 37,7% to 43,1% during the examined years).

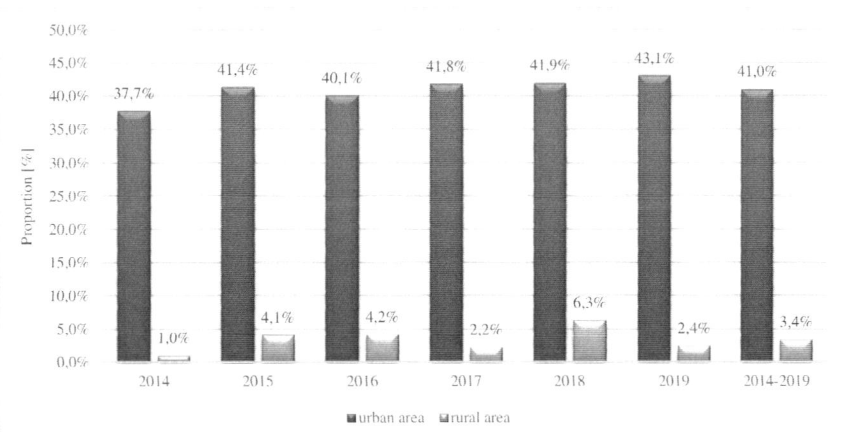

Figure 2. Proportion of pedestrian crashes at designated pedestrian crossings in Hungary (2014-2019)

Accidents at designated pedestrian crossings between 2014 and 2019 resulted in 179 deaths, 1999 serious injuries and 3878 slight injuries in Hungary. In addition to personal tragedies, taking into account the values of road accident losses, all of this caused HUF 194 billion in economic losses to the country in 6 years.

In line with the above presented problems, the purpose of our research was to identify potential risk factors influencing the safety level of uncontrolled (i.e. no traffic lights) pedestrian crossings in a systematic approach. Based on that, factors have been assessed and ranked by statistical methods to identify the most critical risks which have to get the prime focus when the road safety level of pedestrian crossings are to be evaluated or improved.

Template: Anonymized Document

1. Identification and assessment of risk factors

The potential risk factors that influence the road safety level of uncontrolled pedestrian crossings have been revealed and categorized into four main groups. The factors have been notated by f_j^i, where the upper index refers to one of the determined four main groups, and lower index identifies the individual factor within the group. The main groups were:

- Factors related to the technical design and control ($i = 1$);
- Factors related to the construction design and close environment ($i = 2$);
- Factors related to the traffic engineering design ($i = 3$)
- Traffic characteristics, distracting situations and objects ($i = 4$).

When determining the individual risk factors (indicated in Table 1), we have relied on the domestic regulation that deals with the principles of road traffic management and the placement of road signs (20/1984. (XII. 21.) Government Decree), the Hungarian Road Technical Specification prescribing the design of transport facilities for pedestrian traffic (ÚT 2-1-211), as well as on the results of international studies (Antov, Rõivas, Antso, & Sürje, 2011; Pashkevich & Nowak, 2017; Pashkevich, Krasilnikova & Antov, 2016) assessing the risks of pedestrian crossings.

Table 1. The list of the identified risk factors influencing the safety level of pedestrian crossings without traffic lights

Notation	Description of risk factor
f_1^1	The need of crossing more than 2 traffic lanes without the possibility to safely interrupt the crossing movement (e.g. by the presence of pedestrian refugee)
f_2^1	Crossing of parallel traffic lanes (with vehicle traffic towards the same direction)
f_3^1	The absence of adequate separation from the road, or not adequate pedestrian waiting areas (e.g. too narrow, not barrier-free)
f_4^1	The uncontrolled pedestrian crossing is within 100m to an intersection/crossing controlled by traffic light, or can be found on a route controlled by coordinated traffic lights
f_5^1	The minimum width of the designated pedestrian crossing is not ensured, or the angle with the road axis is not perpendicular
f_6^1	The absence of public lighting, or the lighting is not adequate
f_1^2	The pedestrian crossing cannot, or can hardly be recognized from a distance of at least 50 metres
f_2^2	Mutual detection of pedestrians and drivers is obstructed due to obscuring elements
f_1^3	Bad condition of road markings and traffic signs designing the pedestrian crossing
f_2^3	The absence of road markings or traffic sign predicting the pedestrian crossing where it would be justified
f_3^3	Bad condition of road markings and traffic signs predicting the pedestrian crossing
f_4^3	The absence of the prohibition of overtaking in front of the pedestrian crossing
f_1^4	The presence of a situation that divides the attention of drivers right in front of to the pedestrian crossing
f_2^4	The presence of a situation that divides the attention of drivers right behind the pedestrian crossing
f_3^4	The presence of an object that divides the attention of drivers in the vicinity of the pedestrian crossing
f_4^4	Generally high traffic volume in the vicinity of the pedestrian crossing
f_5^4	Generally high traffic speed in the vicinity of the pedestrian crossing
f_6^4	The pedestrian traffic is characterized by an increased risk (e.g. low pedestrian traffic, high share of elderly or children)

The identified risk factors may affect the behavior of drivers and pedestrians according to several aspects, and therefore have effects on road safety, on the willingness to give priority for pedestrians as well as on the level of attention.

A questionnaire survey has been carried out among transport and traffic engineering experts assessing the potential effect of the identified risk factors. The experts evaluated the listed factors according to the following criteria, on a scale of 5:

1. C_1: Impact on the risk of accidents due to driver error (1- does not increase at all, 5- increases significantly);
2. C_2: Impact on the risk of accidents due to pedestrian error (1- does not increase at all, 5- increases significantly);
3. C_3: Impact on the rate of giving priority for the pedestrians (1- does not affect at all, 5- affects strongly (whether in a negative or positive direction));
4. C_4: Incidence frequency on the road network (1- does not occur at all, 5- occurs extremely frequently);
5. C_5: Effect on achieving increased attention of drivers (1- does not inhibit at all, 5- strongly inhibits).

Template: Anonymized Document

The experts evaluated each influencing factor according to the 5 criteria. During this evaluation, average vehicle and pedestrian traffic have been assumed. For the analysis, the evaluation scores of experts have been averaged and summarized in Table 2. The higher values refer to higher risks, as explained by the scales of the criteria.

Table 2. Average score of risk factors according to the defined evaluation criteria

	C_1	C_2	C_3	C_4	C_5	Relative standard deviation
f_1^1	3,429	3,143	2,857	2,429	3,286	0,159
f_2^1	4,429	2,714	3,571	3,571	3,143	0,136
f_3^1	2,714	3,429	2,286	2,429	2,143	0,226
f_4^1	4,143	3,000	3,429	2,571	3,429	0,203
f_5^1	2,286	2,857	2,429	2,143	2,000	0,250
f_6^1	4,571	2,714	4,286	3,714	4,000	0,196
f_1^2	4,143	2,571	4,143	2,429	4,143	0,178
f_2^2	3,857	3,571	3,571	3,286	3,571	0,082
f_1^3	3,571	1,857	3,143	2,857	3,571	0,163
f_2^3	3,429	1,286	3,000	2,000	3,000	0,126
f_3^3	3,143	1,571	2,857	2,714	2,857	0,149
f_4^3	2,857	1,143	1,857	3,857	2,143	0,133
f_1^4	4,429	2,143	3,857	3,429	4,286	0,133
f_2^4	4,000	1,857	3,571	3,000	3,714	0,104
f_3^4	3,714	2,143	3,571	3,571	3,571	0,098
f_4^4	3,000	3,000	3,000	3,286	3,571	0,085
f_5^4	4,571	3,429	4,143	3,143	3,571	0,174
f_6^4	3,429	3,429	3,286	2,857	1,714	0,140

The highest average scores (4,571) were attributed to deficiencies in public lighting and high traffic speeds in the vicinity of the crossing in terms of the effects on the risk of accidents due to driver error. The lowest average score (1,143) was assigned for the impact of the absence of prohibition of overtaking on the risk of accidents due to pedestrian errors.

Based on the evaluation of experts, the risk factors have been ranked using the KIPA method. The factors have also been classified using cluster analysis, according to different aspects. Our aim was to find the group of factors that can be considered particularly risky according to some criteria.

IOP Conf. Series: Materials Science and Engineering 1202 (2021) 012032 doi:10.1088/1757-899X/1202/1/012032

Template: Anonymized Document

2. Ranking of the risk factors based on the KIPA analysis

As first step of the analysis, the relative standard deviations of the average scores of experts have been calculated to confirm the adequacy of the data (see in the last column of Table 2). The relative standard deviations were higher than the expected 0,2 threshold in only three cases, and even in these cases the difference was small, and no outlier data could be observed. This shows the heterogeneous structure of the scoring. Thus, in our primary studies the sample has been considered to be well characterized by the means of the values, and the KIPA method was found to be applicable.

The KIPA method is suitable for comparing complex systems, with the help of which the identified 18 potential risk factors can be ranked taking into account all the examined aspects (Kindler & Papp, 1977). Using the method, the most critical risk factors can be identified based on the experts' opinion.

The alternatives have been characterized on scales based on the weight of the evaluation criteria, by pairwise comparison. The main steps of the procedure are the following:

1. construction of scales for measuring evaluation criteria (taking into account weights);
2. preparation of the basic table of the KIPA method (transforming the scores of experts to the previous scales);
3. preparation of the KIPA matrix (pairwise comparison);
4. setting preference and disqualification thresholds;
5. determining the order of preference (ranking of risk factors).

The KIPA method also provides an opportunity to take into account weights of the evaluation criteria. In case of our research, all 5 defined evaluation criteria were considered equally important (with equal weight of value 1). Thus, the measurement scales of the first step were the same as the original evaluation scales (ranging from 1 to 5), and the basic table of the KIPA analysis included the average scores given by the experts.

To produce the KIPA matrix, the preference (c_{ij}) and disqualification (d_{ij}) indicators have been calculated based on pairwise comparisons. The preference indicator provides information on the advantage of the i-th alternative over the j-th, and is calculated for each relationship. Its value is obtained by summing the weights of the evaluation criteria for which the given alternative is preferred or indifferent (the value assigned to it in the basic table is greater than or equal to) compared to the other alternatives. The disqualification indicators have also been calculated in all respects, but only the evaluation criterion for which the preference intensity was highest had to be taken into account. Thus, in order to determine the value of the disqualification indicators, the evaluation criterion that meets the following two conditions has been selected:

- according to the investigated criterion, the value assigned to the j-th alternative is higher than the value assigned to the i-th;
- the absolute value of the difference between the values assigned to the i-th and j-th alternatives is the largest.

The absolute value of the difference between the values selected in this way is the largest scale difference where the examined i-th alternative is at a disadvantage compared to the j-th alternative. To calculate the disqualification indicator, this value has been divided by the size of the largest scale. Note that in our case the size of each scale was the same since all evaluation criteria were given the same weight. The result has been transformed to the percentage form.

With the use of the calculated preference and disqualification indicators, the KIPA matrix has been created. The matrix compares the alternatives (risk factors) to each other, therefore the size of it is 18x18. The intersection of the i-th row and the j-th column contains both the c_{ij} and d_{ij} indicators. Due to space constraints, the matrix has not been visualized in the article.

The determination of the order of preference (the ranking of the risk factors) has been performed based on the data of the matrix. The preference threshold was considered at 70% ($c_{ij} > 0{,}7$) and the disqualification threshold at 20% ($d_{ij} < 0{,}2$). A total of 66 pairwise comparisons met both of these criteria.

According to the results of the pairwise comparisons, an assortment graph have been elaborated. Nodes of the graph represent the examined risk factors. The edges between the compared risk factors are directed, starting from the preferred alternative and pointing towards the other node (representing the results of the comparison of the given alternatives). Based on the directed edges, the preference order of the compared alternatives (risk factors) has been determined. An alternative is more favorable the more arrows start from it, and the more unfavorable the more arrows pointing there.

The graph illustrates the results well in case of less compared alternatives, but the presentation of it has been excluded from our article due to the high number of considered risk factors. In Figure 3 however the ranking of the alternatives are well illustrated (the most preferred alternatives are on the top of the pyramid), and the structure of the graph has also been represented by the numbers in the brackets. The first number in the bracket indicates the number of alternatives over which the given factor is preferred (number of edges starting from that node), while the second

Template: Anonymized Document

number represents the opposite (number of alternatives that are preferred over the given factor, i.e. the number of edges arriving to that node).

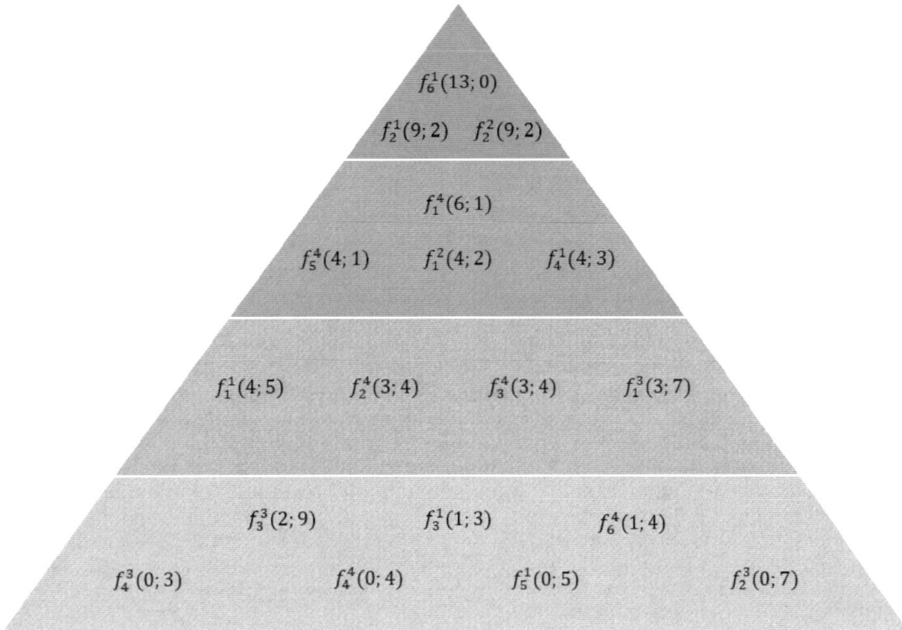

Figure 3. Ranking of the examined potential risk factors

The ranking was based on the level of safety risk assigned to the alternatives. According to the results of the KIPA analysis based on the evaluation of the experts, the factors at the top of the pyramid can be considered to be the most critical risks at urban pedestrian crossings. The less preferred alternatives at the bottom of the pyramid represent the factors with the lowest risks. These were preferred in less than 20% of their pairwise comparisons.

The results showed, that – according to the experts - the absence or inadequacy of public lighting was considered the most critical problem in the case of designated pedestrian crossings within urban area (taking into account all criteria). In the ranking, this is followed by risks due to the possible difficulties in detecting pedestrians or drivers (covering effect of parallel traffic in the same direction; obscuring roadside elements like trees, vegetation, parking vehicles). Each of the three most critical risks is therefore directly related to the issue of visibility at the pedestrian crossings.

The following factors have also been found to be among the most critical risks based on the analysis (second level of the pyramid):

- the presence of a situation that divides the attention of drivers right in front of to the pedestrian crossing;
- generally high traffic speed in the vicinity of the pedestrian crossing;
- the pedestrian crossing cannot, or can hardly be recognized from a distance of at least 50 metres;
- the uncontrolled pedestrian crossing is within 100m to an intersection/crossing controlled by traffic light, or can be found on a route controlled by coordinated traffic lights.

It is interesting to note, that all of the factors related to the traffic engineering design were in the back half of the ranking so can be considered less critical from the road safety point of view. The average scores of the four main groups of risk factors were:

- 3.11 points in case of factors related to the technical design and control ($i = 1$);
- 3.53 points in case of factors related to the construction design and close environment ($i = 2$);

Template: Anonymized Document

- 2.64 points in case of factors related to the traffic engineering design ($i = 3$);
- 3.34 points in case of factors related to traffic characteristics, distracting situations and objects ($i = 4$): 3.34 points.

That is, based only on the average scores of the experts, the risks related to the construction design of the pedestrian crossings were rated as the most critical.

Based on the KIPA analysis, the following 4 factors were not "preferred" to any of the other risk factors so have been considered to be the least critical:

- the absence of the prohibition of overtaking in front of the pedestrian crossing;
- generally high traffic volume in the vicinity of the pedestrian crossing;
- the minimum width of the designated pedestrian crossing is not ensured, or the angle with the road axis is not perpendicular;
- the absence of road markings or traffic sign predicting the pedestrian crossing where it would be justified.

3. Grouping of the risk factors based on cluster analysis

Following the ranking process, the risk factors have been investigated also with the use of cluster analysis. Clusters have been created based on the average scores given by the experts, using the SPSS statistical program.

To perform the classification, the hierarchical Ward-method was used. This procedure is very common in economic applications, can be interpreted well in practice, and results in groups of roughly the same size (Simon, 2006). The method is based on aggregation. In the first step it considers each element as a separate cluster and connects to each other, while forming larger and larger groups. The mean and the sum of the squared deviations from the mean are calculated for all points within the cluster. For the larger cluster formation, the point or cluster is used with which the difference in the sum of squares of the deviation is the smallest.

During the clustering, the examined 18 risk factors have been classified into 3-8 clusters. Based on the examination of the results, the 5 cluster solution was found to be the most suitable for the characterization. The results are showed in Table 3, where all clusters have been characterized based on the average score (and standard deviation in brackets) given for the elements according to the defined criteria.

The table also shows the average based on all of the factors, which can be used for comparison when analyzing the characteristics of the different clusters. Significant downward deviations from the mean were highlighted by italic numbers (difference was greater than 10%), while the significant positive deviations have been indicated by gray background color. In the last column, the risk factors included in each cluster are presented.

Table 3. Average score (and standard deviation) of the clusters according to the criteria

	Number of elements	C_1	C_2	C_3	C_4	C_5	Cluster elements
All risk factors	18	3,65 (0,68)	2,55 (0,77)	3,27 (0,66)	2,96 (0,56)	3,21 (0,76)	
Cluster 1.	3	3,52 (0,58)	3,05 (0,08)	3,10 (0,30)	2,76 (0,46)	3,43 (0,14)	$f_1^1; f_4^1; f_4^4$
Cluster 2.	6	4,33 (0,28)	2,86 (0,54)	3,93 (0,31)	3,26 (0,46)	3,79 (0,43)	$f_2^1; f_6^1; f_1^2; f_2^2; f_1^4; f_5^4$
Cluster 3.	3	*2,81 (0,58)*	3,24 (0,33)	*2,67 (0,54)*	*2,48 (0,36)*	*1,95 (0,22)*	$f_3^1; f_5^1; f_6^4$
Cluster 4.	5	3,57 (0,32)	*1,74 (0,33)*	3,23 (0,33)	2,83 (0,57)	3,34 (0,39)	$f_1^3; f_2^3; f_3^3; f_2^4; f_3^4$
Cluster 5.	1	*2,86 (-)*	*1,14 (-)*	*1,86 (-)*	3,86 (-)	*2,14 (-)*	f_4^3

The most critical factors, characterized by the highest average scores have been included in Cluster 2. The elements of this cluster were associated with higher risks of accidents, higher effects on the rate of giving the priority, higher probability of distracting drivers' attention, and were also considered to be frequent on the road network. The result is in line with our previous analysis, the elements of the most critical cluster have been assigned from the first elements of the rank determined by the KIPA analysis.

Based on the examination of the further clusters, the following conclusions can be made. There were 3 elements in Cluster 1 that were rated as average by experts according to the most criteria. However, in case of the elements of this cluster, the risk of accidents due to pedestrian fault was considered higher than average. Thus, these are situations

Template: Anonymized Document

where pedestrians can make mistakes more easily than average, while the risks are not higher than the average in terms of the other criteria.

Cluster 3 contains similar elements. Here the risk of accidents due to pedestrian error is also higher than the average, but these factors were characterized by lower-than-average risks in all other respects.

Cluster 4 is the "opposite" of Cluster 1. It contains elements where the risk of accidents due to pedestrian error was considered to be lower than average, while the risk was considered average in all other respects. These factors have less influence on the frequency of pedestrian errors, which seems valid since the factors listed in this cluster are conditions that distract the drivers.

A single risk factor ("the absence of the prohibition of overtaking in front of the pedestrian crossing") forms an independent cluster. According to the experts, this factor is not so significant in terms of influencing traffic safety, the level of giving the priority and the level of attention, however, it was identified as the most common factor on the road network.

Conclusions

Pedestrian crossings are the points of the road network designated to ensure the safe crossing of pedestrians. However, as it has been pointed out by our analysis, a significant proportion of pedestrian accidents occurs at these places. This fact justified the need of assessing the potential risks at designated pedestrian crossings and their environments. During our research, potential risk factors that may negatively affect the level of traffic safety have been identified in a systematic approach. Our investigations focused on the designated pedestrian crossings within residential areas.

The factors were evaluated on the basis of expert opinions, using the KIPA method, which allows the ranking of data by several aspects, as well as with the help of cluster analysis. Our goal was to identify the most critical risk factors, by eliminating which the greatest traffic safety benefits can be achieved.

The results of the ranking and clustering pointed in the same direction, clearly identifying the factors considered by experts to be the most critical:

- crossing of parallel traffic lanes (with vehicle traffic towards the same direction);
- the absence, or inadequacy of public lighting;
- difficult identification of the designated pedestrian crossing from a suitable distance;
- obstruction of the mutual detection of pedestrians and drivers due to obscuring elements;
- the presence of a situation that divides the attention of drivers right in front of to the pedestrian crossing;
- generally high traffic speed in the vicinity of the pedestrian crossing.

These were the factors which were considered by the experts the most critical in terms of the risks of accidents and the distraction of drivers' attention level. In addition they have been evaluated as relatively common risks, having a greater than average impact on the rate of giving the priority for pedestrians. When designing new pedestrian crossings and reviewing existing ones, it is important to focus on eliminating, or reducing the risks arising from the above presented factors in order to increase the safety of pedestrian crossings in the most effective way.

Disclosure Statement

Authors declare that they have no competing financial, professional, or personal interests from other parties.

Template: Anonymized Document

References

20/1984. (XII. 21.) Government Decree. *Management of road traffic and placement of road signs*. Retrieved from: https://net.jogtar.hu/jogszabaly?docid=98400020.KMB

Antov, D., Rõivas, T., Antso, I., & Sürje P. (2011). A Method For Pedestrian Crossing Risk Assessment. *Urban Transport 2011, 116*, 587-598. DOI: 10.2495/UT110501

Department of Transport Infrastructure of the Ministry of Transport, Communications and Energy. (2017) *Road Technical Specification.* ÚT 2-1.211. Design of public transport facilities for pedestrian transport. Budapest, Hungary

Eilert-Petersson, E., & Schelp, L. (1999). An epidemiological study of non-fatal pedestrian injuries. *Safety Science*, *29*(2), 125-141. https://doi.org/10.1016/S0925-7535(98)00014-9

European Commission. (2010). *Communication "Towards a European road safety area: policy orientations on road safety 2011-2020"*. COM(2010) 389 final. Retrieved from https://ec.europa.eu/transport/sites/transport/files/road_safety/pdf/com_20072010_en.pdf

European Commission. (2019). *Commission Staff Working Document "EU Road Safety Policy Framework 2021-2030 - Next steps towards "Vision Zero""*. SWD(2019) 283 final. Retrieved from https://ec.europa.eu/transport/road_safety/sites/roadsafety/files/move-2019-01178-01-00-en-tra-00_3.pdf

European Commission. (2020). *2019 road safety statistics: what is behind the figures?* Retrieved from https://ec.europa.eu/commission/presscorner/detail/HU/qanda_20_1004

Kindler, J., Papp, O. (1977). Komplex rendszerek vizsgálata- Összemérési módszerek. *Műszaki Könyvkiadó: Budapest*, (41), 151-175. ISBN: 963 10 1830 X (in Hungarian)

Pashkevich, A., & Nowak, M. (2017). Road safety risk assessment at pedestrian crossings: a case study from Sułkowice. *Scientific Journal of Silesian University of Technology. Series Transport*, *95*, 159-170. DOI: 10.20858/sjsutst.2017.95.15.

Pashkevich, M., Krasilnikova, A., & Antov, D. (2016). Method for Pedestrian Crossing Risk Assessment and Safety Level Determination: the Case Study of Tallinn. In *Proceedings of the CIT2016 – XII Congreso de Ingenieria del Transporte València, Universitat Politècnica de València*, Valencia, Spain. DOI: http://dx.doi.org/10.4995/CIT2016.2016.4124

Simon, J. (2006). A klaszterezés alkalmazási lehetőségei a marketingkutatásban. *Statisztikai Szemle,* 84(7), 627–651. (in Hungarian)

Stutts, J.C. & Hunter, W.W. (1999). Motor vehicle and roadway factors in pedestrian and bicyclist injuries: an examination based on emergency department data. *Accident Analysis & Prevention*, *31*(5), 505-514. https://doi.org/10.1016/S0001-4575(99)00007-X

HIGH PERFORMANCE PAVEMENT MARKINGS ENHANCING CAMERA AND LIDAR DETECTION

Gernot Sauter, Marcel Döring, Rik Nuyttens

3M Deutschland GmbH, Transportation Safety Laboratory, Neuss, Germany

28th February 2021

Abstract. It is well known that camera and video sensors have limitations in detecting pavement markings under certain conditions e.g. glare from sunlight or other vehicles, rain, fog etc. First generations of lane keeping systems depend on visual light. Erroneous detection is also resulting from irregular road surfaces such as glossy bitumen sealing strips, rain puddles or simply worn asphalt. The role of higher performing markings and better visual camera detection has been studied with Vedecom France. LiDAR (light detection and ranging) technology could help to fill remaining gaps, as it actively sends out IR (infrared) light, that returns reliable images of the road scenario and pavement markings both day and nighttime. In order to evaluate the opportunities of LiDAR technology for the detection of road markings, 3M Company and the University of Applied Sciences in Dresden decided to work together in a joint research project. All-Weather Elements AWE, are the latest development of high-performance optics, using high index beads to provide reflectivity both in dry and wet condition. It could be determined that high performance markings help to increase the level of detection by both camera and LiDAR sensors. The AWE marking was detected from significantly longer distances, especially in wet and rainy conditions. In combination with common camera based LKA and LDW systems, the LiDAR sensors can increase the overall detection rate of pavement markings. This is especially important for vehicles with higher SAE levels of automated driving and can support the overall safety of vehicles. The research also evaluated existing test methods for wet and rain reflectivity in EN 1436 and ASTM E 2832 and how measured performance correlates with LiDAR detection.

Keywords: LIDAR, Pavement Marking, Automated Driving, Wet Reflectivity, Lane Keeping Assistance, Lane Departure Warning.

Introduction

The fast-growing automation of road traffic demands even better accuracy with regard to vehicle localization. Independent systems in the vehicle need to get or generate data to support the degree of automation.

Global Navigation Satellite Systems (GNSS) are used to get a localization information, but the accuracy is not high enough for automated driving. Camera systems are broadly used, creating an environmental model, but in various situations these systems fail. Radar systems are a good complement to get more detailed information of the environment and will increase the needed level of redundancy. Specially the velocity information of dynamic objects could be read out very easily.

Nowadays LiDAR systems are used in cars to complete the range of sensors which are able to provide additional information for the automated driving function. As it is sending out an own signal in a defined wavelength, it could easily calculate the distance to an object by measuring the time the signal needs back and forth to the sensor.

Here two approaches are described to support the sensors. A camera research project with Vedecom France and a planned LiDAR research project at HTW university of Dresden, of which first experiments are already presented.

1. Background on machine vision of pavement markings

In many vehicles, cameras are installed to recognize pavement markings. These basic driver assistance systems are commonly known as lane departure warning or lane keep assist. New cars in the European Union will soon have such systems as mandatory fitment. As mentioned in the introduction, camera-based systems may fail in various situations. Critical situations are e.g. glaring sun, bitumen stripes or shadows around tunnels and bridges. Also, in bad weather conditions it is difficult for cameras to detect the pavement markings. In order to quantify the effectiveness of camera systems under varying environmental conditions, 3M Company and Vedecom partnered to assess the performance of different road marking products under various conditions by evaluating the detectability with camera based Advanced Driver Assistance Systems.

In addition to that study, another research project is requested at the BMWI (Federal Ministry for Economic Affairs and Energy) to investigate pavement markings under different environmental conditions for the detection performance with LiDAR systems. Preliminary examinations of different markings showed high differences in detectability as retroreflection is the most important feature for LiDAR sensors. Specially in rain and nighttime conditions this will play an important role for the effectiveness of LiDAR systems.

2. LiDAR systems detecting pavement marking

Mono or stereo cameras with suitable image processing algorithms have been used as sensors for lane detection up to now. For use in highly and fully automated driving functions, these sensors will be one of the sources of information to manouvre the car in the center of the traffic lane. However, there are various situations in which this sensor technology fails.

In the existing driver assistance systems of SAE Level 2, the driver must immediately deactivate the systems himself (if not switched off automatically) and take over control of the vehicle. This is state-of-the-art for actual vehicles, e.g. in Tesla's 'autopilot' system. The driver remains responsible and must always stay in control. In the case of higher automation level (SAE level 3) of an automated vehicle, a period of at least 4 seconds is provided for the changeover; in the case of fully automated vehicles (SAE level 4), it is several minutes. This gives a completely new challenge for the detection systems for the unambiguous and permanent recognition of the lane.

This cannot be solved by means of camera or video sensors alone, especially since this is not a measurement process from a physical point of view. A particularly well-suited sensor for this application is a laser scanner. Here, the travel time of infrared radiation (time of flight, ToF) is measured, so that a physical connection exists. This is a particularly important circumstance for the development of safety critical technology. In the following graphic, the potential is illustrated by means of a measurement on the HTW (University of Applied Sciences Dresden) test site.

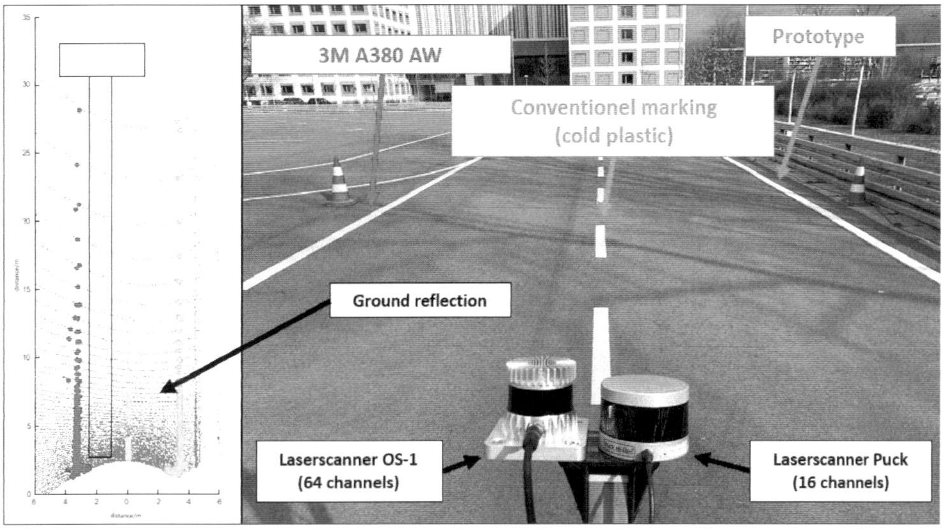

Figure 1: LiDAR Retroreflection intensities of different pavement markings

If we look at the intensities of the signals, we see clear differences between a conventional marking (cold plastic) and the new 3M marking (3M Stamark A380 AW). High retroreflectivity leads to significantly higher backscatter intensities and thus to much better detection of the marking. Based on this, new requirements can be placed on markings when they are used on road sections for highly and fully automated vehicles.

Since the established test methods for the performance evaluation of pavement markings according to EN 1436 are exclusively done in the visible range of light, the retroreflective effectiveness in the infrared spectral range, as needed for the use of laser scanners is unknown. Here it is necessary to evaluate whether the current test procedures are sufficient and meaningful or whether further test methods and requirements must be included. For example, the reflective characteristics in the infrared light range could be different from the reflective characteristics in the visible light range.

3. Camera systems for pavement marking detection

The aim of Vedecom study (by *BARES Victor*, *REDONDIN Maxime*) is to estimate the performance of different pavement marking products under various conditions by evaluating the detectability with camera based Advanced Driver Assistance Systems (ADAS), We specifically measure the contrast between the road surface and the pavement marking using a camera as an indicator of how easily or not they would be detected by an ADAS such as a Lane Keeping Assist system (LKA) in real operation.

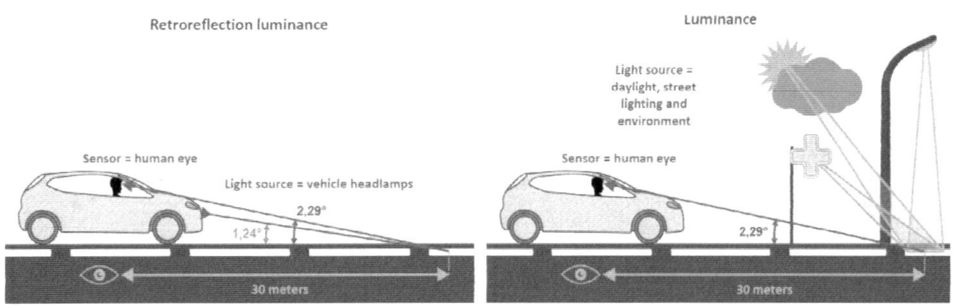

Figure 2: Retroreflection luminance and luminance of a pavement marking according to the CEN standard EN 1436

In this project, different products are compared and analysed. Seven pavement marking products were chosen, provided by 3M and installed at Versailles Satory on a test track near VEDECOM. The chosen products are four preformed pavement marking tapes and three different variants of liquid markings. The pavement marking tapes are Stamark A380AW, A380ESD, A711 (yellow marking) and A715 (black marking). The liquid markings are retroreflective white paint, non-retroreflective white paint and aged non-retroreflective white paint. Several different use cases were chosen to represent different environmental conditions (day, night, weather) encountered by a vehicle.

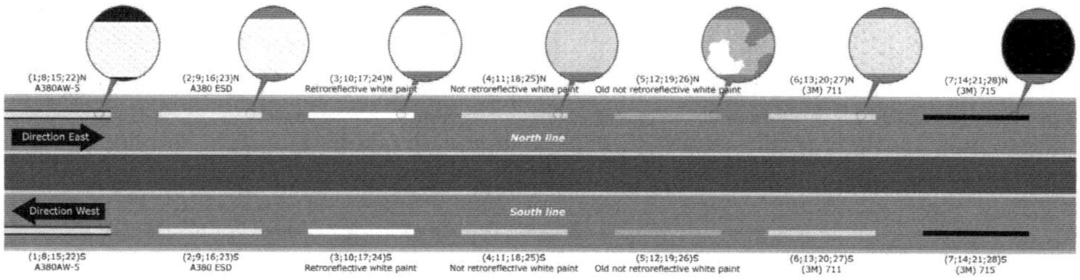

Figure 3: Schematic view of the Satory test set up (above section 3 x repeated)

Figure 4: The Satory test track after its installation (2nd December 2019)

The VEDECOM perception vehicle was chosen to perform this study and more specifically the camera system embedded in the vehicle, as it is widely used in typical ADAS in various standard vehicles. A dedicated methodology has been developed in order to maximize the detection of the pavement marking section by the on-board camera so as to measure accurate contrast values. The vehicle camera system and the algorithms were modified in order to acquire data for this study. Datasets were recorded with the perception vehicle over several months under a variety of environmental conditions.

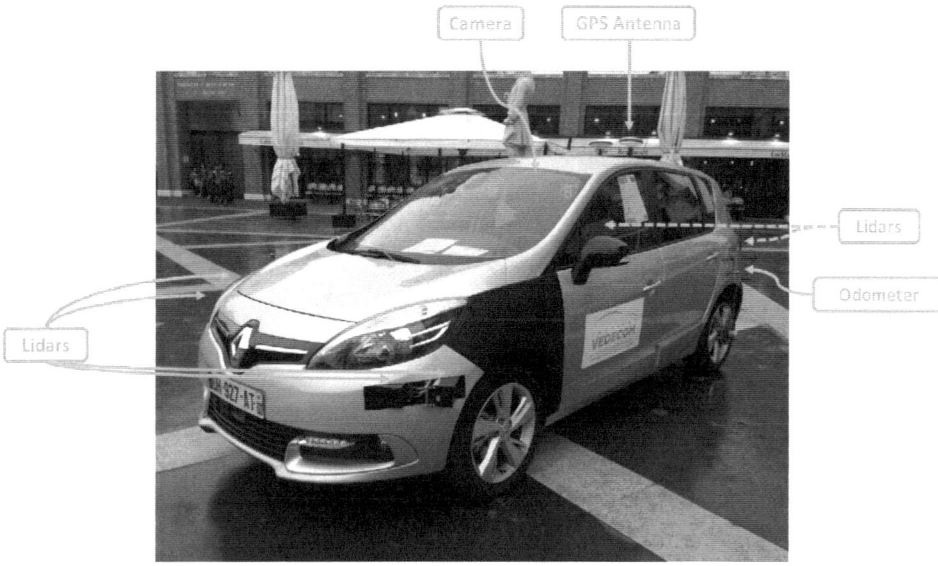

Figure 5: Picture of the perception vehicle

All processing, performed in order to analyze the detectability of the different pavement markings, were done offline. The pavement markings are detected and analysed using a line detection algorithm developed at VEDECOM. It allows us to extract the lines around the vehicle from the previously recorded image. The extraction process begins by removing the obstacles, detecting the lines and then the pavement markings composing them. It continues by extracting the identified markings and eventually evaluating them using the contrast with the surrounding road surface. The contrast values are used to fill a database specifically set up for this project. This database is finally used to extract results on the different use cases in order to analyse them.

The analysis is first done separately by use case and then makes a general assessment of the best performing product over all use cases. Detectability for some of the different use cases is more challenging and this is detailed in the report. Some of the general conclusions of the study regarding the detectability by an ADAS-like system (image-based detection) are given thereafter. The preformed pavement marking Stamark A380ESD is best adapted for the different use cases and obtained the best overall performance. The A380AW and the A711 are able to handle the more challenging use cases very well but have shown weaker performances than other products in standard daylight condition. The retroreflective white paint was detected in most use cases but at lower levels than A380ESD. The non-retroreflective white paint is only functional at dry and daylight condition. The black A715, intended as masking tape for the permanent white marking during road works, remains undetected throughout this study, except when the vehicle is facing low sunlight, which could lead to a false positive in such situations.

Conclusions

Camera detection of pavement marking in normal daylight condition does work well. The different pavement markings in the Vedecom project could be correctly detected and there were no big differences between them in terms of contrast. Even the old, worn pavement marking was correctly detected in most daylight conditions and the non-reflective white paint performed best.

On the contrary, during nighttime conditions, the different pavement markings gave totally different performances. Overall, the three high performance preformed pavement marking tapes (3M Stamark A380AW, A380ESD and A711) obtained the best performances on the night use cases, both in terms of contrast and detection rate. These three pavement markings are clearly the best ones in challenging night conditions (with rain). Both white paint pavement markings clearly get poorer results than the 3M white tapes.

Overall, the most versatile product over all use cases is the Stamark A380ESD tape. It is adapted for all environmental conditions. Only in challenging night conditions with heavy rain, the A380AW and the A711 performed better.

The special use case of glaring sun showed that, in specific configurations (sun in front), the black masking tape (A715) becomes visible through its gloss level and is detected as marking, which does lead to false positives.

In the Vedecom project, it could be shown that nighttime and rain are the most challenging conditions for the camera detection of pavement marking. In order to quantify the respective performance, EN 1436 offers two distinct test methods. The first is called 'wet reflectivity' (sometimes called the bucket method), in which water is poured on the road marking and retroreflective measurements are performed after letting the water drain for 1 minute. You could argue that you measure 'after the rain'. The performance class is WR for 'Wet Reflectivity'. The second test method for 'continuous rain', measures retroreflectivity during constant artificial rainfall with a rain rate of 20 mm / hour. The performance class is RR for 'Rain Reflectivity'. Although this test method is more meaningful for the real performance, it is seldomly used or specified, as the test is not reliable and very time consuming. No standard device for the artificial rain exists, in reality, the few test labs that measure for the RR class use spray nozzles, creating mist or fog, but not realistic raindrops, as required in the test method.

Proposals have been made to the respective CEN TC 226 / WG2 working group, to develop a more realistic and practical test method for rain reflectivity. Work is ongoing at this stage, but it will take some years before a new test method is validated and published.

Author Contributions

The authors would like to thank Professor Dr. Trautmann from the HTW Dresden University of Applied Sciences for his valuable input on the chapter for Lidar detection.

Disclosure Statement

The authors are full time employees of the 3M Company, Transportation Safety Division, a manufacturer of retroreflective pavement marking.

References

BARES Victor, REDONDIN Maxime (2020). Image-based performance analysis of different road marking materials under various environmental conditions, TRIICA Project, Vedecom, France.

CEN, EUROPEAN COMMITTEE FOR STANDARDIZATION. European Standard EN 1436 (2018), Road marking materials – Road marking performance for road users and test methods

IOP Conf. Series: Materials Science and Engineering 1202 (2021) 012034 doi:10.1088/1757-899X/1202/1/012034

ROAD TRAFFIC SAFETY ANALYSIS OF DIFFERENT JUNCTION TYPES ON THE STATE ROADS

Valentina AMARE[1], Juris SMIRNOVS[2]

[1]*Department of Roads and Bridges, Civil Engineering Faculty, Riga Technical University, Riga, Latvia*
[2]*Department of Roads and Bridges, Civil Engineering Faculty, Riga Technical University, Riga,*
Corresponding author: Valentina Amare, email: valentina.amare@lvceli.lv

Abstract. The highest number of road accidents occurs at junctions. One of the aims of traffic organisation is to improve traffic safety in these areas. Based on a variety of indices – road capacity, points of conflict, number, and severity of road traffic accidents – different alternatives for junctions are evaluated. However, the road network has many junctions and roads serve to travel from point "A" to point "B" at a given time. Therefore, one of the most important tasks when addressing the issue of road safety is to find a rational way of improving the safety without losing the importance of the road. The aim of this paper is to analyse the impact of different junctions on the road network and basing on actual data develop a method for the evaluation of different types of junctions with respect to road class.

Keywords: traffic safety, accidents, roads, junction types, signs, injury.

Introduction

Traffic safety is a priority issue in developing the road network. Traffic organisation serves as a basis for the driver to understand the situation and make the right decisions when driving. Therefore traffic organisation has to be unified and understandable. Research of traffic safety level includes analysis to identify problem locations in the road network with the help of different methods. One of the most applied criteria in analysing the existing road network in Europe is the absolute number of road traffic accidents. Concentration of accidents shows that a problem exists in specific location. Deeper analysis helps to identify common aspects and implement safety improvements by eliminating the cause of accidents.

When analysing the network of state main roads in Latvia and studying road traffic accidents in order to identify dangerous locations, it was determined that road traffic accidents concentrate at specific locations, mostly different types of intersections. Statistical data shows that traffic accindents concentrate both in intersections at grade and in intersections in separate grades. The character of traffic accidents differs and these differences are related to points of conflict, geometrical parametres of intersections, as well as traffic organization equipment.

The aim of this report is to identify common aspects in accident concentration locations by analysing the safety of road intersections basing on related criteria. Accident concentration locations are determined according to the Regulations of the Cabinet of Ministers of December 28, 2010, No. 1240 “Order of classifying road sections wih frequent occurrence of accidents and road network safety in the European road network”, namely, such locations are considered as dangerous if at least three road traffic accidents with injured have occurred, or at least eight road traffic accidents have occurred in one kilometre long road section.

This report includes the analysis of frequent road traffic accidents in road interesections in different time periods and on roads of different class: state main roads, state regional roads and state local roads. Causes of accidents in intersections with different parametres and on roads with different classes are mostly depending on traffic intensity and type of vehicles involved, for example, share of heavy transit traffic which greatly inluences the character of accidents.

Several previous studies on dangerous locations have been reviewed, for example, the study of Latvian state roads and accident concentration locations carried out by Vitalijs Jurenoks, Vladimirs Jansons, Konstantins Didenko “INVESTIGATION OF ACCIDENT BLACK SPOTS ON LATVIAN ROADS USING SCAN STATISTICS METHOD”. Scan statistics method was applied in the analysis of road network in order to study road traffic accidents on Latvian state main roads. The basic method applied was Monte Carlo method to identify so called “black spots” in the Latvian road network with the help of stachostic P-value modelling. Scan methodology allows to identify the most dangerous locations on Latvian state roads in time and space and in its essence it differs substantially from traditional statistical methods.

IOP Conf. Series: Materials Science and Engineering 1202 (2021) 012034 doi:10.1088/1757-899X/1202/1/012034

1. General regulations

Traffic safety is topical issue around the world. The European Union has adopted its DIRECTIVE 2008/96/EC OF THE EUROPEAN PARLIAMENT AND OF THE COUNCIL of 19 November 2008 on road infrastructure safety management. The Regulations of the Cabinet of Ministers of December 28, 2010, No. 1240 "Order of classifying road sections whith frequent occurrence of accidents and road network safety in the European road network" are based on this Directive. In 2019 these Regulations were amended. According to the new Regulations State Limited Liability Company "Latvian State Roads" (further in the text – LSR) carries out the analysis of TEN-T road network once every three years and prepares the list of dangerous road sections or intersections. To evaluate traffic safety the comparison with the previous period is carried out. The network of TEN-T roads exceeds 1600 km and includes urban road sections. Only one dangerous intersection in urban area in the TEN-T road network managed by LSR was identified and it is a cross type intersection.

The biggest share of road traffic accidents occurs on a small proportion of the road network, namely on roads with high traffic intensity and high speeds where different types of vehicles drive at different speeds. The most recent "List of dangerous road sections and intersections" covers the period of 2017 – 2019 and it includes 48 locations in the network of state main roads (Figure 1 and Figure 2).

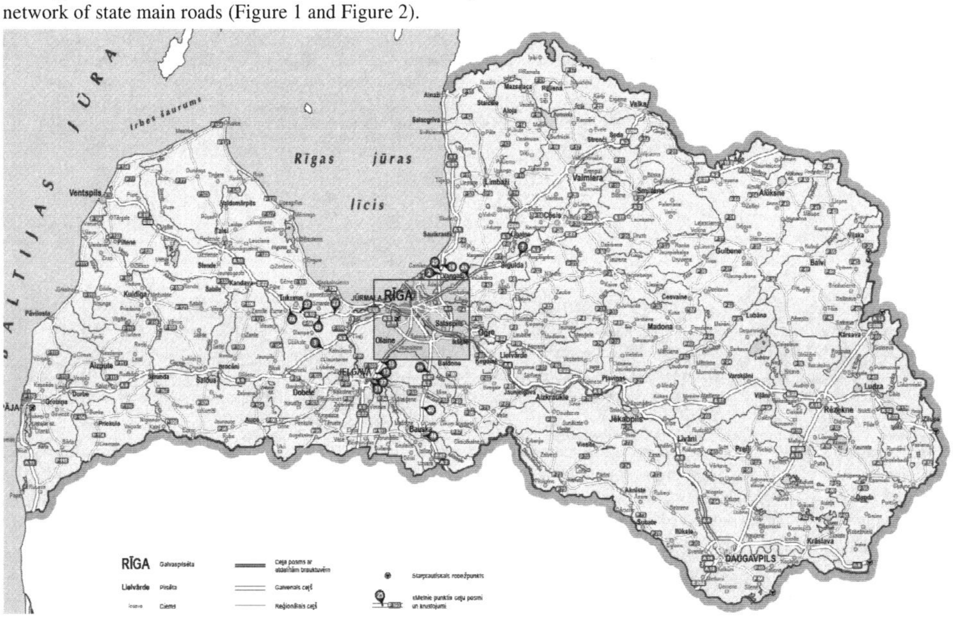

Figure 1. Dangerous sections and intersections on Latvian state main roads in 2017-2019 (map lvceli.lv).

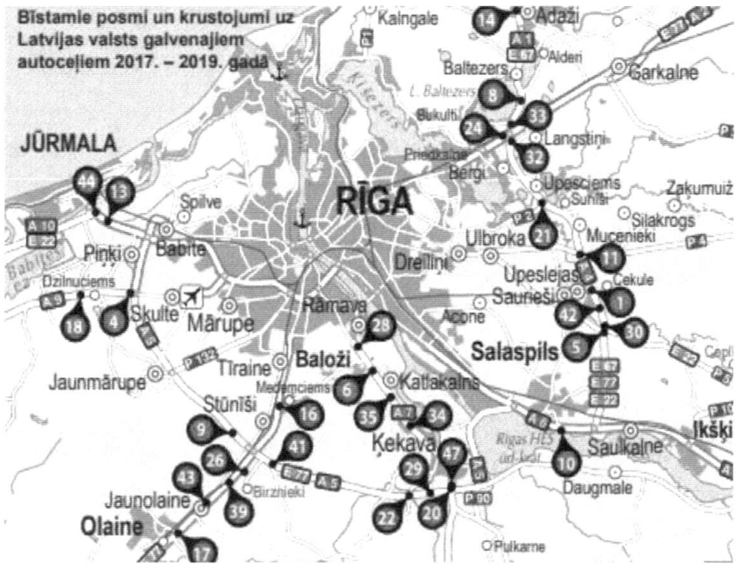

Figure 2. Dangerous sections and intersections on Latvian state main roads in 2017-2019 around Riga (map lvceli.lv).

1.1. Analysis of dangerous intersections

48 dangerous sections and intersections were identified in the analysis of TEN-T road network in 2017 -2019. In the previous period of 2014 – 2016 the state main road network had 99 dangerous road sections and intersections. When comparing these two periods from geographical point of view, it may be determined that only three years ago such locations were dispersed in the whole road network. The analysis of identified sections in the most recent period shows the concentration of such locations in the central part of Latvia, namely the Riga city region. One of related causes is recently constructed and reconstructed road sections.

Summary and evaluation of data on dangerous road sections or intersections without any doubt shows that the majority of such sections are road sections with numerous access roads, as well as intersections. From the road geometry point of view dangerous sections are road sections with poor visibility, and in particular, with poor longitudinal visibility. The biggest share of dangerous locations is on roads with accesses, mostly on high intensity roads with accesses where insignificant accesses create lots of risks as shown in Chart 1. Out of previously mentioned 48 road sections, road sections with the above listed characteristics amount up to 26 dangerous road sections. 22 dangerous locations are intersections with state main, regional and local roads, which constitutes 46 % out of total number in the most recent period.

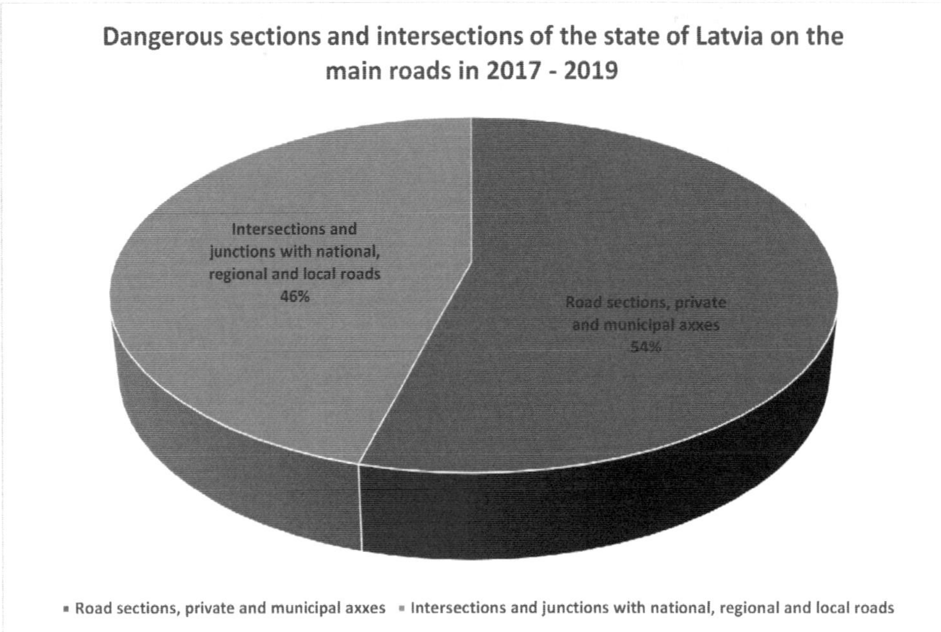

Chart 1. Dangerous sections and intersections on Latvian main roads in 2017-2019 around Riga

Traffic intensity on state main roads differs greatly, and the highest intensity is around the Riga city. The territory around Riga develops fast, the existing TEN-T road network is being renewed but the capacity of these roads has literally reached its limit. Each road access at such high traffic intensity creates risks for accidents as it was clearly seen in the period of 2017 -2019. If dangerous road sections and intersections are considered, mostly the area around the Riga city has the majority of road sections with accesses in close distance to each other which mostly are the cause for road traffic accidents.

Chart 1 clearly shows that insignificant accesses on high intensity roads cause accidents. REGULATION OF THE EUROPEAN PARLIAMENT AND OF THE COUNCIL on Union guidelines for the development of the Trans-European Transport Network stipulates that TEN-T core network has to be developed as motorways or expressways. The only exception is roads with low traffic intensity, the upgrading of which to motorways or expressways is not economically feasible. TEN-T road network should not have any insignificant accesses as it serves as international road network. However, the TEN-T road network managed by LSR has numerous insignificant accesses, as well as numerous intersections with regional and local roads at close distances. Figures 1 and 2 show that dangerous road sections concentrate around Riga where the traffic intensity is the highest. All state main roads in fact start at the Riga city and the concentration of accidents around Riga is closely connected to the centralisation of the whole state main road network in one location. Around Riga not a single road intersection is able to provide sufficient capacity and this is the cause of accidents (luckily there are no accidents with injured in peak hours). Accident statistics in road sections with average traffic intensity and with intersections at grade or road accesses at the allowed speed of 90km/h or even 70km/h unfortunately is much poorer.

1.1.1 Dangerous road sections by road category

The analysis of distribution of dangerous road sections managed by LSR by road category shows that the distribution of regional and local road junctions to state main roads and regional and local roads intersections with state main roads is equal (see Chart 2). The number of dangerous road intersections on roads with the same category – on state main roads - is much smaller. The number of dangerous intersections or junctions on state regional roads and local roads is the same – 16 in total.

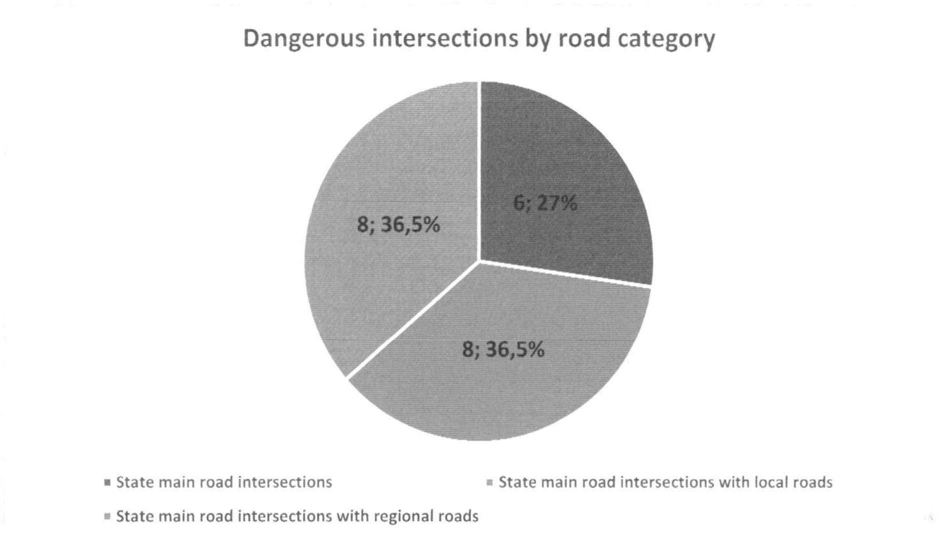

Chart 2. Dangerous sections and intersections

The development of Riga region directly influences the fast increase of intensity on local and regional roads, however the reconstruction of intersections is not implemented so quickly. Therefore road accesses with high intensity mostly are identified as dangerous intersections.

Table 1 shows detailed data per year. Column 1 shows the number of identified dangerous intersections in the previous period. 15 out of previously identified 22 locations have been classified as dangerous also in the next period. This leads to a conclusion that traffic safety improvements in these locations are urgently needed. One of the reasons why this was not done previously is the lack of funding as traffic safety level in dangerous intersections may not be significantly improved with insignificant investments. All the necessary traffic organization devices are installed mostly in every intersection, therefore more significant improvements of infrastructure, namely, reconstruction is needed in order to reduce the number of conflict points and to improve road capacity.

Another problem related to the previously mentioned dangerous road intersections is the increase in driving time. For example, dangerous road intersections on road A4 / Riga bypass are located only in the distance of 3 km from each other. If a driver driving in normal conditions at 90 km/h could pass this section in 15 minutes, then with all regulated road intersections this travel time may increase up to 30 minutes or more. If an accident occurs in this road section, this will lead even to further increase in driving time. The function of a bypass is to divert traffic from the city and to ensure the reduction of driving time. However, at the moment the road A4 as a state main road has unanimously lost its primary function as now it provides accesses to settlements and private properties located closely to the road. In addition to that, it has 5 regulated intersections in a 21 km long road section.

Similar situation may be seen on the road A5 / another part of Riga bypass, where the increase in traffic intensity and the type of the existing intersections lead to insufficient road capacity. It has a lot of cross type intersections identified as dangerous. From the uniformity point of view the road A5 has every imaginable type of intersections – it has a dangerous roundabout, dangerous road interchange in separate grades and dangerous cross type intersection.

Table 1. Dangerous intersections on Latvian state main roads

2014-2016 No.	Road index	km	Accidents				Accidents with injured				Fatalities				Injuried			
			2017	2018	2019	total	2017	2018	2019	total	2017	2018	2019	total	2017	2018	2019	total
45	**A1**	**1**	4	3	2	9	1	1	0	2	0	0	0	0	3	1	0	4
66	**A2**	**14**	2	5	7	14	0	2	2	4	0	0	0	0	0	2	2	4
4	**A4**	**1**	2	3	6	11	0	1	0	1	0	0	0	0	0	3	0	3
	A4	**5**	1	2	4	7	1	1	4	6	0	0	0	0	1	1	7	9
	A4	**10**	2	5	3	10	1	2	1	4	0	0	1	1	1	2	1	4
	A4	**15**	3	1	3	7	2	1	2	5	0	0	1	1	3	1	12	16
	A5	**9**	8	1	3	12	2	1	1	4	0	0	0	0	3	2	2	7
	A5	**12**	3	4	3	10	0	3	2	5	0	0	0	0	0	4	3	7
	A5	**21**	3	3	2	8	1	0	0	1	0	0	0	0	3	0	0	3
A9-86	**A5/A9**	**35/1**	5	5	8	18	1	3	5	9	0	0	1	1	5	9	2	16
34	**A7**	**9**	5	4	5	14	1	0	1	2	0	0	0	0	1	0	2	3
13	**A7**	**10**	6	5	5	16	2	2	2	6	0	1	0	1	4	1	5	10
11	**A7**	**20**	5	1	3	9	0	0	0	0	0	0	0	0	0	0	0	0
69km, no.85	**A7**	**66**	5	1	4	10	0	0	0	0	0	0	0	0	0	0	0	0
62	**A8**	**23**	7	6	6	19	3	0	2	5	0	0	0	0	4	0	2	6
97	**A8**	**32**	9	6	3	18	2	2	1	5	0	0	0	0	3	5	1	9
38 km, no.69	**A8**	**37**	3	4	3	10	1	1	1	3	1	0	0	1	0	2	1	3
60	**A8**	**42**	6	4	3	13	0	1	0	1	0	0	0	0	0	1	0	1
41km, no.95	**A9**	**39**	7	2	4	13	6	1	1	8	1	1	0	2	7	1	1	9
	A10	**54**	3	4	2	9	0	0	0	0	0	0	0	0	0	0	0	0
40	**A10**	**63**	2	6	3	11	2	0	1	3	0	0	0	0	3	0	2	5
89	**A10**	**69**	1	1	9	11	0	0	2	2	0	0	0	0	0	0	2	2

The severity of road accidents serves as an evidence of dangerous intersection, as well. In total, 7 people were killed in intersections shown in Table 1 within the period of three years. The fact that in the previous period there were no fatalities at all also serves as an indication of dangerous situation. In total 19 people were killed in dangerous road sections or intersections identified in the most recent period.

Statistical data shows that the most dangerous road intersections with the majority of fatalities are cross type intersections. The biggest share of cross type intersections with fatalities is located outside urban areas where the allowed driving speed is 90km/h. Despite the fact that speed cameras are installed at two such intersections, accidents with fatalities still took place and this fact shows problems in infrastructure.

1.1.2 Types of dangerous intersections

Chart 3 shows the dangerous intersections managed by LSR by type. It has to be noted that the biggest share is regulated intersections followed by intersections in different grades and intersections at grade. In the analysed time period there has been only one fatality - a cyclist - in intersections in different grades.

Cross type intersections may be characterized with significant severity of road traffic acidents. If causes of road traffic accidents are analysed, they are mostly the combination of two factors - absent-mindedness of drivers and high traffic intensity. Left turn, merging into the traffic flow or even turning off from road become practically fatal at high traffic intensity. Cross type intersections ensure high driving speed on the main road but they definitely do not ensure traffic safety.

The fourth type of intersection is roundabout. There were no fatalities in roundabouts and this shows that roundabouts are safe solution of intersections. The main advantage of roundabouts is the possibility to control driving physically.

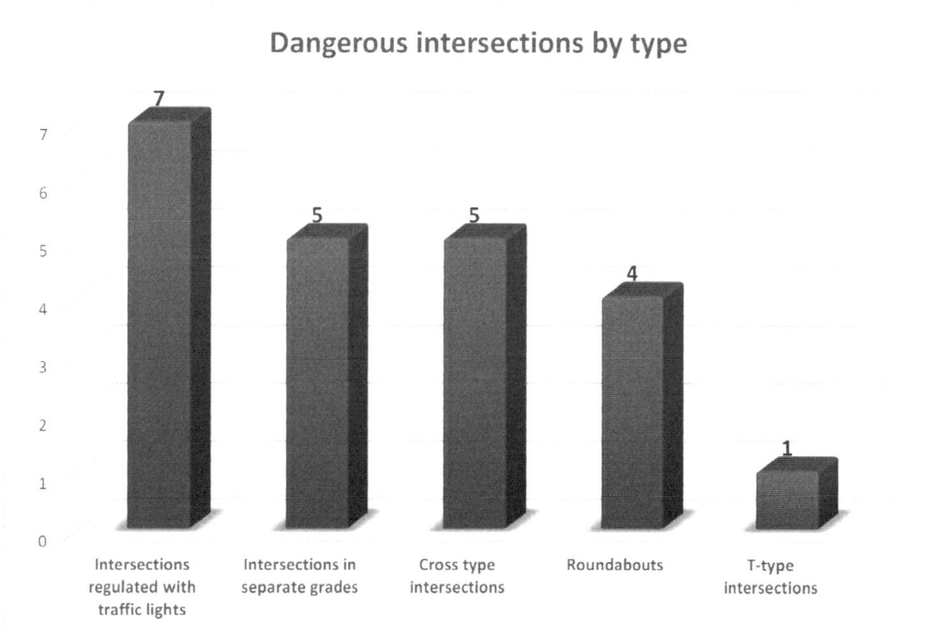

Chart 3. Dangerous intersections on Latvian state main roads

The fifth type of intersections is T-type intersection and there is one such intersection identified as dangerous in the sate main road network. It is also the only one intersection managed by LSR which has a speed camera that is an effective tool for speed control but that unfortunately does not prevent accidents.

Figure 3 cleary shows the number of conflict points in different types of intersections. The indicated conflict points actually show the level of danger in cross type intersections in comparison with other types of intersections.

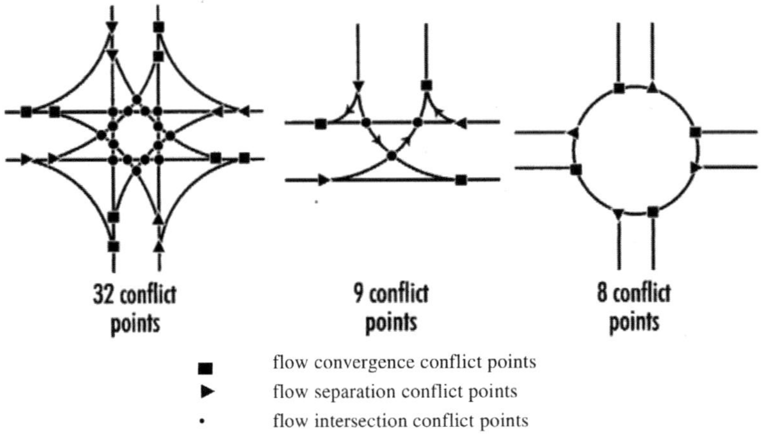

Figure 3. Conflict points at different types of intersections

The number of accidents in T-type intersections on Latvian roads shows that this type of intersections is the safest, but at a specific traffic intensity. In fact, the number of T-type of intersections in the Latvian road network is the biggest. A study [Рунэ Эльвик, Аннэ Боргер Мюсен, Трулс Ваа. Справочник по безопасности дорожного движения/Пер. с норв. Под редакцией проф. В. В. Сильянова. М.: МАДИ (ГТУ),] shows the efficiency of transforming a cross type intersection in two T-type intersections at a specific traffic intensity and taking into consideration other specific aspects.

Further analysis shows that the most dangerous type of intersections are regulated intersections. Data shown in Table 2 is clear evidence that 7 out of 10 intersections managed by LSR may be regarded as dangerous. The mentioned dangerous intersections are located outside urban areas. For example, the Riga bypass which is a little longer than 20 km has 5 regulated intersections and 3 intersections out of this number may be regarded as dangerous. There were fatalities in two of them. The analysis of road traffic accidents shows that the most frequent cause is absent-mindedness of drivers as they do not observe the red traffic light when entering the intersection.

Regulated intersections in urban areas on the other hand may be regarded as positive examples as none of these intersections were identified as dangerous.

Table 2. Dangerous intersections regulated with traffic lights.

Number	TEN-T route	Dangerous
1	A6 (236.2km) / road A13	
2	A4 (2.9 km) / Skolas Street (Berģi)	
3	A4 intersection with P2	Black spot
4	A4 intersection with P4	Black spot
5	A4 intersection with P5	Back spot
6	A7 (8.5 km) / turn to Valdlauči	Black spot
7	A7 (9.2 km) / turn to Baloži	Black spot
8	A8 (22.8 km) / turn to Olaine	Black spot
9	A8 (31.5 km) / turn to Ozolnieki	Black spot
10	A4 (7.8 km) Mucenieki	

Conclusions

1) The analysis shows that proportionally the most dangerous intersections in the road network managed by LSR are regulated intersections outside urban areas. This is the evidence that intersections with traffic lights do not ensure traffic safety.
2) The number of conflict points in cross type intersections with every possible permitted manoeuvres directly influences the level of danger in the intersection and this fact is also approved by statistical data used in the analysis.
3) Road category defines the main function of a road. Every roundabout in the network of regional roads will most probably save some lives as the level of severity of accidents in roundabouts is much smaller. As each road has its function, every type of road intersection in the specific category of roads has to retain certain subordination so that each specific road would serve its function and would be safe for all drivers both local or transit drivers.
4) Common properties of the most dangerous intersections are hight traffic intensity, diversity of different intersection types on the same road, as well as intersection equipment.
5) The importance of TEN-T core network is lost if the traffic intensity is high and there are too many road accesses that are the cause of additional manoeuvres that directly influence road capacity.

Acknowledgements

The author gratefully acknowledges State Limited Liability Company "Latvian State Roads" for providing the data for analysis carried out in this report.

Funding

This research/publication was supported by Riga Technical University's Doctoral Grant programme.

Author Contributions

Dr. ing. Juris Smirnovs, Riga Technical University, has developed the methodology for analytical identification of dangerous road sections in Latvia. In particular, this methodology may be used to evaluate the necessity of installing lighting in such road sections.

Disclosure Statement

No potential conflict of interest was reported by the author.

References

Jurēnoks, Vitālijs (2008). *Investigation of Accident Black Spots on Latvian Roads Using Scan Statistics Method.* http://www.scs-europe.net/conf/ecms2008/ecms2008%20CD/ecms2008%20pdf/lt-ECMS2008_0021.pdf
Smirnovs, J. (1996) *Bīstamo autoceļu posmu noteikšanas metodikai. 1*- 35
Рунэ Эльвик, Аннэ Боргер Мюсен, Трулс Ваа. (2001) *Справочник по безопасности дорожного движения*/Пер. с норв. Под редакцией проф. В. В. Сильянова. М.: МАДИ(ГТУ), 753
Eiropas Parlamenta un Padomes Direktīva 2008/96/EK (2008. gada 19. novembris) par ceļu infrastruktūras drošības pārvaldību https://eur-lex.europa.eu/legal-content/LV/TXT/PDF/?uri=CELEX:32019L1936&qid=1605781014742&from=en
Ministru Kabineta noteikumos Nr.1240 "Kārtība, kādā klasificē ceļu posmus, kuros bieži notiek ceļu satiksmes negadījumi, un ceļu tīkla drošību Eiropas ceļu tīklā" (2010.gada 28. Decembra,)
Road traffic Accidents. https://gis.ic.iem.gov.lv/giswebcais/
Road Traffic Accidents. https://lvceli.lv/celu-tikls/celu-kartes/melnie-punkti/2017-2019/
https://www.trafikverket.se/en/startpage/operations/Operations-road/vision-zero-academy/Background-Vision-Zero/
https://ec.europa.eu/transport/road_safety/sites/roadsafety/files/specialist/knowledge/pdf/speed_enforcement.pdf

Krumina et al., 2021

THE DEVELOPMENT OF BRIGTHNESS EVALUATION METHOD FOR DIGITAL BILLBOARDS

Gunta KRUMINA ORCID iD (0000-0002-5726-3819)[1], Liva VOLBERGA[1], Gatis IKAUNIEKS ORCID iD (0000-0001-9328-7810)[1], Lauris NAUMOVS[2]

[1]*Department of Optometry and Vision Science, Faculty of Physics, Mathematics and Optometry, University of Latvia, Riga, Latvia*
[2]*Department of Road Equipment and System, The Latvian State Roads, Riga, Latvia*
Corresponding author: Gunta Krumina, email: Gunta.Krumina@lu.lv

Received 28 February 2021; accepted date (____)

Abstract. LED-based billboards are becoming increasingly popular due to their low cost, flexibility, and efficiency. On another side, the too-bright digital billboard may pose a potential threat to road safety. The driver's perception of advertising depends on the different factors, including the billboard's luminance value. The aim of the study was to develop the assessment method for the digital screen parameters (luminance or illumination) which could be applied by supervising authorities, using budget-friendly devices, such as a lux meter and could be used during the billboard active operation. We tested the recommended method of OAAA and concluded that it is not applied for large billboards (over 25 square-meter) due to the calculated far distance and low sensitivity of lux meter. We have developed the new method based on driver discomfort glare and De Boer ratio. The ratio helps to determine how likely a display is to cause discomfort to those around it. The lower the value, the less discomfort the driver will experience from the billboard. The new method is applied for the active and different size billboard and using lux meter that is not an expensive device. Moreover, we take into account the glare effect of billboard on drivers' vision.

Keywords: driver, brightness, discomfort glare, luminance, illuminance, De Boer ratio, photometer, lux meter.

Introduction

Outdoor advertising has developed simultaneously with the development of imaging technologies. From Egyptian stone obelisks to publicize laws and treaties to paper posters, neon signs, plastic panels, vinyl posters and, quite recently – digital billboards. Advertisements are strategically placed in popular and busy locations. The advertisements aim to communicate and disseminate information, use it as a means of visual expression, and raise thoughts, ideas, and awareness (Chien, 2011). Over the course of a thousand years, the ways of advertising have surely changed, but speaking in terms of conveying information to the public, the nature and purpose of the advertisements have remained the same.

Nowadays, outdoor advertising is all around us; Huge advertising stands have become an integral part of modern culture. New technologies have made it possible to display different images on any surface. Today, light-emitting diode (LED) screens are increasingly used in a variety of devices and objects, including billboards. The main advantages of LED displays compared to displays from previous generations are the low cost, compactivity and energy efficiency. Using LEDs to create advertising stands opens new possibilities: it is possible to place on the screen not only text and graphics, but also animations and videos that make the advertisements more entertaining and attract attention of more people. The content displayed on LED billboards can be changed quickly, and the displays can be clearly seen in daylight as well as in dark, by virtue of the ability to set the required brightness.

In terms of increased use of LED technologies in various fields, two aspects have changed compared to previous light technologies: (1) increased light pollution; (2) changed temperature of lighting colours: lighting has shifted from white to cool white or bluish, particularly on the city streets. In past, advertisement displaying was carried out using various light sources such as fluorescent, halogen, metal halide lamps. If the display bench surface was around 5 m^2, it produced relatively low lighting from 0.5 lx to 5 lx (Domke et al., 2012). Today, advertising stands are getting larger and implementation of LED technologies are becoming more popular; It is essential to take into account that the brightness of the screens is getting much higher, which may have a serious impact on drivers and perhaps be a potential threat to road traffic safety (Domke et al., 2011).

The LED billboards are much brighter than the traditional ones and the usual advertising stands. The maximum luminance of white light in currently available LED advertising stands can reach 7 000-8 300 cd/m^2 (Zalesinska & Wandachowicz, 2012; Zalesinska, 2018). The luminance of LED screen is much higher compared to the traditional outside-lit advertising stand (7-67 cd/m^2) (Zalesinska, 2018) and the interior illuminated screen or light box (106-360 cd/m^2) (Zalesinska & Wandachowicz, 2013).

However, these screens have also caused new problems. The screens are relatively bright, therefore if the LED displays are not properly adjusted, they can cause glare for drivers. The bright, large, and attractive LED advertising screens can cause visual discomfort for drivers as they can cause distraction or even interfere with ability to see what is happening on the road, especially during the dark (Young et al., 2009). A study (Lee et al., 2007) showed that drivers at night pay more attention to digital billboards than road signs or traditional advertising stands. Since international common guidelines have not yet been developed, it is important for each country to develop its own guidelines and laws for setting up LED billboards to take care of road traffic safety and the well-being of people living in an alluring city.

The main concept of billboards is to draw the attention of road users, particularly passengers. However, it is not uncommon that also the drivers can get distracted for a short moment of time (Lee et al., 2007; Young et al., 2009), which can lead to dangerous situations. According to the World Health Organisation, around 1.35 million people worldwide die in car accidents each year (WHO, 2018). At night, the number and consequences of accidents are much more due to limited visibility (Regev et al., 2018). A large part of the accidents is linked to drivers' lack of attention to the road. For example, when analysing major car accidents in Australia between 2000 and 2011, the majority of accidents are associated with driver negligence (Beanland et al., 2013).

A study Née et al., (2019) carried out in France included interviews with car drivers who were hospitalised after car accidents. The injured drivers were asked whether the car accident was related to negligence and, if the answer was affirmative, the driver was asked to clarify exactly what had caused the distraction. Most commonly, participants mentioned visual distractions (the driver had briefly turned his eye off the road). Other studies have shown that the distraction for looking off the road increases the risk of going out of current driving lane (Peng et al., 2013) and delayed responses to unexpected events on the road, such as an unexpected car braking or the appearance of a pedestrian (Dozza, 2013).

As mentioned above, one of the visual distractions for drivers is the billboards situated to the side of the road. The digital billboards may have a negative impact on the drivers: (1) digital billboards prevent drivers from paying more attention than conventional road signs (Dukic et al., 2013), so it is important to assess the content of advertising and the location of the screen at the side of the road; (2) an overly bright advertising screen can dazzle the driver (Zalesinska, 2018).

Given the high luminance of LED screens, many countries have developed regulations that set the maximum luminance of LED advertising displays. Most commonly, these regulations refer to the maximum luminance of LED screens in daylight and darkness, and in some countries allowed screen brightness depends on whether advertising is placed in a residential area. The regulations may also limit how close to the road LED displays should be placed. In some countries, even maximum size of the screens is limited (Zalesinska, 2018).

How can the luminance of LED billboards affect drivers? Laboratory studies have shown that the increasing of screen's brightness in peripheral vision increases the response time of drivers to various simulated events on the road (pedestrian appearance, etc.), as it is more difficult for the driver to see what happens on the road (Zalesinska, 2018).

LED billboards may not only dazzle the driver, but also cause discomfort glare. In Taiwan, scientists conducted a study of how drivers' vision comfort feelings change depending on the angular size of the LED screen, its brightness, and image flicker rate (Lin et al., 2014b). The measurements were performed under laboratory conditions and the drivers had to assess their feelings subjectively at different measurement conditions. To assess the senses, they used De Boer scale, according to which feelings are considered comfortable if the score is 5 points and more. The assessment for participants was limited to the luminance of the LED screen used in the lower measurements (505 cd/m^2) and the angular size of the smaller screen (9.2 degrees). Such angular size shall be for a screen situated at 30 m from the observer and of a width of ~5 m. Moreover, the larger discomfort in the study participants was at a screen flashing at a frequency of 16 Hz (period of 0.06 s). Consequently, it is important to keep an eye on moving advertising, it does not create an unpleasant flicker of light. A study by Lin et al. (2014b) showed that it is important for LED billboards to limit both their maximum luminance, their size, and the speed of animation movement.

One of the negative aspects directly posed by LED billboards is the dazzling of drivers. Drivers' complaints about blinding created by LED advertising screens have appeared in Latvia too, especially during the dark hours. Driver's glare is classified in various ways. One of the breakdowns is whether it inhibits vision or merely causes visual discomfort. There are three main types of glare identified as follows (Vos, 2003; Schreuder, 2008) (see Figure 1): (1) discomfort glare which is distracting or uncomfortable, but which does not significantly reduce the ability to see information needed for activities; (2) disability glare which reduces the ability to perceive the visual information and which is caused by light scattered within the eye, causing a haze of veiling luminance that decreases contrast and reduces visibility; (3) dazzling or blinding glare which is so intense that for an appreciable length of time after it has been removed, no visual perception is possible. These three types of glaring affect traffic safety, particularly blinding the visual function.

Krumina et al., 2021

Discomfort glare is caused by too large or rapidly changing lights. Discomfort glare makes person uncomfortable, they can start to squint, blink more often and the eyes may start to tear. It may weaken a person's concentration, so it is also important to assess this type of glare for drivers. The intensity of discomfort glare depends on the illumination in the plane of the eye, the background illuminance, and the angle at which the light source is located from the observer (Lin et al., 2014a). Increasing the distance and angle to the discomfort source will significantly reduce the driver's level of discomfort (Mehri et al., 2017).

(a) (b) (c)

Figure 1. Examples of three main types of glare – (a) discomfort glare; (b) disability glare; (c) dazzling or blinding glare.

Discomfort glare, as well as blinding of visual function, depends to what lighting conditions the human eye have adapted. Human glare threshold depends on the adaptation to brightness (Schreuder, 2008). The studies point out that the recommended luminance of billboards should be no more than 10 to 40 times the luminance of the surrounding background so that it does not dazzle drivers (Oviedo-Trespalacios et al., 2019).

Discomfort glare is more difficult to assess as the blinding effect of visual function. The most common way it is assessed is by asking a person to measure his or her feelings and by putting a level of discomfort on the De Boer scale (Fekete et al., 2010). The intensity of discomfort on this scale is divided from 1 to 9, where 9 is the lowest level of discomfort and 1 is the most severe discomfort. The intensity of discomfort to light is sometimes assessed by making the person choose the smallest brightness at which he or she begins to experience discomfort (Zivcevska et al., 2018).

Fekete et al. (2010) assessed the level of discomfort glare depending on the length of the light waves in more detail than in the above study at different wavelengths. The results showed that the most intense discomfort was felt by humans at wavelengths around 500 nm, which can yet be considered by the short waves of light. Studies (Fekete et al., 2010; Zivcevska et al., 2018) showed that short light waves have led to more intense visual discomfort as long light waves. Consequently, LED billboards, emitting an increased amount of blue light, can amplify the sense of dazzling for drivers.

Demand for such screens in the advertising sector is increasing everywhere including in Latvia, and their number is forecast to grow rapidly by 2025 (Stone, 2021). It means that the billboards will replace the traditional stands and create additional distraction for drivers, especially during the dark. As complaints from car drivers about LED billboards on the roadside in Latvia are being increasingly received, a study was commissioned by VSIA The Latvian State Roads to make proposals for amendments to Cabinet regulations regarding the luminance norms for digital advertising screens, and the aim of the study was to develop the assessment method for the digital screen parameters (luminance or illumination) which could be applied by supervising authorities, using budget friendly devices, such as a lux meter or in-built feature in the phone for illuminance assessment.

1. Method

The study methodology was divided into four objectives: (1) to study other national guidelines and legislations on screen luminance standards; (2) to study the literature on human discomfort glare that is impacted by large-size screens or light objects, e.g., billboards; (3) to evaluate the methodologies and the most appropriate methods for the measuring billboards luminance or illuminance; (4) if the relevant method is not adaptable, to develop a new approach for assessment of illuminance produced by large billboards during the processing of advertising (this objective was developed during the study, as it was found that the objective (3) cannot be applied in Latvia because our legislation is closer to European legislation).

The materials available in the web sources are covered by 23 national guidelines or laws describing and regulating the placement of road ads, a variety of parameters, including the maximum luminance standards. Boets et al. (2016) project summarized the guidelines and laws on digital billboards placement opportunities near the road, technical parameters, including the regulations of luminance in Australia, New Zealand and 18 European countries. The description of the Mississauga (2017) summarises the guidelines for the Canadian city's advertising recommendations and the OMA (2014) document for the United States. By studying the standards of different countries, we conclude that in Europe most countries the luminance norms are lower than those of the United States. To propose the standards of digital billboards luminance and design parameters for the Regulations of Latvian Cabinet of Ministers, the guidelines and laws were structured and divided into categories based on advertising location and design criteria set by Roberts et al. (2013) and luminance by Boets et al (2016).

The evaluation of theoretical discomfort glare produced by light sources

The assessment of the drivers' dazzling produced by billboard is a relatively complex process. This can be seen by surveying many subjects and by assessing the intensity of their discomfort glare on one of the developed subjective discomfort scales. One of the most known is the De Boer scale (De Boer, 1967). On this scale, the visual discomfort is divided from 1 to 9, where 9 is the lowest level of discomfort glare and 1 is the most severe discomfort (De Boer, 1967). However, such scales have a number of shortcomings, since the dazzling caused by light sources may vary depending on the complexity of driving conditions and how well road infrastructure is established. Consequently, these scales do not necessarily describe exactly how disruptive light sources can be in the driver's field of vision (Theeuwes et al., 2002).

There are a number of models in the literature that can calculate how much visual discomfort can be caused by a light source on the De Boer scale. One of the following models is proposed by Bullough et al. (2008) and Bullough & Hickcox (2012) based on illuminance quantities:

$$DG = log(E_L + E_S) + 0.6\, log\left[\frac{E_L}{E_S}\right] - 0.5\, log\,(E_A) \qquad (1)$$

where DG – the characteristic value of discomfort glare intensity; E_L – illuminance from the light source (lx); E_S – illuminance from the area surrounding the source (lx); E_A – illuminance of the ambient environment (lx). The quantity DG can be related to values along the De Boer ratio (DB) scale of discomfort glare (1=unbearable, 3=disturbing, 5=just acceptable, 7=satisfactory, 9=unnoticeable glare) according to the following equation:

$$DB = 6.6 - 6.4\, log\,(DG) \qquad (2)$$

The Pearson correlation coefficient between the measured and predicted values was 0.87, which suggested that the trend in measured De Boer ratio was sufficiently accurately predicted by the model (Lin et al., 2014b).

A specific calculation used to calculate human discomfort glare from billboards is not available in the literature, since the various calculations used in the past are based on point light sources. Billboards are screens that are not comparable to a point light source, but its total amount of source emitted light (luminous intensity in candelas) from large square unit is enough to dazzle a driver.

Other glaring calculations, such as the indoor glare of light objects, should also be used for assessments. UGR (Unified Glare Rating) is a method of calculating glare from luminaires, light through windows and bright light sources. The UGR rating helps to determine how likely a luminaire is to cause discomfort to those around it. The UGR could also be partly applied to assessing drivers' discomfort glare, as the size of the light source is taken into account, in the solid angle of the luminaire (steradian), which can be calculated if the area of the light is known. It is calculated using the below stated formula:

$$UGR = 8\, log\left[\frac{0.25}{L_B}\sum\frac{L^2\omega}{P^2}\right] \qquad (3)$$

where: L_B – the average background luminance (cd/m^2), L – the luminance of glare source in the direction of the observer's eye (cd/m^2), ω – the solid angle of the glare source seen from the observer's eye (sr), P – the position index (Guth's index) based on the likelihood of glare, also known as visual comfort probability (Sawicki & Wolska, 2016). Within an office setting, for the luminaire to be classified as "low glare" it must have the UGR below 19 at desk level. Anything above this may cause discomfort – this further enforces the need for high quality interior lighting that is rated UGR<19.

The literature (Lin et al., 2014b) also contains models adapted to the assessment of perceived discomfort glare expressed by means of the De Boer ratio that is related to the illuminance at the eye, the background luminance, and the angle between the glare source and the line of sight:

$$DB = 3.45 - log\left[\frac{(L\,\omega)^{2.21}}{L_B{}^{1.02}\theta^{1.62}}\right] \qquad (4)$$

where DB – the intensity of discomfort glare on the De Boer scale, L – luminance of the light source (cd/m^2), LB – background luminance (cd/m^2), ω – solid angle of the light source (sr), determined from the observer position, θ – the angle between the glare source and the line of sight (degree). The developers of this model compared real measured

Krumina et al., 2021

De Boer values to different light sources with theoretical values using the formula (4). This model resulted in an adjusted R^2-value of 0.929 (Lin et al., 2014b).

2. Results

The methods for the determining the parameters of digital billboards

This chapter describes three methods that can be applied to assess the luminance and illuminance of road advertising screens. A photometer is performing the most accurate measurements for luminance and we recommend this method as a basic method and during the day the photometer can be used. However, the other two methods are carried out with a lux meter and only during the dark. Their accuracy is lower, but methods can be used to assess the specific situation where complaints have come from people or drivers around them. Each method lists both the advantages and disadvantages to be considered by specialists in evaluating the luminance of advertising screens.

The first two methods are more suitable for setting up digital billboards and light objects, while a third method can be used to control screen brilliance.

Method 1: Assessment of screen luminance with a photometer

Photometry is the most accurate to assess the luminance of the screen. The operating range of photometers, the accuracy and quality of the measurements depends on the manufacturer. Based on CIE calibration standards, Konica Minolta LS-100/110, LS-150/160 (Japan) are considered the most precise. The devices are hand-held, easy to use and usable outside 00 to 400 C, preferably at a humidity of not more than 80%.

As drivers' dazzling may result from a bright advertising, it is important to assess the maximum screen luminance and to adjust it accordingly following developed guidelines or legislations. Adjustment should be carried out by setting up an LED display and at least every six months to make sure that the pure white colour does not exceed the prescribed norms. When testing with a photometer, the screen can be measured during both the light and dark days. Ambient background lighting does not affect the measurement of the photometer significantly.

To correctly assess the screen's maximum luminance, screen must display a white colour (R 255, G 255, B 255) and turn on the maximum screen brightness. If the screen's luminance exceeds the requirements of the guidelines or legislations, the screen's luminance should be reduced so far as it corresponds to the prescribed values.

The measurements obtained by the photometer will most accurately describe the screen's luminance in cd/m^2. However, you should remember, the larger the screen, the total screen luminous intensity (cd) will be many times larger than in the case of a small screen size. Consequently, at the same luminance, the larger screen per square meter will more dazzle the driver than the small screen. Therefore, we suggest that the maximum values for screen brilliance, depending on the size of the screen and the level of ambient lighting, should be incorporated into the guidelines and legislations.

Method 1 Advantages

- Measurements can be performed at any time of the day.
- The screen luminance (cd/m^2), as defined in the guidelines and legislations, is assessed at once.
- No specific calculations are required.
- Measurements can be performed at different distances, depending on the recommended distance of the device and preferably perpendicular to the surface of the screen.

Method 1 Disadvantages

- If the measurements are performed at the time of the advertisement imaging, it is difficult to capture a moment that shows a pure white colour, even though the human eye perceives it as the brightest colour.
- If it is not possible for the device to change the measurement time, it may not be possible to include white colour and to measure the average intensity of several colours.
- The measurement target is in small motion if the measurement performer puts focus on the screen from a long distance and without using the device stand.

Method 2: Assessment of lighting from the screen with a lux meter (OAAA and ISA method)

Outdoor Advertising Association of America (OAAA) and International Sign Association (ISA) have developed and published the method that is applied for the assessment of Electronic message centres' (EMCs) brightness using a lux meter. In order to simplify the recommendation of Illuminating Engineering Society of North America (IES) former Member of the Board I.Lewin and to take a more reasonable approach to ensure that EMCs are sufficiently visible but not overly bright, it is recommended that EMCs not exceed 0.3 foot-candles (~3.2 lux) over ambient lighting conditions when measured at the recommended distance, based on the EMC size (Lewin, Undated; Wachtel, 2014; ISA, 2016). The following steps shall be taken to obtain the measurements:

Krumina et al., 2021

- The illuminance of an EMC shall be measured with an illuminance meter set to measure foot-candles accurate to at least two decimals.
- Illuminance shall be measured with the EMC off, and again with the EMC displaying a white image for a full colour-capable EMC, or a solid message for a single-color EMC.
- Determine the square footage of the face of the electronic message sign (EMC) by multiplying the height and width of the EMC.
- Determine the distance to measure the brightness of the EMC by applying formula:

$$distance\ [ft] = \sqrt{S\ x\ 100} \tag{5}$$

where S – area of billboard (sq. ft.).

- The distance should be measured perpendicular to the EMC sign face. The use of a measuring wheel, laser finder or a smartphone app are the most convenient ways to measure the distance.
- As the display alternates between a solid white message and an "off" message, note the range of values on the illuminance meter. If the difference between the readings is less than 0.3 foot-candles, then the brightness of the display is in compliance. If not, the display will need to be adjusted to a lower brightness level using the manufacturer's recommended procedures.
- If the municipality is unable to obtain access to the sign controls or attempting to take the measurement after business hours, this method should be followed.
- A helper should position themselves about 7 ft. to 10 ft. in front of the light meter and hold up an opaque, black sheet of material. The sheet should be positioned so it blocks all light from the EMC, but still allows the remaining ambient light to register on the illuminance meter.
- The illuminance meter should be held at a height of about 5 ft. (which is approximately eye level) and aimed directly at the EMC. The illuminance meter will account for surrounding sources of light or the absence thereof.
- At this point, readings should be taken from the illuminance meter to establish a baseline illumination level, add 0.3 foot-candles to the baseline level to calculate the max brightness limit.

Suggested limits for eye illuminance depending on the lighting zone, E1 to E4 (see Table 1). To determine the maximum billboard average luminance that can be allowed so as to meet a given illuminance limit at the viewer's eye (E_V) the following must be know:

$$L_{max}\ [cd/m^2] = \frac{d^2\ x\ E_V}{S} \tag{6}$$

where d – distance from viewer's eye plane to the billboard (m); E_V – illuminance limit at the viewer's eye (see Table 1); S – area of billboard (m^2).

Table 1. Four different eye illuminance limits depending on the environmental zones ranging from very low ambient light zone to high ambient light.

Zone	Eye illuminance limit (fc)	Eye illuminance limit (lx)
E1	0.1	1.1
E2	0.3	3.2
E3	0.8	8.6
E4	1.5	16.2

Method 2 Advantages

- The price of the illuminance meter (or lux meter) is lower compared to the photometer.
- At an off-screen, it is possible to assess more accurately the level of lighting generated by the environment.

Method 2 Disadvantages

- The method is inaccurate because the luminance level (cd/m^2) is not determined, but the level of illumination produced is determined, depending on the measuring distance.
- Specific calculations must be made: (1) specify the exact area of the screen, (2) calculate the measuring distance.
- The lux meter must be mounted on a stable surface as the small movements of the hand affect the acquisition of measurements.

Krumina et al., 2021

- If measurements are performed at the time of the display, it is difficult to capture the moment in which the brightest advertising appears, and measurements will depend on the colour composition of the ad.
- If it is not possible to change the measurement time for your device, it may not be possible to include the lightest display and measure the average intensity of several colours.
- The measurement target is in small motion if you focus on the screen from a long distance and without using the device stand.
- The measurement conditions are affected by adjacent light objects, such as lanterns.

Method 3: Assessment of billboard produced illumination using lux meter (developed by our team)

By carrying out measurements on large billboards (above 50 m^2) nearby Riga as well as making theoretical calculations, we concluded that OAAA method also has its disadvantages on theoretical background i.e., when calculating measurement distances depending on screen dimensions (see formula (5)), they are not equal for every advertisement displayed on the screen. On the other hand, when calculating the maximal luminance of the billboard, it is constant and does not depend on the billboard's size when assessing the difference between billboard produced illuminance at calculated distance and environmental illuminance level that should be 0.3 foot-candles. If the billboard is installed in environment E4 (assuming that the background light is 5 lx in the dark), theoretically the maximum billboard luminance is 827 cd/m^2 (see Table 2).

Table 2. The calculated maximum luminance (L) of billboard applying OAAA method that the difference between illuminance at the viewer's eye plane and ambient lighting; approximately illuminance of environment (E) by IESNA (2000).

Zone	L (cd/m^2)	E (lx)
E1	322	0
E2	423	1
E3	524	2
E4	827	5

The larger the digital billboard area, the more it dazzles a person if the luminance remains constant. Consequently, we do not see this method suitable for Latvia, where dark period of the day reaches 18 hours during winter period. Knowledge of the visual system helped to develop a new approach to both development of standards for roadway billboards brightness (cd/m^2) and determination of produced lighting using lux meter.

The ambient lighting level forms overall scene of the billboard. The lighting around the billboard is influenced by lantern placements on the roadside and the lighting of surrounding buildings. The darker the surroundings, the brighter and more noticeable the luminance of the billboards will appear; thus, it will draw more attention of drivers. The recommended environmental lighting (IESNA, 1999) to avoid light pollution is shown in Figure 2.

Zone	Environement	Lighting	Example
E1	Nature	Intrinsically dark	
E2	Rural	Low district brightness	
E3	Suburban	Medium district brightness	
E4	Urban	High district brightness	

Figure 2. Environmental zones and examples of ambient lighting levels.

We also carried out theoretical calculations prior to the experimental testing of the method. We used the Netherlands approach (Boets et al., 2018), but the relationship between screen area and luminance were developed based on mathematical calculations (see Figure 3):

$$L\,[cd/m^2] = k/\sqrt{S} \quad (7)$$

where L – the maximal luminance of the billboard (cd/m^2), k – the ratio related to environmental zones (see Table 3), S – area of billboard (m^2).

In order to check if the calculated luminance levels would not, at least theoretically, result in a discomfort glare, we assessed De Boer ratio for critical distances by applying the formulae (1) and (2). De Boer ratio is 4.7–4.9 at the borderline between comfort and discomfort, being dose to the label "5-feel admissible" in the de Boer rating scale (Sammarco et al., 2010; Lin et al., 2014b). Assuming that on suburban roads, the billboard (50 m^2) is 15–17 m from the side of the road and is the nearest distance when it is on the periphery of driver's visual field. Although the bright object projects on the periphery of the retina, it still creates the retinal illuminance, and in this case De Boer discomfort ratio will be 5.1 if the driver is driving in zone E3 (ambient lighting – 2 lx, billboard luminance – 75 cd/m^2). However, the De Boer ratio is critically low in cities because the distances at which billboards can be located are much closer to the driver's vision plane. Consequently, large-size advertising stands should be placed above driver's viewing angle so that the care frame would overshadow the eyes and driver would not dazzle when approaching.

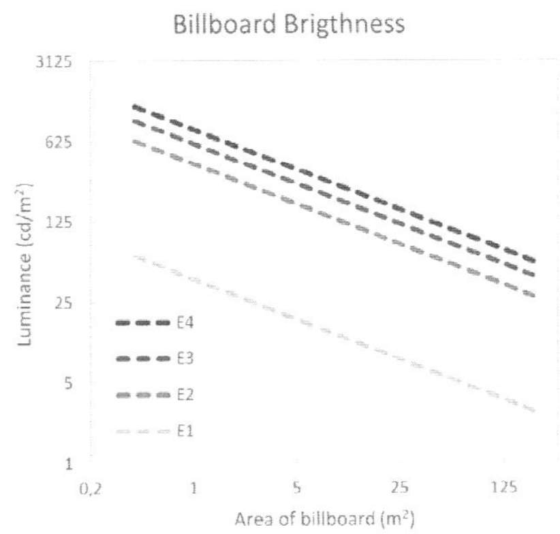

Figure 3. A theoretical calculation of the luminance of billboards according to the environmental zones. Logarithmic scales (x and y) at base 5 have been used to better reflect results.

Table 3. The coefficients that are used to calculate the maximum luminance of billboards are appropriated to specific environmental zones.

Zone	k
E4	800
E4	600
E2	400
E1	40

The new method can be used to estimate the approximate luminance of billboards during the advertisements' operating time and the amount of illumination produced by the screen at a given distance can be used as a measurand (see Formula (9)). The method is not as accurate as measuring luminance using photometer, but it is possible to assess whether at some point the prescribed standards are exceeded and whether complaints about the brightness of the screen can be verified.

To perform measurements, preliminary work must be carried out first. A small pyramid shaped mold should be made (see Figure 4) of black, opaque, matte material. Preferably waterproof, so it would not soak during rain. As the measurement is carried out during advertisements' operating time, it will only be possible to determine the maximum illumination of the screen generated by advertisements which have more white colours. Therefore, the measurement will only be approximate as it also depends on the duration of the brighter advertisement and whether the lux meter is able to perform the measurement in the particular timing.

To determine the measurement of LED advertising screen lighting, following steps shall be performed:

1. Measurements shall be performed only during dark, one hour after sunset marked in the astronomical calendar or one hour before sunrise.
2. Measure the height and width of the billboard using a laser meter or other suitable device prior to lighting measurements. Calculate the area of billboard:

$$S\,[m^2] = b\ x\ c \tag{8}$$

where S – area of billboard (m^2), b – width (m) and c – height (m).

Krumina et al., 2021

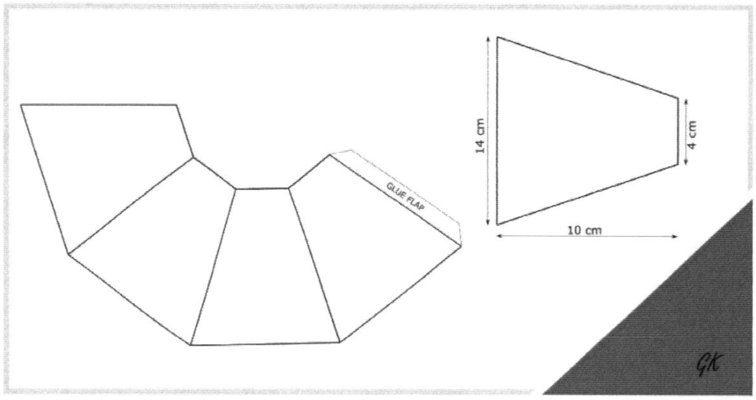

Figure 4. Template of black pyramid shaped mold for reducing illuminance coming from lanterns and car headlamps. The mold pass for billboards that high or with not exceed 12 m.

3. As the illuminance changes depending on the measuring distance, it is important to determine the appropriate distance at which the illuminance measurements will be taken. Formula for calculations:

$$d\,[m] = 4\sqrt[4]{S} \tag{9}$$

where d – the measuring distance (m), S – area of billboard (m^2).

4. When looking through a pyramid shaped mold, the observer should avoid from directly confronting lights, such as car headlights and lanterns closer to an advertising screen or above the head of the observer. A smooth background should be reached through the mold "window" that shows only the billboard on and the illuminated background.
5. Before measurements, the lux meter sensor should be placed as parallel as possible to the surface of the screen and the small hole of the pyramid shaped mold. Make sure that lighting objects on side such as lanterns and car lights do not shine in the vision field. If possible, the lux meter sensor or device shall be placed on a stand or stable surface so that measurements can be taken, and it would be possible to read the resulting figures. Preferably at the height of an eye or 1.5 m from the ground (see Figure 5).

Figure 5. Measurement: Acquisition of a measurement with a pyramid shaped mold and a lux meter – billboard illuminance + environmental illuminance.

6. You have to see device display and look at the numbers that change when billboard shows different advertising. The largest number shown in lux must be recorded. The measurement shall be carried out for at least 2-5 minutes but must be fixed or all advertising will be re-displayed at that time.
7. Record the highest number that was obtained with the lux meter (E_{BB}, lx). The highest value characterise the billboard produced illuminance and environmental illuminance or background.
8. Then turn around 180 degrees and measure the environmental illuminance to avoid getting the light objects into the tube's "window". This number should be recorded (E_B, lx). Approximately expected levels for measurements obtained with lux meter are shown in Table 4.

Table 4. Approximately expected levels of lighting in different residential areas (IESNA, 2000) and our suggested limits of illuminance in different environmental zones.

	E1	E2	E3	E4
Twilight*	2 lx	5 lx	10 lx	25 lx
Darkness **	0 lx	1 lx	2 lx	5 lx
E_{BB} (max)	**3 lx**	**25 lx**	**38 lx**	**50 lx**

Note: * twilight – time period between (astronomical) dawn and sunrise, and between sunset and (astronomical) dusk; or one hour before sunset or one hour after sunrise; ** looking at an astronomical calendar the time period after one hour after sunset or before one hour before sunrise.

When measuring with a lux meter, the sensor must be positioned as parallel as possible to the billboard surface so that the maximum sensor illumination can be obtained. Valid measurements can also be obtained at different angles from the advertising bench, but the accuracy will decrease.

Method 3 Advantages

- The price of the lux meter is lower compared to the photometer.
- The illuminance of brighter advertising is possible assess.
- A more constant light effect on eye retina, regardless of the size of the screen.

Method 3 Disadvantages

- Method is inaccurate because the luminance level (cd/m^2) is not determined, but the level of illuminance produced by billboard is determined, depending on the measuring distance.
- Specific calculations must be made: (1) area of billboard, (2) the distance for measuring.
- Additional materials – black, opaque, matte pyramid shaped mold.
- The lux meter must be mounted on a stable surface as the small movements of the hand affect the acquisition of measurements.
- If measurements are performed at the time of the advertising, it is difficult to capture the moment in which the brightest advertising appears, and measurements will depend on the colour composition of the billboard too.

The pyramid shaped mold may also not be used, but you should critically assess if there are no lanterns or car headlamps that can affect the measurements because the lights can be much brighter than billboard produced illuminance.

Considering theoretical aspects and measurements taken, we have also developed the recommended luminance standards for roadway billboards, based on both the physiological features of human vision and the assessment of De Boer ratio for critical situations where the billboard is closest to the driver's eyes and can lead to a dazzling (see Table 5).

Table 5. Developed and suggested luminance limits (cd/m^2) of different billboards' areas and in the different environmental zones.

Area (m^2)	**E1**	**E2**	**E3**	**E4**
< 0.4	60	600	900	1200
0.4 – 0.8	45	450	600	900
0.8 – 1.5	30	300	450	600
1.5 – 3	20	200	300	450
3 – 6	15	150	200	300
6 – 12	10	100	150	200
12 – 25	7	75	100	150
25 – 50	5	50	75	100
50 – 100	0	40	50	75
100 – 200	0	25	40	50

The distance from billboard to viewer in formula (6) has a significant effect on maximum billboard luminance. Billboards are typically viewed over a range of distances, so the "d" value choice will be somewhat arbitrary. We have

Krumina et al., 2021

calculated theoretical luminance and measured billboard's luminance for real situation nearby Riga (see Figure 6 and Table 6)

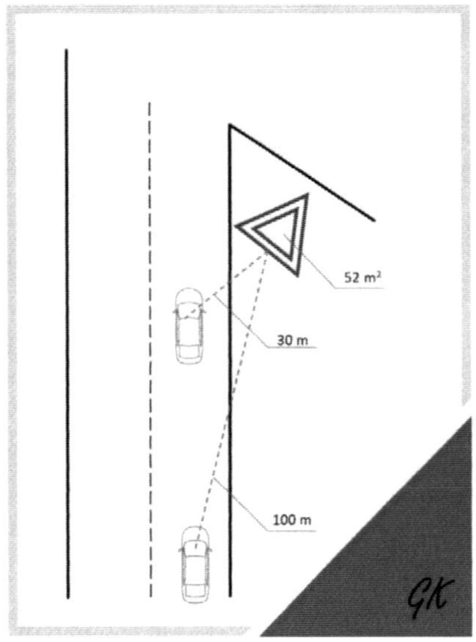

Figure 6. The crossroad on Lielirbes street nearby Riga. Area of billboard is 52 m^2. The location of billboard from the road border line is up to 15-19 m. Two different situations were analysed theoretically and experimentally.

Table 6. Calculated De Boer coefficient that is one of the characteristics value of discomfort glare. L_{max} (cd/m^2) was calculated for area of billboard 52 m^2. De Boer coefficient marked in red means that it is out of criterion (De Boer, 1967).

Zone	Lewin & OAAA method (Lewin, Undated)				The Netherland normative (Boets et al., 2018)				Our method			
	Distance 30 m		Distance 100 m		Distance 30 m		Distance 100 m		Distance 30 m		Distance 100 m	
	L_{max}	De Boer	L_{max}	De Boer	L_{max}	De Boer	L_{max}	De Boer	L_{max}	De Boer	L_{max}	De Boer
E1	19	4.7	215	3.9	0	-	0	-	0	-	0	-
E2	58	5.2	646	4.3	100	4.6	100	6.0	40	5.5	40	6.5
E3	155	4.5	1722	3.8	150	4.5	150	5.5	50	5.4	50	6.1
E4	291	4.2	3229	3.5	200	4.4	200	5.1	75	5.2	75	5.6

3. Discussion

Developed luminance standards for roadway billboards and the method for assessing the illumination produced by billboard are based on the specific characteristics of human vision. Roadway advertisements must not blind or distract drivers from the road, especially during the dark, and physiological characteristics of drivers' visual functions in different ages should be considered. We have developed our method taking into account the standards of visual comfort/discomfort, blinding glare and impact on visual functions, proposed disability glare and ratio by De Boer, as well as the UGR ratio when setting people's workplaces, public places and have been taken into account for the determination of the light objects' placement and luminance for the reduction the impact of human visual discomfort.

One of the main characteristics of digital advertising stands is luminance. The radiant brightness plays key role in how advertising screens affect the surroundings, especially drivers' vision and night peace. During daylight, the level of light emitted by billboard may be much higher than in dark. There is a big difference between the billboard's illuminance and ambient lighting at night, which can lead to the distraction of drivers or can cause sleep disorders for people who live near the advertising screens and who receive screen produced lighting through their home windows.

Krumina et al., 2021

Driving is a complex task which involves extensive interactions between road users and the other components of the transport system such as drivers, vehicles, and road traffic environment. Studies (Oviedo-Trespalacios et al., 2019) have shown roadside advertising signs impact on drivers' behaviour and road safety. Driving performance is influenced by a wide range of factors, including fatigue (Filtness et al., 2012), distraction (Regan et al., 2011), mood (Rhodes et al., 2015), etc. Due to changed optical transparency of eye structures that produce larger dazzling in human vision field for older people, the bright objects distract drivers, especially during the dark (van den Berg et al., 2009; van der Berg et al., 2010). In order not to pose a threat to human safety on road, it would be advisable to reduce the luminance levels or even turn off billboards on roads and advertising screens in cities (Illinois Coalition for Responsible Outdoor Lights, 2010) at night.

Analysing information on the guidelines and laws of different countries gathered by Roberts et al. (2013), it can be concluded that one important restriction on the roadside advertising is the luminance: the amount of light emitted by the advertising screen and the produced illuminance at driver's eye plane, the ambient light or background of billboards, the reflection of billboards, and the use of retroreflectors when the beams are reflected in the same direction back. One of all mentioned parameters as luminance of digital billboard had the largest impact on the driver's response time. A luminance of 200 cd/m^2 already had some deteriorating effects on driving performance (Zalesinska, 2018). When billboard is brighter than environmental lighting or other objects, e.g., pedestrians, it increases the risk of distracting the driver (Sendek-Matysiak, 2017). Studies (Dukic et al., 2013; Wachtel, 2018) showed that drivers more frequently looked at the digital billboards than conventional billboards, and traffic road signs and had more driving errors when digital billboards were present (Edquist et al., 2011). For that reason, the luminance of digital billboards must be less than the luminance of traffic road signs.

The adverse effect of LED screens on driving are mentioned by majority of drivers. Surveys show that approximately 80% of drivers have responded that digital advertising stands distract them from driving (Domke et al., 2013). Another study showed that the impact of LED billboards on drivers' attention depends on the information displayed on screen e. g. graphic elements or text. The least attention is paid to advertising stands, which have mostly graphical elements and small text (Marciano, 2020).

The issue of large-sized billboards remains topical, as not only luminance, but also the contents and presentation play a major role, which can draw drivers' attention and thus distract them from looking down the road. As the number of digital roadside advertising signs is only increasing, it will certainly be important in future to focus on how much and how far to place one advertisement from another, so that they do not affect road safety, especially at night.

Once the luminance standards in Latvia will be approved and the digital advertising control system will be introduced, our method will be assessed. If necessary, the values will be reviewed and reconciled with Latvian regulations until the proposed environmental zones (E1, E2, etc.) will be introduced. Currently, there is a discussion about whether digital road advertising billboards should be allowed in nature areas because there are many of them in Latvia. It is important to remember that setting up a roadside advertising billboard may pose a threat not only to road safety but also to nature.

Conclusions

The deployment of digital billboards must be strictly assessed on roads where cars travel at maximum permitted speed of 90 km/h and on roads that are not illuminated. Although the proposals mention the billboards maximum luminance during dark, it should be taken in mind that every digital billboard is a bright light element in driver's vision field and driver unknowingly draws attention to it. The larger the area of digital billboard or light object, the more it dazzles the driver even at a low screen brightness level.

Acknowledgements

We would like to express our gratitude to the people involved in the project, who brought in valuable discussions that made it possible to think about various things on the road, when the driver is travelling and unknowingly paying attention to roadside advertising, sometimes avoiding a longer-term view, without even being aware of it by the driver himself. Special thanks go to Maris Zaļaiskalns, Director of the Transport Organisation, who participated in the discussions and helped to perform measurements of the luminance and illuminance of digital billboards in Riga. We must also thank Lita Holcmane, the head of the State Language and Urban Environment Management of the Riga Building Board, for taking part in the conversations regarding the developed proposals.

Funding

The work was supported by the VSIA The Latvian State Roads under Grant No LVC/2020/1.10/AC/61) and the University of Latvia under Grant No Y5-AZ77 and Y9-B003.

Krumina et al., 2021

References

Beanland, V., Fitzharris, M., Young, K. L., & Lenné, M. G. (2013). Driver Inattention and Driver Distraction in Serious Casualty Crashes: Data from the Australian National Crash In-depth Study. *Accident Analysis and Prevention, 54*, 99–107. https://doi.org/10.1016/j.aap.2012.12.043

Boets, S., Vandemeulebroken, F., & Daniels, S. (2018). *Roadside Advertising: An Inventory of Current Practices and Future Trends*. ADVERTS project, 94 p. Retrieved from https://www.cedr-adverts.eu/storage/minisites/adverts-d1.1b-current-practices-and-future-trends.pdf

Bullough, J., & Hickcox, K. S. (2012). International among Light Source Luminance, Illuminance and Size on Discomfort Glare. *SAE International Journal of Passengers Cars – Mechanical Systems, 5*(1), 199–202. https://doi.org/10.4271/2012-01-0269

Bullough, J., Brons, J., Qi, R., & Rea, M. (2008). Predicting Discomfort Glare from Outdoor Lighting Installations. *Lighting Research & Technology, 40*(3), 225–242. https://doi.org/10.1177/1477153508094048

Chien, C. (2011). Billboard Evolution. *The Four Peaks Review, 1*(1), 86–97. http://dx.doi.org/10.7152/fpr.v1i1.11424

De Boer, J. B. (1967). Visual Perception in Road Traffic and the Field of Vision of the Motorist. In J. B. de Boer (Ed.), *Public Lighting* (pp.11–96). Philips Technical Library, Netherlands: Eindhoven.

Domke, K., Wandachowicz, K., Zalesinska, M., Mroczkowska, S., & Skrzypczak, P. (2012). Large-Sized Digital Billboards Hazard. *International Journal of Design and Nature and Ecodynamics, 7*(4), 367–380. https://doi.org/10.2495/DNE-V7-N4-367-380

Domke, K., Wandachowicz, K., Zalesińska, M., Mroczkowska, S., & Skrzypczak, P. (2011). Digital Billboards and Road Safety. *WIT Transactions on the Built Environment, 121*(244286), 119–131. https://doi.org/10.2495/LIGHT110111

Domke, K., Zalesinska, M., Wandachowicz, K., Skrzypczak, P., Mroczkowska, S. (2013). *The Evaluation of Impact of Large LED Billboards on the Drivers' Visual Conditions in Road Traffic*. Research project financed by the ministry of science and higher education in 2011-2013, No. N 510 666140. The Technical Report (pl). Poznan, Poland.

Dozza, M. (2013). What Factors Influence Drivers' Response Time for Evasive Maneuvers in Real Traffic? *Accident Analysis and Prevention, 58*, 299–308. https://doi.org/10.1016/j.aap.2012.06.00

Dukic, T., Ahlstrom, C., Patten, C., Kettwich, C., & Kircher, K. (2013). Effects of Electronic Billboards on Driver Distraction. *Traffic Injury Prevention, 14*(5), 469–476. https://doi.org/10.1080/15389588.2012.731546

Edquist, J., Horberry, T., Hosking, S., & Johnston, I. (2011). Effects of Advertising during Simulated Driving. *Applied Ergonomics, 42*, 619–626. https://doi.org/10.1016/j.apergo.2010.08.013

Fekete, J., Sik-Lányi, C., & Schanda, J. (2010). Spectral Discomfort Glare Sensitivity Investigations. *Ophthalmic and Physiological Optics, 30*(2), 182–187. https://doi.org/10.1111/j.1475-1313.2009.00696.x

Filtness, A. J., Reyner, L. A., & Horne, J. A. (2012). Driver Sleepiness – Comparisons Between Young and Older Men During a Monotonous Afternoon Simulated Drive. *Biological Psychology, 89*(3), 580–583. https://doi.org/10.1016/j.biopsycho.2012.01.002

IESNA (1999). *Recommended Practice for Outdoor and Environmental Lighting*. Illuminating Engineering Society of North America, Report RP-33-99, New York: Illuminating Engineering Society of North America.

IESNA (2000). *IES TM-11-00 Light Trespass: Research, Results and Recommendations*. Illuminating Engineering Society of North America. IESNA Publication no. TM-11-00. New York (NY): IES. Retrieved from https://civilnode.com/download-standard/10630464835876/ies-tm-11-00-light-trespass-research-results-and-recommendations

Illinois Coalition for Responsible Outdoor Lighting (2010). *Digital Billboards: New Regulations for New Technology*. Retrieved from www.illinoislighting.org/bill-boards.html

ISA (2016). *Night-time Brightness Level Recommendations for On-Premise Electronic Message Centers*. ISA Research. Retrieved from https://www.signs.org/media/files/ISA_EMC_Recommendations_Refresh_FINAL.pdf

Lee, S. E., McElheny, M. J., & Gibbons, R. (2007). *Driving Performance and Digital Billboards*. Final Report prepared for Foundation for Outdoor Advertising Research and Education. Virginia Tech Transportation Institute. Retrieved from http://omac-website.s3.amazonaws.com/wp-content/uploads/2016/03/Driving-Performance-Digital-Billboards-VTTI-Study.pdf

Lewin, I. (Undated). *Digital Billboard Recommendations and Comparisons to Conventional Billboards*. Retrieved from https://scenicnevada.org/wp-content/uploads/Condensed-Verstion-Part-I-Wachtel-Study-1.pdf

Lin, M.-W., Hsieh, P.-H., Chang, E. C., & Chen, Y.-C. (2014a). Flicker-Glare and Visual-Comfort Assessments of Light Emitting Diode Billboards. *Applied Optics, 53*(22), E61. https://doi.org/10.1364/AO.53.000E61

Lin, Y., Liu, Y., Sun, Y., Zhu, X., Lai, J., & Heynderickx, I. (2014b). Model Predicting Discomfort Glare Caused by LED Road Lights. *Optics Express, 22*(15), 18056. https://doi.org/10.1364/OE.22.018056

Marciano, H. (2020). The Effect of Billboard Design Specifications on Driving: A Driving Simulator Study. *Accident Analysis and Prevention, 138*, 105479. https://doi.org/10.1016/j.aap.2020.105479

Mehri, A., Dehghan, S. F., Hajizadeh, R., Zakerian, S. A., Mohammadi, H., & Abbasi, M. (2017). Survey of Discomfort Glare from the Headlamps of Cars Widely Used in Iran. *Traffic Injury Prevention, 18*(7), 711–715. https://doi.org/10.1080/15389588.2017.1310379

Mississauga (2017). *Guidelines for the Review of Sign Variance Applications for Billboard Signs with Electronic Changing Copy*. Retrieved from http://www7.mississauga.ca/Departments/PB/Building/Electronic%20Billboards.pdf

Née, M., Contrand, B., Orriols, L., Gil-Jardiné, C., Galéra, C., & Lagarde, E. (2019). Road Safety and Distraction, Results from a Responsibility Case-Control Study Among a Sample of Road Users Interviewed at the Emergency Room. *Accident Analysis and Prevention, 122*, 19–24. https://doi.org/10.1016/j.aap.2018.09.032

OMA (2014). *Digital Billboards and Road Safety: An Analysis of Policy and Research Findings in 2013–14*. Discussion Paper. Outdoor Media Association. Retrieved from http://docplayer.net/25954698-Discussion-paper-digital-billboards-and-road-safety-an-analysis-of-policy-and-research-findings-in.html

Oviedo-Trespalacios, O., Truelove, V., Watson, B., & Hinton, J. A. (2019). The Impact of Road Advertising Signs on Driver Behaviour and Implications for Road Safety: A Critical Systematic Review. *Transportation Research Part A: Policy and Practice, 122*, 85–98. https://doi.org/10.1016/j.tra.2019.01.012

Peng, Y., Boyle, L. N., & Hallmark, S. L. (2013). Driver's Lane Keeping Ability with Eyes off Road: Insights from a Naturalistic Study. *Accident Analysis and Prevention, 50*, 628–634. https://doi.org/10.1016/j.aap.2012.06.013

Regan, M. A., Hallett, C., & Gordon, C. P. (2011). Driver Distraction and Driver Inattention: Definition, Relationship and Taxonomy. *Accident Analysis & Prevention, 43*(5), 1771–1781. https://doi.org/10.1016/j.aap.2011.04.008

Regev, R., Rolison, J. J., & Moutari, S. (2018). Crash Risk by Driver Age, Gender, and Time of Day Using a New Exposure Methodology. *Journal of Safety Research, 66*, 131–140. https://doi.org/10.1016/j.jsr.2018.07.002

Rhodes, N., Pivik, K., & Sutton, M. (2015). Risky Driving Among Young Male Drivers: The Effects of Mood and Passengers. *Transportation Research Part F: Traffic Psychology and Behaviour, 28*, 65–76. https://doi.org/10.1016/j.trf.2014.11.005

Roberts, P., Boddington, K., & Rodwell, L. (2013). *Impact of Roadside Advertising on Road Safety*. Austroads Ltd, ISBN 9781921991721, 56 p. Retrieved from https://takedowntheadsdotcom.files.wordpress.com/2016/02/austroads-research-report-on-impact-of-roadside-advertising-on-road-safety.pdf

Sammarco, J. J., Mayton, A. G., Lutz, T., & Gallagher, S. (2010). Discomfort Glare Comparison for Various LED Cap Lamps. *IEEE Transactions on Industry Applications, 47*(3), 1168–1174. https://doi.org/10.1109/IAS.2010.5615978

Sawicki, D., & Wolska, A. (2016). The Unified Semantic Glare Scale for GR and UGR Indexes. *Proceedings of 2016 IEEE Lighting Conference of the Visegrad Countries* (Lumen V4) https://doi.org/10.1109/LUMENV.2016.7745536

Schreuder, D. (2008). Outdoor Lighting: Physics, Vision and Perception. Springer, 448 p.

Sendek-Matysiak, E. (2017). Influence of Roadside Illuminated Advertising on Drivers' Behavior. *The Archives of Automotive Engineering – Archiwum Motoryzacji, 77*(3), 149–162. https://doi.org/10.14669/AM.VOL77.ART10

Stone, M. (2021). *Global LED Billboard Market 2020 Emerging Trend, Top Companies, Industry Demand, Business Review and Regional Analysis by 2025*. Press Release. Market Watch. Published: Feb.24, 2021 at 2:51 p.m. Retrieved from https://www.marketwatch.com/press-release/global-led-billboard-market-2020-emerging-trend-top-companies-industry-demand-business-review-and-regional-analysis-by-2025-2021-02-24

Theeuwes, J., Alferdinck, J. W. A. M., & Perel, M. (2002). Relation Between Glare and Driving Performance. *Human Factors, 44*(1), 95–107. https://doi.org/110.1518/0018720024494775

van den Berg, T. J. T. P., Franssen, L., & Coppens, J. E. (2010). Ocular Media Clarity and Straylight. In: Darlene A. Dartt, Ed., *Encyclopaedia of the Eye*, Vol 3. Oxford: Academic Press; 173–183. https://doi.org/10.1016/B978-0-12-374203-2.00230-X

van den Berg, T. J. T. P., van Rijn, L. J., Kaper-Bongers, R., Vonhoff, D. J., Völker-Dieben, H. J., Grabner, G., Nischler, C., Emesz, M., Wilhelm, H., Gamer, D., Schuster, A., Franssen, L., de Wit, G. C., & Coppens, J. E. (2009). Disability Glare in the Aging Eye. Assessment and Impact on Driving. *Journal of Optometry, 2*(3), 112–118. https://doi.org/10.3921/joptom.2009.112

Vos, J. J. (2003). Reflections on Glare. *Lighting Research and Technology, 35*(2), 163–176. https://doi.org/10.1191/1477153503li083oa

Wachtel, J. (2014). *Report on Digital Sign Brightness*. Prepared for the Nevada State Department of Transportations, Washoe County, City of Reno and City of Sparks. Retrieved from https://scenicnevada.org/wp-content/uploads/Condensed-Verstion-Part-I-Wachtel-Study-1.pdf

Wachtel, J. (2018). *Compendium of Recent Research Studies on Distraction from Commercial Electronic Variable Message Signs (CEVMS)*. Veridian Group, Berkeley, California. Retrieved from https://www.scenic.org/wp-content/uploads/2019/09/billboard-safety-study-compendium-updated-february-2018.pdf

WHO (2018). *Global Status Report on Road Safety 2018*. World Health Organization. Retrieved from https://www.who.int/violence_injury_prevention/road_safety_sta-tus/2018/en/

Young, M. S., Mahfoud, J. M., Stanton, N. A., Salmon, P. M., Jenkins, D. P., & Walker, G. H. (2009). Conflicts of Interest: The Implications of Roadside Advertising for Driver Attention. *Transportation Research Part F: Traffic Psychology and Behaviour, 12*, 381–388. https://doi.org/10.1016/j.trf.2009.05.004

Zalesinska, M. (2018). The Impact of the Luminance, Size and Location of LED Billboards on Drivers' Visual Performance-Laboratory Tests. *Accident Analysis and Prevention, 117*, 439–448. https://doi.org/10.1016/j.aap.2018.02.005

Zalesinska, M., & Wandachowicz, K. (2012). The Research of Outdoor Billboards with Light Emitting Diode Using Luminance Measuring Camera. Poznan Univerisity of Technology Academic Journals. *Electrical Engineering, 69*, 275–282. Retrieved from http://www.iee.put.poznan.pl/wydawnictwa/wydawnictwaFiles/2012/21_Academic%20Journals%20Poznan%20University%20of%20Technology_69/!275_34_Zalesi%C5%84ska.pdf

Zalesinska, M., & Wandachowicz, K. (2013). Examination of Luminance Distributions in the Field of Vision of Drivers in Locations with LED Billboards. *Przeglad Elektrotechniczny, 8*, 270–273. Retrieved from http://www.red.pe.org.pl/articles/2013/8/58.pdf

Zivcevska, M., Lei, S., Blakeman, A., Goltz, H. C., & Wong, A. M. F. (2018). A Novel Visual Psychometric Test for Light-Induced Discomfort Using Red and Blue Light Stimuli under Binocular and Monocular Viewing Conditions. *Investigative Ophthalmology and Visual Science, 59*(3), 1467–1474. https://doi.org/10.1167/iovs.17-23526

Ikaunieks et al., 2021

THE EFFECT OF DISPLAY BRIGHTNESS ON VISUAL FUNCTION FOR YOUNG AND OLD DRIVERS

Gatis IKAUNIEKS ORCID iD (0000-0001-9328-7810)[1], Gunta KRUMINA ORCID iD (0000-0002-5726-3819)[1], Liva VOLBERGA[1], Lauris NAUMOVS[2]

[1]*Department of Optometry and Vision Science, Faculty of Physics, Mathematics and Optometry, University of Latvia, Riga, Latvia*
[2]*Department of Road Equipment and System, The Latvian State Roads, Riga, Latvia*
Corresponding author: Gatis Ikaunieks, email: gatisik@lu.lv

Received 3 May 2021; accepted date

Abstract. Nowadays outdoor advertising displays have become popular. Bright displays near the roads could cause drivers to experience disability or discomfort glare, especially at night. Disability glare increases with age, but discomfort glare thresholds are independent of age. The aim of the study was to assess a luminance level of displays, which causes glare for younger and older subjects. 24 young subjects age of 20 to 24 years and 13 older subjects age of 55 to 69 years participated in the study. The task was by using the method of adjustment to find out the acceptable level of display brightness when the recognition of high (>90%) contrast objects was comfortable. Measurements were done in a photopic and mesopic lighting conditions. Results showed that discomfort glare were larger in mesopic than in photopic lighting conditions ($p < 0.001$) for both age groups. Preferred display brightness in both lighting conditions did not significantly differ between age groups ($p > 0.05$). We can conclude that discomfort glare thresholds for displays with textual elements are independent of age.

Keywords: display brightness, discomfort glare, light scattering, age, drivers, illuminance.

Introduction

Due to developments in lighting technologies more digital billboards appear in locations visible from the roadway. Digital billboards could be potential danger to road traffic safety (Oviedo-Trespalacios, Truelove, Watson & Hinton, 2019) because these billboards attract more and longer glances than regular traffic signs (Dukic, Ahlstrom, Patten, Kettwich & Kircher, 2013). Marciano (2020) showed that billboards with many textual elements attract more attention than billboards with graphic elements.

Digital billboards could also be source of glare. They are much brighter than traditional billboards (Zalesinska, 2018). Glare effects are usually classified into disability glare and discomfort glare. Disability glare is the loss of retinal image contrast as a result of intraocular light scatter. Disability glare impairs the vision of objects without necessarily causing discomfort. This type of glare increases with age (Van Den Berg et al. 2007).

Discomfort glare causes discomfort without necessarily impairing the vision (Vos, 2003). Perceived discomfort glare is related to the illuminance at the eye, the background luminance and the angle between the glare source and the line of sight (Lin, Liu, Sun, Zhu, Lai & Heynderickx, 2014). Discomfort glare is less dependent on age (Bargary, Jia & Barbur, 2014).

Discomfort glare, has a psychological response, and therefore, is difficult to evaluate. This type of glare is usually assessed by asking observers to rate their discomfort sensation. The most often used scale to evaluate discomfort glare is de Boer rating scale (Lin et al., 2014). Sometimes psychophysical methods is also used to assess perceived discomfort glare (Zivcevska, Lei, Blakeman, Goltz & Wong, 2018).

This study's objective was to assess maximum comfortable level of display brightness for participants with different age in different lighting conditions. We hypothesize that comfortable level of brightness (discomfort glare threshold) for display with text will be lower for older than younger participants.

1. Method

24 young adults (3 males and 21 females; mean age 22.8±1.5 years) and 13 older adults (2 males and 11 females; mean age 66.2±7.8 years) who did not have ocular diseases participated in this study. A spherical equivalent refraction for participants was between -5.50D and +2.75D.

Written consent of participants was obtained before research. Ethical approval was given by the Ethical Committee of The Institute of Cardiology and Regenerative Medicine, University of Latvia.

23-inch LED monitor was used for the research. To evaluate preferred display brightness for participants, MS PowerPoint presentation was used. Presentation consisted of slides which background luminance changed from 10 to

Ikaunieks et al., 2021

300 cd/m^2 with step 15 cd/m^2. There was a text in each slide which letters size correspond to visual acuity 0.6 (decimal units) and Weber contrast was ~0.99. Subject task during measurement was to choose slide with maximum comfortable level of brightness (discomfort glare threshold). Preferred display brightness measurements were done in three conditions:

- Photopic conditions (room illuminance ~220 lux; Konica Minolta Illuminance Meter T-10);
- mesopic conditions (room illuminance ~1 lux);
- mesopic conditions with light scattering filter BPM 2 (this measurement was done only for 18 younger participants).

BPM 2 (Tiffen Black Pro Mist) filter is a good early-cataract-simulating filter (De Wit, Franssen, Coppens, & Van Den Berg, 2006).

Measurements in each condition were repeated 5 times. Test distance was 3 m. Average values between the data groups were compared using a one-tailed dependent samples t-test. A significance level of 0.05 was used for the statistical analysis, and all data were processed using MS Excel.

2. Results

Figure 1 shows the relation between the preferred display brightness in photopic and mesopic lighting conditions for young participants (mean age 22.8 ± 1.5 years). Preferred brightness obtained in photopic conditions (mean [M] = 206.2, SE = 15.9 cd/m^2) was significantly larger compared to selected brightness in mesopic conditions (M = 115.4, SE = 13.9 cd/m^2), $t(23) = 6.48$, $p < 0.001$. Pearson's correlation coefficient was R = 0.56, p = 0.004.

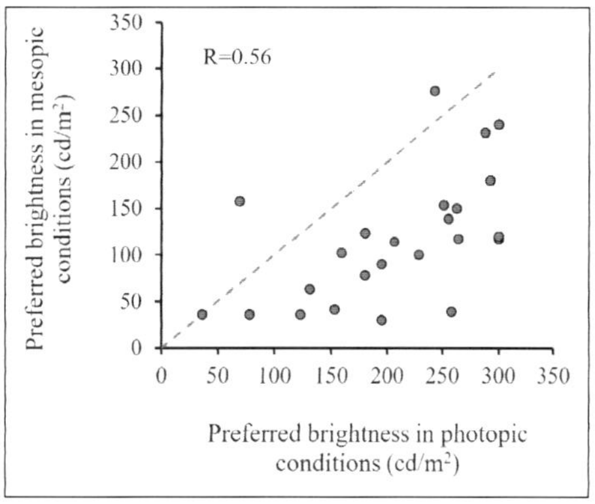

Figure 1. The relation between the preferred brightness in photopic and mesopic conditions for young participants. Dashed line represents perfect agreement. R is Pearson's correlation coefficient.

The relation between preferred brightness in photopic and mesopic lighting conditions for older participants (mean age 66.2 ± 7.8 years) is presented in figure 2. Preferred brightness obtained in photopic conditions (M = 189.0, SE = 20.9 cd/m^2) was statistically larger compared to preferred brightness in mesopic conditions (M = 137.2, SE = 24.6 cd/m^2), $t(23) = 4.30$, $p < 0.001$. Pearson's correlation coefficient was R = 0.86, p < 0.001.

Ikaunieks et al., 2021

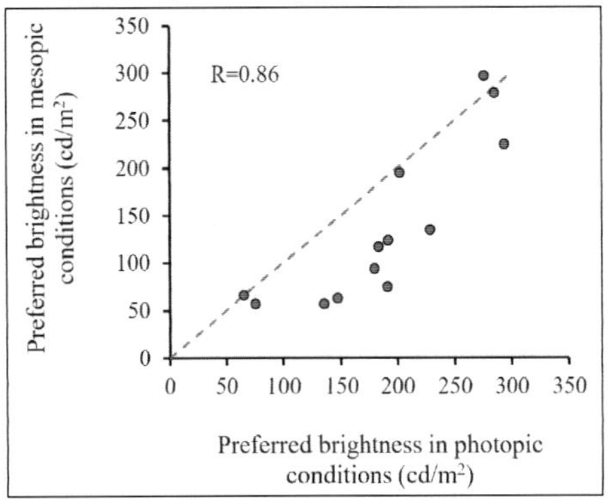

Figure 2. The relation between the preferred display brightness in photopic and mesopic lighting conditions for older participants. Dashed line represents perfect agreement. R is Pearson's correlation coefficient.

Figure 3 shows the mean preferred display brightness values in photopic and mesopic lighting conditions for all participants. Preferred brightness obtained in mesopic conditions (M = 123.1, SE = 12.2 cd/m^2) was statistically smaller than preferred brightness in photopic conditions (M = 200.2, SE = 12.4 cd/m^2), $t(36) = 7.41$ $p < 0.001$.

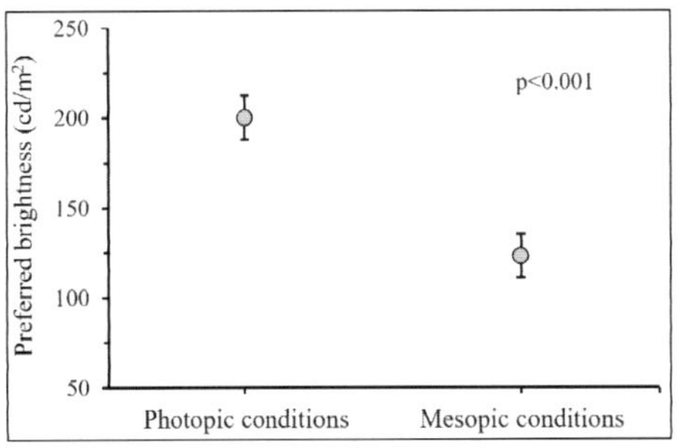

Figure 3. Mean preferred display brightness values in photopic and mesopic lighting conditions for all participants. Standard error (SE) is shown for each data point.

For 18 younger participants additional preferred display brightness measurements were done in mesopic lighting conditions. These measurements were done with light scattering filter BPM 2 in front of the eye. Preferred display brightness with BPM 2 filter was significantly higher (M = 152.4, SE = 19.2 cd/m^2) than without filter (M = 101.9, SE = 14.4 cd/m^2), $t(17) = 3.02$ $p = 0.004$.

3. Discussion

Results showed that preferred display brightness was not statistically different for both age groups in photopic ($t(35) = 0.66$, $p > 0.05$ and mesopic lighting conditions ($t(35) = 0.85$, $p > 0.05$). This is in accordance with the findings of Bargary et al. (2014), which reveal that age has no significant effect on discomfort glare thresholds for stimuli with uniform luminance. We expected that for display with textual elements maximum comfortable level of display brightness will be lower for older than younger participants. Older participants have increased light scattering level in the eye (Van Den Berg et al. 2007), therefore brighter stimulus background should reduce perceived text contrast for these participants more than for younger subjects. Results did not confirm our hypothesis. Future research should apply the method to larger groups of subjects.

All participants who participated in our study had no cataract (opacities in the lens of the eye), which could significantly increase disability glare (De Wit et al., 2006). To simulate vision of drivers with cataract, additional maximum comfortable level of display brightness measurements were done for 18 young with BPM 2 filter in front of the eye. These measurements were performed in mesopic lighting conditions. Preferred display brightness with light scattering

Ikaunieks et al., 2021

filter was significantly higher (p = 0.004) than without filter. Such results could be explained by reduced retinal illuminance when participants did task with filter. Light transmittance of filter BPM 2 is ~66% (de Wit et al., 2006). These results also did not confirm our hypothesis that discomfort glare thresholds will be lower for participants with increased light scattering level in the eye compared to participant with normal level of retinal straylight.

As expected from previous studies (Lin et al., 2014), maximum comfortable level of display brightness for all participants was significantly lower ($p < 0.001$) in mesopic than in photopic lighting conditions. These values are higher than permissible values of luminance of LED billboards in many countries (Zalesinska, 2018). Our task was not to simulate real driving conditions, in which drivers usually see LED billboards with peripheral vision, therefore perceive less discomfort glare than from displays, located in central part of visual field (Domke, Wandachowicz, Zalesinska, Mroczkowska, & Skrzypczak, 2010).

Conclusions

This study showed that discomfort glare thresholds for displays with textual elements are independent of age.

Funding

The work was supported by the VSIA Latvian State Roads under Grant No LVC/2020/1.10/AC/61) and the University of Latvia under Grant No Y5-AZ77 and Y9-B003.

References

Bargary, G., Jia, Y., & Barbur, J. L. (2014). Mechanisms for discomfort glare in central vision. *Investigative ophthalmology & visual science*, *56*(1), 464–471. https://doi.org/10.1167/iovs.14-15707

De Wit, G. C., Franssen, L., Coppens, J. E., & Van Den Berg, T. J. (2006). Simulating the straylight effects of cataracts. *Journal of cataract and refractive surgery*, *32*(2), 294–300. https://doi.org/10.1016/j.jcrs.2006.01.048

Domke, K., Wandachowicz, K., Zalesinska, M., Mroczkowska, S., & Skrzypczak, P. (2010). Assessment of hazards in road traffic caused by large-surface LED billboards. Grant of City Hall Poznan No. RoM.III/3420-50/10, The Technical Report (pl). Poznan, Poland.

Dukic, T., Ahlstrom, C., Patten, C., Kettwich, C., & Kircher, K. (2013). Effects of electronic billboards on driver distraction. *Traffic Injury Prevention*, *14*(5), 469–476. https://doi.org/10.1080/15389588.2012.731546

Lin, Y., Liu, Y., Sun, Y., Zhu, X., Lai, J., & Heynderickx, I. (2014). Model predicting discomfort glare caused by LED road lights. *Optics Express*, *22*(15), 18056. https://doi.org/10.1364/oe.22.018056

Marciano, H. (2020). The effect of billboard design specifications on driving: A driving simulator study. *Accident Analysis and Prevention*, *138*, 105479. https://doi.org/10.1016/j.aap.2020.105479

Oviedo-Trespalacios, O., Truelove, V., Watson, B., & Hinton, J. A. (2019). The impact of road advertising signs on driver behaviour and implications for road safety: A critical systematic review. *Transportation Research Part A: Policy and Practice*, *122*, 85–98. https://doi.org/10.1016/j.tra.2019.01.012

Van Den Berg, T. J., Van Rijn, L. J., Michael, R., Heine, C., Coeckelbergh, T., Nischler, C., Wilhelm, H., Grabner, G., Emesz, M., Barraquer, R. I., Coppens, J. E., & Franssen, L. (2007). Straylight effects with aging and lens extraction. *American journal of ophthalmology*, 144(3), 358–363. https://doi.org/10.1016/j.ajo.2007.05.037

Vos, J. J. (2003). Reflections on glare. *Lighting Research and Technology*, *35*(2), 163–176. https://doi.org/10.1191/1477153503li083oa

Zalesinska, M. (2018). The impact of the luminance, size and location of LED billboards on drivers' visual performance—Laboratory tests. *Accident Analysis and Prevention*, *117*, 439–448. https://doi.org/10.1016/j.aap.2018.02.005

Zivcevska, M., Lei, S., Blakeman, A., Goltz, H. C., & Wong, A. M. F. (2018). A novel visual psychometric test for light-induced discomfort using red and blue light stimuli under binocular and monocular viewing conditions. *Investigative Ophthalmology and Visual Science*, *59*(3), 1467–1474. https://doi.org/10.1167/iovs.17-23526

A COMPACT ROAD WEATHER STATION

Taisto Haavasoja, Pauli Nylander and Leena Puhakka

Teconer Oy, Helsinki, Finland
Corresponding author: Taisto Haavasoja, taisto.haavasoja(at)teconer.fi

Received 12.03.2021

Abstract. Road Weather Stations (RWS) have been extensively used for collecting information about slippery road conditions during the last thirty years. Recently, vehicle based mobile road condition monitors have challenged the fixed RWS. Both approaches have their advantages and can complete each other. Fixed RWS can provide accurate trend data whereas mobile condition monitors can cover the road sections between RWS. Nevertheless, a traditional RWS is a fairly expensive investment, typically 30 000 € or much more, and often includes a number of components not essential for the purpose of measuring and predicting road conditions. To reduce the total cost we have developed a bare minimum of a fixed RWS including only the essential sensor information like road surface state, friction, water and frozen layer thickness, air temperature, road surface temperature, dew point temperature, atmospheric pressure, wind speed and estimated ground temperature at -6 cm. The targeted end user price of the station is one third of the traditional price level. We report experiences with the first installations during 2020-2021.

Keywords: friction, surface condition, non-invasive measurement, road weather, mobile road condition, optical sensors, radiation thermometry, infrared measurement, water layer thickness, RWIS

Introduction

Using of road weather stations to observe and predict slippery winter road conditions started expanding on a global level during the 1990's. The first stations were based on invasive road sensors, which required a laborious installation procedure with special tools and materials. The installation cost could reach to a big fraction of the whole investment. The first commercial non-invasive optical road condition sensors appeared nearly 20 years ago (Pilli-Sihvola, Toivonen, Haavasoja, Haavisto & Nylander, 2006). This technology is based on optical sensing of water and ice absorption at near infrared wavelengths to detect road surface condition and on radiation thermometry to measure road surface temperature remotely. The same technology was applied later to mobile detection of road surface conditions with vehicles (Haavasoja, J. Nylander & P. Nylander, 2012a).

The data of fixed road weather stations is intended for assessing required proper winter maintenance actions, like plowing and salting. Another important aspect is to provide initial information to prediction models. Currently, the Metro model (Crevier & Delage, 2001) has become a popular open source choice in prediction models. To cover the needs to observe the current surface conditions and to provide initial information to prediction models the measured parameters in road weather stations consist typically of road surface condition, water and frozen layer depths, road surface temperature, estimated coefficient of friction, road temperature at a depth, wind speed and direction, precipitation, visibility, present weather, short and long wave radiation, concentration and amount of deicer chemical, and depression of freezing point. The long list of measurands has implications on the cost. Fixed road weather stations tend to cost 30 000 € and much more, if all the measurands are involved, and this may not even include the cost of installation.

Due to the high cost the density of installed road weather stations is fairly scarce so that the distance from station to station is often tens of kilometers in most cases. On varying weather conditions a much denser network would be beneficial for a reliable view of surface conditions in a given area. However, the cost would double for every division of distance to half. Using mobile measurements instead of increasing the number of fixed stations is one approach. Nevertheless, organizing of mobile measurements frequently enough on a given section of road is challenging (Haavasoja & Karki, 2018). By combining fixed and mobile measurements the limitations of temporal resolution in mobile measurements can be substantially reduced.

In this work we report about experiences obtained in developing a compact fixed weather station. The target has been to include a minimum set of measurements and a relatively low cost of equipment and installation. Then it would be possible to install a compact station to selected locations, where following an accurate trend on weather development could pay off the additional cost.

A Compact Road Weather Station

1. Sensors of the Road Weather Station

For winter maintenance purposes we need to know at least road surface condition and surface temperature to be able to decide required operations in avoiding slippery road surfaces. From the point of view of available technologies optical remote detection is an appealing approach, since the installation would be non-invasive and fairly straightforward by exploiting existing road infrastructure, like masts, poles, traffic signs and gantries. Sensors of optical surface condition detectors are currently still fairly expensive, on the order of 5000 €, but the cost is expected to become lower with higher volumes in near future.

It is feasible to measure surface condition, i.e. type of contamination, and layer thickness of the contamination with optical sensors. These sensors send infrared radiation beams to the road surface and measure back-reflected radiation. The wavelengths have been selected so that absorption due to water and ice or other frozen contaminations can be detected individually. Possibility to distinguish frozen and nonfrozen water enables estimation of braking friction as discussed in Haavasoja, J. Nylander & P. Nylander, 2012b. Modeled friction reading is more valuable than concentration of deicer chemicals, since friction is a direct, absolute and unique measure of slippery conditions.

Radiation thermometry enables optical surface temperature measurement. The challenges in radiation thermometry come from varying emission properties of road surfaces and from the unknown temperature of the objects seen in the angle of reflection. Fortunately, the emissivity of a typical dry road surface or even contaminated (water, ice, snow/frost) surface is rather high, about 0.90 or more (Engineering Toolbox, 2021). Thus, the implied error due to emissivity is fairly small, when no hot or cold objects are in the reflected view. The accuracy may get smaller, if open skies are seen at the angle of reflection at nighttime. Luckily, in this case the deviation tends to cause measured surface temperature to appear lower than actual. This deviation is towards the same direction where the road surface temperature is heading during open sky cases. It can therefore be treated as a short-term forecast of actual surface temperature.

An existing commercial mobile sensor branded as Road Condition Monitor RCM411 has been redesigned for measuring surface conditions in the compact fixed station. The redesign includes rearrangement of optical components to obtain maximum possible signal levels up to distances of nearly 10 m. The firmware has been reworked as well to enable integrating the small signal levels without getting confused by the traffic passing under the sensor view.

Figure 1. The sensors of the compact weather station installed on a wooden street light pole. The lower sensor is the optical surface condition monitor RCM411 installed inside a black dust protection tube. The upper sensor is the optical surface temperature and dew point sensor RTD411 installed inside a radiation shield.

A surface temperature and dew point sensor RTD411 was selected for temperature measurements. This sensor has a number of other measurements bringing added value to the fixed road weather station. Figure 1 shows the sensors installed on a wooden street light pole.

A Compact Road Weather Station

The two sensors provide the following list of measurands:

- road surface state, including dry, moist, wet, slush, ice, snow or frost
- contaminant layer thickness, ice or frozen, in mm
- friction, modelled estimate of braking friction, from 0.10 to 0.80
- air temperature
- surface temperature
- dew point temperature
- atmospheric pressure.

The list of measurands contains the most important parameters to enable decision-making of winter maintenance operations and short-term forecasting of surface conditions by linear extrapolation. Longer term forecasting models require additional ground temperature values at some depths. We can make a fair estimate of ground temperature at a given depth by averaging the history values of surface temperature. This is feasible, since the thermal diffusion length is proportional to the square root of time and diffusivity. The thermal diffusivity of an asphalt pavement is about $10^{-3}\sqrt{(m^2/s)}$ (Bai, Park, Vo, Dessouky & Im, 2015). Thus, thermal history extends to about 60 mm in one hour and twice as deep in four hours. By calculating an average surface temperature over a proper time we can estimate the pavement temperatures at a given depth and save laborious invasive installation of depth temperature sensors.

A precipitation sensor would bring in added value to the data set. Precipitation information will be valuable for nowcasting. However, an optical surface condition measurement is a very sensitive way to observe the effect of precipitation on current surface condition and then knowing the actual type of precipitation is not a must in fixed weather stations.

2. Road Weather Station

In addition to the sensors we need an enclosure for powering hardware and a communication unit. The power unit consists of a 7 Ah rechargeable 12 V battery and a commercial smart charger. If power is available continuously, the charger keeps up the voltage at nominal level without overcharging the battery. The station may be powered from street lights as well. The battery will cover the daytime without street lighting and is charged over the night-time for the next day. This seems to run smoothly without problems for the first year of testing. At very northern locations the street lights may not come on at all during the midsummer time, and then we need to consider other powering options.

If street light powering is not available, a solar panel power system is a feasible solution. In a continuous mode the power consumption level is approximately 3 W. This may require a fair size of panel and batteries in northern locations to cover the time during December – January, if we want to have continuous data from the station. Arranging a sleep mode to the system could reduce the size of the solar power system appreciably. A fuel cell solution is also viable. However, then the price tag would increase by over 50 %, since there seems not be available low power fuel cell systems.

The communication unit is a 3G or LTE modem with a low throughput data plan supporting internet connection and SMS. The SMS support is needed to be able to command the station remotely without a need of a static IP address at the station. It is also feasible to set remotely sensor parameters and update the communication unit or sensor firmware. Figure 2 shows and internal view of the enclosure cabinet for power and communication units.

An overview of an installation to a wooden street light pole is shown on the Figure 3. Full installation of this setup took less than an hour from the start to the finish with data flowing to a server and to a web based user interface. The sensors report data roughly every 10 seconds and the data is transmitted to a cloud server once every minute.

A Compact Road Weather Station

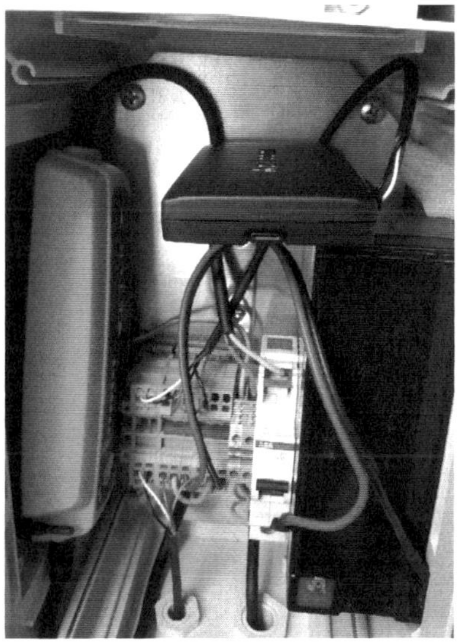

Figure 2. An inside view of the cabinet. The communication unit is on the top, the smart charger on the left and the 12 V battery on the right. The external cabling with feedthroughs consists of the sensor cable and the power cable.

Figure 3. The road weather station installed on a wooden street light pole. The power is taken from the street light mains line. The cabinet (300x200x150 mm) is on the other side of the pole between the two traffic signs.

3. Field Testing

The fixed compact optical road weather station has been tested during the winters starting from 2019. The setup shown

A Compact Road Weather Station

in Figure 3 has been running over a year without any service after an initial fixing of a circuit breaker and after changing the charging mode of the smart charger to an optimal setting.

An open web based user interface is available at https://roadweather.online/rwsdemo. The user interface is originally designed for mobile data, but can be used also with fixed stations. Figure 4 shows a general view of the interface with the test road weather station located on a zoomable map. On the right side of the view there is a browsable daily based list of the available data files.

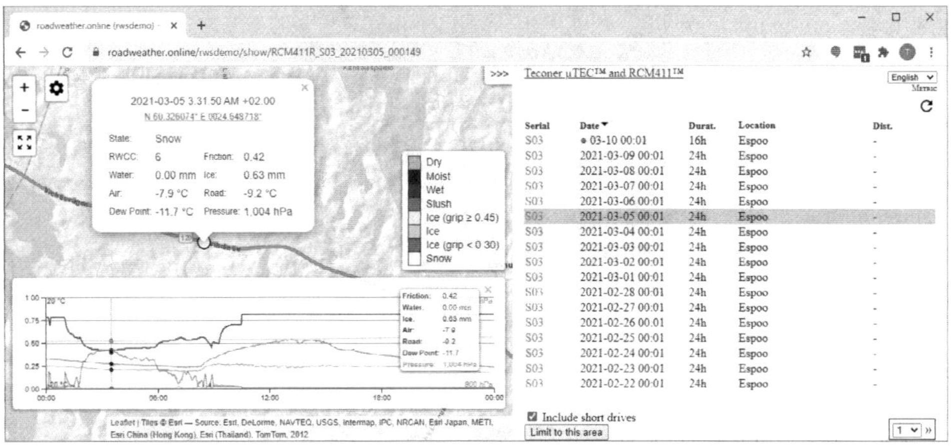

Figure 4. A web user interface for the data of the road weather station. The interface is zoomable and can be used globally for mobile and fixed station data.

Sensitivity to rapid changes in weather and surface conditions are easily detectable and can be used for making warnings or short-term nowcasting. When a salty wet surface approaches freezing to a slippery state, it is feasible to follow the evolution of friction and decide, whether more salting is needed. Formally, the required additional amount of salt can be calculated by using the layer thickness and surface temperature, while not allowing the ice fraction of the binary solution to reach well over 60 %, which is a limit for a safe level of friction.

Figure 4 shows development of surface condition during a light snow fall. At about 1 o'clock the snow starts to slowly accumulate on the surface and friction drops finally to roughly 0.40. The station is in a rural area, but there is a fair amount of traffic in the morning. The traffic and studded tires eat the snow away and the surface is dry again before noon. There are a number of similar cases during the last winter concerning freezing and snowing.

Annual maintenance of the station has turned out minimal. The sensor windows remain uncontaminated for extensive periods of time. Finally, it may turn out that annual maintenance is only needed for fixing a failing component. The expected lifetime of the road condition sensor is on the order of 10 years. Service period of the humidity sensor is not yet known.

Conclusions

The designed compact road weather station contains a minimum number of measured parameters, which have been selected to enable following optimally development of road surface conditions for winter maintenance purposes. The most important measured data include surface state, friction, contaminant layer thickness, surface temperature and dew point temperature. Other targets have been low cost of the equipment, easiness of the installation and long survivability without on spot maintenance. The field testing indicates that these targets have been adequately achieved.

Disclosure Statement

The authors do not have any competing financial, professional or personal interests from other parties than the affiliation.

A Compact Road Weather Station

References

Crevier, L.-P. & Delage, Y. (2001). METRo: A New Model for Road-Condition Forecasting in Canada. *J. Appl. Meteor.*, 40, 2026-2037.

Engineering Toolbox (2021). Retrieved from https://www.engineeringtoolbox.com/emissivity-coefficients-d_447.html (09.03.2021).

Haavasoja, T. & Karki, O. (2018). Using Buses for Road Weather Data Collection. *PIARC XV International Winter Road Congress, 20-23 February 2018, Gdansk, Poland.*

Haavasoja, T., Nylander, J. & Nylander, P. (2012a). Experiences of Mobile Road Condition Monitoring. *Proceedings of SIRWEC 2012, Helsinki, Finland.*

Haavasoja, T., Nylander, J. & Nylander, P. (2012b). Relation of Road Surface Friction and Salt Concentration. *Proceedings of SIRWEC 2012, Helsinki, Finland.*

Bai, B.C., Park, D.-W., Vo, H.V., Dessouky, S. & Im, J.S. (2015). Thermal Properties of Asphalt Mixtures Modified with Conductive Fillers. *Hindawi Journal of Nanomaterials, Vol 2015, Article ID 926809.*

Pilli-Sihvola, Y., Toivonen, K., Haavasoja, T., Haavisto, V., & Nylander, P. (2006). New Approach to Road Weather: Measuring Slipperiness. *Proceedings of SIRWEC 2006, Turin, Italy.*

Template: Anonymized Document

WHY IT IS IMPORTANT TO HAVE A CORRECT ROAD MAINTENANCE TERMSTOCK?

KAI SIMSON

Linguistic and Language Editing, School of Humanities, Tallinn University, Tallinn, Estonia

Abstract. This article is based on my Master's Thesis entitled „Planning of the Road Maintenance Termstock in Estonian", which examines the current situation of the road maintenance terms. Road maintenance influences every single person and therefore it is important that terms and concepts are clear and precise. I will give a review of the condition of the Estonian road maintenance terms, using legislation, instructions and other materials which are used to regulate the area, and a web questionnaire was carried out among this field's experts. This is the first time when such a thorough study has been carried out about the road maintenance termstock. All the results indicate that there is a need to plan road maintenance termstock in Estonian. Experts who had been questioned, pointed out that there is a need for common termstsock that all parties would have a common understanding of concept definitions and what terms denote them. Most frequently emerging problems were following: terms do not have clear concept definitions; many synonymous terms for one concept; no consistent use of terms in legislation nor in the guidelines; terms systematization has not been taken fully into account; ambiguous terms are in use; lack of termstock that would cover needed terms and have correct concept definitions.

Keywords: road maintenance, termstock, road conditions, termstock planning, winter road maintenance, term.

Introduction

All people are exposed to road maintenance every day. It surrounds us all the time and all the work done or undone is immediately visible. It is therefore a subject that is under lot of attention and widely spoken and written about. When we speak about such important issues like safe driving during winter or what does road maintenance means, it becomes a problem when there is no clear terminology and it is understood differently by various parties.

Ülle Sihver has written that concepts need to be named and the conceptual diversity to be organized. As the society is constantly changing, so there is a continuous search for terminology to describe it. The concept helps to explain phenomena and objects, and connections between them. The term is a name for the concept or in other words technical word to express thoughts. A term is a language unit and a concept is a knowledge unit that can be defined. (Sihver 2018)

The issue is important as the road maintenance terminology has not been a subject for studies. I have been engaged in media communication for road maintenance over four years. During this period, several doubts have arisen about the use of terminology and the clarity of their concept. Also, the ambiguous terms used in regulations are confusing. This creates further misunderstandings when several terms are used for one concept and it is not clear which one should be preferred. There are problems with the expediency and clarity of the terms used, and with the development of language expressions that are understandable only to people who work in this area.

1. Theories of terminology

The terminology was discussed for the first time scientifically by the Austrian engineer Eugen Wüster (1898–1977). He developed a theory of terminology. In 1968 he published the interlingual dictionary that systematically arranged French and English standardised terms (with a German supplement) intended as a model for future technical dictionaries. (Cabré 2003)

As M. Teresa Cabré has written that the development of theoretical and applied terminology that started in the second third of the 20th century occurred thanks to the interest of scientists and technicians. Subject matter and methodology develop when there is a need and are pursued to the extent that they are the result of clear social needs. (Cabré 1999)

There are also disagreements between experts whether terminology constitutes a separate discipline. Some consider it as a practice dealing with social needs that are often related to political and/or commercial ends. Others consider terminology as a scientific discipline that owes much to the other subject fields from which it borrows fundamental concepts. (Cabré 1999)

Template: Anonymized Document

Juan C. Sager has taken a position that we see terminology as a number of practices that have evolved around the creation of terms, their collection and explication and finally their presentation in various printed and electronic media. But practices do not constitute a discipline. Disciplines establish knowledge about things and as such are justified in their own right; methodologies are only means to an end, in the case of terminology, how to do things. (Sager 1990)

Terminology can only be understood in relation to special languages and communication and addresses a variety of purposes, all of which are related to communication and information. The two main groups of people who use terminology as a communicative tool are direct users and intermediaries. The direct users of terminology are the specialists in each subject field, in this case people who are connected to the road maintenance and intermediaries use terminology to facilitate communication for other users. (Cabré 1999)

It is important to understand that specialists use terminology regardless the term is appropriate within a particular linguistic system or not. They have a knowledge of a concept that needs to be communicated; their interest in terminology focuses on concepts and how they can be named clearly and unambiguously. Experts use terminology to transfer specialized knowledge in one or more languages and to structure the information contained in specialized texts. (Cabré 1999)

The use of standardized terminology helps to make communication between specialists themselves and others more efficient.

There are five different theories of terminology: General Terminology Theory (GTT), Socioterminology, Sociocognitive Terminology, Frame-based Terminology and Communicative Theory of Terminology (CTT). Although all of them have certain aspects that could be used, still the communicative theory seems to be the most applicable.

The GTT focused on specialized knowledge concepts for the description and organization of terminological information. Within this framework concepts were viewed as being separate from their linguistic designation (terms). Concepts were conceived as abstract cognitive entities that refer to objects in the real world, and terms were merely their linguistic labels. (Faber 2009)

Teija Pihkala has described socioterminology as an approach which objective is that besides cognitive aspect, social aspects are also acknowledged and considered in all terminology theory and practice does not try to be an independent discipline under terminology, but its objective is that, besides cognitive aspects, social aspects are also acknowledged and considered in all terminology theory and practice. (Pihkala 2001)

Temmerman added that socioterminology tries to get the study of terminology back to the study of real language usage (Temmerman 2000).

Sociocognitive terminology differs from other theories by emphasizing on conceptual organization, and focusing on category structure from the perspective of cognitive linguistics approaches. (Faber 2009)

Communicative Theory of Terminology was established by M. Theresa Cabré it started to draw linguistics and terminology closer. She has pointed out that terminology can be understood when it is set into relationship with language and communication.

The basis of this theory were the notions that terminology is not a separate discipline and communicators do not differentiate special language from general language. (Taukar, Tavast 2013)

Cabré stated that terms belong to general lexicology and act as a term if this is allowed by pragmatic and communication parameters. Terms can be used in different specialties and therefore have different meanings. The concepts are organized into concept system via different relationships. (Taukar, Tavast 2013)

2. Which are the most problematic issues in the terminology?

In order to understand the situation of the road maintenance terminology, I studied legislation, instructions and other materials which are used to regulate the area. In addition, also dissertations from the students of TTK University of Applied Sciences have been studied to find which terms are used for concepts.

In addition to the quantitative study also a qualitative study was carried out via web questionnaire to find out how what experts consider as a problem in road maintenance terminology, how they assess the present condition of it and what they suggest that should be done to improve the situation.

For the analysis all road maintenance terms and definitions were assembled and entered in a terminological database in Excel, which contains 282 terms. Only the terms that are used or are directly connected to this area, were entered to the database, as the road maintenance is one part of the road field. The road maintenance was divided into five concept fields: road, road maintenance, regular and periodic road maintenance, winter road maintenance and road conditions (see Figure 1).

IOP Conf. Series: Materials Science and Engineering 1202 (2021) 012038 doi:10.1088/1757-899X/1202/1/012038

Template: Anonymized Document

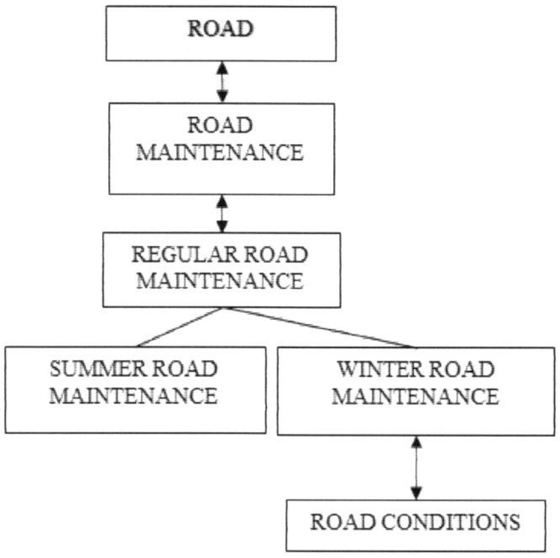

Figure 1. Concept fields of road maintenance.

This division was justified, as due to this it became clear that the use of certain terms is neither necessary nor justified. Several concept diagrams also encountered difficulties in defining superordinate concepts and subordinate concepts. More comprehensive view of the difficulties of generic concept and specific concept.

The qualitative analysis showed that experts consider the situation of road maintenance terminology to be medium (3) or good (4) at scale 1 (very bad) to 5 (very good). The experts answered that the terminology is mostly clear, and specialists understand each other, but the problems arise when you try to communicate with people outside the subject. Another problem that was brought up was the situation with legislation and materials which are not unified, and different terms are used for concepts that might rise problems and misunderstandings.

While studying the terms representing road maintenance concepts it became clear there are problems with clarity of terms. There are terms which do not have clearly defined concepts or do not have a related concept at all. For example, terms like time of the maintenance cycle, active snow clearance, passive snow clearance etc. These terms are in active use, but they do not have a concept to them. Therefore, these terms are used under the assumption that the other party understands it the same way. But this is already a ground for misunderstandings, especially in contracts.

Another problem that occurred was with terms that do not represent its concept accurately. As an example, last couple of years the Estonian Road Administration (since 01.01.2021 Estonian Transport Administration) has declared *severe weather conditions*, although it means severe road and driving conditions. This term is taken from the weather agency but in such situations, it would be wise to consider thoroughly if it is needed and suitable to use terms from other subjects. It might be better to create own term.

Terms that might have different concepts in different situations also were noted as a problem. Especially when users are not sure whether the terms are synonyms or not. This creates hesitation which term should be used or do they even mean the same concept. If we use *korrashoid* or *hooldus* (in English both have same translation – maintenance), do they represent the same concept or are they superordinate and subordinate concepts.

The same problem was with synonyms that mean the same term but there is no indication which one is preferred. While studying materials it was clear the author of the document decided which terms to use according to his/her preference. It was pointed out that this area should be clarified, and preferred terms should be stated.

The biggest problem is that there is no term database for road maintenance in Estonian. The situation in English language is much better due to PIARC dictionary. But while talking about native language, it is not good to use foreign language terms, but to find or create native language terms.

This could be achieved through a terminology committee that should be established with experts of the field and linguist to plan the road maintenance termstock.

Conclusions

It is important to agree on preferred terms that will be used. As Cabré has stated – specialists will use terms to communicate the knowledge of a concept. They do not thing if this term is appropriate within a particular linguistic system or not. Therefore, it is important to have clear and precise terms.

Template: Anonymized Document

The results of a qualitative analysis show that problems with terms and concepts exist. Most frequently emerging problems were following:

- terms do not have clear concept definitions;
- many synonymous terms for one concept;
- no consistent use of terms in legislation nor in the guidelines;
- terms systematization has not been taken fully into account;
- ambiguous terms are in use;
- lack of termstock that would cover needed terms and have correct concept definitions.

Experts pointed out that there is a need for common termstsock that all parties would have a common understanding of concept definitions and what terms denote them. This avoids problems in understanding what is actually meant.

Acknowledgements

I thank my instructor Peep Nemvalts, who is linguist and senior research fellow from the Tallinn University. Also, many colleagues from the Estonian Road Administration (since 01.01.2021 it is Estonian Transport Administration) helped with gathering and reviewing the terms database and concept diagrams.

Disclosure Statement

I do not have any competing financial, professional, or personal interests from other parties.

Template: Anonymized Document

References

Cabré, M. Teresa 1999. *Terminology. Theory, methods and applications*, 6, 11-12, 165. John Benjamins Publishing Company. Amsterdam/Philadelphia

Cabré Castellví, M. Theresa (2003). *Theories of terminology. Their description, prescription and explanation,* 2, 6. Terminology 9:2. John Benjamins Publishing Company.

Faber Benítez, Pamela 2009. *The Cognitive Shift in Terminology and Specialized Translation*, 111, 117. MonTI. Monografías de Traducción e Interpretación, 117 http://www.erevistes.uji.es/index.php/monti/article/view/1646

Pihkala, Teija 2001. *Sosioterminologia: lyhyt oppimäärä*, 6-7. Terminfo 1/2001 Nordterm 2001, Seija Suonuuti, sosioterminologia, ympäristösanasto EnDic 2000, termipalveluselvitys, hakemisto 2000, multimedia...

Sager, Juan C. 1990. *A Practical Course in Terminology Processing*. John Benjamins Publishing Company. Amsterdam/Philadelphia

Sihver, Ülle 2018. *Eestikeelse terminivara hoole ja kasutus on terminoloogiakomisjoni eesmärk ja mõte*, 62. Oma Keel number 36 / kevad 2018. Emakeele Selts.

Tavast, Arvi; Taukar, Marju 2013 *Mitmekeelne oskussuhtlus*, 61-62. Valgus. Tallinn

Temmerman, Rita 2000. *Towards New Ways of Terminology Description: The sociocognitive approach*, 31. John Benjamins Publishing Company

COMPARISON OF WINTER MAINTENANCE REQUIREMENTS (ESTONIA, LATVIA, LITHUANIA)

Jānis Kastanovskis
Head of Maintenance Supervision Department
State Limited Liability Company "Latvian State Roads"
Janis.Kastanovskis@lvceli.lv

March 1, 2021

Abstract.

Three Baltic states are located next to each other on the coast of the Baltic Sea. Similarly, to Finland, density of population in these countries is low. All three countries have large amount of state roads with low traffic intensity.
Climatic differences are minimal. Milder climate is in coastal area but inland the weather conditions are more severe. The average air temperature in January in Estonia is from -2 to -7 degrees Celsius but in Lithuania from -1 to -5 degrees Celsius. The number of snowy days fluctuates from 50 to 120.
All three countries have common contracts for summer and winter maintenance where the responsibility for maintenance activities lays on the contractors. Despite common history the methods of the assignment of road maintenance contracts are different in every country. Estonia awards contracts in open tenders and has performance-based contracts. Latvia has the contract awarded to the State Joint Stock Company «Latvijas autoceļu uzturētājs» ("Latvian Road Maintainer") by law and the contract is unit price based. Lithuania has the contract awarded to the state-owned company « Kelių priežiūra» the contract is performance-based.
The requirements for road conditions are quite similar in all three countries. They have three levels of maintenance.
High service level means snow and ice-free surface of the road pavement during winter in constant weather conditions and quick response in case of worsening road conditions.
Medium service level means that snow and ice is allowed on the surface, but activities must be undertaken to improve skid resistance.
The lowest service level means that snow and ice is allowed on the surface and activities to improve skid resistance (mostly only snow cleaning) may be undertaken in some spots only.
In Estonia the expenses for winter maintenance are lower than in the neighbouring countries but it does not have impact on traffic safety during winter.

Keywords: Snow, ice, winter maintenance, weather, the Baltic states, maintenance level

Introduction

More than 20 years within the Baltic Road Association a special Technical Committee on Road Maintenance is working actively. The Committee improves information exchange between road maintenance experts in three Baltic countries and below the names of involved experts who have contributed to this report are listed:

Mr Aldis Lācis,

Mr Jānis Kastanovskis,

Mr Meelis Saat,

Mr Taavi Umal,

Mr Modestas Lukošiunas,

Mr Tomas Ratkevičius

1. Background

Three Baltic states are located next to each other on the coast of the Baltic Sea. Similarly, to Finland, density of population in these countries is low; All three countries have large share of state roads with low traffic intensity.

Climate

Long term average climate data from1981 to 2010 are shown in Figure 1. The climatic differences are minimal. Milder climate is on the coastal area but inland the conditions are more severe. The climate in Estonia is a little colder than Latvia and Lithuania. Precipitation is a little more severe in Latvia and Lithuania. The number of snowy days fluctuates from 50 to 120. The Meteorological offices in the Baltic states have admitted that the average temperature and amount of precipitation are increasing and we will have less snow in the future.

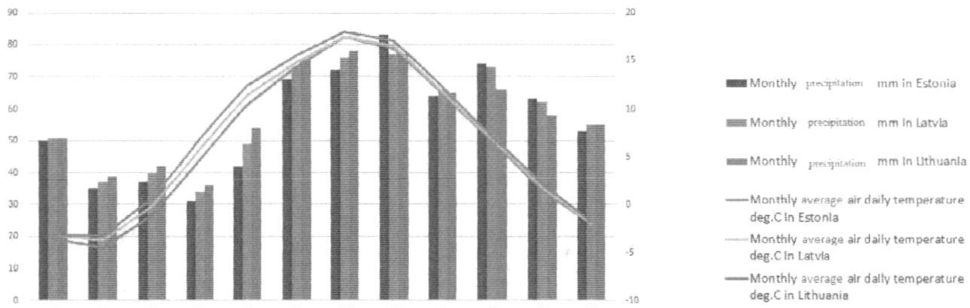

Figure 1. Average amount of precipitation and average temperature in the Baltic states

Maintenance Organization

All three countries have common contracts for summer and winter maintenance where the responsibility for the activities lays on the contractors. Despite common history the assignment of road maintenance contracts is different in every country. One of such differences is the determination of winter maintenance period. Estonia does not determine a special winter maintenance period but pays for 5 months. Latvia has winter maintenance season from November 1 until March 31. Lithuania has winter maintenance season from October 15 to April 15.

Estonian Maintenance Organization

The contracts are awarded in open tender and there are 18 contract areas and 9 contractors. All contractors are private contractors. The contracts are performance-based. The customer is the Estonian Transport Administration which has four Road Maintenance Divisions.

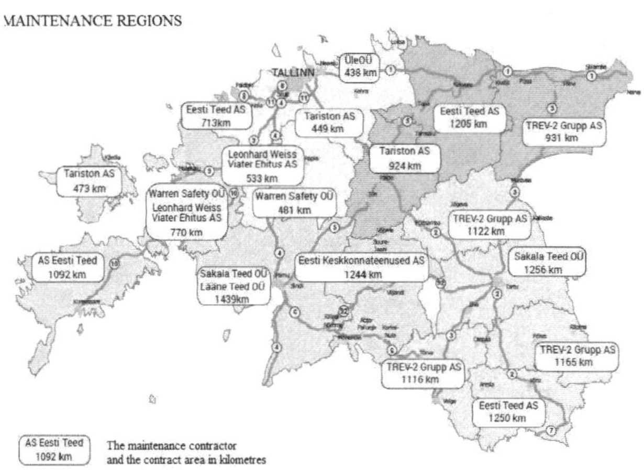

Figure 2. Estonian Maintenance organization.

Latvian Maintenance Organization until May 1, 2022

Latvia has the contract awarded to the State Joint Stock Company "Latvijas autoceļu uzturētājs" by law. The contract is unit price based and there are three parties involved in the contract. The customer is the Ministry of Transport, road network manager is State Limited Liability Company "Latvian State Roads". LSR supervises the state-owned contractor according to the contract with the Ministry.

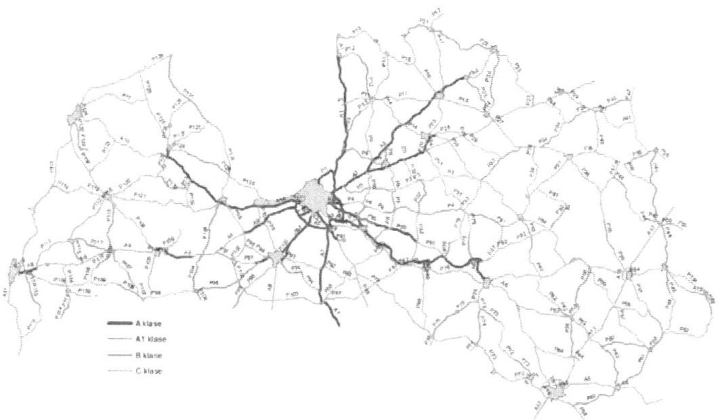

Figure 3. Latvian maintenance classes in the winter 2020 /2021.

Latvian Maintenance Organization after May 1, 2022

Customer for the road maintenance works will be State Limited Liability Company "Latvian State Roads". The execution of road maintenance works will be awarded in open tender in 19 contract areas. There will be more than one contractor. But the share of works performed by state owned company will still be large. Contracts will still be unit price based except one which will have fixed payment for maintenance performance on main roads. Will the maintenance be in same quality as before and cheaper? We will see.

Figure 4. Maintenance contracts after May 1, 2022

IOP Conf. Series: Materials Science and Engineering 1202 (2021) 012039 doi:10.1088/1757-899X/1202/1/012039

Lithuanian Maintenance Organisation

Lithuanian contractual arrangements are like Latvian, although with few important differences. Lithuania has the contract awarded to the state-owned company Public limited liability company « Kelių priežiūra». The contract is performance based with some works with unit prices. Instead of the Ministry the Customer is the State Enterprise "Lithuanian Road Administration". The supervision is carried out by independent Supervisors according to the Contract with the Lithuanian Road Administration.

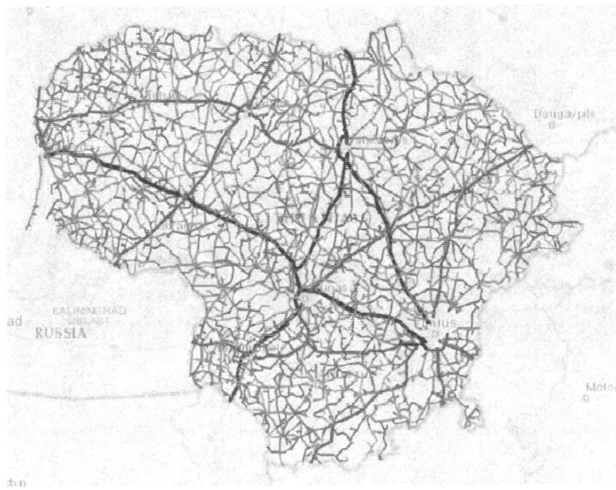

Figure 5. Lithuanian state road network with different classes.

Service standards for road winter maintenance

The requirements for road conditions are quite similar in all three countries, namely, all three countries have three standards of maintenance. The high standard means that the pavement surface is free from snow and ice during winter in constant weather and quick response is expected in worsening road conditions. The medium standard means that snow and ice may be on the surface, but activities have be done to improve friction. The lowest standard is that snow and ice may be on the surface and the activities to improve friction have to be done in some spots only. And this means only snow cleaning mostly.

Country	High Standard	Medium Standard	Low Standard
Estonia	Level 3 and 3+	Level 2	Level 1
Latvia	Level A	Level B	Level C, D and E
Lithuania	Level 1 and 2	Level 3 and 4	Level 5
Carriageway	Ice and Snow free	Partly snowy and icy	Snowy and icy
Anti-skidding	As soon as possible	Long reaction time	Only spots

Table 1. Maintenance service standards in the Baltic.

Lowest Service Standard

The lowest service standard in Estonia is Level 1 and it is applied on roads with average traffic intensity less than 250 vehicles per day.

Latvia has 3 levels: Class D on roads with traffic less than 100 vehicles per day and class C on roads with traffic from 100 to 499 vehicles per day, and Class E for collapsed roads. Class D and E means no requirements and activities are done occasionally.

In Lithuania the lowest service level 5 is applied only for unpaved roads and paved roads with traffic intensity less than 500 vehicles per day.

Country	Estonia	Latvia from April 2021	Lithuania
Service level (traffic volume)	Level 1 (<250)	Class C (100 -499)	Service level 5 and 5g on gravel roads (0 - 500)
		Class D (<100)	
		Class E collapsed road	
Time for snow removal	24 hours after snowfall	Class C 24 hours after snowfall	Service level 5: 9 hours after snowfall; Service level 5g: 18 hours after snowfall.
		Class D snow removal at least 2 times per season	
		Class E snow removal at least once per season	
Permitted snow depth	10 cm (loose snow), 5 cm (slush)	Class C 10 cm (20 cm in snow drift places)	Service level 5: snow removal is done when snow thickness exceeds 15cm; Service level 5 on gravel road: snow removal is done when snow thickness exceeds 20 cm.
		Classes D, E no requirements	
Skid - resistance coefficient	0.20 (0.25 on unsafe spots)	No requirements, spreading and grading of unsafe spots	No requirements, spreading and grading of unsafe spots
Time for de-icing	12 hours	No requirements	Service level 5: de-icing* with sand in 9 hours; Service level 5g: de-icing* with sand in 18 hours. *only in case of extremely slippery road surfaces (e.g., freezing rain, wet ice, etc.)
Hours when requirements are applied	Round the clock	6:00-20:00 no requirements in other time	9:00-18:00 no requirements in other time

Table 2. Requirements for the lowest service standard.

Medium Service Standard

The medium service standard in Estonia is Level 2 with traffic intensity of 250 to 1000 vehicles per day. For Latvia it is Class B with traffic intensity of 500 to 999 vehicles per day. The medium standard in Lithuania is Level 3 and 4 with the average traffic of 500 -2000 vehicles per day.

Country	Estonia	Latvia from April 2021	Lithuania
Service level name (traffic volume)	Level 2 (250 -1000)	Class B (500 -1000)	Level 3 and 4 (500 -2000)
Time for snow removal	12 hours after snowfall	8 hours after snowfall	Service level 3: 4 hours after snowfall;
			Service level 4: 6 hours after snowfall.
Time for de-icing	8 hours	8 hours, 24 hours mechanical treatment (grooves)	Service level 3: 4 hours after snowfall
			Service level 4: 6 hours after snowfall

Permitted snow depth	8 cm (loose snow), 4 cm (slush)	4 cm loose snow, during snowfall (8 cm, 5 cm slush, 16 cm on snow drift places)	Service level 3: in normal conditions - 3cm (loose snow), 2cm (slush), during snowfall or sweep - 10 cm (loose snow), 5 cm (slush), allowed depth of unevenness - 2 cm;
			Service level 4: in normal conditions - 3cm (loose snow), during snowfall or sweep - 10 cm (loose snow), allowed depth of unevenness - 4 cm.
Skid - resistance coefficient	0.25	No requirements	No requirements
Hours when requirements are applied	7:00-21:00 in other time service level 1	6:00-20:00 In other time service class C	Service level 3: 4:00-19:00
			Service level 4: 6:00-18:00.

Table 3. Requirements for medium service standard.

Highest Service Standard

The highest service standard means no snow and ice on carriageway during normal weather. Estonia has level 3 and level 3+, Latvia - class A, Lithuania has level 1 and 2. The average traffic intensity is 1000 and more vehicles per day in Estonia and Latvia and more than 2000 vehicles per day in Lithuania.

Country	Estonia	Latvia from April 2021	Lithuania
Service level name (traffic volume)	Level 3+(>3000 on most important main roads)	Level A (>1000)	Service level 1 (≥10000)
	Level 3 (>1000)		Service level 2 (2000 - 10000)
Time for snow removal	5 hours after snowfall, for 3+ in 2 hours	3 hours after snowfall	Service level 1: 2 hours after snowfall
			Service level 2: 3 hours after snowfall
Pavement surface during normal weather	Snow and ice not allowed	1cm of snow and ice	Service level 1 and 2: snow or ice not allowed
Permitted snow depth during snowfall	4 cm (loose snow), 2 cm slush	6cm loose snow (12 cm on snow drift places) 3 cm slush	Service level 1: 4 cm (loose snow), 0 cm (compact);
			Service level 2: 7 cm (loose snow), 2 cm (compact).
Skid - resistance coefficient	0.3 in wheel tracks and 0.28 in other spots	No requirements	No requirements
Time for de-icing	Level L3+ 2 hours. Sidewalk and wheel tracks free of snow and ice. Preventive de-icing required	Slippery roads must be spread with de-icing or abrasive within 3 hours.	Service level 1: 2 hours
	Level L3 4 hours. Sidewalk and wheel tracks free of snow and ice. Preventive de-icing required		Service level 2: 3 hours
Hours when requirements are applied	Level L3+ 24 hours	Level A 5:00 - 23:00, level B required in other time	Service level 1: 24 hours
	Level L3 6:00 - 22:00 L2 required in other time		Service level 2: 04:00 - 22:00, no requirements in other time

Table 4. Requirements for the highest service standard.

Conclusions

Requirements for winter road maintenance in the Baltic states are not controversial, they are ideologically similar. Maintenance level depends on traffic intensity on roads. It seems that Latvia has less requirements for busy roads. There are no requirements for snow and ice-free carriageway during normal weather although this so far was provided by state owned maintenance company.

The table below shows maintenance requirements for the road E67 during night. During daytime difference is not significant. There is opportunity to harmonize the requirements for the maintenance of E roads in the future.

	Estonia	Latvia from April 2021	Lithuania
Allowed snow thickness	free	4 cm	free
Time for snow removal	2 hours	8 hours	2 hours
Time for de-icing	2 hours	8 hours	2 hours

Table 5. Requirements for the maintenance of E67 VIA BALTICA during nighttime.

Disclosure Statement

I have no competing financial, professional, or personal interests from other parties.

References

Saat M. (2021). *Winter maintenance in Estonia.* Chief Engineer, Western Road Maintenance Department, Estonian Transport Administration. Presentation in the Seminar of BRA Maintenance Committee, February 3, 2021.

Ratkevičius T. (2021) *Road maintenance management in Lithuania.* Project Manager, Department of Transport Infrastructure, Transport Infrastructure Maintenance Division, State Enterprise "Lithuanian Road Administration". Presentation in the Seminar of BRA Maintenance Committee, February 3, 2021.

AUTOMATION OF ISSUING PROCESS OF PERMIT TO USE ROADS BY ABNORMAL TRANSPORT

Modestas LUKOŠIŪNAS

State Enterprise Lithuanian Road Administration, Vilnius, Lithuania
Corresponding author: Rolandas Mazaliauskas, e-mail rolandas.mazaliauskas@gmail.com

26 February 2021

Abstract. Since the end of 2017 the Lithuanian Road Administration (LRA) has taken over the authority to issue permits for oversized and heavy vehicles using the national significance roads from the former State Road Transport Inspectorate. LRA has launched a pilot project for automated authorising system, in which application is submitted by filling electronic form and providing a driving route on a map. The route in most cases is automatically approved by the system. If there are any restrictions on the route or in case of larger parameters of a vehicle, an LRA specialist shall review it. Upon coordination, the system automatically calculates a charge for the use of roads by abnormal transport and informs the applicant. The applicant pays this charge via electronic banking, and the system automatically issues a permit (such payments make up to 90 %). If the charge is paid by standard transfer, the system automatically issues a permit after LRA employee registers payment details.

Keywords: abnormal transport, oversize and heavy transport, application for permit, permit to use roads by abnormal transport.

Introduction

Permits for the use of national significance roads used to be issued by the former State Road Transport Inspectorate (later – SRTI) which had 5 branches across the country. SRTI had to coordinate all the driving routes with the Lithuanian Road Administration (later – LRA), thus making it a slow process with all the data processed manually. LRA employees had to review all the driving routes, check if they are correct, calculate and provide the length of the route on national roads, including the part of the route on the highest category roads, which is necessary for the calculation of the charge for the use of national significance roads. To facilitate this process, LRA has started to develop a web application in which all the required data about the vehicle is filled in and the driving route on the map is planned. According to this data and various restrictions on the roads, this system can automatically measure and approve the driving route. Later, a political decision was made, so that LRA not only had to measure and approve the route, but also to calculate the charge and issue the permit. The application system was expanded with new features and now it can do all these processes automatically.

1. Legal Framework

The use of roads by abnormal transport in Lithuania is mainly regulated by 6 legal acts:

1. The Lithuanian Republic Law on Roads, which sets out the basic requirements and principles for issuing permits;
2. The Law on the Financing of the Road Maintenance and Development Program, establishing the principles of road financing;
3. Government Resolution No 447 establishing specific tariffs, procedure for calculation and administration of the charge for use of national significance roads;
4. Government Resolution No 757 approving the list of national significance roads;
5. Order No 3-289 of the Minister of Transport and Communications, which regulates the procedure for issuing and using the permit;
6. Order No 3-66 of the Minister of Transport and Communications, which regulates the maximum authorized parameters of multiple vehicles.

Roads in Lithuania are divided into two categories depending on their significance (Figure 1):

- roads of nation significance;
- roads of local significance.

Roads of national significance belong to the state, they are managed by LRA which also issues permits to use them.

Roads of local significance are mainly owned and managed by municipalities. Permits to use roads of local significance are issued by municipal administrations.

The owner of abnormal transport must obtain the permit of the road owner to use it. Therefore, depending on the significance of the road, the owner of abnormal transport must obtain LRA and/or municipal permits.

LRA is responsible and manages website www.eismoinfo.lt in which detailed information about road traffic conditions and restrictions on roads of national significance is provided (Figure 2).

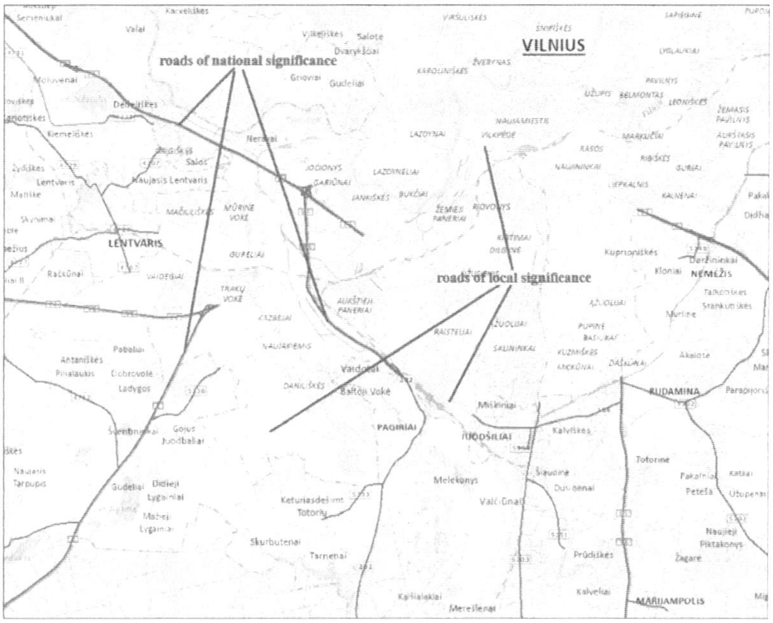

Figure 1. Roads of national and local significance on map (www.eismoinfo.lt with a layer of road surface course)

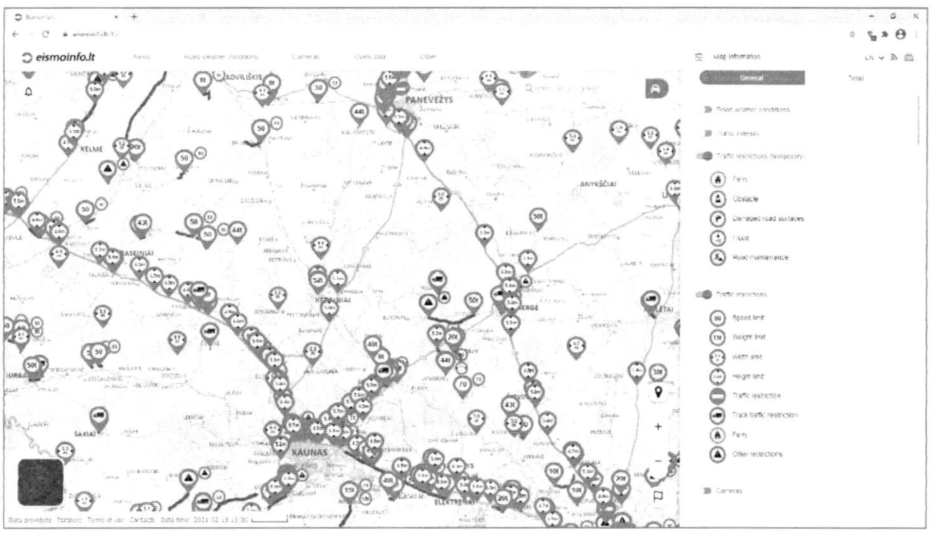

Figure 2. Information about traffic restrictions on roads of national significance (www.eismoinfo.lt with layers for temporary and permanent traffic restrictions)

IOP Conf. Series: Materials Science and Engineering 1202 (2021) 012040 doi:10.1088/1757-899X/1202/1/012040

2. Application Process

The owner of abnormal transport must submit an application request for a permit to LRA, which contains:

1. vehicle and cargo data;
2. driving route.

The application request for a permit can be submitted to LRA only by electronic means on the website https://eismoinfo.lt/leidimai. The applicant fills in the application form, which contains the following necessary data:

- the name and address of the vehicle owner;
- e-mail address for further communication;
- the type of permit (one-time, monthly or annual) and the start (validity) date of it;
- details of the axles, including the distances between the axles and the actual axle weight;
- the dimensions of the vehicle. The gross weight is calculated by summing the specified axles weights;
- data of the load, its dimensions, weight, including the type of load: divisible or indivisible;
- the driving route, which is presented on a map created on the basis of ArcGIS;
- required documents attached to the application.

The applicant is responsible for the accuracy and correctness of the data. When entering the data to the application form, the system carries out an initial verification of the data on the possibility of issuing long-term permits (monthly or annual).

According to the vehicle parameters specified in the application and traffic restrictions on the route, the system automatically assesses the possibility to take the intended route and also a possibility to issue the permit. The system is allowed to automatically approve the route if the vehicle's dimensions, weight or axle weight do not exceed the permitted norms significantly (Table 1) and there are no restrictions on the route that could prevent the vehicle from following the planned route (Figure 3). If the parameters of the vehicle exceed the norms specified in the Table 1 below, or there are some restrictions on the route that could prevent the vehicle from driving, requests must be reviewed by an LRA employee.

Table 1. Parameters that the system can approve automatically

Objects	Parameters
All vehicles except for tractors / self-propelled machines and car transporters	–
Length (all except vehicles with trailer)	≤ 30,00 m
Length (vehicles with trailer)	≤ 18,75 m
Width (all vehicles)	≤ 4,00 m
Height (all vehicles)	≤ 5,00 m
Weight (all vehicles)	≤ 55,0 t
Axle weight (all vehicles)	≤ 12,0 t

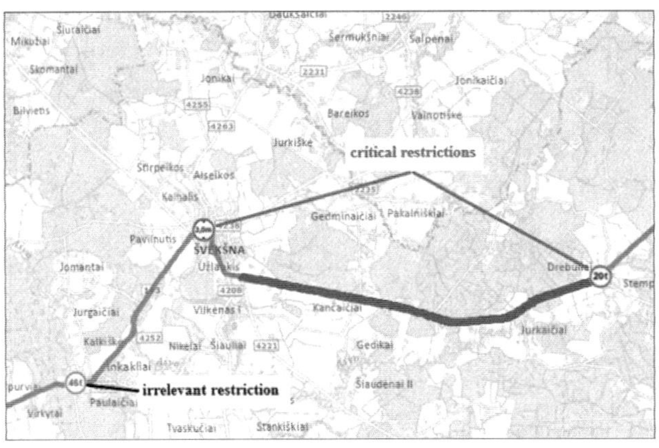

Figure 3. Barriers on the planned driving route.

When the route is approved, the system automatically calculates the charge, which depends on classifier values, and informs the applicant about the amount of the charge to be paid. The applicant has the option to pay it using the same system or to make a bank transfer. When the charge is paid through the system, the process continues automatically. After receiving the bank's approval, the system forms a permit based on the application data. If the charge is paid through the bank transfer, the applicant must send a payment confirmation to the LRA employee, who registers it and only then the system can automatically issue the permit (Figure 4).

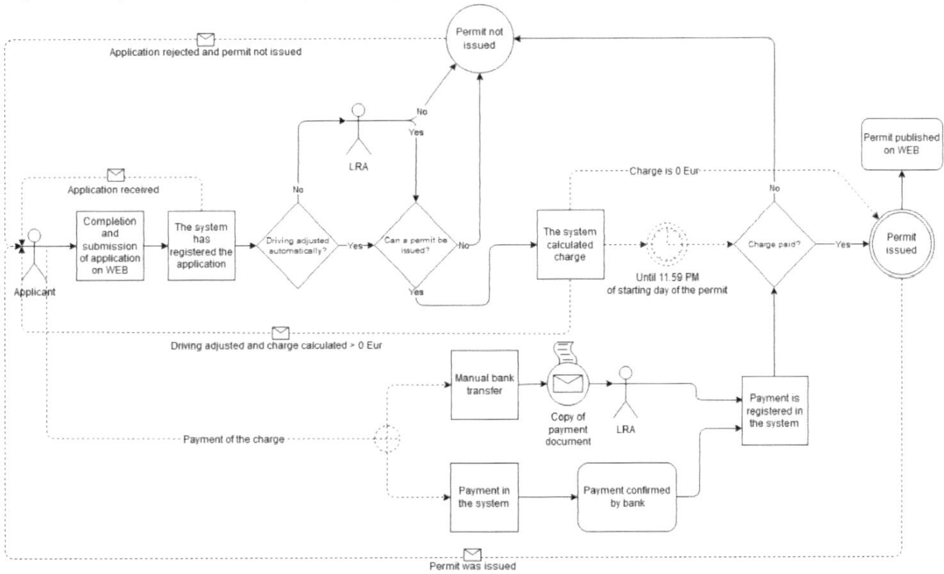

Figure 4. Typical scheme of application process.

The system sends automatically generated emails throughout all of the application process, to the email address specified in the application form. If additional driving conditions are specified during manual review of the route, these conditions shall be automatically recorded by the system and sent in the message informing about the approval of the route and in the permit issued.

All email messages sent by the system contain a link through which the application form can be opened and the details of the submitted application can be adjusted, in such way a new application can be submitted.

In addition, the owner of vehicle can change the vehicles (mark, model and plate number) specified in the permit. In the email message about the issued permit sent by the system, there is a link to the data change window, in which the applicant himself can change the data, after the system forms an amended permit, without the participation of an LRA employee (figure 5).

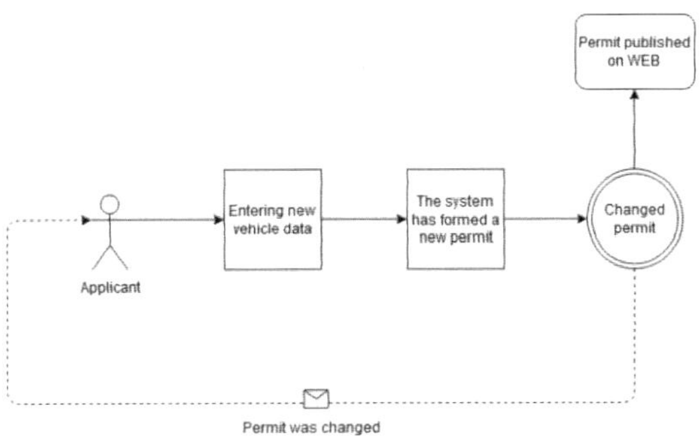

Figure 5. Typical scheme of vehicle change process.

The issued permit message notification also includes a link to download the permit data file, but this is not an official document. All issued permits are automatically published on the Internet, where they can be found and viewed by all interested parties: an applicant, a driver, a consignor, a consignee, a controlling officer, etc.

3. Results of automation

In principle, the intervention of an LRA employee in the permitting process is required only in the following cases:

1. the route was not approved automatically by system because of the vehicle's exceeding parameters or critical traffic restrictions (barriers) on the planned route;
2. the charge was paid not through the system, but by the bank transfer.

The share value of charges paid not through the system is about 8% of all permits issued. Manual registration of payments does not require particularly high time cost. However, it can be reduced by implementing a wider range of billing options. A payment card has recently been introduced as an option to pay the charge, which further increases the number of payments made through the system and reduces the burden on LRA employees.

The share of manually coordinated and approved applications is about 30%. Some of them are only a formal confirmation of an LRA employee, which can also be reduced by expanding the system features of automatic assessment. However, some of them require careful and responsible analysis, such as the load-bearing capacity of bridges, the height of viaducts and so on.

It should be also noted that permits for under one quarter of the applications received are not issued. These applications have to be either adjusted or rejected, or is applicants just wanted to check the driving possibilities and to find out the amount of the charge.

The implementation of this system by LRA since the end of 2017 has significantly increased the number of permits issued (Figure 6).

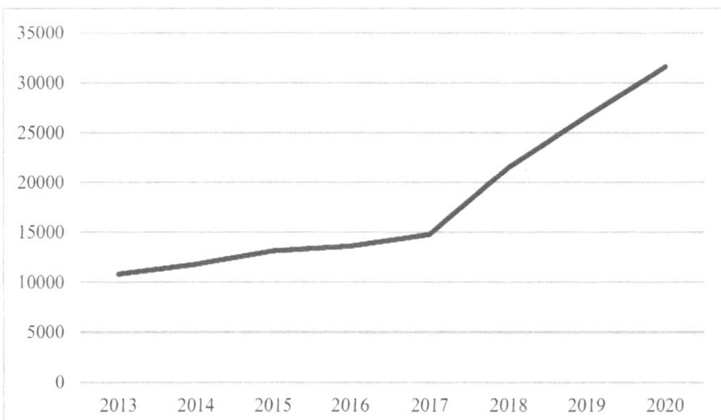

Figure 6. Permits issued per year.

Conclusions

Here, in LRA, we are observing such trends that the cargo loads are increasing and becoming heavier each year, especially the equipment and structures used on construction sites. However, we also believe, that a relatively fast way in which a permit can been obtained has pulled other owners of vehicles out of the shadow, who took the risk of driving without a permit because of the long-time spawn form permit to be issued earlier in the days.

Advantages of the automated system:

- authorization process period was reduced from 2–5 days to 5–10 minutes in case of the system's automatic approval and payment via system, or to 1–2 days in case an LRA employee has to review it;
- a number of employees required to manage this process was reduced from 4–5 in SRTI plus 1–2 in LRA to 1–2 employees in LRA (one – full-time position, second – consultant on bridges and a substitute for the first one);
- the number of permits issued and charge paid has increased significantly;
- key data of application is saved and can be used multiple times;

- there is no direct contact between an applicant and an LRA employee (anti-corruption measure);
- all issued permits are electronic and published on the Internet, so there is no need for paper copy;
- permit data is saved and stored in a georeferenced database;
- there is a possibility to incorporate municipalities into one permit system, if the data about local significance roads were stored in the same manner as the data about the state significance roads.

Disadvantages of the automated system:

- representing the driving route on map is quite a challenge for unskilful applicants;
- there is no possibility to correct some of the submitted data in the application during the authorization process.

Further possibilities for using georeferenced data of permits:

- to plan repairs or development of roads and bridges assessing the accumulated data of abnormal transport;
- implementation of automated tracking (control) system of abnormal transport (the system can immediately identify if a vehicle has a permit to drive on a particular road).

In summary, majority of customers are satisfied with the current fast operating system in Lithuania.

Moreover, we have reduced human resource and time costs required to manage this process and accumulated georeferenced data about the routes of abnormal transport.

E-WAYBILL USING EXPERIENCE & DEVELOPMENT IN ESTONIAN STATE ROAD BUILDING

E-WAYBILL USING EXPERIENCE & DEVELOPMENT IN ESTONIAN STATE ROAD BUILDING

Taavi TÕNTS & Aivo SALUM #66, Smart Road Solutions & ITS

Road Management Department, Leading Engineer, Estonian Transport Administration, Tallinn, Estonia

Received day month year; accepted date: 17.01.21

Abstract. Estonian Transport Administration (ETA) has since 2010y developed digital solutions for monitoring abnormal 52t transport heavy vehicles (HV). Since 2020y we signed the memorandum between 8 different parties for developing bulk material transport digital solutions (e-waybill system) for road building. The focus is to make the logistic more transparent since beginning of the loading point - for the different authorities. The second focus is to make the truck movement corridor visible for the traffic control, avoiding week roads, bridges etc. The final, and the most difficult, is to develop the mass control system, so that there is automated weight info in the e-waybill system visible for the traffic police and for building supervisors etc. We have met with our Association of Estonian Cities and Municipalities and many others, and everyone is very interested of going from paper waybills for faster, cloud based, e-waybill systems, what is also more $C0_2$ friendly. This digital e-waybill allows single data entry, and all the rest data with statistics is visible for concerned people. ETA is planning to pilot in 2021y also many road building projects with e-waybill demand. So far, the feedback has been mainly positive from different parties. We have started with our Estonian Ministry of Economic Affairs and Communications (EMEAC) also wider digitalisation projects concerning the new regulation (EU) 2020/1056 of eFTI for the gross-border transport logistics digitalisation, what must be applied in every member state 21.08.24.

Keywords: e-waybill; eFTI; Intelligent Access; on-board-weighing (OBW); CEDR RFT WG; $C0_2$ reduction in transport; contact free ITS; VELUB;

Introduction

Following paper gives an overview about Estonian's digital research and developments in road building and transportation area, since 2010y.

As there are many new EU regulations which ones demands digital freight and transport information systems to develop for the EU member states, it might be useful to chare the development overview with other EU member states.

Some pilot projects have already been with the international e-waybills (e-CMR), made by EMEAC with the neighbour countries, where also ETA was a partner (Figure 1).

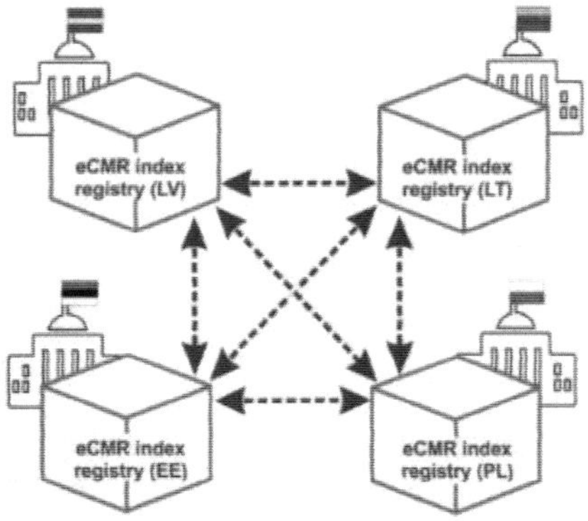

Figure 1. Schema. The index registries exchange data with each other using DLT technology /1/

IOP Conf. Series: Materials Science and Engineering 1202 (2021) 012041 doi:10.1088/1757-899X/1202/1/012041

E-WAYBILL USING EXPERIENCE & DEVELOPMENT IN ESTONIAN STATE ROAD BUILDING

1. EU new regulations for the transport sector

EU's Electronic Freight Transport Information (eFTI) regulation demands EU member states to provide the eFTI service starting from 21. August 2024y /2/. Based on that regulation EMEAC is piloting eFTI / eCMR together with the neighbor states and ETA.

Since 2020y ETA is piloting e-waybill in the bulk material transport in building contracts with the target to demand it since 2022y in all building contracts. In the beginning of the 2021y EMEAC joined to our Bulk Material Transport Memorandum and declared the interest to develop also the internal state e-waybill (internal eCMR) - together with international eCMR, according to the eFTI directive. Also, Estonian Association of Information Technology and Telecommunications (ITL) joined as 10. member to the memorandum.

ETA has developed since 2010y also Intelligent Access, with the VELUB System which one allows to control on the Smart Road map, 52t abnormal transport vehicles (with GNSS) for more effective and greener transport, protecting the roads constructions at the same time. /3/

Greener transport vision is detailly described in EU's new Sustainable and Smart Mobility Strategy until 2050y. It describes how greener transport should look like and how it should be developed in the EU member states etc. /4/

EU and EFTA Ministers of Transport declared last year in the Passau Declaration that digitalization is the Smart Deal for Mobility and it shapes the mobility of the future – sustainable, safe, secure and efficient. The Passau Declaration talks about Building Information Modelling (BIM) and eFTI and other digital tools for making the whole transport sector more efficient in coming years. /5/

Different type of e-waybills and digital information (Figure 2) can be very likely developed in one, digital, cloud-based system.

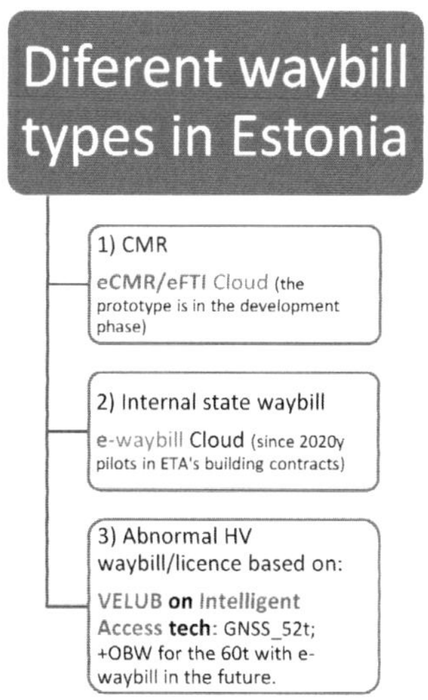

Figure 2 Different waybill types.

In the following paragraphs, the short overview about the Estonian different e-waybill connected developments and researches will be given.

IOP Conf. Series: Materials Science and Engineering 1202 (2021) 012041 doi:10.1088/1757-899X/1202/1/012041

2. eCMR pilots and ongoing process

Estonia supports the fast adoption of eFTI to achieve the goals of the road transport digitalisation as early as possible. There are e-CMR service providers and there are governments capable and interested to receive the data that they are entitled to check. Still, the governmental institutions cannot easily accept the electronic e-CMR, even if the legal ground for it exists. The missing link is the secure and trustful way to exchange e-CMR data between governments and e-CMR service providers.

eCMR solution, which meets the market´s expectations, can only function as an internationally connected digital ecosystem which would allow the data exchange on transport documents along the logistical corridor. Therefore, it is reasonable to carry out pilot projects on road transport digitalisation as cross-border service development in cooperation with neighbouring countries.

Under the leadership of EMEAC, the cross-border eCMR prototype between Estonia, Latvia, Lithuania and Poland (Figure 3) was introduced in September 2020 /1/.

The objective of prototyping was to create an e-governance compatible eCMR indexing scheme that will allow controlling institutions of partner countries to check the availability and validity of CMR transport documents in a secure and trustful way. Every authorized governmental institution will be able to access minimum available indexing information via a specialized application programming interface (API) and get a link to request original documents relevant data directory from e-CMR service from the county of origin.

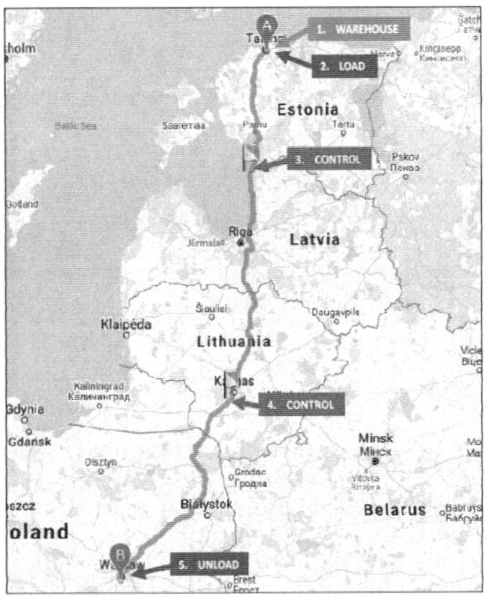

Figure 3 eCMR prototype piloting scheme.

For the public sector and authorities, it will create conditions for the more effective transport sector and trade supervision. One of the outcome examples is reducing the time spent on stopping trucks for CMR control, but also minimizing the number of offences related to tax avoidance and transport safety.

The prototype development team includes governmental organisations (ministries, road administrations, tax and customs, police, statistics etc.), logistics and road carriers' associations, consulting experts and leading software developers with an international background.

EMEAC will analyse in 2021y (Figure 4) NAP system architecture and how internal e-waybills can be linked (Figure 1). Also, will be analysed Cost-Benefits, the ownership alternatives and different risks etc. ETA is involved to the development as a road owner, whose interest is to increase construction and traffic safety etc. /6/

E-WAYBILL USING EXPERIENCE & DEVELOPMENT IN ESTONIAN STATE ROAD BUILDING

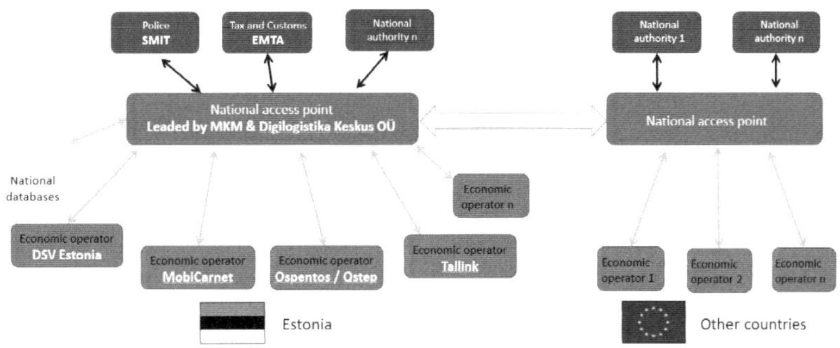

Figure 4 National Access Point (NAP) as part of national and global ecosystem (Estonian possible version)

EMEAC will also introduce in EU in 2021y, the Position Paper Towards Interoperable Electronic Freight Information Exchange in Europe. In this position paper the road owners' vison is included for extending the vision to road safety, road maintenance, taxation, etc.

The European Commission's vision to digitize G2B, B2G transport information flow across Europe with eFTI regulation is undoubtedly ambitious. However, it addresses only the information related to cargo, but not the vehicle or a driver. At the same time, there is a strong need for public authorities to handle this information together to achieve greater transparency, safety and efficiency.

Therefore, the eCMR concept can be expanded using the same data exchange channel and technology in:

- **Reducing risks in road traffic** (automated validity of driving license and health certificate; inspection of compliance with working, driving and rest time rules; technical inspection of truck and trailer; payment of fines and tolls; insurance of truck and goods, etc.).
- **Improving road constructions safety** (truck weight information from On-Board Weighing Systems).
- **Handling of dangerous goods, special cargo and contract deliveries** (GPS tracking, automated fee payment control).
- **Increasing the transparency of tax revenues** (payment of customs duties, automatic VAT refunds).
- **Other areas** (rescue, statistics, insurance, etc.).

Handling this information in a single channel has the potential to significantly improve the efficiency of transport and, therefore, to contribute to the smooth functioning of the Single Market.

Taking into consideration the quick need for a positive transformation in the electronic freight transport information sector, Estonia and others highlight the need to swiftly start discussions by setting up the joint development of eCMR data exchange model in Europe.

3. Internal state e-waybill using experience

In 2020, ETA carried out three procurements, where it was mandatory to use an e-waybill of lading for the transport of bulk materials. The special requirements for documenting, the works applied to both: the contractor and the owner's supervision. All consignment notes for bulk materials and the summary tables, compiled based on them, had to be prepared in an electronic data exchange platform. /7/

The pilot projects assumed the use of either the Waybiller environment developed in Estonia or an analogous electronic data exchange platform. The procurement required that the digital platform allow the creation of separate objects and GNSS location-based tracking of each load (Figure 5). The vehicle and/or trailer number information had to be generated automatically from the traffic register database and had to reach the electronic environment on the e-waybill. If the truck had a special cargo permit of 48 or 52 tons, its data, permit number and the validity period had to be included. Supervisors and subscribers needed to have access to the environment, so that they could control the information through the cloud.

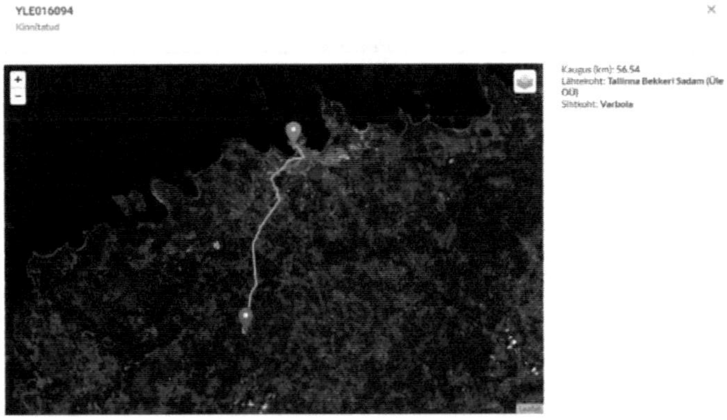

Figure 5 GNSS makes material transport, and strong road corridor usage, is in road building more transparent.

The e-waybill had to contain at least the number of the truck and/or trailer, the number of axles, the permissible weight / load capacity of the truck; the mass and name of the material, the name of the driver, the haulier, the owner of the load and the quarry. If the transport was on public roads, an e-waybill had to be created for the transport from the intermediate warehouses to the site (except for the intermediate warehouses immediately adjacent to the site). An indication had to be made on the e-waybill, whether the material came from a quarry or from an intermediate warehouse.

As mentioned, the owner supervision was also obliged to use the electronic environment of the e-waybill. For example, he had to check the e-waybills for bulk materials provided by the contractor and confirm receipt of the load in the digital environment. Also, during the asphalting works, the engineer used a data exchange platform to validate the e-waybill of asphalt loads arriving at the site.

This year, the ETA is already planning several times more (14) pilot procurements in order to prepare for the full transition to e-waybills in 2022y, for the transport of bulk materials. One procurement in Pärnu area, will require the use of a weighing station, loader or OBW interfaced with the service provider - to protect the weaker road constructions and bridges (Figure 6).

E-WAYBILL USING EXPERIENCE & DEVELOPMENT IN ESTONIAN STATE ROAD BUILDING

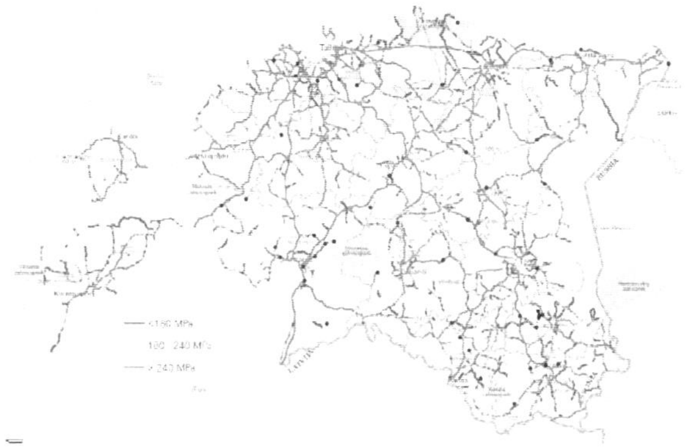

Figure 6 Strength of the Estonian state roads (black dots, weaker bridges).

For 2022y, a separate contract annex (2 pages) has been prepared for contractors, which contains detailed minimum technical requirements for the e-waybill. This will make it easier for the new service providers to bring data exchange platforms to market and will be better traceable to the all parties.

In 2020y OÜ Üle won the public procurement of surface works in Harju and Rapla County, who, despite the initial critical attitude, finally found that the use of the e-waybill is simple, compact and allows real-time monitoring of the data. The problem was caused by the delivery note, which was required for the forwarding of the material from the intermediate warehouse to the object, although it is also required as a sheet of paper, according to the Road Transport Act.

AS Järvak, which performed the transportation work, highlighted the time savings and the possibility to adapt quickly and expressed the wish that the e-waybill would be fully switched soon. In total, about 2,000 e-waybills were completed during the project. Lääne Teed OÜ turned out to be the repairer of Rapla gravel roads. At the end of the pilot project, their project manager assessed both the ease of use of the e-waybill and the time saved that would otherwise have been spent on paper sheets and the subsequent precise compilation of volumes and shipments. Taital Trans OÜ, which made transports within the project and who already had previous experience, added that the e-waybill is convenient to use. Approximately 1,700 e-waybills were used at the sites.

The largest test object in 2020y was the reconstruction of the state highway No. 11412 (Liikva-Rannamõisa). The contractor of the project was AS TREV-2 Grupp (VINCI Concern), whose project manager said that the environment of Waybiller (Figure 7), the e-waybill service provider they used, worked quite well, although it should be further developed. In the course of the project, about 2,100 e-waybills were issued.

Figure 7 User view of the Waybiller app.

IOP Conf. Series: Materials Science and Engineering 1202 (2021) 012041 doi:10.1088/1757-899X/1202/1/012041

4. VELUB Intelligent Access

ETA has since 2010y developed digital Intelligent Access solutions in the VELUB platform for the abnormal 52t transport monitoring (Figure 8). This system can be used in the future for controlling also heavier trucks, adding more controlling criteriums etc. At the moment, 60t and EMS longer and heavier vehicles CBA is ongoing in ETA.

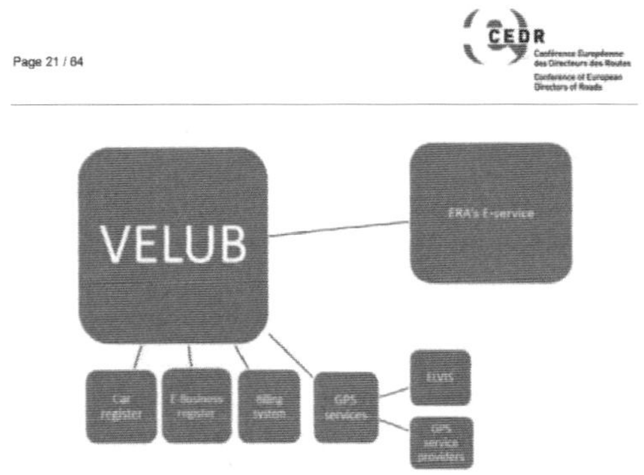

Figure 8 VELUB– ETA’s special permit application system; ELVIS – State Forest Management's electronic cargo list information system. /8/

Study of the HV impact to the bridges showed that, less than 7% of the total bridges, are not OK for 52t /9/
Those weaker bridges are not marked inside the strong, digital Smart Road (Figure 9) corridor. /10/

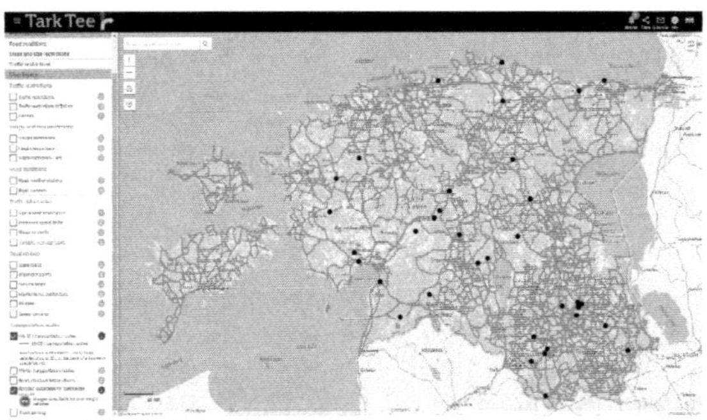

Figure 9 Smart Road (ETA, Tark Tee) strong corridors marked with the purple lines (dots - weaker bridges) 28.02.21

The next step is to develop the VELUB further - automatic control, OBW in the future, etc. Overview about Stakeholders in the VELUB and different waybill developments in Estonia, was given in Aeroflex/i4DF workgroup /11/.

Detail overview about the VELUB development and its digital possibilities was given in HVTT15 Symposium. / 12/

In 2020y, OBW telemetry (Figure 10) tests were completed, with 5 different HV. Good results were achieved for continuing in coming years. This study with OBW tests with HV’s, was carried out to determine the possibility of using OBW equipment to reliably and accurately monitor weights of HV’s. /13/

Estonia has long been the leading innovator in managing heavy vehicles via new technologies such as telematics, GIS and ITS. Previous developments such as current road networks for HV („green roads“ for winter time and „purple roads“ for whole year (Figure 4.2) have been widely accepted by logistics companies and road administrators. These developments give strong background for new innovations with OBW equipment.

E-WAYBILL USING EXPERIENCE & DEVELOPMENT IN ESTONIAN STATE ROAD BUILDING

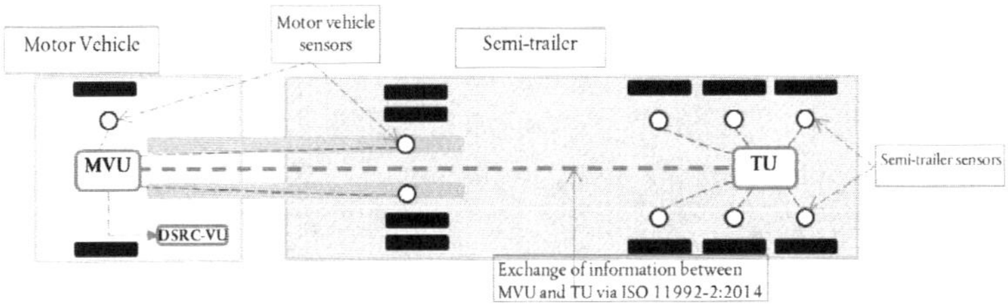

Figure 10 Example of layout for OBW in a stage 1 truck/semi-trailer vehicle combination / 14/

For this purpose, five trucks were selected, connected and monitored through fleet management platform provided by FleetComplete. Necessary additional hardware and software developments were done to allow to read the weighting data by telematics devices from vehicles CAN bus using FMS interface and display the data through web interface (Figure 11& 12). For verification purposes, weighing of fully loaded HGV vehicles was carried out with portable scales.

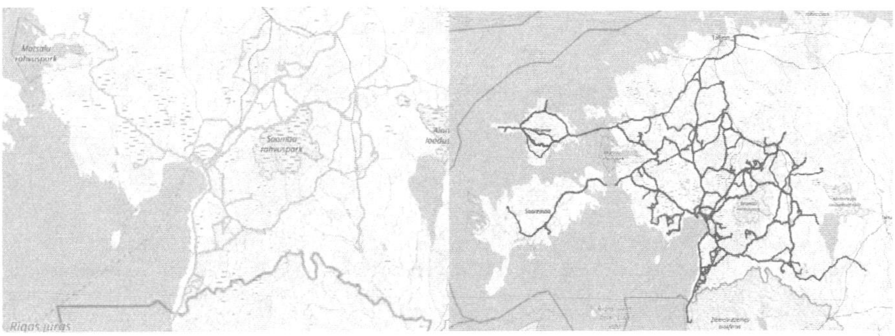

Figure 11 HV „Volvo 02 timber truck 4+" moving paths 10.09.2019 – 21.10.2019 (left) & 21.10.2019 – 15.01.2020.

Besides enforcement capabilities, OBW systems provide logistics managers with robust way of optimizing the usage of trucks. As our analysis showed, fuel usage per kilometre does not increase significantly at higher loads. Therefore, loading trucks to maximum safe limit for roads allow saving fuel and thus reduce carbon dioxide emissions.

In order to determine said safe limit, all data must be integrated with road database. In this study, only pavement and IRI were used, providing insight that pavement type does affect the measured values, but road roughness has no effect. In future studies, other road parameters can be used with similar methodology. Also, the current road networks for heavy vehicles („green roads" for winter time and „purple roads" for whole year) can be linked to weight data providing even more insight for better logistics management and when necessary, also enforcement. In terms of pavement management, more detailed information about actual weights on the road will yield more accurate predictions than current methods using standardized axles and vehicles.

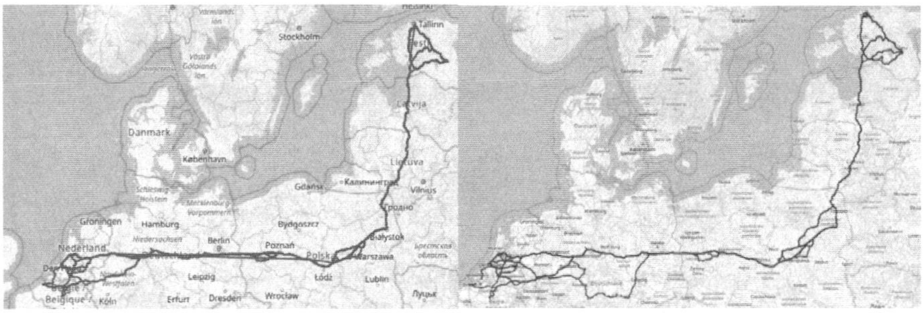

Figure 12 HV „Scania 02 saddle 2+“ moving paths 02.09.2019 – 21.10.2019 (left) & 21.10.2019 – 15.01.2020

Integration with road tolling system provides road administrators an opportunity to promote logistics solutions better for road structures and environment. This is also valuable information for procurement of logistics services allowing to choose most effective provider with smallest carbon footprint. In order to achieve this, a follow-up study with more vehicles included is needed to better understand the relations between fuel usage, road deterioration and load weight.

OBW data gives for the traffic control good overview where are the most likely overloaded HV. With that data traffic control points can be decided much quicker - where to go to do the static weighing with traffic control van.

CEDR new RFT (Road Freight Transport) work group has also started Intelligent Access subgroup, to study further possibilities what new, digital cloud-based system allows: road construction protection for aging infra, higher traffic safety and at the same time greener transport etc.

Conclusions

E-waybill development in Estonian state road building and other areas, has given us already a lot of savings in CO_2 emissions, making transport more efficient in the same time - with the single data entry to the cloud. Defiantly there is still a lot to do and develop in coming years, to achieve EU’s climate targets in the transport sector, using digitalisations as a tool for it.

Hopefully our neighbour Baltic countries, and other EU states, are continuing in coming years, the EU digitalization goals to make more efficient our freight transport between the states, and why not, inside state as well.

ETA’s bulk material e-waybill piloting and usage is just a small part of internal transport (ca 5%), but it is a huge step to digitize all internal waybills (over 2mln/y) together with EMEAC, and together with international eCMR, based on the eFTI regulation /2/.

In 2022y ETA is planning to demand e-waybills in all building contracts. There are already now many interested ITS companies with the big interest to develop e-waybills in Estonia, as it’s not so complicated in 21. century anymore. If there is interest, our e-waybill providers can help to pilot other EU road owners as well - for the greener future.

Digitized cloud-based transport and road information systems are allowing to develop further Intelligent Access to protect road construction and increase traffic safety. In the same time transport transparency increases, and ca 50mln EUR/y socio-economic benefit is produced in small Estonia.

IOP Conf. Series: Materials Science and Engineering 1202 (2021) 012041 doi:10.1088/1757-899X/1202/1/012041

E-WAYBILL USING EXPERIENCE & DEVELOPMENT IN ESTONIAN STATE ROAD BUILDING

References

1. Final Report eCMR index registry prototype, 2020y, Estonian Ministry of Economic Affairs and Communications https://www.diginnobsr.eu/ecmr-testing-results
2. Regulation (EU) 2020/1056 of the European Parliament and of the Council of 15 July 2020 on electronic freight transport information https://eur-lex.europa.eu/eli/reg/2020/1056/oj
3. ETA's VELUB page, https://eteenindus.mnt.ee/maantee.jsf?lang=en
4. EUROPEAN COMMISSION, Brussels, 9.12.2020, *Sustainable and Smart Mobility Strategy – putting European transport on track for the future*. https://eur-lex.europa.eu/legal-content/EN/TXT/?uri=COM:2020:789:FIN
5. PASSAU DECLARATION on the occasion of the virtual Informal Meeting of EU and EFTA Ministers of Transport under the auspices of the German EU Council Presidency on 29th October 2020 https://www.bmvi.de/SharedDocs/EN/Documents/K/passau-declaration.pdf
6. E-waybill future possibilities in road owners point of view (E-veoselehe väljavaated taristuomaniku pilgu läbi), Roadpaper (Teeleht), pg 32-33 nr 100, 2020y, Taavi Tõnts, ETA https://www.mnt.ee/sites/default/files/road_paper_pdf/mnt_teeleht_suvi2020_nr100_veeb_1_1.pdf
7. EU sees "YES" for eCMR (Euroopa ütles e-veoselehele „jah"), Roadpaper (Teeleht), pg 19-21 nr 102, 2021y, Taavi Tõnts, ETA https://www.mnt.ee/sites/default/files/road_paper_pdf/teeleht_nr_102_veeb.pdf
8. Conditions for efficient road transport in Europe, CEDR 2017y https://www.cedr.eu/download/Publications/2017/2017-5-Conditions-for-efficient-road-transport-report.pdf
9. ETA's Smart Road information system https://tarktee.mnt.ee/#/en
10. HV Impact to the bridges, 2020y, ETA (Raskeveokite mõju sildadele. Nende koormuste valideerimine reaalsetel koormustel ning mõju analüüs). https://www.mnt.ee/sites/default/files/survey/raskeveokite_moju_sildadele_lopparuanne_2020-03-23.pdf
11. Experiences with the stakeholder process from Estonia, T.Tonts (2019), Aeroflex/i4DF, Special Focus Workshop: Intelligent Access Part II, Paris.
12. Estonian VELUB system for more efficient and greener transport VELUB System is a small part of the much bigger, theoretical, MVMC System (E^n dimensional movement controlling system), T.Tonts (2018), HVTT15, https://hvttforum.org/wp-content/uploads/2019/11/Taavi-Tonts-ESTONIAN-VELUB-SYSTEM-FOR-MORE-EFFICIENT-AND-GREENER-TRANSPORT.pdf
13. OBW tests with HV's (Tänapäeva ja tuleviku rakendused sõidukite kaalumisel, sõidukite masside jälgimissüsteemi loomine ja jälgimine reaalajas), Maanteeamet, ETA, AS Teede Tehnokeskus, Fleet Complete Eesti OÜ, Tallinn 2020y. https://www.mnt.ee/sites/default/files/survey/pohiaruanne_06_02_2020.pdf
14. Commission Implementing Regulation (EU) 2019/1213, Laying down detailed provisions ensuring uniform conditions for the implementation of interoperability and compatibility of on-board weighing equipment pursuant to Council Directive 96/53/EC https://eur-lex.europa.eu/legal-content/EN/TXT/?uri=CELEX%3A32019R1213

LESS TEMPTATION TO EXCEED THE SPEED LIMIT OR TOWARDS *VISION ZERO*

Viktoras Lapinas[1], Mantas Kišonas[2]

[1,2] *State Enterprise Lithuanian Road Administration, J. Basanavičiaus str. 36, 03109 Vilnius, Lithuania*
E-mails: [1]viktoras.lapinas@lakd.lt; [2]mantas.kisonas@lakd.lt

Abstract. In 2020 the Lithuanian Government approved traffic safety programme *Vision Zero*. One of the integral measures applied to improve traffic safety and to reduce the number of road traffic infringements is the development of average speed cameras' network on state significance roads. It is planned that the network of average speed cameras will cover more than 800 km of state significance road network in Lithuania in 2020-2021. Initially, it was planned to implement these measures only on rural roads. However, taking into consideration the principles of road eligibility for average speed camera installation, some road sections crossing the so-called linear settlements were selected to test the impact of such systems on driving habits as well. It is presumed that from the beginning of exploitation of these systems the reduction in the consequences of severe traffic accidents on the selected most dangerous state significance road sections will be observed.

Keywords: Average speed, speed control, distance control, driving habits, driving rules infringements

Introduction

According to the EU Road Safety Framework 2021–2030 – Next Steps Towards *Vision Zero*[1], about one third of fatal crashes are (partly) caused by excessive or inappropriate speed. According to the conducted study, the risk to be involved in a crash when speeding is 12.8 times higher. Moreover, higher speed crashes cause far more damage than lower speed ones. Based on the research results, the European Transport Safety Council (ETSC) has calculated that if average speeds dropped by only 1 km/h on all roads across the EU, more than 2,200 of road deaths could be prevented every year. It is expected that covering certain road sections with average speed cameras will give a significant positive effect in the reduction of fatal road accidents caused by overspeeding.

1. Key Principles of Average Speed Camera Installation

During the last 20 years Lithuania as well as other EU member countries have implemented various traffic safety programmes, soft and hard traffic safety improvement measures, transposed and implemented EU law into national legislation, set an aim to reduce the number of road fatalities and injuries. It has resulted in safer road infrastructure, safer vehicles, more responsible road users. A lot of state budgetary, EU and other funds have been allocated to traffic safety; therefore, the efficiency of their use shall be clearly substantiated and measured. Even though in the long run a considerable drop in the number of fatalities per 1 million inhabitants from 173 (in 2000) to 63 (in 2020) has been observed in the Republic of Lithuania (Figure 1). In the period of last four years the number of fatalities remains the same and fluctuates from 62 to 68 fatalities per 1 million inhabitants.

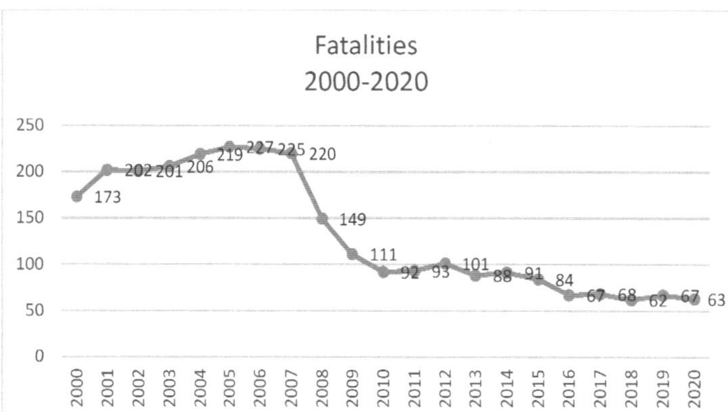

Figure 1.

It is obvious that earlier implemented measures were effective; however, the current situation shows that it is high time to review traffic safety components and look for further effective traffic safety improvement solutions. Therefore, in 2020 the Government of the Republic of Lithuania approved the new traffic safety programme. One of the integral measures implemented by SE Lithuanian Road Administration is the new traffic safety strategy, i.e. the implementation of the system of average speed cameras on state significance roads.

Key principles of implementation:

The methodology of locating average speed cameras has been prepared. According to the set criteria and conditions the whole state significance road network has been evaluated. Homogeneous road sections of the state road network were evaluated according to the four criteria (average annual daily traffic volumes (AADT), consequences of road accidents, unsafe driving speed, which is considered to be the main cause of accidents, limited overtaking). The obtained findings enabled to compile a priority list of 4,003 homogeneous road sections, according to which cameras are installed. It is planned to review the compiled list every three years.

In 2021 it is planned to set up 106 average speed control road sections (Figure 2). In 2017–2020, 687 injured and 66 fatalities were registered on these sections (Figure 3). The length of sections are usually the same as those of homogeneous ones. The length of the first road sections varies from 0.9 to 20.1 km.

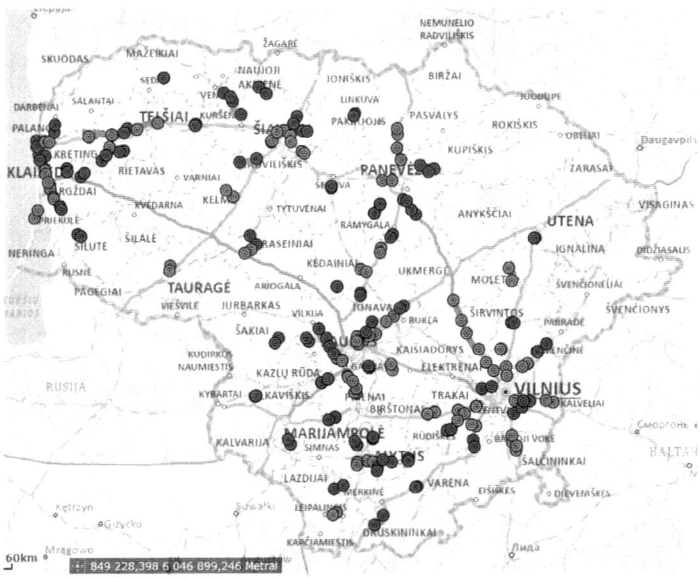

Figure 2. Location of average speed sections in 2020-2021

IOP Conf. Series: Materials Science and Engineering 1202 (2021) 012042 doi:10.1088/1757-899X/1202/1/012042

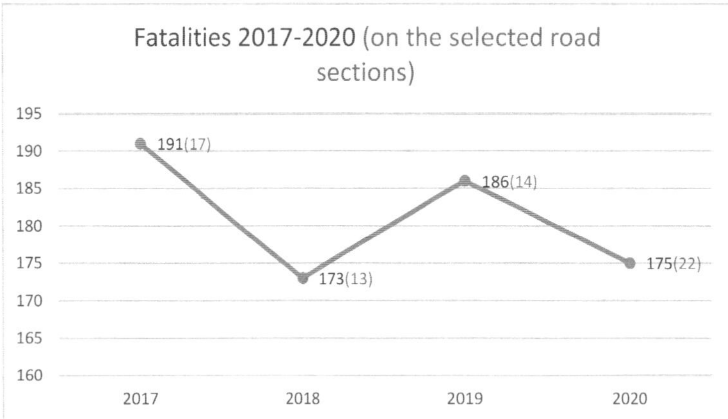

Figure 3. The number of road fatalities in Lithuania in 2017–2020 (value in brackets shows cases falling under 106 selected first average speed system implementation sections).

The aim of the average speed control system implementation on the most accident-prone road sections, ca 850 km, which would make up 13 per cent of the state significance main and national road network, is to alleviate the consequences of road accidents and to eliminate fatal road accidents on these road sections altogether. In 2018, the first installed 25 average speed control sections (Figure 4) showed that the measure influenced on the drivers' speed on this road section and reduced the number of overtakings.

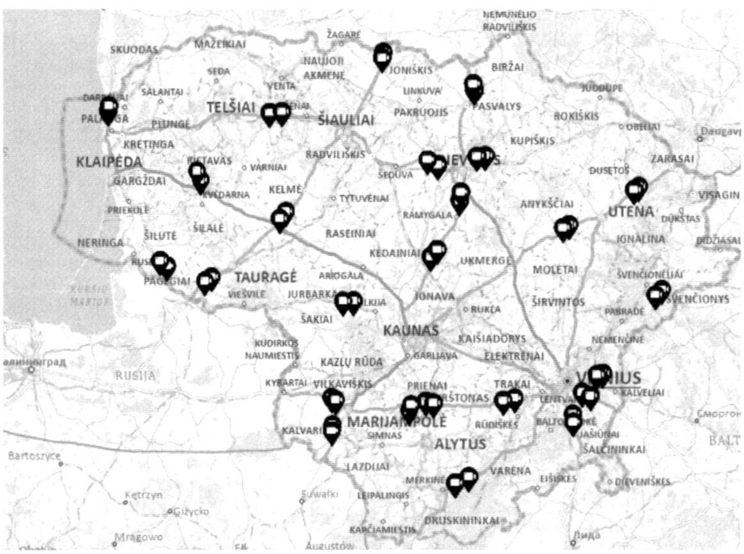

Figure 4. Current average speed control sections

The system will have a surplus value: new-generation technological solutions will enable to automatically check if transit vehicles carry compulsory insurance and valid technical inspection documents as well as if the road user charge has been paid.

2. Driving Habits on Average Speed Sections

Average speed control sections contribute to educating drivers and changing their driving habits, which is confirmed by the available statistical data (Table 1). To illustrate the situation, 3 average speed control sections on different roads have been selected. For example, the average speed of vehicles driving on A9 road on a selected day was 83.52km/h, when the permitted speed limit is 90 km/h. Drivers exceeding the average speed limit by 1km/h accounted for 8.28 per cent. The average speed of vehicles on A5 road on a selected day was 83.1km/h, when the permitted speed limit is 90 km/h. The number of those exceeding the speed limit by 1km/h accounted for ca 4 per cent. The average speed of vehicles on A13 on a selected day was 79.71km/h, when the speed limit is 90 km/h. The amount of those exceeding the average speed by 1km/h on this section accounted for 3 per cent only.

Average speed section	Speed limit (km/h)	Traffic volumes	Average speed (km/h)	Average speed violations from 1 to 10km/h (%)	Average speed violations from 10 to 20km/h (%)	Average speed violations from 20 to 30km/h (%)	Average speed violations above 30 km/h (%)
A9 14.03 – 20.555 km	90	3717	83.52	297 (7.99)	9 (0.4)	0	2 (0.05)
A5 76.936 – 80.368 km	90	4428	83.1	145 (3.27)	28 (0.63)	3 (0.07)	1 (0.02)
A13 33.309 – 36.184 km	90	2580	79.71	78 (3.02)	0	0	0

Figure 5. Mean of average speed and exceeded limit on a selected day

3. Average Speed Enforcement Systems in Lithuania

To achieve the best effect of average speed sections, high requirements for average speed system must be set. In Lithuania, we require at least 90 % of passed vehicle detection and at least 95 % for correct number plate recognition from suppliers. We perform data quality check to make sure that the systems operate properly. At the moment JENOPTIK VECTOR P2P system for average speed enforcement system is employed. After expanding the existing network of average speed enforcement systems, PASSCAM average speed enforcement system will be introduced as well. Both systems will use 2 ANPR cameras: one at the beginning and the second at the end of the section, without any additional triggers like radars or inductive loops. High quality of average speed enforcement system is required because after a vehicle passes one of the sections, an ANPR camera sends the data to the software that checks the validity of insurance, technical inspection and road user charge (if applicable).

IOP Conf. Series: Materials Science and Engineering 1202 (2021) 012042 doi:10.1088/1757-899X/1202/1/012042

Another important thing to take into consideration is the type of equipment that meets local requirements for recorded violations. For example, if you need to identify only number plate, additional infrared lighting is not required. If you need to identify a car or a driver, additional infrared lighting is required.

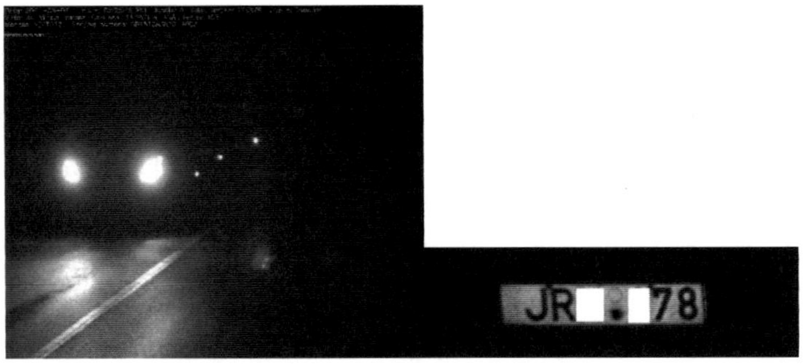

Figure 6. Example 1: When just number plate is enough

Figure 7. Example 2: When you need to identify the car as well

PERSONAL DATA: In Lithuania personal data (photos, number plates) are stored until different registers are checked and infringements are detected. Then the data are forwarded to relevant institutions and all personal data is deleted. It is only depersonalized data that are stored for statistics.

Conclusions

Exceeding speed limit remains one of the most frequent violations of road traffic regulations. For some drivers it has become a driving norm. Such driving frequently results in extremely severe road accidents. The average speed control on homogeneous road sections obliges drivers to observe the permitted speed limit on the whole road section. It is expected that compiling a priority list of road sections according to the developed methodology, where the main accident cause is overspeeding, will bring about give a significant positive effect. Presumably, it will surpass the current international experience [3] in reducing the consequences of the most severe road accidents. It is also very important to select an efficient system complying with the legal acts and the environment. In the long run, it is believed that this measure together with other strategic traffic safety measures will change driving habits.

1. Commission staff working document SWD(2019) 283 final. EU Road Safety Policy Framework 2021-2030 - Next steps towards "Vision Zero".
2. Kelių eismo taisyklių phttps://lkpt.policija.lrv.lt/lt/statistika/keliu-eismo-taisykliu-pazeidimu-statistikaažeidimų statistika | Lietuvos kelių policijos tarnyba (lrv.lt)
3. De Ceunynck, T. (2017), Installation of section control & speed cameras, European Road Safety Decision Support System, developed by the H2020 project SafetyCube. Retrieved from www.roadsafety-dss.eu on 1/2/2021

THE ROLLING OUT OF INTERNATIONAL PROJECT SMART E263/E77 FOR ADVANCED TRAFFIC MANAGEMENT

Boriss JELISEJEVS[1], Kristjan DUUBAS[2]

[1]*Latvian State Roads, Riga, Latvia*
[2]*Estonian Transport Administration, Tallinn, Estonia*
Corresponding author: Boriss Jelisejevs, boriss.jelisejevs@lvceli.lv

Abstract. Intelligent transportation systems (ITS) provide significant added value to road transportation, making the related investments distinctively effective and long-lasting. Moreover, some ITS activities may be eligible for financial support of the European union (EU). That was the way how Estonian Transport Administration and *Latvian State Roads* worked on the project proposal "Smart corridor Tallinn-Tartu-Luhamaa-Riga E263/E77" (acronym – SMART E263/E77), which was approved by EU program Interreg Central Baltics as CB891 project. The project started on June 1, 2020, and its implementation will last till the end of 2022 according to quite challenging schedule. Project activities primarily include numerous installations or road telemetry and telematics devices (especially, variable message signs) for advanced traffic management to be supported by cross-border traffic plans and improvements of traffic control centers. Project target is to provide general travel time savings at least by 0.88% across the whole corridor, however for the motorway-type sections it should reach more than 5.5%. Expected project results will establish new and improve existing functions on the E263 and E77 road transport corridors, namely: traffic management adaptive to variable road conditions; gathering and dissemination of traffic information; decision-making support for road maintenance operations (especially in winter). This report will summarize the information on project progress with emphasis on traffic management considerations.

Keywords: intelligent transportation system, adaptive traffic management, variable message sign, cross-border traffic management plans, traffic information, road weather information, traffic signal.

Introduction

European transport policy has defined highly ambitious goals to be achieved in 2050, but intermediate progress indicators are to be reached already in 2030. (European Commission, 2011). One of prerequisites is consequent rolling-out and further development of Intelligent Transportation Systems (ITS). Now ITS is a set of quite well-defined scope of services and supportive measures in the European Union (EU) level, which is aimed to improve all facets of mobility, including safety, network throughput and environmental impact.

The primary involvement of national road administrations in the ITS domain is covered by the EU Directive 2010/40/EU (emergent support of critical traffic situations and dissemination of traffic information) and European ITS action plan (European Commission, 2011). A set of supportive delegated acts was adopted to make such prescribed ITS services interoperable and seamless between the EU member states at least for the Trans European network (TEN).

But such infra-oriented ITS offering intelligent road network operations, namely – adaptive traffic management, also may be suited for improving the mobility, as well. That is why national priorities and EU supportive measures provide opportunities for international ITS projects. Close cross-border co-operation allows to achieve even more than direct project deliverables only: knowledge transfer, harmonisation of methodologies and co-operation procedures are the proper contact points between the regional partners. Latvian and Estonian national road authorities were able to form a successful partnership (had the roots in the recently implemented project SmartE67) aimed to reach these targets within the on-going EU co-financed project SmartE263/E77. This report outlines the main deployment aspects and features of traffic management within the project.

General information

As mentioned before, both partners from Latvia and Estonia already had success story and good cooperation practice to deploy joint ITS project (Smart E67) with the support of the same EU programme (Jelisejevs, 2017). Near the end of the period of Interreg Central Baltic Programme for 2014-2020 of the European Regional Development Fund, the partners decided to put together their experience and intentions in the project proposal for the last (fifth) call of the Programme, aiming to define even higher targeted key performance indicators (KPIs) in comparison to the previous one. The partners changed their roles for the on-going project, thus Estonian implementing body is the lead partner.

Overall objective of the project is improving the efficiency and safety of cargo and passenger transport flow on 403 km long road sections in total (localized and section-like pattern of roadside ITS equipment placement) of the crossed TEN corridors E263 (in Estonia) and E77 (in Latvia) by introducing broad ITS pattern (see Figure 1). According to proposal evaluation criteria, the main effect to be reached is the decrease of travel time for both passenger and cargo transportation within the whole section by at least 0.88%, comparatively to the present situation (ERC, 2019). Additionally, road safety is to be improved and CO_2 emissions reduced. According to the cost-benefit analysis done especially for the project it will have significant macroeconomic effectiveness ratio for the planned investments of 3.62 for the project estimated lifecycle of 15 years. The realistic average traffic forecast for TEN corridors in the respective period showed the annual growth of 1.5-2% before. However, the current long-lasting emergent situation of Covid-19 pandemic changed the traffic pattern considerably (decrease of 13-17% from 2020 to 2021), probably affecting also societal habits in general so that this data is to be updated. Co-operative systems (C-ITS) and autonomous driving also are to be taken into account in the future, although such topics are still out of the planning horizon. Cost-benefit analysis has included HDM-4 methodology of traffic flow distribution for the most congested sections in the project (made in late 2019) and it showed that peak hours (carrying about 20% of all traffic) occurred for up to 800 hours, but busy flow – 61% or up to 3500 hours annually, which are to be primarily treated by adaptive traffic management approach (Figure 2).

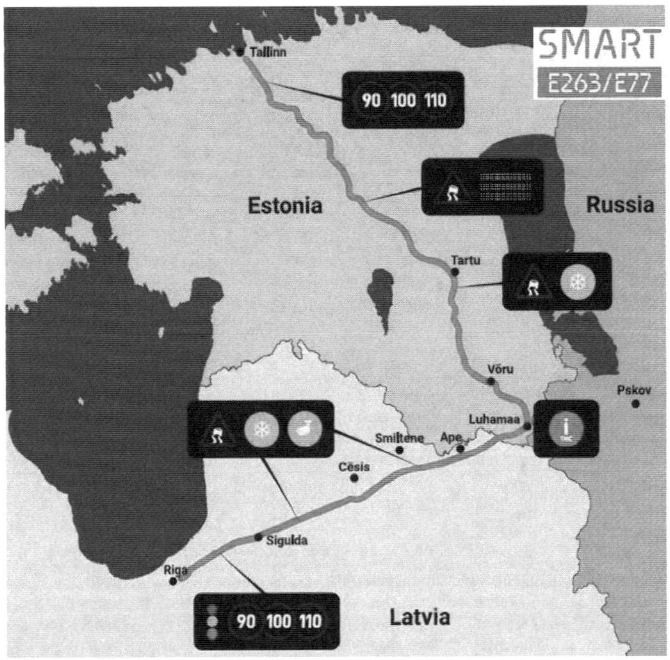

Figure 1. Promotional information of the project Smart E263/E77 (infographics and logo).

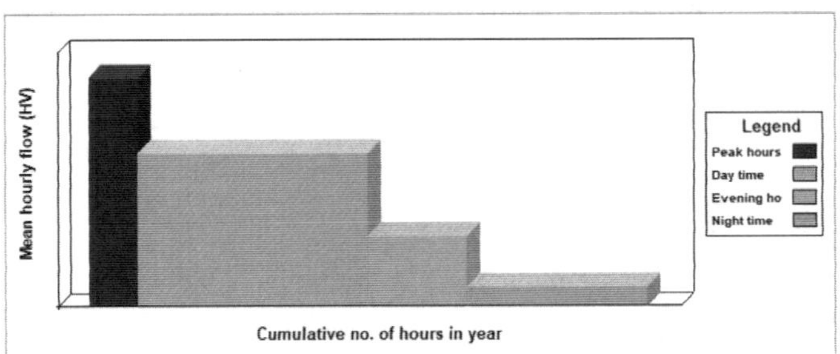

Figure 2. Average traffic flow distribution for the most congested sections of the project (from ERC, 2019)

The partners estimate notable co-effects, giving opportunity for further ITS implementations, transposable to the rest of the network later (for instance, Latvia has announced highway development program 2020-2040, where variable speed limits up to 130km/h. will be introduced). The project aims to introduce new or to modernise the existing roadside ITS equipment and to ensure proper solutions for traffic information exchange between traffic information centres (TIC) of the partners, joint technological approach and best-practice examples, available in the region (Jelisejevs, 2010).

The total budget is almost 2.5 million EUR, where the EU will co-finance 81% of the eligible costs, with almost equal distribution between the partners. There are a lot of ambitious activities within the project, including specific studies, interventions in roadside ITS equipment, TIC adaptation, communication to road users and society, as well, as administrative costs. This is one of the so-called whole implementation life-cycle projects, where target measures (investments in ITS deployment) are combined with a broad list of supportive measures. The project started on June 1, 2020 (4 months after the initial plans due to organizational issues which did not depend on the partners), and it has to be completed not later, than on December 31, 2022 (whereas extension is not allowed at all, due to deadline of certain EU programme). That means a lot of additional challenges for the partners to provide smooth deployment activities in time.

Both partners see not only direct and measurable effect of SmartE263/E77 to be reached, but also significant potential of multiplication, while the project results (traffic control scenarios, technical solutions, evaluation methodology, etc.) might be applicable for other road sections and improve operational ITS to a better linkage at regional (cross-border) and local (interurban/urban) levels (Ehrlich & Jelisejevs, 2019). The project has also aimed to increase professional competence in ITS technical and organisational issues of partners' personnel.

Project scope

Technological harmonisation is one of the project key goals, as in reality the activities of partners are not absolutely equal due to their differences in the road infrastructure, covered by the planned ITS and domestic approach to traffic management priorities (Figure 3, Table 1). That is why just some core activities are to be covered by joint procurement (staff training session and information campaign), while deployment tasks are to be solved separately, but still based on specific common principles.

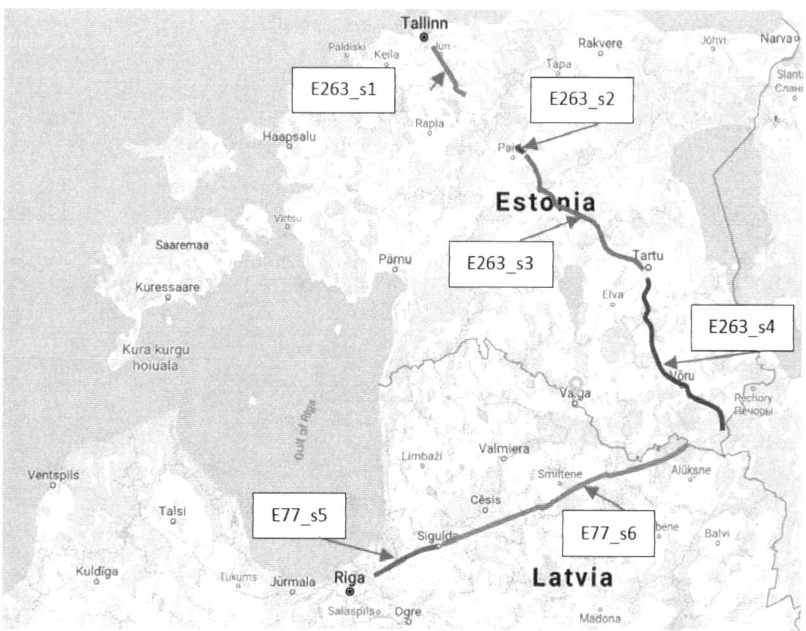

Figure 3. Map of road sections covered by project activities.

Table 1. Detailed data on road sections covered by project activities

Section	TEN route	Length, km	Type	Pavement overall condition	Pavement condition, IRI, mm/m	AADT, veh. per day	HGV, %	Start location	End location
E263_s1	E263	34.3	4-lane	Good	1,05	14 609	11	Tallinn	Kose
E263_s2*	E263	5.9	4-lane	Good	1,10	8 917	17	Tarbja	Maekula
E263_s3	E263	87.5	2-lane	Good	0,87	7 466	16	Maekula	Tartu
E263_s4	E263	97.1	2-lane	Good	0,86	3 482	11	Tartu	Luhamaa
E77_s5	E77	24.2	4-lane	Good	0,75	21 735	16	Baltezers	Senite
E77_s6	E77	150.6	2-lane	Moderate	2,10	4 094	20	Sigulda	Veclaicene

Remark: * this section of adaptive traffic management will be implemented as a part of another road construction project.

According to the initial application, there are 6 working packages with subsequent and parallel activities to be done:

- external expertise, attracted to detailed decision making on the planned investments and evaluation of project results,
- deployment of roadside ITS elements (variable message signs (VMS), road weather stations, video surveillance etc.), including preparatory works (for instance, on-site wiring),
- TIC adaptation foresees the functional improvements of traffic management systems (central software) and also training of TIC staff to be ready for the new functionality (scenarios applicable for adaptive traffic management);
- communication measures, which will cover general project PR issues and bring together the main stakeholders and target groups,
- provision of electricity supply to roadside ITS installations where necessary,
- management package, provided by the project team, as the main supportive effort to all above mentioned topics (for instance: procurements, reporting, etc.).

The main differences between the planned activities of the partners are the following:

- LV firstly puts maximum effort on the 2+2 section Riga-Senite to provide variable speed management for 24km of roads in total (divided in 3 co-ordinated sections) and localized warning VMS installations similar to SMARTE67 farther on the route E77, and secondly to unifying all traffic management scenarios to a centralized advanced traffic management system (ATMS),
- EE expands their 2+2 adaptive traffic management to additional 40km of roads and develops light VMS installations on the peripheral part of E263 to be joined to the existing ATMS.

In fact, the partners have different approaches to VMS performance, namely:

- EE uses wide range of VMS (5 types) and auxiliary equipment in the classic way (side-mounted and overhead installations),
- LV in its turn plans to deploy 3 types of side-mounted VMS for speed limits, traffic warnings and additional information that do not support broad textual notifications and have limited re-routing functionality but are fully usable and upgradeable to the next level when such a necessity will appear.

Organizational approach towards the planned ITS deployment also differs, since EE prefers to split the corresponding services to several phases (technical design, supply and construction works of road ITS installations, inclusion of the new devices in the existing ATMS), but LV (having public consultations with ITS market actors) decided to put them together in a single complex procurement.

Progress status

Since the starting point (the first project period ended in November, 2020) numerous preparatory works were completed and the following main project activities are on-going according to the schedule:

- administrative efforts and preparations to launch the project implementation (were based on similar previous experience),
- working meetings (including kick-off), that involve project management team and steering committee,
- reporting to the first level control institution and the Programme institution for the first project period,
- development and implementation of project communication plan,
- preparation of project modification request which is related to some formal (changes in the status of EE partner, cost reallocation between the budget lines according to the programme flexibility rules), as well as, more serious content-related issues, since the detailed latest data gives the opportunity to improve some of the initial assumptions by principle "more results with less costs".

IOP Conf. Series: Materials Science and Engineering 1202 (2021) 012043 doi:10.1088/1757-899X/1202/1/012043

The completed content-oriented measures of the project cover:

- provision of cost-benefit analysis which proved the feasibility and full compliance of the project with the awarding criteria of travel time savings (was prepared in advance and linked to the application),
- provision of pre-design activities where necessary (for instance, LV had such separate study to define proper VMS placement for 2+2 road section),
- provision of supportive and on-site preparatory services (design and provision of electricity connections for road ITS equipment, a.o.),
- final preparation of procurement documentation for main investment packages to be launched in the second quarter of 2021,
- provision of some minor services that were offset from the main packages (for instance, adaptation of 2 traffic lights across E77 2+2 road section in Latvia).

The partners are currently doing all the best to implement the project activities in the planned timeframe. We hope that also the market is ready to respond adequately, since this is not the first VMS deployment project in both countries. Therefore technical requirements for procurement procedures are balanced to provide good performance for possibly quite broad product range for better competition.

Notable features of the planned ATMS and VMS

All the decisions on adaptive traffic management will be initiated centrally within the ATMS based on customized operational scenarios that includ triggering thresholds and their settings, priorities etc. (Figure 4). Such scenarios consist of 5-6 levels of priority (relative significance of situations) and a number of VMS status combinations and their relative weight, prescribed within the ATMS engine. The following ATMS operational modes are to be generally provided:

- automated (third priority) – direct trigger from road weather station to display certain road sign pictograms (speed limit, traffic warning, additional plate) on VMS in specific locations according to ATMS settings, referring to adaptation of speed limits at the step of 10km/h (50-110 km/h);
- semiautomated (second priority) – the same principle as automated, but needed to be approved by TIC operator (the system parameter is critical and has to be validated by human intervention);
- manual (highest priority) – TIC operator provides highly customized (by VMS status and localization including editable textual notifications) settings based on wide range of available traffic information, whereas only few basic variable speed limits are in use (50, 70, 90 km/h);
- background (lowest priority) – possibility of showing the binding traffic information (for instance, road temperature data from road weather station or any general traffic related notification) in VMS passive mode, when there is no ATMS prescribed command to be "on".

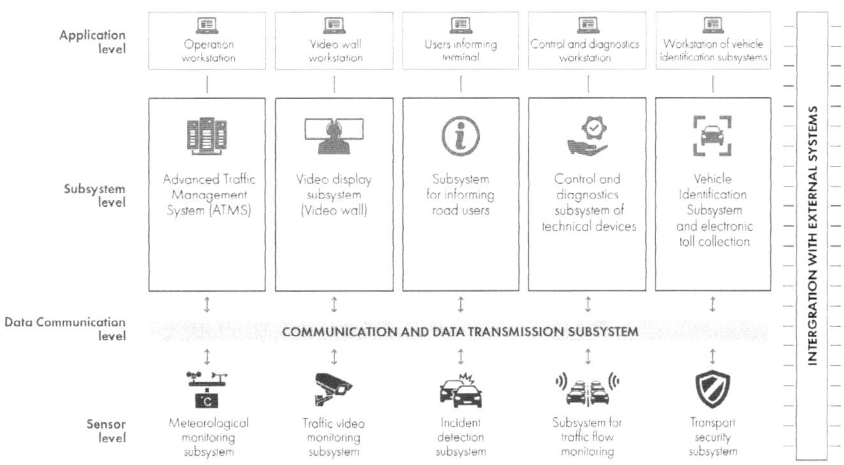

Figure 4. ATMS principal scheme (with functional modules, including VMS control).

ATMS asks for in-depth functional integration between different types of road equipment and seamless data flow, also to semantically support a broad scope external feeds in perspective (Guevara & Jelisejevs, 2016). TIC operators should have maximum friendly user interface (profiling, functional dashboards for direct interaction through workstations and video wall) to react on real-time basis and to support their actions by the embedded set of traffic management scenarios. The related data communication is to be solved internally within the system (NTCIP protocol for VMS) and by the outer exchange hub (machine readable, according to DatexII specification). The partners will develop the detailed cross-border traffic management plans for co-ordinated handling of the respective traffic situations.
The result of professional discussion was that maximum variable speed limit for 2+2 road sections would not exceed 110km/h in summer season and not more than 100km/h in winter conditions. It depends on the technical performance of such roads (since some parameters are not applicable at higher speed levels) and ATMS limitations. As it was already mentioned VMS types and placement principles may differ between the partners. Harmonization is aimed here to technological coherence and visual alignment (CEN, 2014). Thus, VMS screens for pictograms of road signs will be required to be of pre-defined matrix and just multifunctional parts of VMS will have to be freely programmed. The same approach is for harmonized visual performance allowing colour inversion of black and white, as well as similar matrix resolution. Some interesting things coming from the domestic approach are:

- LV plans to use only side-mounted duplicated VMS that means a pair per location per direction of carriageway for 2+2 road section, whereas up to 4 conventional zones per each VMS exist (Figure 5),
- regarding the optimization of VMS posts for 2+2 road sections LV decided to mix ATMS with stationary road signs where appropriate (Figure 6, ProVia, 2020),
- EE plans to use mainly side-mounted duplicated variable speed limits linked where necessary with big complex VMS boards above the carriageway, mounted on the structures of consoles or gantries.

Figure 5. VMS general composition for LV 2+2 road sections (on the right – complex installation in the beginning of ATMS segment, on the left – variable speed limits for continuous traffic management within the ATMS segment).

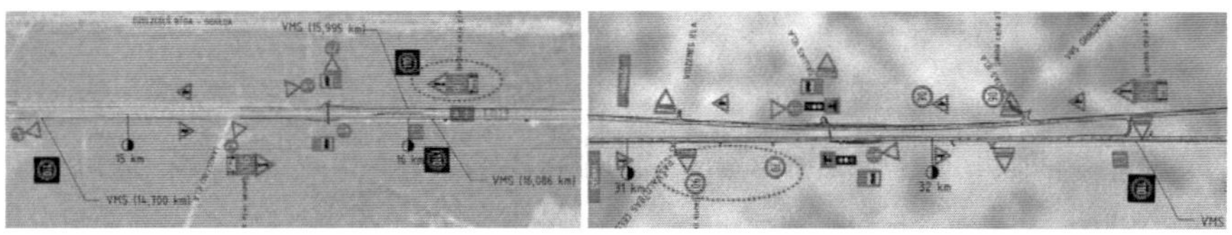

Figure 6. Use cases of VMS mix with stationary road signs for LV 2+2 road sections (on the left – ATMS zone continuation and traffic speed suggestion at U-turns, on the right – traffic calming before controlled pedestrian crossings).

Figure 7. One of typical VMS composition for EE 2+2 sections (on the right – variable speed limit at the road side and complex VMS board partly above the carriageway in couple hundred metres away, on the left – duplicated variable speed limit in road median).

Conclusions

The project still is in the rolling-out phase when the main activities to be implemented are planned. The partners have a lot of challenges, but also experience and professional awareness to deal with them, hopefully, in the most effective way.

The main conclusions so far are the following:

- There are indications of forming professional ITS community in the Baltic region, whereas general awareness in the field and capabilities of service providers are considerably growing,
- EU supports investments in mobility, whereas proposals on ITS projects are highly feasible, announcing turn from "hard" to "soft" infrastructural measures,
- the planned actions within SmartE263/E77 are quite ambitious, pointing out development of complex approach for corridor based adaptive traffic management,
- some changes are possible in the initial application due to project complexity and data detailing,
- harmonized ITS services do not mean the deployment of completely the same equipment due to complex site-specific considerations,
- investments in road ITS let to achieve not only project goals but also bring new functionality and wider coverage of other services (for instance, decision-making for road routine maintenance and facilitation of emergent response on traffic incidents) (Jelisejevs 2018);
- adaptive traffic management reflects broad involvement of public and stakeholders to be solved in the most careful manner (operational scenarios, road user tolerance, enforcement methods etc.),
- general approach to SmartE263/E77 and parallel activities build up the framework for expertise and technology transfer to the rest of the national road network,
- ATMS provision at the arterial entrances to the cities of Riga and Tallinn gives an opportunity to build up direct interlocks with the subsequent urban ITS solutions.

A lot of practical benefits are awaited within the road sections covered by ATMS, namely:

- road authorities will have a new communication tool to affect the road user behaviour in real time (meeting their actual expectations),
- the on-trip info availability of traffic conditions will motivate the road maintenance contractors for better performance,
- VMS will serve drivers with actual traffic prescriptions and advices in the era where cars are quite intelligent and may create illusion of safety (the physics still remains and skid resistance factor and surface temperatures are good to be noted).

Acknowledgements

The authors express their acknowledgements to the international (Latvian and Estonian) project Smart E263/E77 management team, steering committee and Interreg CB Secretariat, who gave their collegial input to the proceeding.

Funding

This work (directly related to the EU co-financed project and to be done, informing the stakeholders and professionals of road transportation industry) was supported by the <ERDF Interreg Central Baltic Programme of 2014.-2020.> under Grant [number CB891].

Disclosure Statement

Authors declare that they do not have any competing financial, professional or personal interests from other parties, regarding the information, included in the report and that the published information is for public needs and is not confidential at any extent.

References

CEN (2014) *EN 12966: Variable message signs.*

Directive 2010/40/EU of the European parliament and of the council of 7 July 2010 on the framework for the deployment of Intelligent Transport Systems in the field of road transport and for interfaces with other modes of transport. Retrieved from http://eur-lex.europa.eu/LexUriServ/LexUriServ.do?uri=OJ:L:2010:207:0001:0013:EN:PDF

Ehrlich, J., Jelisejevs, B., a.o. (2019) *Low cost ITS.* PIARC TC report, 1 122

European commission (2008) *Action Plan for the Deployment of Intelligent Transport Systems in Europe*. Retrieved from https://ec.europa.eu/transport/sites/transport/files/themes/strategies/doc/2011_white_paper/white-paper-illustrated-brochure_en.pdf

European comission (2011) *White paper on transport.* Retrieved from https://ec.europa.eu/transport/sites/transport/files/themes/strategies/doc/2011_white_paper/white-paper-illustrated-brochure_en.pdf
Guevara, G., Jelisejevs, B., a.o. (2016) *Advanced technology for data collection and information to users and operators.* PIARC TC report, 1–45
Jeļisejevs, B., (2010) *ITS tools for winter road maintenance in Latvia.* Scientific Proceedings of RTU, vol. 11(2), 83–87
Jelisejevs, B., (2017) *Implementation of SMART E67 project.* Proceedings of 29th International Baltic road conference, 1–8
Jelisejevs B., (2018) *The Main Challenges of Winter Road Service to be Solved Within the Framework of Intelligent Transportation System.* Springer notes: Reliability and Statistics in Transportation and Communication,349-359
ERC (2019) *E263 and E77 cost-benefit analysis of a prospective cross-border (Estonia-Latvia) IT solutions.* Final report
ProVia (2019) *Feasibility study on variable message signs placing at E77 section Riga-Senite.* Final report

BENEFITS AND PERFORMANCE OF THE ELECTRONIC CONSTRUCTION JOURNAL

Paulius Bautrėnas[1], Marius Čereška[2]

[1,2] *State Enterprise Lithuanian Road Administration, J. Basanavičiaus str. 36, 03109 Vilnius, Lithuania*
E-mails: [1]*paulius.bautrenas@lakd.lt;* [2]*marius.cereska@lakd.lt*

Abstract. Electronic construction journal (later – e-journal) has been developed to replace a former paper construction journal. The purpose of e-journal is to ensure effective supervision of road construction and reconstruction works and to avoid potential abuse. Records made by contractors, supervisors and controlling institutions in e-journal enable to manage the work process in each road construction and reconstruction building site. All road construction process participants have access to the data, comments and findings entered into e-journal. E-journal accurately records all deadlines, technical supervision or control performance time. It is a significantly faster and more effective means of control and quality assurance of performed works.

Keywords: Asset Management Information System, electronic construction journal, e-journal, road construction works, road construction process monitoring and control.

Introduction

To ensure effective record and supervision of road construction works, to prevent potential abuse and to control construction processes, the electronic construction journal has been developed. It is considered that in the long run it will replace paper documents and will become the key instrument of monitoring and controlling the road construction process.

1. Road Asset Management Information System

To enhance administration, management and maintenance of road assets and provision of public e-services, in 2017 the State Enterprise Lithuanian Road Administration (later – the Road Administration) started the implementation of the project Development of Road Data Electronic Service (later – the Project). The Project was financed from 2014–2020 EU funds. The Project aimed to develop advanced e-services for road data providers and economic entities using the road data via the centralized state and local significance road data source, where the data would be handled from the beginning until the end of the road asset life cycle. This would enable to develop higher quality and user-friendlier e-services in the transport sector on the Lithuanian and European level as well as to inform the society about Lithuanian roads by providing relevant and timely data, which in its own turn will improve traffic safety, comfort and efficiency of driving as well as will create conditions to develop other e-services in the transport sector.

Based on the best practices and international standards, the instrument and the database have been developed and implemented under the umbrella of the Project. The Asset Management Information System was adjusted to the needs of the Road Administration. The system enables to manage all types of assets on one software platform. The solution is based on the service-oriented architecture and carries comprehensive information on all types of assets, their environment, location, supporting work processes, which will enable optimal planning, control, audit and compliance. The database provides crucial information on assets, including their main attributes, configuration and physical as well as logical interactions with other resources. The developed system has been integrated into other state registers and external systems via administrative and public e-services portal E-Government Gateway.

2. Construction E-journal

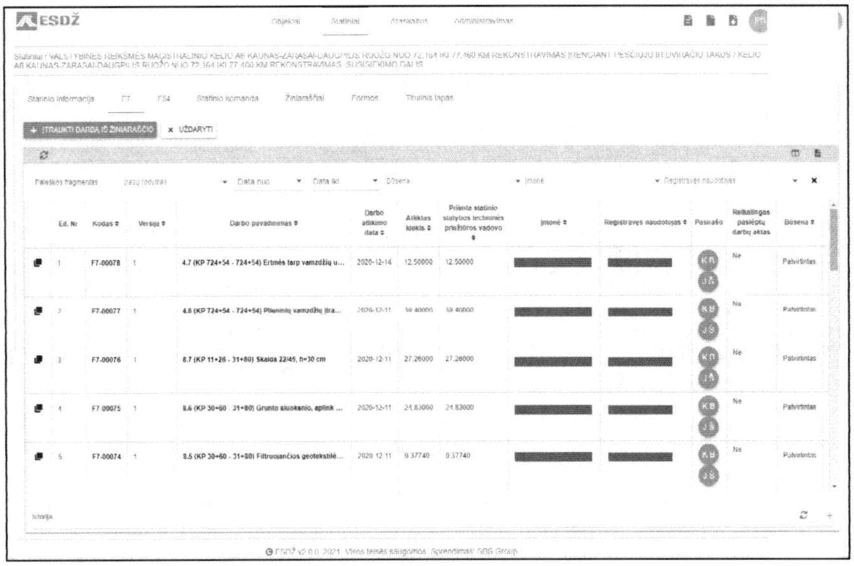

Figure 1. E-journal user interface

The electronic construction journal (Figure 1) of the Road Administration Asset Management Information System (later – the e-journal) has been developed in order to replace the existing paper construction journal. It is believed that the e-journal will ensure effective record and supervision of road construction works and will prevent potential abuse.

The e-journal is a compulsory document of work performance where the process of construction works, the quality of performed works as well as commissioning of separate works to the client are described. The necessity of such document is regulated by the Technical Regulation of Construction of the Republic of Lithuania. Contractors are responsible for filling-in the journal; however, records may be entered by technical supervisors, project implementation supervision managers, representatives of the client, controlling institutions. So far paper construction journals have been used. The aim of developing the electronic version of this document was to replace and optimize the so-called paper process of construction work management.

Records (Figure 2) made by contractors, technical supervisors, project implementation supervisors and controlling institutions enable to manage the process effectively in each road construction object. All data, notes and recommendations are available to all road construction process participants. The e-journal records the data of work performance and commissioning. Tracking of amendments and other records is available, which guarantees faster and more effective work performance control and quality assurance.

Key Features of the System:

- All forms of Technical Regulation of Construction STR 1.06.01:2016 "Construction Works. Construction Maintenance" and other additional blanks integrated into the system;
- Internet-based: different users can work with the same e-journal at the same time;
- Simple and interactive: easy to attach and to review related documents, to search for and to filter records;
- Imports sheets of work quantities submitted by designers and contractors;
- Easy tracking of the history of user's actions and data entry;
- Different versions of input information and documents by saving all actions of users;
- Transfer of the e-journal data to the information systems of the Road Administration and other institutions via the Internet.

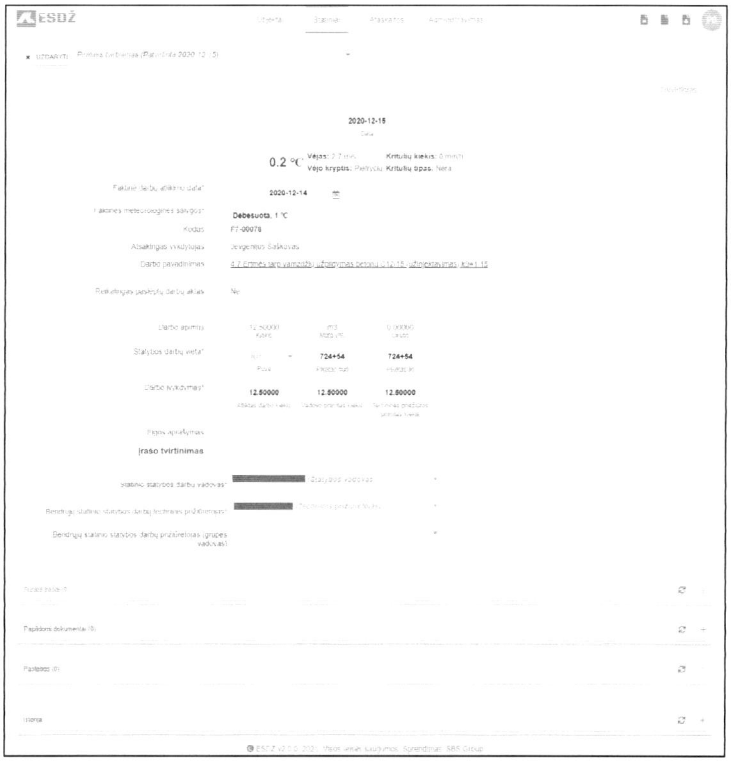

Figure 2. E-journal entry example

3. E- journal Development

The idea to use e-journals came up in 2017. It took a long and complicated public procurement process to sign the contract on the e-journal development with JSC SBS GROUP on 9 May 2018. During the development of the e-journal, a number of meetings, discussions, presentations and workshops with contractors, technical and project implementation supervision specialists were held in cooperation with the Road Association *Lietuvos keliai*. Having identified the needs of potential users of e-journals and implemented new and reviewed requirements, the system was developed and put into operation. In September and October of 2020, three pilot e-journals were tested to verify the functionality and expedience of the system. Remote MS Teams training sessions on using e-journals were held (and later periodically updated) for the participants of road construction process. Since November 2020, e-journals are issued for all new contract works. For example, on 9 March 2021, 22 contract items were registered and 65 e-journals were issued. Currently, the system has 100 registered users.

Conclusions

The innovative modulus of electronic construction journal of the Road Administration Asset Management Information System will enable to take another important step towards technological progress when a paper journal has been replaced. This will ensure effective record and supervision of road construction works, will prevent potential abuse, will enable to control the processes of ongoing constructions works and will become the key instrument of road construction process monitoring and control.

USE OF BIM IN SLLC "LATVIAN STATE ROADS" FOR PROCESS MANAGMENT

Zigmārs KRŪMIŅŠ[1], Juris SMIRNOVS[2], Valentina ĀMARE[3], Roberts AUZIŅŠ[4]

[1]*Department of Roads and Bridges, Civil Engineering Faculty, Riga Technical University, Riga, Latvia*
[2]*Department of Roads and Bridges, Civil Engineering Faculty, Riga Technical University, Riga,*
Corresponding author: Zigmars Krumins, email: zigmars.krumins@edu.rtu.lv

Abstract. BIM is an integrated set of building design, construction and management processes, technologies and regulations that allow all parties involved to jointly design, build and manage a building in a digital environment.

In Europe, a common approach to the use and adaptation of technologies is currently being developed with the aim of providing a common regulatory framework that would promote the efficiency of technology application between the countries of the European Union. The use of BIM technologies is based on economic (financial) benefits. The financial benefits are estimated in the long run, and the main factors for financial savings are productivity and quality, which can be achieved by applying BIM technology and appropriate processes during the construction cycle (design, construction and management). Studies show that savings can reach up to 10% of construction costs.

Currently underway is the process of implementing and configuring the Microsoft Dynamics platform for processing and using work information of SJSC "Latvian State Roads" (LSR). Microsoft Dynamics is a line of enterprise resource planning and customer relationship management software.

In general, LSR intends to use the system in the following stages of the construction cycle:

- Design stage
- Construction procurement stage
- Construction stage
- Warranty period

Microsoft Dynamic has both pros and cons for processing LSR information. Only after full implementation it will be possible to assess the effectiveness of this platform.

Objective of the article is to explore how the use of BIM can improve the LSR work process and what happening at this moment in LSR.

Keywords: BIM, LSR, road management, building design, construction process, financial benefits.

Introduction

The construction industry is one of the largest industries in the world, accounting for significant ten percent of the total GDP in the world (approximately 10 trillion USD of construction related expenditure worldwide) and employing nearly seven percent of the global population. The European construction industry accounts for nine percent of GDP (approximately 1.3 trillion EUR of construction related expenditures worldwide), providing eighteen million jobs. The situation in Latvia is similar - construction sector expenditures amount up to ten percent of total expenditures (almost 2 billion EUR of construction related expenditures), and the sector employs almost seven percent of all taxpayers, indicating that the Latvian construction industry follows global trends and is considered one of the largest economic driving forces in the country.

In 2020 and 2021, the global economy, and in particular the Latvian economy, was affected by the crisis caused by the Covid-19 virus. The Latvian government on April 28, 2020, responded by allocating 75 million EUR for the rehabilitation of pavements on state main roads in 2020 and on December 8, 2020, it allocated additional 100 million EUR for the renovation of roads in 2021. The justification was that investments in road infrastructure would both secure orders for the local construction industry and have a positive effect on road users and stimulate the economic development in general. [2]

In Europe, a common approach to the use and adaptation of technologies is currently being developed with the aim of providing a common regulatory framework that would promote the efficiency of technology application in the countries of the European Union. An EU Task Group has been set up to develop a common policy for using technology to boost productivity and quality in the construction sector. The EU Task Group has developed the document "BIM Handbook", which is available also in Latvian.

1. General regulations

In Latvia, the current political situation and politically set goals are favorable for streamlining construction processes. According to the Ministry of Economics, it is important for Latvia with its open economy in conditions of high international competition to continuously increase the efficiency and competitiveness of construction companies both locally and internationally, using the opportunities provided by digitalisation [3]. At the end of 2019 the Ministry in co-operation with other state institutions, public companies, educational institutions and construction clients jointly developed the Building Information Modeling Roadmap as a step in the digitalisation of the construction industry.
The Ministry of Transport has not yet defined concrete steps in the field of digitalisation, but the need for process efficiency and productivity improvements is emphasized, and this may be interpreted as a positive signal in the use of technology to improve process efficiency.
An important political factor is the digital transformation guidelines developed by the Ministry of Environment and Regional Development, which determine Latvia's digital transformation (information society development) policy for the period from 2021 to 2027. The guidelines elaborate the provisdion, action directions and tasks in the digital transformation policy defined in the National Development Plan for 2021-2027.
The use of BIM technologies is justified by economic benefits. The financial benefits are estimated in the long run, and the main factors for the financial savings are productivity and quality, which may be achieved by applying BIM technology and appropriate processes during the construction cycle (design, construction and management). Studies show that savings may reach up to 10% of construction costs [6].

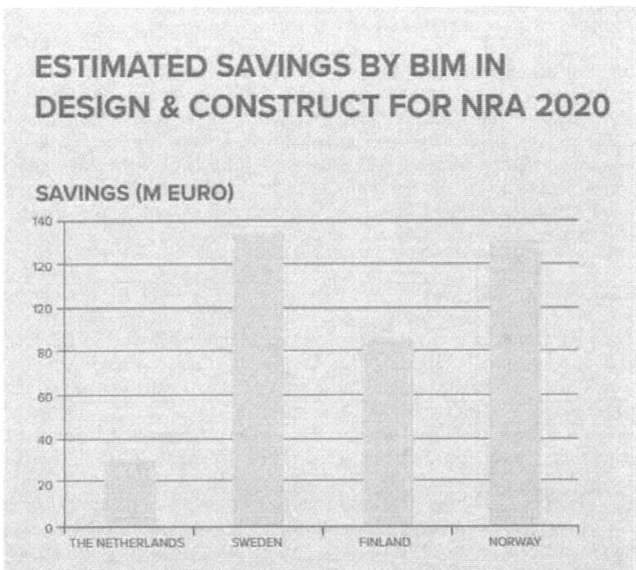

Figure No.1. CEDR estimated benefits from the use of BIM technologies. [6]

CEDR has determined the following main benefits:

- reduction of construction costs,
- improved communication with stakeholders,
- solid information base for asset management and construction cycle management,
- more transparent process and division of responsibilities.

1.1. BIM application technologies

The development trends of digital technologies in the construction industry in recent years indicate a rapid development of technologies. One of the reasons is the low level of technology application in the construction industry compared to other industries. [3] The rapid development of technology is a positive indicator, as the use of technology may significantly contribute to the efficiency of construction and design process.
BIM is one of the most innovative and promising technologies in the construction industry. BIM is an integrated set of building design, construction and management processes, technologies and regulations that allow all parties involved to jointly design, build and manage a building in a digital environment.[7] In addition to BIM technologies, there are a number of other technologies that complement the acquisition, use and processing of information during the construction cycle.
Four international companies dominate among digital technology developers: Bentley, Trimble, Autodesk and Grafisoft. In addition, there are a number of other companies that develop digital BIM tools for preparing BIM information. These companies invest significant financial resources in technology development, but there is a problem of data incompatibility. Currently, a number of different data types are used in the implementation of infrastructure projects (Appendix No. 4 "Overview of data types used in the BIM process"). Among BIM digital technologies, data types are divided into closed and open data types. Closed data types may be opened only with specialized software, open data (otherwise described as OpenBIM) is types of data that may be opened by other (similarly configured, often competing) software and free viewers. Data diversity (and the slow pace of development of open data standards) encourages software developers to develop technologies that nay handle closed data types, creating a unified and open information processing environment. Such technologies enable data users (clients) to maintain an independent approach to the application of technologies, thus promoting an open environment for service providers, with an emphasis on the end result.
In Latvia, several platforms have been developed with the aim of digitizing construction processes. The BIS system has a significant impact in the context of processes of SLLC "Latvian State Roads" (LSR), as data entry and retrieval from this system must be performed on a regular basis. The second phase of the Construction Process and Information System Development (BIS) project has been launched. The aim of the project is to improve BIS, including further process optimization and automation (design submission, coordination, construction process control, commissioning and operation stages in the system), reduction of administrative burden, development of an automated analytical decision tool (agent) and improvement of the system to there would be a unified platform for the life cycle management of any building from concept to the end of operation, as well as a basis for the implementation of BIM processes in Latvia and provide the necessary platform on the part of the state information system. [15]

1.1.1. Current situation in Latvia for the use of BIM

One of the important economic factors in Latvia is the implementation of the Rail Baltica project and the use of digital technologies is of great importance in the implementation of this project. The technology is based on BIM, for the use of which a strategy framework, a detailed strategy, customer requirements and a manual with several annexes and additional documentation have been developed, which is integrated in the technical regulations for the evaluation of tenderers. The purchasing ability of RB Rail company has a significant impact on market participants - it is observed that Latvia's leading construction companies invest resources in the implementation of technologies in order to meet the requirements set by RB Rail. [5]
SJSC „State Real Estate" (SRE) has developed BIM documentation, which is used to organize its procurements, and it explains the need for BIM, its meaning, principles, as well as provides documentary framework for including BIM information requirements in procurement procedures (including a document describing requirements for BIM specialists).
In Latvia, several documents related to BIM technologies and processes have been developed and adapted, including documents developed by RB Rail and SRE, which help both public and private capital companies to adapt technologies.
The implementation of the Rail Baltica project creates a significant economic impact in the market with its purchasing power. RB Rail places significant emphasis on the use of technology for efficient project management and has developed a set of digital technology (BIM) documents (including technical requirements for procurement), facilitating the adaptation of market participants to the BIM guidelines developed by RB Rail. From the point of view of the technological factor, significant technological development trends have been observed, and there is active work on the development of data standards, which may significantly facilitate the exchange and development of information, which is the basis for streamlining construction processes.
If we talk about traffic safety, Latvia has a high number of fatalities in accidents compared to other EU countries for a long time. One of the factors that could have a positive effect on reducing the number of fatalities is improving infrastructure. Therefore, for the development of the Latvian road network, the issue of traffic safety is a priority.
The study of the level of traffic safety includes analytical activity, which allows to identify problematic places in the road network by applying various methods. Traffic intensity on state main roads is very different and it is much higher around Riga. The development of the area around Riga is indeed very fast, the existing TEN-T road network (state main roiad network) is currently being resurfaced, but the road capacity is literally exhausted. At such high traffic intensity, each

road connection poses an increased risk of an accident, which was also clearly seen in the period 2017-2019. If dangerous road sections and intersections are compared, there are many short-distance road connections around Riga which are a cause of road traffic accidents. Therefore, in reconstruction projects and new projects the compliance with traffic safety requirements, the improvement of intersections and the elimination of so called "black spots" must be considered in relation to the environment and the surrounding infrastructure. The use of BIM makes it easier to process large amounts of information for a larger number of stakeholders, with the largest amount of information available. These preconditions make it possible to achieve the best possible results in improving road safety.

1.1.2. Situation and preconditions for the implementation of BIM in the daily use of LSR

LSR manages the state road network, administers the funding allocated to roads, plans the development of roads, organizes procurement for works on the state road network, administers the state road design and construction process, ensures supervision of road traffic organization in the state road and municipal road network, performs the inspection of the fulfillment of daily maintenance requirements of state roads and supervision of works, as well as supervises the compliance of maintenance of local government roads with the requirements of legal acts. LSR also provides public services and consultations, performs expertise and testing of roads and artificial structures, develops draft legal acts for the industry, manages information management systems necessary for road maintenance and development, and organizes trainings for road specialists. The certificate issued by Bureau Veritas Latvia confirms that LSR operations comply with the international quality management system standard ISO 9001: 2015.

Quality management plays an important role in internal processes, as it brings together all business processes and procedures. However, it should be noted that LSR department directly responsible for quality management has limited capacity (limited human resources), and for this reason part of quality and process documentation is developed and maintained in each individual department.

LSR has specialists who are able to develop construction projects (simple projects) independently. Designs are prepared under the management of the construction control division, where each respective project is managed by a responsible project manager. LSR also performs the function of the construction board for the fulfillment of requirements. The fulfillment of project conditions is also checked.

For the purposes of design and construction process management, the process of implementing and configuring the Microsoft Dynamics platform is currently underway. Microsoft Dynamics is a line of enterprise resource planning and customer relationship management software. Microsoft markets Dynamics applications through a network of reseller partners that provide specialized services. Microsoft Dynamics is part of Microsoft Business Solutions.

Microsoft Dynamics will replace the existing project management system, but it is not planned to integrate the existing HDM-4 system into it at present.

LSR intends to use the system in the following stages of construction cycle:

- Approval of the financial plan / programme,
- Design stage,
- Construction procurement stage,
- Construction stage,
- Warranty stage.

System development plan integrating all stages (combining all cycles) is 2-3 years. LSR currently has 14 Dynamics licenses. Additional licenses will be purchased as the number of users and their required number of licenses increases. 3x concurrent user licenses for 10 employees are for Bentley Microstation software. For system integration purposes, it is possible to use the OpenApi protocol.

The current strengths are the following:

- A large amount of information related to construction processes has been accumulated over the years,
- Use of the FIDIC contract form,
- Efficient use of resources in project management,
- Internal design resources,
- Application of HDM-4 in the programme planning stage,
- Some technologies are already being used in practice,
- A specialist for adaptation and implementation of BIM technology is appointed in the company.

Weaknesses are the following:

- LSR has a lot of unstructured data (registers, documents, etc.) that is not used in practice,
- No clear vision for asset management digitalisation,
- "Lowest price" principle in procurement,
- Diversified competence in the application of technologies,
- Insufficient communication within the company,
- Some project data from State Construction Control Bureau of Latvia is not available to project managers.

The opportunities are the following:

- By digitizing the existing documentation, it can be used in planning and management;

• A register of standardized hours can be used to determine construction deadlines based on BIM model information;

• LSR design specialists can learn BIM design capabilities in depth to be able to define BIM requirements and control BIM deliverables.

The present threats are:

• LSR construction project information is not stored in LSR (BIS work diary and EIS procurement data);

• Possible unsuitability of the current MS Dynamics platform for BIM needs;

One of the most significant shortcomings observed in the study "Framework of the Strategy for Improving the Efficiency of Construction Lifecycle Processes Management" is the organization of procurement according to the lowest price principle, which can potentially create difficulties for LSR to request information according to detailed and specific requirements. One of the strengths of LSR was the efficient use of resources in project management, as well as LSR ability and opportunities to develop small projects without the involvement of external specialists, which enables the development of internal competencies for validation, documentation and verification of deliverables. [1]

1.1.3. Study "Framework for a strategy for streamlining construction life cycle process management"

Taking into account internal and external factors, three alternatives for streamlining the construction life cycle process management were considered, based on BIM technologies and processes. The three proposed alternatives were compared taking into account advantages and disadvantages, costs and time required for each alternative.

Recommendations were developed for the implementation of the strategy, which envisages such factors as the need to develop documentation for the development of recommendation processes, expected BIM technologies and IT infrastructure, competence development procedures, necessary resources, the need for pilot projects and their implementation. In addition to that, the risks associated with the implementation of the strategy are considered. The developed recommendations serve as a basis for the strategy implementation roadmap.

The result of this research, or the strategic framework for streamlining the construction life cycle process management serves as a starting point for the implementation of BIM technologies and processes in LSR, based on the provided recommendations and the developed road map. It is important to continue working on aspects of BIM implementation, providing resources for the development of BIM requirements, competence building, implementation of pilot projects, market education and other activities related to new technologies and their accompanying processes.

Alternative No.1 is to adopt the BIM requirements developed by RB Rail with all the accompanying documentation (including the already developed detailed strategy, customer requirements, BIM manual), as well as used technologies and principles.

Alternative No. 2 is to develop independent BIM strategy details, customer requirements, BIM manual and other accompanying documentation.

Alternative No.3 is to develop a detailed BIM strategy, customer requirements, BIM manual and other accompanying documentation taking into account RB Rail, SRE and other BIM documents already developed in Latvia (including RB Rail BIM manual, SRE BIM templates, LVS translated standards, etc. available documentation), using the documentation already developed as much as possible by adapting it to internal factors.

When developing the recommendations, it was concluded that there was a need for planned activities for implementing the strategy gradually. An important aspect in the implementation of a successful strategy is the human factor, to which it is recommended to place great emphasis and allocate all the necessary resources. It is recommended to gather information from CEDR members with the aim of learning from the mistakes and successes of other similar organizations through a structured survey.

The strategy formulated as a result of the research envisages integrating the advantages provided by BIM technologies and processes in streamlining LSR construction life cycle management based on BIM documentation developed by RB Rail.

BIM Roadmap has been prepared with the aim to envisage a plan for the introduction of BIM technologies for the efficiency of LSR construction processes. The form of a calendar schedule is used to display the Roadmap. The BIM Roadmap is divided into five blocks. The strategy, which is a Roadmap for the implementation of BIM, is planned to be integrated into public procurement in 2025. The BIM Roadmap is based on the minimum of necessary activities to be carried out in order to streamline the management of the construction life cycle processes. The blocks and measurable final results of the Roadmap are summarized below. [1]

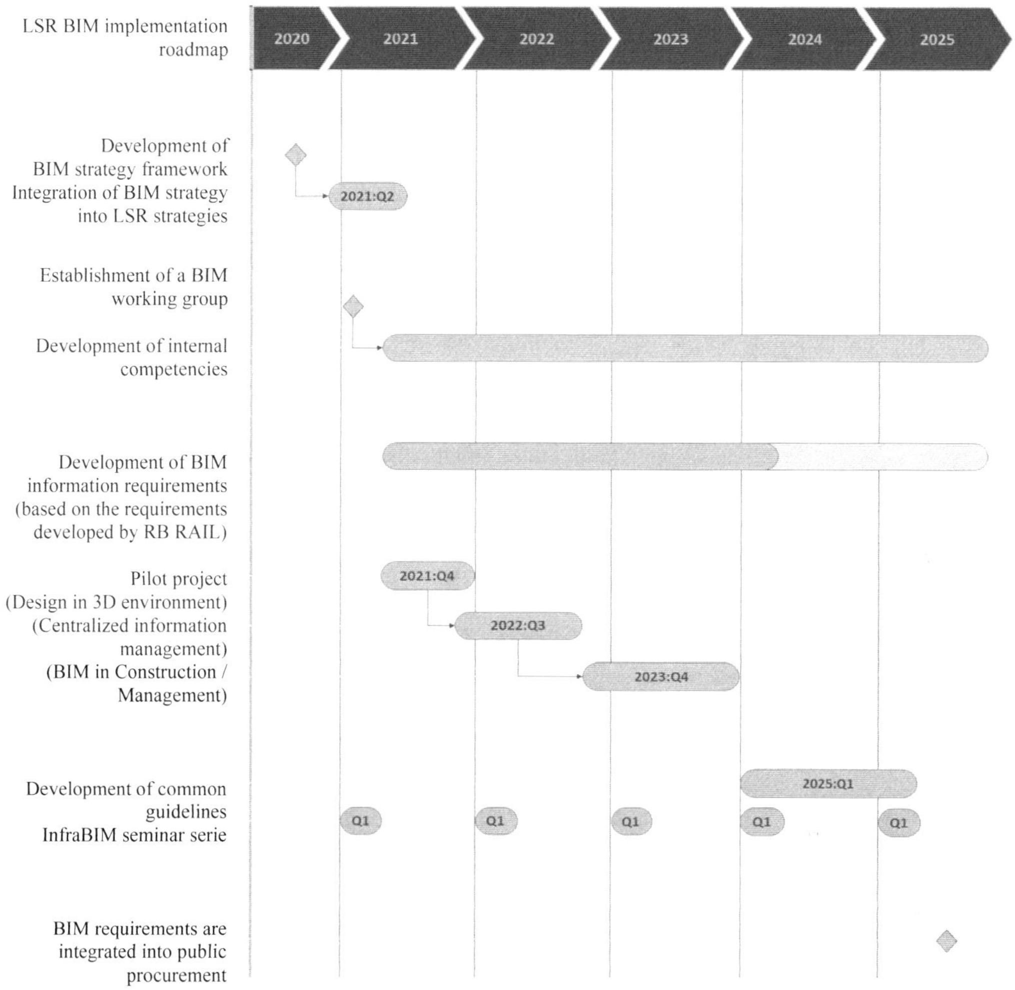
LSR BIM implementation roadmap
2020
2021
2022
2023
2024
2025
Development of BIM strategy framework
Integration of BIM strategy into LSR strategies
2021:Q2
Establishment of a BIM working group
Development of internal competencies
Development of BIM information requirements (based on the requirements developed by RB RAIL)
Pilot project (Design in 3D environment) (Centralized information management) (BIM in Construction / Management)
2021:Q4
2022:Q3
2023:Q4
Development of common guidelines
InfraBIM seminar serie
2025:Q1
Q1
Q1
Q1
Q1
Q1
BIM requirements are integrated into public procurement

Conclusions

1. From a technological point of view, LSR is currently implementing the MS Dynamics platform, which is potentially also planned to be used in asset management. A number of different platforms are used and thus there is a challenge for the exchange, input and quality of information (there are data registers the information of which is practically not used).
2. Currently, before the development of pilot project I (according to the Strategy Implementation Map), a requirement is already included in the design procurements of new bridges, which provides for additional points by submitting a 3D model of the bridge load bearing structure according to the LOD 300 level of detail based on BIM Forum (https://bimforum.org/ lod /) "Level of Development (LOD) Specification Part I & Commentary, December 2020". A 3D model of the bridge load bearing structures must be submitted in IFC and in the original file format.
3. The implementation of BIM in LVC is at an early stage. Following the study "Framework for a strategy for streamlining construction life cycle process management", an internal working group has been set up to coordinate BIM issues, but it should be noted that the current members are not BIM specialists but specialists in various technical issues of road and bridge life cycle who are ready to provide advice on the technical side of the BIM implementation process. In order to fully start work on the implementation of BIM, new specialists with BIM competence in various fields should be involved in the working group and work process, which could form the new LSR BIM structure and train the existing employees.

Acknowledgements

The author gratefully acknowledges State Limited Liability Company "Latvian State Roads" for providing the data for analysis carried out in this report.

Author Contributions

Dr. ing. Juris Smirnovs, Riga Technical University, report management

Valentīna Āmare, Riga Technical University, PhD student, LSR traffic organization engineer studied and provided information on the use of BIM in road design and construction that could be used to improve accident statistics,

Roberts Auziņš, Riga Technical University, PhD student, LSR leading bridge project manager, studied and provided information on the use of BIM in design and construction of static structures, gave his assessment of the current data processing in LSR.

Disclosure Statement

No potential conflict of interest was reported by the author.

References

[1] Study "Framework for a strategy for streamlining construction life cycle process management" made by society with limited responsibility "EVISOL".
[2] Cabinet of Ministers Republic of transport https. ://www.mk.gov.lv/en
[3] McKinsey Global Institute, "Reinventing Construction: A Route to Higher Productivity," 2017.
[4] "Par mums," *Latvijas Valsts ceļi*. https://lvceli.lv/par-mums/
[5] Ekonomikas Ministrija, "BIM Ieviešanas Ceļakarte." 2019.
[6] CEDR, "Open information exchange (BIM) of National Road Authorities is of vital importance for Asset management."
[7] "Būvniecības informācijas modelēšana - BIM - Ekonomikas Ministrija," 2019. https://em.gov.lv/lv/nozares_politika/buvnieciba/buvniecibas_informacijas_modelesana/ (accessed Jul. 21, 2020).

[8] Digitālās transformācijas pamatnostādnes 2021-2027. gadam (informatīvā daļa), *VARAM, 2020.*
[9] "Open information exchange (BIM) of National Road Authorities is of vital importance for Asset management."
[10] EU Task Group , Rokasgrāmata Būves informācijas modelēšanas ieviešanai Eiropas publiskajā sektorā 2016.
[11] RB Rail AS BIM Stratēģija '"BIM Strategy", https://www.railbaltica.org/rb-rail-as-bim-documentation/
[12] RB Rail AS BIM Pasūtītāja prasības "BIM Employer's Information Requirements (EIR)", https://www.railbaltica.org/rb-rail-as-bim-documentation/
[13] RB Rail AS BIM Rokasgrāmata "BIM Manual", https://www.railbaltica.org/rb-rail-as-bim-documentation/
[14] RB Rail AS BIP sagatave "BEP Template", https://www.railbaltica.org/rb-rail-as-bim-documentation/
[15] Būvniecības valsts kontroles birojs, "Būvniecības procesu un IS attīstība (2.kārta) Detalizētais projekta apraksts."

IMPLEMENTING BUILDING INFORMATION MANAGEMENT (BIM) IN ESTONIAN TRANSPORT ADMINISTRATION

Erko PUUSAAG[1], Anti PALMI[1]

[1]*Estonian Transport Administration, Tallinn, Estonia*

Received 28 February 2021; accepted date

Abstract. Digitalisation is the key to efficiency in the road construction Industry. Building Information Management (BIM), being the main developments in the architecture, engineering, and construction (AEC) Industry, offers a technology as well as processes to bring traditional design and construction practices to the digital era. Estonian Road Administration started the process of BIM implementation to the Estonian road Industry in 2017, when the first planning of pilot projects began. Today, the utilisation of BIM is set as a strategic goal for Estonian Transport Administration (formerly Estonian Road Administration) and a structured approach has been developed in order to overcome such a complex shift in the ways, how road information is being managed in the whole life cycle of a road. Development of principles of data movement, setting technical guidelines and requirements, managing legal issues and communication within the organisation, as well as outside, training of personnel are all aspects that need to be taken into consideration. Successful adaption also needs good communication with the Industry, which is mainly done with Estonian Digital Construction Cluster – a collaboration which brings together the main stakeholders in the sector. In a few years time most of the main sections of the road life cycle will hopefully utilise BIM successfully in Estonia, but the key lies within a well developed and excecuted implementation. The paper describes the development of such a implementation plan and also brings out the main issues and success factors, that are relevant for a successful shift towards digitalisation in the road construction industry.

Keywords: Road Construction, Digitalisation, BIM, InfraBIM, information models, AEC

Introduction

Efficiency in the Architectural, Engineering and Construction (AEC) sector is lagging behind compared to other sectors. The key for solving the issue is to utilise the benefits of digitalisation. The main tool or methodology used in the AEC sector for digitalisation is called Building Information Management (BIM), which is often also referred to as Model or Modelling instead of Management. Estonian Transport Administration (formerly Estonian Road Administration) has been implementing BIM as a tool to be used in the whole life-cycle for several years now, but as every change which disrupts the ways of how people are used to work and manage information, the process is ongoing.

1. What is BIM and how it helps to make road management more efficient

The perception of BIM is quite different across different actors in the AEC sector. According to Vallimaa, Puusaag and Ivask, stakeholders in the Estonian AEC sector see BIM as a process, especially in the planning phase. Later in the life-cycle of an asset, BIM is seen more as a tool for managing information. Contractors, supervisors and facility managers also see BIM simply as a model.

The main core of the methodology is a digital representation of an existing asset or object, the digital representation has geometry in 3D form and there is also information attached to these elements. These 3D geometrical representations with added associated information are called BIM Models. The main benefit or purpose of these models is to gather and transport data, which is relevant to a specific construction element and the most important part is that, the data is in a machine readable format. This digital data makes it easily storable and also accessible, enables statistics and also allows to manage the design and construction processes, but also the whole construction life-cycle more efficiently.

2. BIM implementation started in 2017

In 2017 Estonian Ministry of Economic Affairs and Communications initiated an agreement among Estonian public sector clients, that are ordering design and construction works, as well as do facility management and use public money to do so. The purpose of the agreement was to initiate the usage of BIM among the public sector clients. Estonian Transport Administration was one of the actors in the agreement and Road Management Division was made responsible for the implementation in the organization.

3. Estonian Transport Administration's Road Management Division

Road Management Division, which is responsible of master design, construction and maintenance of state owned public roads is divided into four regions: northern, southern, eastern and western. The regions are responsible of managing master design and construction contracts and also maintenance contracts. The Division has central departments as well, with the main purpose coordinating the works, updating requirements and regulations, manage research and development projects related to road management etc.

4. BIM Working group

In order to tackle the problem of BIM implementation a working group was formed. The group included personnel from all the regions as well as other departments, including development, road registry, planning etc and the woking group has a routine monthly gathering for several years now. The initial size of the active group was around 10 persons and the first task was to manage the first pilot project which were chosen as the starting point for the implementation. The purpose of such a broad working group was two-fold. Firstly, to gather as much knowledge across the organization and include all the relevant departments from day one. The second aim was to start the information sharing as early as possible. Including people from all the relevant departments enabled those persons to be the advocates of BIM among their colleagues. The approach has been to work as a team, which means to tackle problems and decide on solutions as one group, so no one would feel left out and would feel as part of the implementation process. Instead of having a dedicated development team with very little practice the working group consists of people with actual practical experience with design and construction contracts, and know there is no need to sell the idea to the managers of contracts, because they themselves are part of the process.

After first pilot projects were carried out the working group had a seminar dedicated to the results and lessons learned. The outcome of the seminar was an action plan for further actions, with deadlines and responsible persons. The five main goals for BIM in Estonian Transport Administration were defined as: ease design and construction processes, ease data collection and usage, quality BIM models, common information exchange, transparent design, construction and maintenance processes . The action plan was then used to give input for the strategy of the organization. The action plan was later made more detailed and all the relevant factors for BIM implementation were mapped out. Also, two additional sub-working groups were formed for developing technical requirements for BIM design projects and for BIM construction projects.

5. BIM as a strategic goal

The high-level action plan for BIM implementation, which was formed in the BIM working group was used to set strategic goals for Estonian Road Administration in 2020. The timeline was set for four years: by the end of 2021 all the design projects should end with geometrical models (without attributes), by end of 2022 all the construction projects should end with as-built models (including attribute data), by the end of 2023 design models should also incorporate attribute data and by the end of 2024 road maintenance should utilize BIM maintenance models.

In order to fulfil the goals set in the strategy, several problem must be tackled. Such problems include: setting principles of data movement, developing technical guidelines and requirements, tackling legal issues, manage communication – inhouse as well as outside, providing hardware tools for personnel, training of personnel. The BIM working group is currently tackling the mentioned problems.

6. Working together with the industry

Successful implementation of BIM has to be done in close cooperation with the partners from the private sector. If the contractual partners do not grasp what we, as clients, are actually trying to achieve and/or the requirements set are not feasible or efficient to fulfil, then the risk might realise that we will not be able to actually become more efficient in the sector with digitalisation and BIM, but we might just add byrocracy and problems, which do not add any value at the end. To avoid such risks close cooperation has been developed with Estonian Construction Cluster and also with Estonian Asphalt Pavement Association.

IOP Conf. Series: Materials Science and Engineering 1202 (2021) 012046 doi:10.1088/1757-899X/1202/1/012046

Estonian road construction sector is overall quite innovative. In most cases, designs include 3D geometry, but probably not at the best condition for delivering fast machine control models. Machine control is extremely popular on construction sites, because it makes construction more efficient. Some sites have also a common data environment implemented, but it is mostly in use among the construction team and not shared with other participants. As-built data is given over digitally, but it almost never finds a way directly to maintenance contractor.

Estonian Transport Administration as the main paymaster in Estonian road construction sector, is trying to make its BIM process compatible with the whole life-cycle and for all the participants.

Conclusions

There are several approaches to manage change, which digitalisation and BIM implementation most certainly is. Estonian Transport Administration has chosen the path of using internal team, consisting of practitioners with real life experience form inside the organisation. On the other hand the approach has been to use a step-by-step method by implementing BIM model requirements in smaller steps and allowing the organization itself and also the road design and construction sector in general to keep up with the pace of changes.

Disclosure Statement

The authors of this paper have no competing financial, professional, or personal interests from other parties.

References

Puusaag, E., Vallimaa, O. & Ivask, F. (2019). *Building Information Modelling (BIM) process challenges and development of future business models in Estonia.* MSc Thesis, Estonian Business School, Tallinn, Estonia.

IOP Conf. Series: Materials Science and Engineering 1202 (2021) 012047 doi:10.1088/1757-899X/1202/1/012047

Template: Anonymized Document

PRELIMINARY RESEARCH ON WASTE RUBBER APPLICATION IN CEMENT BOUND BASE LAYER

Matija ZVONARIĆ ORCID iD (0000-0002-7604-3002)[1], Ivana BARIŠIĆ ORCID iD (0000-0002-8680-977X)[1], Tihomir DOKŠANOVIĆ ORCID iD (0000-0003-3396-8176)[2], Martina ZAGVOZDA ORCID iD (0000-0002-5334-3731)[1]

[1]*Department for geotechnics, road construction and geodesy; Faculty of Civil Engineering and Architecture Osijek, Josip Juraj Strossmayer University of Osijek, Osijek, Croatia*

[2] *Department for Materials and Structures; Faculty of Civil Engineering and Architecture Osijek, Josip Juraj Strossmayer University of Osijek, Osijek, Croatia*

Corresponding author: Matija, Zvonarić, mzvonaric@gfos
Received day month year; accepted date (use style Received dates, or Alt + Ctrl + R)

Abstract. Besides all the positive characteristics of cement bound courses (CBC), it has some detrimental effects on the pavement wearing courses. Due to cement hydration, this mixture is affected by shrinkage. Shrinkage induces cracks in the whole layer which along with weather conditions propagate through asphalt layers in a short period. Also, it's stiffness negatively affects cracks propagation without providing elastic support for upper layers. As a result, roads are covered with various damages which reduces driving comfort and safety and demand new financial investments. The focus is on reducing the detrimental effect of CBC on the pavement. Nowadays, large quantities of recycled rubber can be found on the market. Wasted rubber is a large ecological problem due to its long decomposition period. On the other hand, by mechanical grinding and separation process, suitable fractions of rubber can be obtained for use in construction. Consequently, the replacement of conventional material by crumb rubber reduces the consumption of natural material and energy for its exploitation. Appropriate amounts and fractions of recycled rubber have the potential to reduce shrinkage and increase the elasticity of CBC. Within this paper, preliminary research results will be presented on the possibilities of crumb rubber implementation in CBC and its effects on mechanical characteristics. By using recycled materials in construction processes we undertake a major step in the sustainable management of natural resources.

Keywords: cement bound base layer, pavement, rubber, compressive strength, ultrasound pulse velocity.

1. Introduction

Nowadays, when we are witnesses to the rapid development of technology and science, the emphasis is on the pronounced mobility of people, goods and services. That entails a need for fast development of transportation infrastructure in terms of railways, air transport, sea traffic and, the most, road infrastructure. Under the TEN – T project, European Union plans to build nine network corridors that connect the most important nodes of the European Union until 2030 (Trans-European Transport Network (TEN-T) n.d.). Enormous quantities of natural raw materials are needed for such an infrastructure project which conflicts with our efforts to establish greater sustainability and nature preservation. Construction of such infrastructure is a great opportunity to investigate and implement some new, sustainable, technics and materials in road infrastructure. While on one hand, large amounts of natural raw materials are needed, on the other hand, large amounts of long-time decomposing waste are disposed of in nature. End of life road vehicles tires is such a kind of waste. According to (Mashiri et al. 2015) about 1.5 million scrap tires reach their end of life every year. Considering the population growth and monetary status increase, this number also rises through recent years. Waste tires in their original shape are useless for construction, so they go through a six-step process where cutting, separation of metal and textile particles and final grinding to desired granulation is carried out (Dobrota, Dobrota, and Dobrescu 2020). Shredded and crumb rubber of different granulations can be found on the market. In the construction industry, recycled rubber has its part in several areas. Crumb rubber can be added to asphalt mixtures where it takes a dual role depending on the type of production process. There are three possible production processes for rubber modified asphalt: dry, wet and terminal blend. In the dry process, crumb rubber is added to the asphalt mixture before the asphalt binder, which represents a partial substitute for fine coarse aggregate. Contrary, finer rubber particles are added into hot liquid asphalt before mixing with aggregate in the wet process. While in the dry process, crumb rubber plays its role as an aggregate, in the wet process rubber particles melts under high temperatures and unites with a binder. The terminal blend process can be interpreted as a type of wet process where the crumb rubber is blended with the asphalt binder at the asphalt terminal (Bakheit and Xiaoming 2019; X. S. B. Huang n.d.). Ding et al. (2019) claim that asphalt mixture composed of recycled asphalt and stable crumb rubber made by wet process improves asphalt's high temperature and low-temperature performance,

Template: Anonymized Document

moisture stability and fatigue resistance comparing to virgin asphalt mixture. Also, this research states that a stable crumb rubber can increase the proportion of recycled asphalt in the asphalt mixture. Da Silva et al. (2018) investigated the influence of crumb rubber on the propagation of reflective cracks and fatigue life of dry mix asphalt mixtures. Their experiment shows that crumb rubber in the asphalt mixtures increases resistance to reflection cracking and fatigue resistance. Also, in their previous study (da Silva, Benta, and Picado-Santos 2018) they concluded that crumb rubber asphalt produced by both, dry and wet process, develop the same performance in terms of fatigue life and rutting resistance. Chen et al. (2020) state that asphalt mixtures with ground tire rubber can increase rutting resistance. From those researches, the advantage of using recycled rubber in asphalt mixture is obvious. Contrary, Yang et. al (2018) have focused their research on the harmful impact of the use of crumb rubber in hot mix asphalt mixtures and found out that its use largely contributes to hazardous emissions such as xylene and toluene. Furthermore, recycled rubber can be used as a replacement for aggregate in cement bound base course (CBC). Some papers which investigate the use of rubber in CBC have been published. In their extensive researches, Farhan et al. (2020) investigated the possible use of recycled rubber and steel fibres from old tires in subbase layers and concluded that implementation of recycled rubber in cement bound aggregates affect the density of mixture due to low specific gravity of rubber and damping action of rubber on compaction. Furthermore, those papers revealed that recycled rubber has a positive effect on the appearance and propagation of cracks. Namely, by examining the internal structure of failed specimens, they noticed that cracks propagated through rubber particles, which implies that rubber particles absorb energy. By conducting an Ultrasonic Pulse Velocity (UPV) test, researchers measured lower values than for virgin mixtures, which are characteristic of less stiff materials. In papers (Jie Li, Mohammad Saberian 2018; Saberian et al. 2018, 2020; Saberian, Li, and Setunga 2019), recycled concrete aggregate and crushed rock mixed with coarse and fine rubber and, in some mixtures, with crushed glass were prepared and tested in a triaxial testing machine. It was concluded that recycled aggregates with a small amount of rubber could be used in subbase and base pavement layers. Considering that recycled concrete has inferior characteristics compared to natural materials, it is likely that higher amounts of crumb rubber could be used in mixtures with natural raw materials. In their paper, Wei et al. (2015) examined the effect of freeze-thaw cycles on silty clay modified with fly ash and crumb rubber under dynamic loading. The conclusion of their research claims that rubber and ash modified mixtures are superior to unmodified ones. Nevertheless, crumb rubber can be used in some other construction projects. Firstly, and the most noticeable application of RC is at playgrounds and sporting surfaces where this surface type provides better protection against serious injuries (T. J. Huang and Chang 2009). Waste rubber also finds its application in railway construction. In their research, Asgharzadeh et al. (2018) concluded that rubberized asphalt mixtures show better performance under cyclic traffic load than virgin asphalt mixtures and they would provide better support for railways in the track – bed structure.

Given all possible applications and previous researches, this study aims to examine the influence of various amounts of crumb rubber on the mechanical properties of cement bound base course (CBC) mixtures for heavy traffic load pavement application. Here are presented preliminary results of ongoing research to define optimal rubber and binder content in CBC to produce a crack-resistant material for asphalt pavement reflective cracking reduction.

Template: Anonymized Document

2. Materials and methods

2.1. *Materials*

For the preparation of CBC test specimens, sand and three different fractions of gravel from the Sava river were used, 0-4 mm, 4-8 mm and 8-16 mm. In accordance with European norm (European committee for standardization 2013a), solid particle densities of natural aggregate were determined in a pycnometer. The density of sand is 2,86 g/cm^3, while densities of 0-4 mm, 4-8 mm and 8-16 mm gravel fractions are 2,96, 2,63 and 2,70 g/cm^3 respectively. Besides the aggregates for virgin CBC mixtures, crumb rubber of nominal grain size 0-0,5 mm was used for rubber-modified mixtures. In accordance with the above-mentioned norm, the density of used rubber was determined in ethanol instead of water due to its low density. The density of crumb rubber is 1,12 cm^3. As a binder, Portland cement of grade 32,5 (CEM II B/M (P-S) 32,5R) was used comprising 5% of aggregate mass for the reference mixture (C5R0). The density of used cement is 2,92 g/cm^3, determined in a pycnometer using petroleum, according to (European committee for standardization 2018). Virgin mixtures consist of the same mass amount of all four fractions (25% of sand and each gravel fractions), cement and optimum moisture content determined using the Proctor compaction method in accordance with standard EN 13286-2 (European commitee for standardization 2010). The gradation curve of the used aggregate is presented in Figure 1 and it can be classified as a cement bound granular mixture 4 (CBGM 4), according to EN 14227-1(European commitee for standardization 2013b). Cement content was determined according to previous research results based on 28-day compressive test results (Barišić 2012). Due to their similar gradation curve, sand was replaced by rubber, while the gravel fractions kept their mass ratio in mixtures. Considering that sand has a 260.71% higher density than rubber, the rubber replaced sand by volume. Three different rubber modified mixtures were designed, replacing 20, 30 and 60 vol.% of sand (C5R20, C5R30 and C5R60 respectively) with rubber.

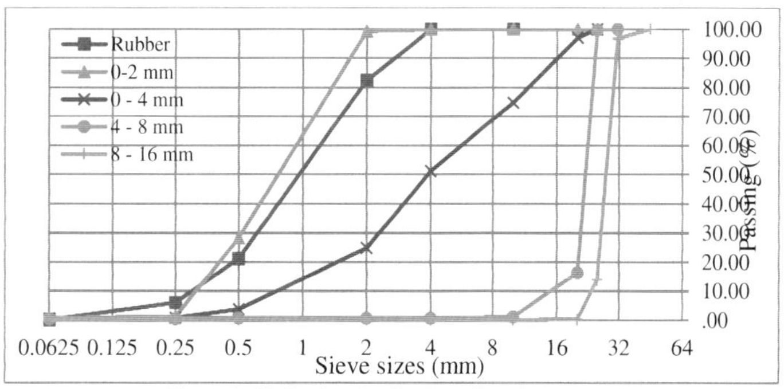

Figure 1. Grain size distribution of used aggregate

2.2. *Specimen preparation and testing*

The maximum dry density (MDD) and optimum water content (OWC) for every mixture were determined following (European committee for standardization 2010). Three of a kind specimens were prepared using the modified Proctor compaction method according to recommendations of (European committee for standardization 2004) (Figure 2.). Three samples were made for each mixture to eliminate possible deviations of laboratory results. After compaction, specimens were demoulded and wrapped into cling film to prevent loss of moisture during the curing period. Wrapped specimens were cured in a climate chamber for 28 days at a temperature of 20°C and humidity of 90%. After the curing period, specimens were unwrapped, measured and weighed. Ultrasonic pulse velocity (UPV) was measured according to (CEN/TC104 2004), to calculate the dynamic modulus of elasticity (Figure 3). This test is usually performed for concrete samples, but some researches prove that UPV measurement gives results of great significance for CBC mixtures as well (Barišić, Dimter, and Rukavina 2016; Dimter, Rukavina, and Barišić 2011). From the obtained time and height of the specimen, ultrasonic pulse velocity can be calculated according to the following equation:

IOP Conf. Series: Materials Science and Engineering 1202 (2021) 012047 doi:10.1088/1757-899X/1202/1/012047

Template: Anonymized Document

$$UPV = \frac{L}{T}$$

Where:
UPV – the pulse velocity [km/s]
L – the path length [mm]
T – the time taken by the pulse to transverse the length [µm].

Figure 2. Compacted and demoulded specimen

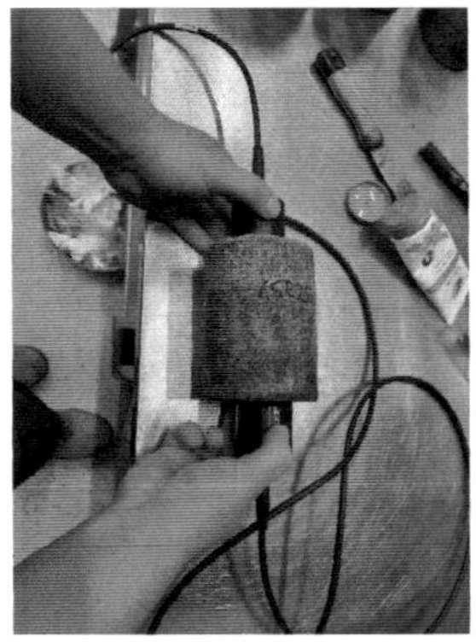
Figure 3. UPV measurement

Based on ultrasonic pulse velocity, the density of the specimens and Poisson's ratio, the dynamic modulus of elasticity can be calculated. Poisson's ratio is adopted as ν = 0,25. The equation for the dynamic modulus of elasticity calculation is the following:

$$E = \rho \times v^2 \frac{(1 + \nu)(1 - 2\nu)}{1 - \nu}$$

Where:
E – the dynamic modulus of elasticity [MN/m2]
ρ – the density of the specimen [kg/m3]
v – the pulse velocity [km/s]
ν – the Poisson's ratio

After non-destructive measurements, specimens were tested in a universal testing machine to determine their 28-day compressive strength. Testing of compressive strength was conducted in compliance with (European committee for standardization 2003). The load was applied uniformly and the maximum stress of the specimens occurred between the 30th and 60th second of commencement of loading. The compressive strength is defined as the stress at maximum load on a specimen when tested in uniaxial unconfined compression. It is calculated as a ratio of the force to the surface on which the force acts - a round surface with a diameter of 100 mm. The form of an equation is:

$$f_{c,28} = \frac{F}{Ac}$$

Where:
fc,28 – the compressive strength of the specimen of hydraulically bound mixtures [N/m2]
F – maximum force sustained by the specimen of hydraulically bound mixtures [N]
Ac – cross-section area of the specimen of hydraulically bound mixtures [mm2].

Template: Anonymized Document

Figure 4. presents specimen failure under compression strength test with a satisfactory failure pattern according to (European commitee for standardization 2003a).

Figure 4. Broken specimen after compression

3. Results and discussion

Results of conducted tests are presented in Table 1.

Table 1. Dynamic modulus of elasticity and compressive strength results

Mixture/parameter	MDD [g/cm3]	OWC [%]	UPV [km/s]	E [GPa]	fc28[MPa]
C5R0 (reference mix)	2,12	5,87	4,06	31,49	7,60
C5R20	2,06	5,53	2,98	16,57	3,41
C5R30	2,03	5,13	2,40	10,50	2,48
C5R60	1,89	4,94	0,37	0,45	0,94

Addition of rubber results in a drop of all tested mechanical properties. As expected, there is a drop in MDD and OWC with the increase in rubber content due to the hydrophobic nature and lower density of rubber compared to sand. These results are in accordance with (Farhan et al. 2015; Sun et al. 2020).

MDD drop for 20, 30 and 60 vol. % sand replacement by rubber is 2,83%; 4,25% and 10,85% respectively compared to the reference mix. However, E drop is 47,30%; 66,67% and 98,41% respectively comparing to reference mix. This high drop in E value means that replacing sand with rubber results in a less stiff material which could be beneficial for crack occurrence delay within the layer. Also, this high difference between MDD and E drop with rubber addition cannot be only due to the difference in material densities, but to a change in materials behaviour. The same goes for UPV results, dropping by 26,60%; 40,88% and 90,89% respectively. This is going to be investigated within the next research phase.

A piece of important information for the evaluation of cement bound mixture for use in the pavement is the 28-day compressive strength. According to (General technical conditions for road works 2001), the required value of a mentioned parameter for heavy traffic load is within the range of 2,5 MN/m^2 to 6 MN/m^2. From Table 1. it can be seen that all tested mixtures except C5R60 achieve the required minimum value. Rubber addition resulted in a compressive strength decrease which was predicted based on previous researches (Farhan, Dawson, and Thom 2016b). According to Table 1. a great dropdown trend can be observed by an increase of rubber content in the mixture. The drop in compressive strength of C5R20, C5R30 and C5R60 is 59,21%; 67,11% and 88,16% respectively compared to the reference mix. It can be observed that the decreasing nature of compressive strength is consistent with the drop in the dynamic modulus of elasticity. Despite the decrease in compressive strength, obtained results show that the mixture with 20% of rubber satisfies the condition for use in heavy traffic load pavement, while the mixture with 30% of rubber achieved the minimum required value. For higher cement content, higher rubber content could be used if compressive strength criteria should be the most important one. The influence of rubber replacement level depending on different cement content in a view of compressive strength, elastic behaviour and crack occurrence reduction is to be investigated within the next research phase.

Template: Anonymized Document

UPV measurement is a non-destructive test, it is simple to perform and it is in use for classification of the quality of concrete for a long period now (excellent quality concrete with UPV values > 4,5 km/s and very poor quality concrete with UPV < 2 km/s (Guidebook on non-destructive testing of concrete structures 2002)). A similar analogy was used in previous research where UPV ranging 2-3,2 km/s indicated a material of satisfactory compressive strength (2-6 MP) (Barišić, Dimter, and Rukavina 2016). Within this research, a similar result is observed with UPV less than 2 km/s indicating insufficient $f_{c,28}$ (less than 2 MPa) for high traffic purposes. In Figure 5, the correlation between UPV and MDD (a) and compressive strength (b) is presented and a strong correlation is noted. Although only a limited number of samples and results are presented, this is a good basis for further research.

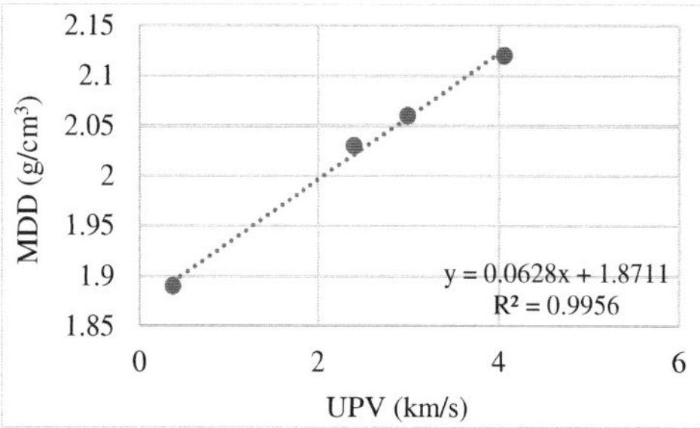

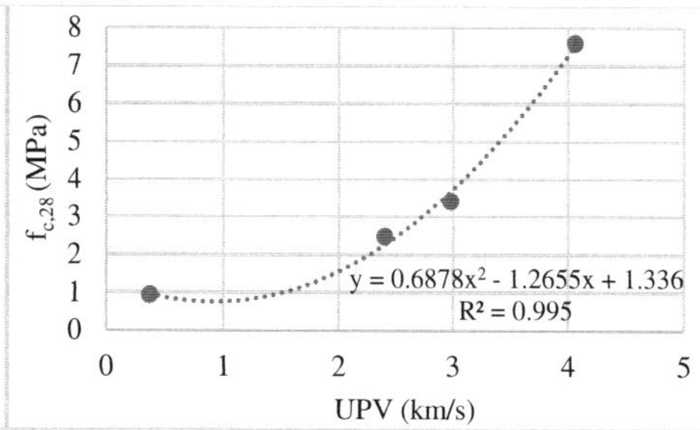

a) b)

Figure 5. Correlation between UPV and maximum dry density (a) and UPV and compressive strength (b)

4. Conclusion

This preliminary study aimed to obtain an initial input on the influence of various amounts of crumb rubber on the mechanical properties of cement bound base course (CBC) mixtures for heavy traffic load pavement application. From the presented laboratory test results, the following conclusions can be made:

- addition of rubber results in a drop of maximum dry density, optimal water content, ultrasound pulse velocity and compressive strength of cement bound aggregate for pavement base layer
- drop in dynamic modulus of elasticity value by rubber addition indicates a less stiff material for CBC construction
- fine aggregate fraction could be replaced by rubber to 30 vol. % while maintaining the minimum value of compressive strength on 2 MPa
- for the next phase of the research, the influence of rubber replacement level depending on different cement content and more detailed insight into elastic behaviour and crack occurrence reduction is to be investigated.

Acknowledgement

This research was supported by Croatian Science Foundation under the project UIP-2019-04-8195 Cement stabilized base courses with waste rubber for sustainable pavements – RubSuPave

Author Contributions

MZ (Matija Zvonarić) and IB conceived this study, MZ (Matija Zvonarić) conducted laboratory tests, TD, IB and MZ (Martina Zagvozda) were responsible for data analysis and article conceptualization.

Disclosure statement

The authors declare no conflict of interest.

Template: Anonymized Document

References:

Ahmed Hilal Farhan, Andrew Robert Dawson, Nicholas Howard Thom. 2018. "Recycled Hybrid Fiber-Reinforced & Cement-Stabilized Pavement Mixtures: Tensile Properties and Cracking Characterization." *Construction and Building Materials* 179: 488–99.

Asgharzadeh, Seyed Mohammad, Javad Sadeghi, Pooneh Peivast, and Makan Pedram. 2018. "Fatigue Properties of Crumb Rubber Asphalt Mixtures Used in Railways." *Construction and Building Materials* 184: 248–57.

Bakheit, Ismail, and Huang Xiaoming. 2019. "Modification of the Dry Method for Mixing Crumb Rubber Modifier with Aggregate and Asphalt Based on the Binder Mix Design." *Construction and Building Materials* 220: 278–84.

Barišić, Ivana. 2012. "Composition Optimization of Pavement Cement Stabilized Base Courses with Slag." University of Zagreb, Zagreb, Croatia.

Barišić, Ivana, Sanja Dimter, and Tatjana Rukavina. 2016. "Characterization of Cement Stabilized Pavement Layers with Ultrasound Testing." *Technical Gazette*: 447–53.

CEN/TC104, Technical Committee. 2004. *EN 12504-4:2004 Testing Concrete - Part 4: Determination of Ultrasonic Pulse Velocity*.

Chen, Siyu et al. 2020. "Investigation of Hot Mixture Asphalt with High Ground Tire Rubber Content." *Journal of Cleaner Production* 277.

Dimter, Sanja, Tatjana Rukavina, and Ivana Barišić. 2011. "Aplication of the Ultrasonic Method in Evaluation of Properties of Stabilized Mixes." *The Baltic Journal of Road and Brigde Engineering* 6(3): 177–84.

Ding, Xunhao et al. 2019. "Laboratory Investigation of the Recycled Asphalt Concrete with Stable Crumb Rubber Asphalt Binder." *Construction and Building Materials* 203: 552–57.

Dobrota, Dan, Gabriela Dobrota, and Tiberiu Dobrescu. 2020. "Improvement of Waste Tyre Recycling Technology Based on a New Tyre Markings." *Journal of Cleaner Production* 260.

European commitee for standardization. 2003a. *HRN EN 13286-41:2003 Unbound and Hydraulically Bound Mixtures-Pari 41: Test Method for the Determination of the Compressive Strenght of Hydraulically Bound Mixtures*.

———. 2003b. "HRN EN Unbound and Hydraulically Bound Mixtures-Part 41: Test Method for the Determination of the Compressive Strenght of Hydraulically Bound Mixtures."

———. 2004. *EN 13286-50 Unbound and Hydraulicylly Bound Mixtures - Part 50: Method for the Manufacture of Test Specimens of Hydraulically Bound Mixtures Using Proctor Equipment or Vibrating Table Compaction*.

———. 2010. "HRN EN 13286-2:2010 Unbound and Hydraulically Bound Mixtures - Part 2: Test Method for Laboratory Dry Density and Water Content - Proctor Compaction."

———. 2013a. *HRN EN 1097-6:2013 Test for Mechanical and Physical Properties of Aggregates - Part 6: Determination of Particle Density and Water Absorption*.

———. 2013b. *HRN EN 14227-1:2013 Hydraulically Bound Mixtures - Specifications - Part 1: Cement Bound Granular Mixtures*.

———. 2018. *HRN EN 196-6:2018 Methods of Testing Cement - Part 6: Determination of Fineness*.

Farhan, Ahmed Hilal et al. 2015. "Flexural Characteristics of Rubberized Cement-Stabilized Crushed Aggregate for Pavement Structure." *Materials and Design* 88: 897–905.

Farhan, Ahmed Hilal, Andrew Robert Dawson, and Nicholas Howard Thom. 2016a. "Characterization of Rubberized Cement Bound Aggregate Mixtures Using Indirect Tensile Testing and Fractal Analysis." *Construction and Building Materials 105*: 94–102.

———. 2016b. "Effect of Cementation Level on Performance of Rubberized Cement-Stabilized Aggregate Mixtures." *Materials and Design* 97: 98–107.

———. 2020. "Compressive Behaviour of Rubberized Cement-Stabilized Aggregate Mixtures." *Construction and Building Materials* 262.

General Technical Conditions for Road Works (In Croatian: Opći tehnički uvjeti za radove na cestama). 2001. Institut građevinarstva Hrvatske, Zagreb, Janka Rakuše 1.

Guidebook on Non-Destructive Testing of Concrete Structures. 2002. Vienna: International Atomic Energy Agency.

Huang, T. J., and L. T. Chang. 2009. "Design and Evaluation of Shock-Absorbing Rubber Tile for Playground Safety." *Materials and Design* 30: 3819–3823.

Huang, Xiang Shu; Baoshan. "Recycling of Waste Tire Rubber in Asphalt and Portland Cement Concrete: An Overview."

Jie Li, Mohammad Saberian, Bao Thach Nguyen. 2018. "Effect of Crumb Rubber on the Mechanical Properties of Crushed Recycled Pavement Materials." *Journal of Environmental Management* 218: 291–99.

Mashiri, M.S., J.S. Vinod, M. Neaz Sheikh, and Hing-Ho Tsang. 2015. "Shear Strength and Dilatancy Behaviour of Sand–Tyre Chip Mixtures." *Soils and Foundations* 55(3): 517–28.

Saberian, Mohammad et al. 2020. "An Experimental Study on the Shear Behaviour of Recycled Concrete Aggregate Incorporating Recycled Tyre Waste." *Construction and Building Materials* 264.

Saberian, Mohammad, Jie Li, Bao Nguyen, and George Wang. 2018. "Permanent Deformation Behaviour of Pavement Base and Subbase Containing Recycle Concrete Aggregate, Coarse and Fine Crumb Rubber." *Construction and Building Materials* 178: 51–58.

Template: Anonymized Document

Saberian, Mohammad, Jie Li, and Sujeeva Setunga. 2019. "Evaluation of Permanent Deformation of a New Pavement Base and Subbase Containing Unbound Granular Materials, Crumb Rubber and Crushed Glass." *Journal of Cleaner Production* 230: 38–45.

da Silva, Luis, Agostinho Benta, and Luis Picado-Santos. 2018. "Asphalt Rubber Concrete Fabricated by the Dry Process: Laboratory Assessment of Resistance against Reflection Cracking." *Construction and Building Materials* 160: 539–50.

Sun, Xiaoyin, Shuyin Wu, Jun Yang, and Rouchong Yang. 2020. "Mechanical Properties and Crack Resistance of Crumb Rubber Modified Cement-Stabilized Macadam." *Construction and Building Materials* 259.

"Trans-European Transport Network (TEN-T)."

Wei, Haibin, Yubo Jiao, and Hanbing Liu. 2015. "Effect of Freeze–Thawcycles on Mechanical Property of Silty Claymodified by Fly Ash and Crumb Rubber." *Cold Regions Science and Technology* 116: 70–77.

Yang, Xu et al. 2018. "Emission Analysis of Recycled Tire Rubber Modified Asphalt in Hot and Warm Mix Conditions." *Journal of Hazardous Materials* 365: 942–51.

CONSTRUCTION OF ECONOMICAL PAVEMENT STRUCTURES WITH WOOD ASH

Sanja Dimter ORCID 0000-0003-0688-8592[1], Martina Zagvozda [1], Tea Tonc [1], Miroslav Šimun ORCID 0000-0002-8410-0249[2]

[1] *Josip Juraj Strossmayer University of Osijek, Faculty of Civil Engineering and Architecture Osijek, Osijek, Croatia;* [2]*Civil Engineering Department, Zagreb University of Applied Sciences, Zagreb, Croatia*
Corresponding author: Sanja Dimter, sdimter@gfos.hr

Received day month year; accepted date (use style Received dates, or Alt + Ctrl + R)

Abstract. Stabilized mixes that are used in pavement structures are composed of aggregate bound with hydraulic binders (cement, lime) or bitumen. The most commonly used for the construction of base layers are mixes stabilized with cement. A long-standing construction practice for pavement structures was based on the use of quality granular materials for the construction of base layers. However, when designing the pavement structure and selecting materials, economy, sustainability, and environmental impact, in addition to their mechanical properties, should also be considered. Clear requirements and guidelines for sustainable development have imposed the need to explore the possibility of using non-standard materials in construction. Wood ash, which is formed as a residue from the combustion of biomass in the production of electricity and heat, is one of the newer and, in Croatia, less researched alternative materials that can be applied in construction. The paper describes compressive strength tests of mixtures of sand from the Drava River and cyclone wood ash stabilized with various contents of cement. The obtained results showed that with wood fly ash (in a content of 30 % mass.) in the stabilization mixture of sand, values of compressive strengths can be achieved within the required limits necessary for the construction of base layers of the pavement structure stabilized by a hydraulic binder.

Keywords: wood ash, sand, pavement structure, base layers, rationalization of construction costs, sustainable road construction

Acknowledgements

This research (scientific-research project IZIP-GrAFOS-2018) entitled "The application of ash from wood biomass in layers of pavement structure—Bio PAV" was funded by the Faculty of Civil Engineering and Architecture Osijek, Croatia. Tests of the constituent materials and physical and mechanical properties of the mixture were conducted in the Geotechnical and Road Construction Laboratory at the Faculty of Civil Engineering and Architecture Osijek. The authors would like to acknowledge Zlatko Treska from Strizivojna Hrast Company in Strizivojna, Croatia, for wood ash used for experiments.

Author Contributions

Conceptualization and methodology, S.D.; investigation, M.Z. and T.T; writing—original draft preparation, S.D., M.Z., and T.T.; writing—review and editing, S.D., M.Z., M.Š., and T.T. All authors have read and agreed to the published version of the manuscript.

Disclosure Statement

The authors declare no conflicts of interest.

Template: Anonymized Document

CONSTRUCTION OF ECONOMICAL PAVEMENT STRUCTURES WITH WOOD ASH

Sanja Dimter ORCID 0000-0003-0688-8592[1], Martina Zagvozda [1], Tea Tonc[1], Miroslav Šimun ORCID 0000-0002-8410-0249[2]

[1] *Josip Juraj Strossmayer University of Osijek, Faculty of Civil Engineering and Architecture Osijek, Osijek, Croatia;*
[2]*Civil Engineering Department, Zagreb University of Applied Sciences, Zagreb, Croatia*
Corresponding author: Sanja Dimter, sdimter@gfos.hr

Received day month year; accepted date (use style Received dates, or Alt + Ctrl + R)

Abstract. Stabilized mixes that are used in pavement structures are composed of aggregate bound with hydraulic binders (cement, lime) or bitumen. The most commonly used for the construction of base layers are mixes stabilized with cement. A long-standing construction practice for pavement structures was based on the use of quality granular materials for the construction of base layers. However, when designing the pavement structure and selecting materials, economy, sustainability, and environmental impact, in addition to their mechanical properties, should also be considered. Clear requirements and guidelines for sustainable development have imposed the need to explore the possibility of using non-standard materials in construction. Wood ash, which is formed as a residue from the combustion of biomass in the production of electricity and heat, is one of the newer and, in Croatia, less researched alternative materials that can be applied in construction. The paper describes compressive strength tests of mixtures of sand from the Drava River and cyclone wood ash stabilized with various contents of cement. The obtained results showed that with wood fly ash (in a content of 30 % mass.) in the stabilization mixture of sand, values of compressive strengths can be achieved within the required limits necessary for the construction of base layers of the pavement structure stabilized by a hydraulic binder.

Keywords: wood ash, sand, pavement structure, base layers, rationalization of construction costs, sustainable road construction

Introduction

Pavement structures are complex engineering structures that comprise 50-70% of the total expense of road construction, particularly under conditions of increased traffic load that affects all modern roads and highways. For these reasons, a special engineering-economic approach is necessary in the design and construction of pavement structures.

Natural aggregates have been used for many years to construct pavement structures. In Croatia, they still represent the greatest source of raw materials. Unlike the rest of the country, eastern Croatia does not have enough quality gravel and stone material available for road construction. Gravel deposits are rare, and materials are of poorer quality, while quarries are far from the place of application. On the other hand, eastern Croatia has considerable quantities of sand materials (river and excavated sands), the application of which was imposed as a necessity to rationalize construction expenses. By years of application in road construction, sand has proven to be a quality and adequate local material for the design and construction of cost-efficient pavement structures and embankment layers. Sand can be applied in its natural state or stabilized by a hydraulic binder (most commonly cement). Such application of a local material enables savings in transport costs that may be as much as 50-60% of the price of the materials. Of course, the application of local materials that is (usually) not covered by existing standards requires detailed laboratory tests, as well as road test sections and the running of the construction technology (Study, 1990.).

In addition to local materials, waste materials and industrial by-products also have a major role in the construction of cost-efficient pavement structures, although in considerably lower quantities. These materials require complex disposal of large quantities at dump sites and are harmful to the environment, but their application is economically justified and is clearly encouraged by the Guidelines for the Sustainable Development (Dimter et al. 2014.). These materials differ from each other by their chemical and physical properties, by their original state, or by the additional processing they must undergo before being used. Specific waste materials, and fly ash produced by the combustion of coal in thermal power plants or metallurgical slag stand out, have good pozzolanic properties, which also makes them useful as independent binders or a binder addition in the stabilization of concrete mixtures and in the production of cement. When designing pavement structures, materials with pozzolanic properties offer multiple benefits and rationalization of expenses that include: (1) reduction in the quantity of a hydraulic binder (cement) in the mixture, (2) improved mechanical properties of stabilization mixtures and greater resistance to the development of cracks from shrinking, (3)

Template: Anonymized Document

reduction in the quantity of waste materials at dump sites that are harmful to the environment, and (4) adding value to the waste material.

The application of fly ash from thermal power plants has been the most researched (Dimter et al. 2014.), and a large part of that research has been dedicated specifically to the pozzolanic effect of fly ash in mixtures. The pozzolanic activity/effect is directly related to the content of CaO in ash, depending on the type of coal from which the ash is produced. Initial research on the application of fly ash in pavement structure layers was conducted on stabilization mixtures that are used for construction of the base layers of the pavement structure (Maher & Balaguru, 1993.; Zenieris & Laguros, 1988.; Sobhan et al. 1999.; Ksaibati, 1995.) In this research fly ashes confirmed their effective properties and the possibility of application in stabilized layers of the pavement structure. Research with fly ash from thermal power plants continued, expanding to application in the subgrade and embankment layers (Dimter et al. 2011.; Simatupang et al. 2020.; Mahvash et al. 2017.). With time, the possible application of other types of ashes began to be researched, such as ash from municipal waste incineration plants, ash from paper mills and bioash of various origins (Zagvozda et al. 2018.; Supancic & Obernberger, 2011.; Šķēls et al. 2016.; Šķēls et al. 2017.). Finer fly ashes (cyclone and filter) and coarser bottom ashes also found a place in the application in the base layers of the pavement structure due to two mechanisms: improvement of the load bearing capacity through mechanical stabilization and improvement of the granulometric composition, or due to the pozzolanic activity of ash when ash can partly or completely replace the hydraulic binder in the mixture. So, it is possible to use ashes during stabilization of the subgrade material with a lower load bearing capacity, in base layers of pavement structures, or in concrete mixtures. These applications are enabled by the required content of CaO and pozzolan in the chemical composition of bioash. Among the various types of bioash, wood ashes are distinguished by a high CaO content. Wood ashes are produced as residue from the incineration of wood biomass to produce electricity and thermal energy. Three types of ash are produced: coarser furnace ash, cyclone fly ash and electrostatic filter fly ash. The chemical composition of wood ash is changeable and depends on many factors: the type of the wood biomass, the parts of a tree that are burned (branches and roots, trunk, bark, and leaves), the season in which the trees are cut down, and their age (Pitman, 2006.).

In Croatia, research on the application of wood ash began relatively recently, spurred by the large quantities of bio ash produced by newly built power plants (Zagvozda et al. 2017.). The construction of biomass power plants in Croatia intensified with the passage of the Strategy for Energy Sector Development of the Republic of Croatia. The first cogeneration biomass power plant began operations in 2011. In eastern Croatia, almost half of the total number of power plants put into operation, including the largest ones, use wood biomass to produce energy. Approximately 110 tons of wood biomass is incinerated daily by the company Strizivojna Hrast d.o.o. The quantities of ash that are produced daily are considerable. At present, there is no systematic way of recovering wood ash; it is mostly deposited. The company has shown interest in scientific research that will uncover and confirm the possibilities of useful application of wood ash. The need for research into the possibilities of recovering bioash at the local level is supported by the fact that two counties of eastern Croatia, Osijek-Baranja and Vukovar-Srijem, are part of the SRCplus Project (Zagvozda et al. 2017.). The project was initiated by the European Commission with the goal of ensuring the production of sufficient quantities of woodchips to satisfy national and European energy goals. The plan of the project was to develop a sustainable supply of local power/cogeneration plants with wood biomass from wood cultures of short rotation grown on agricultural land of poorer quality. Districts from eight European countries were chosen for the project, including the two above-mention Croatian counties.

The goal of the research described in this paper was to ascertain the effect of wood ash on the mechanical properties of the stabilization mixture composed of local materials (sand) and waste material (wood ash) stabilized by cement.

1. Experimental part

The mechanical properties of the cement stabilized materials are most often defined by compressive strength. For cement stabilized materials, testing of compressive strength is usually after 7 and 28 days of sample conditioning, while the period for achieving the required compressive strength for materials with a prolonged time of binding (e.g., pozzolans, such as fly ash) can be even longer.

1.1. Materials

Sand from the Drava River was used as the basic material for preparing of stabilization mixtures, to which cyclone fly ash was added. The mixture of sand and ash was stabilized with 2% and 4% of cement with the addition of an adequate quantity of water necessary for the compacting and hydration of the cement.

Sand from the Drava River has a uniform granulometric composition (Figure 1.) with a medium grain size $D_{50} = 0.3$ mm and a degree of unevenness $U=d_{60}/d_{10}=2$. Due to the uniform granulometric composition of sand, it is necessary to add an adequate quantity of cement. Based on previous research (Study, 1990.) the bearing capacity of sand expressed by the California Bearing Ratio (CBR), was CBR= 9–12%.

Template: Anonymized Document

Figure 1. Photo of used sand and wood ash

Cement CEM II/B-M (P-S) 32.5 N is the cement type that is most frequently used in eastern Croatia. It is produced locally at the Našicecement d.d. cement factory near city Našice.

Cyclone fly ash from the cogeneration plant of Strizivojna Hrast d.o.o. collected on the cyclone deduster was added to the sand. The chemical composition of the cyclone wood ash is shown by the mass content of individual components (mass. %): MgO=3.06, Al_2O_3=0.44, SiO_2=4.05, P_2O_5=2.90, SO_3=1.59, K_2O=2.82, CaO=46.91. As seen from the composition, the main component is calcium oxide, which indicates possibility of binding properties or pozzolanic reaction in combination with suitable material. Testing the mineral composition of ash samples by the X-ray diffraction method (XRD) (Report, 2019.) determined that the main components of wood ash were calcite, quartz, and CaO, and, in smaller quantities, portlandite ($CaOH)_2$) and fairchildite ($K_2Ca(CO_3)_2$).

1.2. Mixture composition

Based on previous tests and the granulometric compositions of materials, the basic composition of the stabilization mixture was designed in a proportion of 70% sand and 30% wood ash. The usual part of cement necessary for stabilization of pure sand from the Drava River and obtaining the required values of compressive strength amounted to 8-12% in the tests carried out thus far (Study, 1990.). Regarding the content of wood ash in the mixture and its potential for pozzolanic activity, and based on previous tests, the content of cement in the mixture was considerably reduced in relation to the usual percentage. Mixtures with the following composition were designed:

1. Mix 1: 70% sand and 30% wood ash, 2% cement.
2. Mix 2: 70% sand and 30% wood ash, 4% cement.

A modified Proctor experiment was conducted according to the norm HRN EN 13286–2. For the research, a Proctor's cylindrical mold A with a 100 mm diameter and 120 mm height was used. Five layers of samples were compacted with the appropriate energy (2,7 MJ/m^3) in an automatic Proctor device. The resulting values of optimal moisture content (OMC) and maximum dry density (MDD) are presented in Table 1. These values were used to create samples to determine compressive strength.

Table 1. Proctor elements for sample preparation

Amount of binder	OMC [%]	MDD [g/cm^3]
2%	12.69	1.73
4%	11.61	1.70

1.3. Testing of compressive strength

Compressive strength is defined as the average stress in a sample exposed to uniaxial pressure at the fracture force (according to HRN EN 13286-41 standard). Compressive strength of samples is usually determined after 7 and 28 days of conditioning. The sample is placed in a press, between two steel plates, and loaded with a constant increase of the force. The load is applied evenly and continuously and without jolts until fracture of the samples occurs. The maximum load on the sample until fracture is recorded and its compressive strength is calculated. The standard HRN EN 13286-41 also prescribes that the fracture of the sample/appearance of cracks occurs within 30 to 60 seconds from the

Template: Anonymized Document

beginning of the application of load. After the fracture, the sample is taken from the press and the types of fractures that can be satisfactory and unsatisfactory are studied.

Figure 2. Compressive strength testing

Testing the compressive strength of samples was carried out after they were conditioned at a temperature of 20°C after 7 and 28 days. During conditioning, the samples were wrapped in adhesive foil. The test results are presented in Figure 3.

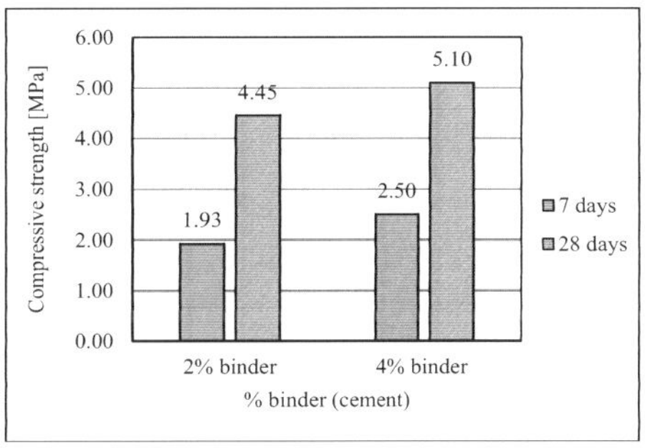

Figure 3. Compressive strength results

2. Results analysis and commentary

2.1. Analysis of compressive strengths results

The values of 7-day and 28-day compressive strengths of mixtures of sand, wood ash and cement are shown in Figure 3. The samples with 2% of cement after 7 days of conditioning achieved a compressive strength of 1.93 MPa, and after 28 days of conditioning, the value of compressive strength was 4.45 MPa. An increase in the content of the binder from 2 %m to 4 %m resulted in an expected increase of compressive strength. The samples with 4% cement after 7 days of conditioning achieve a compressive strength of 2.50 MPa, and after 28 days of conditioning, of 5.10 MPa. The increase of the compressive strength of mixtures with 4% cement compared to mixtures with 2% cement after 7 days of conditioning amounted to 29%, while after 28 days of conditioning, the increase was 14%. The increase of 28-day compressive strength compared to 7-day compressive strength for mixtures with 2% cement was as much as 130%, while for mixtures with 4%, the increase was 104%.

As stated earlier, the usual proportion of cement necessary for stabilization of the sand and achieving the required values of compressive strength ranged from 8-12%. Thus, according to the results of the tests (Study, 1990.) carried out on mixtures of sand from the Drava River and cement, for cement contents of 7%; 9%; 11%; 13% and 15%, the following values of 7-day compressive strength were realized in the mixture: 0.5 MPa; 0.71 MPa; 1.15 MPa; 1.925 MPa and 3.5 MPa. The values for 28-day compressive strength were: 1.62 MPa; 2.43 MPa; 3.18 MPa; 3.75 MPa and 4.67 MPa. The cement used in testing had the same properties as cement that was used in the research of mixtures of sand and ash described in this paper.

Template: Anonymized Document

Comparing the results obtained in this research with the previously stated results from the Study, a significant difference in the values of compressive strengths is apparent, i.e., the addition of wood ash in combination with cement showed binding properties thus increasing compressive strengths. Thus, the mixture of sand and ash with 2% cement achieved the same value of compressive strength as the mixtures of pure sand with 13% cement after conditioning of 7 days (1.925 MPa). The same mixture with ash and 2% cement achieved a higher compressive strength after 28 days of conditioning (4.45 MPa) compared to the mixture of sand stabilized with 13% cement (3.75 MPa). The mixture of sand and ash with 4% cement achieved a 28-day compressive strength (5.10 MPa) greater than the mixture of sand stabilized with as much as 15% of cement (4.67 MPa).

2.2. Comparison of obtained values of compressive strengths with the prescribed values of compressive strengths

For the stabilization mixtures to be applied to the construction of base layers of pavement structures, they must meet the corresponding criteria. Thus, the obtained values of compressive strengths of stabilization mixtures of sand, wood ash and cement were compared with the prescribed compressive strengths according to the HRN U.E9.024 standard and the General Technical Requirements for Road Work (GTR) and with the compressive strengths prescribed in the "Study of the possibilities of application of sand in construction of roads of Slavonia and Baranja region" (1990.). While the HRN U.E9.024 standard and the GTR prescribe a compressive strength of mixtures from stone materials for application in base layers, the minimum compressive strengths according to the Study are prescribed for sand and the criteria are somewhat lower. The prescribed values of the compressive strengths of stabilization mixtures for application in the pavement structure can be seen in Table 2, and the diagram of compressive strengths with the criteria in Figure 4.

Table 2. Prescribed values of compressive strengths

Pavement structure layer	Compressive strength (MPa)		Compressive strength (MPa)	
	according to HRN U.E9.024 and GTR		according to Study (1990)	
	after 7 days	after 28 days	after 7 days	after 28 days
Pavement structure bases	2.0-5.5	3.0-6.5	1.5-3.0	2.5-5.5
Pavement structure sub-bases	1.5-4.5	2.5-6.0	1.0-2.0	1.5-3.0

The set requirements for minimum compressive strengths according to the standard and the GTR are exclusively for stabilization mixtures in which the binder is cement. According to the GTR, construction of the base layer from granular stone material stabilized by the hydraulic binder requires that when using other binders except cement (fly ash, slag), the limits of compressive strength remain the same as for cement, but with a prolonged length of conditioning that should be determined on the basis of laboratory tests and with the consent of the supervisory engineer.

Since there are no precise requirements at present in Croatia about the quality of base layers stabilized by fly ash, it is interesting to compare the obtained results with the requirements prescribed in other countries. Thus, in neighboring Hungary, which has a long tradition of using fly ash in stabilized base layers, and according to the MSZ-07 3703-8271 standard for stabilization of base layers, the minimum compressive strength for mixtures of sand stabilized with fly ash and lime after 60 days amounts to 7.0 MPa.

Template: Anonymized Document

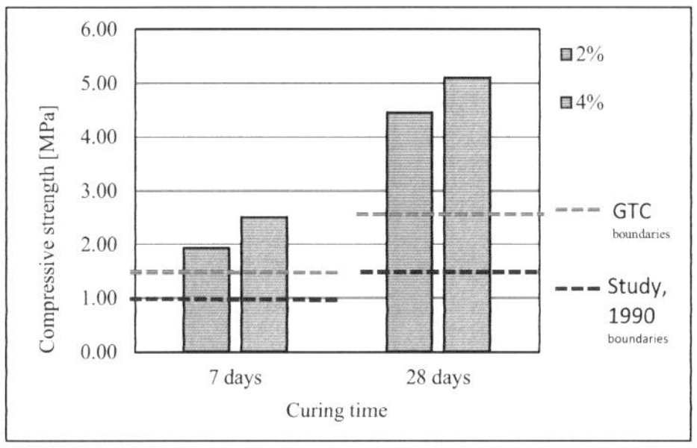

Figure 4. Comparison of mixes compressive strengths with the prescribed criteria

In the diagram in Figure 4, it is evident that both mixtures of sand and ash meet the set requirements, both those defined by the Study and the stricter ones set by the standard and the GTR exclusively for stone materials and cement. The set conditions of compressive strengths were fulfilled for both conditioning types after 7 and 28 days. The obtained results showed that with wood fly ash in the stabilization mixture (up to 30%), the values of compressive strengths in the prescribed limits necessary for construction of base layers stabilized by a hydraulic binder can be achieved. Therefore, the minimum quantity of cement in the stabilization mixture is 2 % mass. It is to be expected that, with the prolonged conditioning of samples and the pozzolanic activity of ash, the values of compressive strengths will continue to grow.

Conclusion

The paper describes compressive strength tests of mixtures of sand from the Drava River and cyclone wood ash stabilized by cement in different contents. Based on previous tests, the composition of stabilization mixtures was designed with the goal of determining the impact of wood ash, i.e. its pozzolanic activity on the development of compressive strength of mixtures. The obtained results showed that, with wood fly ash (in a content of 30 % mass.) in the stabilization mixture of sand, the values of compressive strengths in the prescribed limits necessary for construction of base layers stabilized by the hydraulic binder can be achieved. Thereby, the minimum quantity of cement in the stabilization mixture amounts to 2 % mass. Regarding the necessary quantity of cement and compared to the content of cement that is necessary for stabilization of pure sand from the Drava River (Study, 1990.), cost savings in the construction of stabilized base layers are possible. In addition, all the materials used in the research (natural and waste) are local, and thereby also enable a more rational design of pavement structures.

The obtained results confirmed the application of wood ash in stabilization mixtures for construction of base layers of pavement structures. The application of the results of this research lies in more rational and economical design and construction of pavement structure layers, particularly in areas with large quantities of ash from the wood biomass, such as area of eastern Croatia.

References

Dimter, S., Rukavina, T., & Dragčević, V. (2011). Strength Properties of Fly Ash Stabilized Mixes. *Road materials and pavement design*, 12(3), 687-697. https://doi.org/10.1080/14680629.2011.9695266

Dimter, S., Rukavina, T., & Barišić, I. (2014). Alternative, environmentally acceptable materials in road construction. *Handbook of Research on Advancements in Environmental Engineering*, 557–583.

General technical conditions for road works (GTC), book III: Pavement structures. (2001). Institut IGH, Zagreb, Croatian roads (in Croatian).

HRN EN 13286-2:2010 Unbound and hydraulically bound mixtures - Part 2: Test methods for laboratory reference density and water content - Proctor compaction (EN 13286-2:2010)

HRN EN 13286-41:2003 Unbound and hydraulically bound mixtures - Part 41: Test method for the determination of the compressive strength of hydraulically bound mixtures (EN 13286-41:2003)

HRN U.E9.024 Izrada nosivih slojeva kolničkih konstrukcija cesta od materijala stabiliziranih cementom ili sličnim hidrauličnim vezivima (in Croatian).

Ksaibati K. (1995). Evaluation of Cement Treated Bases with Fly Ash. *Australian Road and Transport Research*, 4(4), December 1995.

Maher, M. H., & Balaguru, P. N. (1993). Properties of Flowable High-Volume Fly Ash-Cement Composite. *Journal of materials in Civil Engineering*, 5(2), 212-225. https://doi.org/10.1061/(ASCE)0899-1561(1993)5:2(212)

Mahvash, S., López-Querol, S., & Bahadori-Jahromi, A. (2017). Effect of class F fly ash on fine sand compaction through soil stabilization. *Heliyon*, 3(3). https://doi.org/10.1016/j.heliyon.2017.e00274

Template: Anonymized Document

Pitman, R. M. (2006). Wood ash use in forestry - A review of the environmental impacts. *Forestry*, 79(5), 563–588. https://doi.org/10.1093/forestry/cpl041

Report on testing the mineral composition of ash samples by X-ray diffraction method (XRD). (2019). 65/19, Croatian Geological Survey Zagreb, December 2019.

Simatupang, M., Mangalla, L. K., Edwin, R. S., Putra, A. A., Azikin, M. T., Aswad, N. H., & Mustika, W. (2020). The Mechanical Properties of Fly-Ash-Stabilized Sands. *Geosciences*, 10. https://doi.org/10.3390/geosciences10040132

Sobhan, K., Jesick, M. R., Dedominicis, E. J., McFadden, J. P., Cooper, K. A., & Roe, J. R. (1999). A Soil-Cement-Fly Ash Pavement base Course Reinforced with Recycled Plastic Fibers. *Transportation Research Board, 78th Annual Meeting*, Washington, DC, 1-12.

Studija mogućnosti primjene pijeska u izgradnji prometnica na području Slavonije i Baranje. (1990). Građevinski institut - Fakultet građevinskih znanosti u Osijeku, Osijek p. 105 (in Croatian).

Supancic, K., & Obernberger, I. (2011). Wood ash utilisation as a stabiliser in road construction – First results of large-scale tests. *19th European Biomass Conference and Exhibition,* 43(0), 859–870.

Šķēls, P., Bondars, K., & Haritonovs, V. (2017). Wood fly ash stabilization of unbound pavement layers. *19th International Conference on Soil Mechanics and Geotechnical Engineering, ICSMGE 2017*, 2014–2017, January 2017.

Šķēls, P., Bondars, K., Plonis, R., Haritonovs, V., & Paeglītis, A. (2016). Usage of Wood Fly Ash in Stabilization of Unbound Pavement Layers and Soils. *In Proceedings of 13th Baltic Sea Geotechnical Conference*, 122–125. http://doi.org/10.3846/13bsgc.2016.017

Zagvozda, M., Dimter, S., & Rukavina, T. (2017). The potential of bio ash for utilization in road construction in eastern Croatia. *Selected Scientific Papers - Journal of Civil Engineering*, 12(2), 121-127. https://doi.org/10.1515/sspjce-2017-0028

Zagvozda, M., Dimter, S., Rukavina, T., & Grubeša, I. N. (2018). Possibilities for using bioash in road construction. *Građevinar*, 70(5), 393–402. https://doi.org/10.14256/JCE.2074.2017

Zenieris, P., & Laguros J. (1987). Fly ash as a binder in aggregate base course. *Materials Research Society Simposium Proceedings*, 113, 231-241. https://doi.org/10.1557/PROC-113-231

AUTHOR INDEX

Institute of Physics
Dirac House, Temple Back
Bristol BS1 6BE UK

ISSN: 1757-8981
ISBN 978-1-7138-4661-1